Animal Behavior in the Tropics

Shakunthala Sridhara •
Akshay Kumar Chakravarthy • Geetha Bali •
Rallapalli Ramamurthi
Editors

Animal Behavior in the Tropics

Vertebrates

Springer

Editors
Shakunthala Sridhara
Department of Zoology
University of Agricultural Sciences
Bengaluru, Karnataka, India

Akshay Kumar Chakravarthy
Indian Institute of Horticultural Research
Division of Entomology and Nematology
Bengaluru, Karnataka, India

Geetha Bali
Karnataka Women's University
Bengaluru, Karnataka, India

Rallapalli Ramamurthi
Sri Venkateswara University
Tirupati, Andhra Pradesh, India

ISBN 978-981-97-7323-7 ISBN 978-981-97-7324-4 (eBook)
https://doi.org/10.1007/978-981-97-7324-4

This Springer imprint is published by the registered company Springer Nature Singapore Pte Ltd.
The registered company address is: 152 Beach Road, #21-01/04 Gateway East, Singapore 189721, Singapore

This book is dedicated to

Late Dr. Smt. Shakunthala Sridhara
(1947–2023)

Foreword

The Ethological Society of India (ESI) has completed 50 years, and to celebrate its Golden Jubilee, it is bringing out a commemorative volume on animal behaviour with the title *Animal Behavior in the Tropics: Vertebrates*. As I have been associated with the ESI since 1977, I am honoured to pen a foreword for this book. Traditionally, animal behaviour has been studied with proximate and ultimate perspectives wherein the proximate factors usually deal with how behaviour arises and the underlying physiological, biochemical, or even psychological mechanisms of behaviour. The ultimate perspective traces the evolutionary origins of behaviour and its function for a species. A comparative study of animal behaviour across species, therefore, is a tool to understand our own evolutionary history. On the application side, studies on animal behaviour have helped conservation management of the threatened species, both in situ and ex situ. For a long time, behavioural biology and conservation biology remained detached from each other, and it is the behavioural ecology perspective that bridged the gap between the two conventional field biology perspectives. We have many species in our country that are threatened with extinction and require attention not only for management in our national parks and wildlife sanctuaries but also need conservation breeding for maintaining self-sustaining captive stocks and stocks for rehabilitation in the wild, if required. Seventy-three species, mostly mammals but also birds and reptiles, have been identified by the Central Zoo Authority of India for conservation breeding in identified zoos and institutions, and experts in animal behaviour are interacting with the managers for creating appropriate ecology of the species through environmental enrichment in a captive environment for a full expression of evolved behaviour patterns that are required for successful breeding. Behavioural studies such as social organisation, mate compatibility and choice, infant rearing, dispersal and migration patterns, xenophobia, etc. have enormous significance for the species' management. Further, animal behaviour research has helped in pest management in the agricultural fields which has much importance for India which is primarily an agrarian economy. Many nongovernmental organisations dealing with animal welfare heavily depend on an understanding of animal behaviour.

This book, edited by Late Shakuntala Sridhara, A. K. Chakravarthy, Geeta Bali and R. Ramamurthi, contains 26 chapters in two parts. Part I deals with general principles underlying behavioural studies, and Part II deals with studies on specific vertebrate species. Chapters on statistical methods, tools and techniques, photography, model-animal for behavioural studies and chronobiology form Part I of this book. Chapters on amphibians and reptiles, pit vipers, bats, bandicoots and squirrels, birds, swamp deer, Himalayan goral, spotted Chevrotain, sloth bear, lion and elephant constitute Part II. In the recent past, advances in neuroethology, neurobiology, socio-biology, communication science, genomics, animal-tracking and monitoring devices, captive breeding, evolutionary animal psychology and animal-husbandry management have facilitated understanding of animal behaviour in depth. Human welfare relies on conservation of nature, of which animals are a vital part. In this context, this book on animal behaviour is important and timely. More than 50 ethologists have contributed to this volume which reflects a wide range of aspects covered on over 20 vertebrate species. It is exciting to read so many new, interesting and intriguing facts about animal behaviour and the incredibly rich behavioural strategies that have been put together in this volume. I sincerely hope that this book will contribute for the cause of animal conservation and spark interest in the young minds and further advance the science of animal behaviour in the tropics.

As animal behaviour is now being offered as a compulsory/optional paper in most biology courses in the country, this book will provide reading material from an Indian perspective for the students who otherwise have to largely depend on the western contents. This book deserves a place on the shelves of Indian biology teachers and students as well as wildlife and animal welfare managers.

Institution of Excellence, Vijnana Bhavan, Mewa Singh
University of Mysore
Mysore, India
March 28, 2024

Preface

We, when still young developed interest in animal behaviour and explored patterns of behaviour in insects, birds, pet and other animals in field explorations in and around our homes. Several people from diverse walks of life are fascinated by the intriguing animal behaviour and people have their own means and approaches to develop understanding of the behaviour in animals. Animal behaviour came of age as a science when profs. Konnard Lorenz, Nicolas Timbergen and Karl Von Frish received the Nobel Prize in 1973 for their contribution to behavioural science. Today the applications of molecular tools and genomics have revolutionised the science of animal behaviour. The advance in imaging and computing has taken behavioural studies to depths unravelling the much needed insights. Behavioural biology is now being incorporated across several universities and institutions in tropical and sub-tropical countries.

It is difficult to present the whole wealth of information on animal behaviour in a book, but chapters have been so selected that well-rounded, understanding of animal behaviour is provided to readers. With contribution from renowned experts in the field, this book covers up-to-date information on animal behaviour, restricted to vertebrates. This book covers small (fish, bandicoot, short-nosed fruit bat, etc.), medium (Himalayan goral, Indian spotted Chevrotain, swamp deer, etc.) and large animals (lion and elephant). Crop raiding, olfaction, diet selection, migration, reproreproductive strategies are some of the aspects covered in this book. Hopefully, information in this book contributes to biodiversity and wildlife conservation. Reduce human–wildlife conflict, functionality and carbon sequestration in tropical forest and enthuse young minds to take off studies on animal behaviour, especially tropics and sub-tropics.

Merely creating green physical cover in afforestation and forest regeneration programmes of the governments will not facilitate conservation of wild animals and other biodiversity elements. Patches of functional forests with three vegetation layers, viz., ground floor (soil+herbs), understory (shrubs) and canopy (trees) will play a crucial role in ecological restoration, recreate natural biodiversity and facilitate ecosystem services. The degraded land, open land and wasteland should also be rendered natural for sustaining biodiversity.

Bengaluru, Karnataka, India — Shakunthala Sridhara
Bengaluru, Karnataka, India — Akshay Kumar Chakravarthy
Bengaluru, Karnataka, India — Geetha Bali
Tirupati, Andhra Pradesh, India — Rallapalli Ramamurthi

Acknowledgments

We would like to acknowledge the immense debt we owe to contributors who over long years were planning to bring out a book on animal behaviour. Our first debt of gratitude is to the late Smt. Shakuntala Sridhara. She did not live to see the publication of this book. Her death at the age of 76 was a great intellectual and administrative loss to the Ethological Society of India and the advancement of science of animal behaviour in the tropics. This book *Animal Behavior in the Tropics: Vertebrates* is based on research conducted on behaviour of animals in India in the early twenty-first century. Biologists are increasingly deploying modern tools and techniques to understand the biology and behaviour of animals. Select studies also use a combination of conventional and modern approaches to unravel hidden behavioural traits in animals. Contributions exhibited a wide spectrum of aspects that have been resolved in two parts in this book, viz. Part I: General Principles Underlying Behavioural Studies in Vertebrates and Part II: Exploring Behaviour in Vertebrates.

There are 26 chapters, consisting of 5 in the first part and 21 in the second part. The general trends, tools and techniques in animal behaviour are presented in the first chapter. This is followed by statistical tools that facilitate interpretation of animal behaviour activities. Dr. Vijay Kumar Koli answered how photography is essential to add to our conservation efforts. Drs. V.V. Sangaraju, Senthil kumar, N.K. Kiran and Y. Chandrashekar dealt with the fish, *Damio rerio* that can be used as a tool for research in neurodegenerative dysfunctions in vertebrates. Drs. Richa Pathak, Nisha Singh and Arti Parganiha highlighted on cancer cells. Drs. Deytima Ghosh, N.A. Arvind and M. Mudke facilitate chapters on behaviour of amphibians and reptiles. Dr. Albert Rajendren contributed a chapter on pit vipers. Prof. J. Vanitarani focused on chemical signalling in bats, while E. Rajan, M. Murugan and D.M. Rajathei dealt with olfaction in short-nosed fruit bats. Dr. Chakravarthy and Ashutosh summarised roosting behaviour in bats. Dr. Anil Kumar contributed a chapter on behaviour of bulbuls in an urbanised set-up. Prof. Arkajyoti Mukharjee studied and contributed information on ducks. Bird migration formed the topic for Dr. S. Sushant and Dr. Manjulaa Menon, who documented adaptive behaviour in an urban set-up by birds. There are also chapters on bandicoots and reproductive strategies in squirrels. Mr. Parvez and colleagues documented behaviour and ecology of swamp deer, Himalayan goral and Indian spotted Chevrotain. This volume also includes a chapter on diet selectivity of sloth bear by Drs. V. Rabari, A. Malik and

N. Dharayia. Dr. Sakuntala Shridhara, A.K. Chakravarthy and Usha Nandini contributed a chapter on lion. A team of scientists, viz. C. Jabili, A. Sanyal, D. Choudhary and T.N.C. Vidya have unravelled social organisation in Asian elephants. Dr. E.A. Jayson and S.K. Govind elaborated on crop raids by elephants.

We would not have been able to publish this book without the continued support of our contributors, who demonstrated enormous skill, precision and warmth despite their hectic schedule. We would like to acknowledge with gratitude, the love and patience of our family members. They all kept us going with the book publishing. Finally, the publisher, Springer Nature has our gratitude for incredible help and matchless care in the formation and development of this book.

Shakunthala Sridhara
Akshay Kumar Chakravarthy
Geetha Bali
Rallapalli Ramamurthi

Contents

Editors and Contributors

About the Editors

Shakunthala Sridhara has published more than 100 papers and has authored a dozen books on rodents, vertebrate pest management, insects and women in agriculture. She has been a frequent sessional chair in international conferences on animal behaviour and rodent pests. Recognising her pioneering work on Indian rodents, Dr. Sridhara has been invited to deliver seminars in prestigious universities all over the world including Oxford, Nottingham, York, Central Scientific Laboratory of the UK, Munich, the Danish Pest Control Laboratory, Spain, Vienna, Tubingen, the World Health Organization (Geneva), UCLA at Berkeley, Davis, Sacramento, Wildlife Research Center at Denver, Bowling Green State University, the National Zoological Park of the Smithsonian Institution at Washington, DC, Monell Chemical Senses Center (Philadelphia), John Hopkins University, the University of Tennessee at Knoxville, etc.

Akshay Kumar Chakravarthy, Head and Principal Scientist (Rtd.,), Division of Entomology and Nematology, is an author of many books and over 320 scientific papers and 60 chapters in entomology and natural history. His interest includes insects, birds, bats, rodents and mammals. With nearly four decades of experience in teaching, research and extension, Dr. Chakravarthy has been the investigator for over 35 research projects and has guided more than 28 students for post-graduation. Recently, he has published books titled, *Innovative Pest Management Approaches for the 21st Century* (foreword by Prof. Dr. Remzi Atlihan, Turkey), *New Horizons in Insect Science:*

Towards Sustainable Pest Management (foreword by Prof. Hiroshi Honda, Japan), *Economic and Ecological Significance of Arthropods in Diversified Ecosystems: Sustaining Regulatory Mechanisms* (foreword by Fani Hatijina, Greece), *Arthropod Diversity and Conservation in the Tropics and Sub-tropics* (foreword by Prof. Madhav Gadgil, India), *The Black Spotted, Yellow Borer, Conogethes punctiferalis Guenee and Allied Species* (foreword by Prof. Alma Solis, USA) and *Experimental Techniques in Host Plant Resistance* (foreword by Prof. Michael C. Smith, USA) by Springer.

Geetha Bali, a well-known scientist, academician and an institution builder, served as the second Vice Chancellor of Karnataka State Akkamahadevi Women's University, Vijayapura, and as the Chairperson, Board of Governors, NIT Bhopal as well as Chairperson of the Standing Committee of the Council of NITs. She is a recipient of Sir Asutosh Mookerjee Fellowship and is currently engaged with research at Cell Space Research Foundation, Bangalore. She is currently a member of the institute body of the National Institute of Mental Health and Neurosciences, Bangalore, India. She served the Ministry of Human Resource Development, Government of India, in various capacities as a Member of the Indian National Commission for UNESCO (Education), Central Advisory Board of Education, Advisory Committee for Central Schools under the Government of India, Steering Committee, Unnat Bharat Abhiyan and governing bodies of a number of institutions of Higher Education in India, as the President's nominee. She was an invited speaker at the International Foundation for Science for Peace, Italy, the World Women Leaders Forum, Beijing, China and the AAAS, Washington, DC, USA, as well as several universities in Europe and the USA. She was elected as the 99th President of the Indian Science Congress Association, and She has to her credit a number of awards and recognitions at both the national and international levels.

Rallapalli Ramamurthi was born on the 27th day of July 1936, at Chennai, India. After a consistently good academic and extra-curricular record, he did his post-graduation in zoology from Sri Venkateswara University, Tirupati, and joined the faculty of the Department of Zoology in 1958 at the same university. Early in his career, his research effort came under the enviable tutelage of Prof. K. Pampapathi Rao, who not only set high standards of research but also carefully tended and incubated the potential in Prof. Ramamurthi. The mid-1960s took him to Oregon in the USA for post-doctoral research, and upon his return, he was elevated to the position of reader and later to Professor of Zoology. He carried out further research and teaching in Bonn, Tubingen, Karlsruhe, Nijmegen, Malta, Jackson, Memphis, Urbana, Los Angeles, Washington, Gainesville, Tallahassee, Columbia, Montreal, Vancouver, Toronto, Quebec City and Tokyo, covering a spread of collaborative research in as variegated fields as chemical regulatory mechanisms in endocrinology, eco-science and environmental toxicology, collagens, bio-energy systems, diabetic and hypertensive nephropathy. The multitude of his research projects were funded by Government of India funding agencies such as UGC, DST, DOEn, CSIR, ICMR, DBT and other overseas agencies.

Contributors

N. A. Aravind Yenepoya Research Centre, Yenepoya (Deemed to be University), Mangalore, India

SM Sehgal Foundation Centre for Biodiversity and Conservation, Ashoka Trust for Research in Ecology and the Environment (ATREE), Bangalore, India

Subhadra Barik Sálim Ali Centre for Ornithology and Natural History, Coimbatore, Tamil Nadu, India

Divya Choudhary Animal Behaviour and Sociogenetics Laboratory, Jawaharlal Nehru Centre for Advanced Scientific Research, Bengaluru, Karnataka, India

Ashutosh Dey Environmental Management and Policy Research Institute, Bengaluru, Karnataka, India

Nishith Dharaiya Wildlife and Conservation Biology Research Lab, Department of Life Sciences, Hemchandracharya North Gujarat University, Patan, Gujarat, India

Centre of Excellence for Wildlife and Conservation Studies, Bhakta Kavi Narsinh Mehta University, Junagadh, Gujarat, India

Deyatima Ghosh Laboratory of Animal Behaviour and Conservation, College of Life Sciences, Nanjing Forestry University, Nanjing, Jiangsu, People's Republic of China

School of Life Sciences, University of Lincoln, Lincoln, UK

SM Sehgal Foundation Centre for Biodiversity and Conservation, Ashoka Trust for Research in Ecology and the Environment (ATREE), Bangalore, India

Suresh K. Govind Department of Psychology, Christ College, Irinjalakuda, Thrissur, Kerala, India

Sameer Gupta Pineal Research Lab, Department of Zoology, Banaras Hindu University, Varanasi, India

Chandana Haldar Pineal Research Lab, Department of Zoology, Banaras Hindu University, Varanasi, India

Chowdari Jabili Animal Behaviour and Sociogenetics Laboratory, Jawaharlal Nehru Centre for Advanced Scientific Research, Bengaluru, Karnataka, India

E. A. Jayson Department of Wildlife, Kerala Forest Research Institute, Peechi, Thrissur, Kerala, India

Neelakanta Sarvashiva Kiran Department of Biotechnology, School of Applied Sciences, REVA University, Bengaluru, Karnataka, India

Vijay Kumar Koli Wildlife Research Laboratory, Department of Zoology, M L Sukhadia University, Udaipur, Rajasthan, India

Anil Kumar Zoological Survey of India, Northern Regional Centre, Dehradun, Uttarakhand, India

Akshay Kumar Chakravarthy Division of Entomology and Nematology, Indian Institute of Horticultural Research, Bengaluru, Karnataka, India

Arzoo Malik Wildlife and Conservation Biology Research Lab, Department of Life Sciences, Hemchandracharya North Gujarat University, Patan, Gujarat, India

Vartika Malik Pineal Research Lab, Department of Zoology, Banaras Hindu University, Varanasi, India

Manjula Menon Department of Environmental Science and Management, Bharathidasan University, Tiruchirappalli, India

Rajesh Kumar Mohapatra Nandankanan Zoological Park, Bhubaneswar, Odisha, India

Madhushri Mudke SM Sehgal Foundation Centre for Biodiversity and Conservation, Ashoka Trust for Research in Ecology and the Environment (ATREE), Bangalore, India

Manipal Academy of Higher Education (MAHE), Manipal, India

Arkajyoti Mukherjee Department of Chemical Engineering, Jadavpur University, Kolkata, West Bengal, India

Ecotechnology Project Lab, Government College of Engineering and Leather Technology, Kolkata, West Bengal, India

Murugan Mukilan Department of Biotechnology, Sri Ramakrishna College of Arts and Science, Coimbatore, Tamil Nadu, India

Arti Parganiha Center for Translational Chronobiology, Pandit Ravishankar Shukla University, Raipur, India

Parvez Society for Science and Technology Applications (SSTA), Bengaluru, Karnataka, India

Richa Pathak School of Studies in Life Science, Pandit Ravishankar Shukla University, Raipur, India

Vasantkumar Rabari Wildlife and Conservation Biology Research Lab, Department of Life Sciences, Hemchandracharya North Gujarat University, Patan, Gujarat, India

Senthilkumar Rajagopal Department of Biotechnology, School of Applied Sciences, REVA University, Bengaluru, Karnataka, India

K. Emmanuvel Rajan Department of Animal Science, School of Life Sciences, Bharathidasan University, Tiruchirappalli, Tamil Nadu, India

D. Mary Rajathei Department of Biotechnology and Bioinformatics, Bishop Heber College, Tiruchirappalli, Tamil Nadu, India

Albert Rajendran Research Department of Zoology, St. John's College, Tirunelveli, Tamil Nadu, India

Varun Varma Sangaraju Department of Information Technology, University of the Cumberlands, Williamsburg, KY, USA

Ankana Sanyal Animal Behaviour and Sociogenetics Laboratory, Jawaharlal Nehru Centre for Advanced Scientific Research, Bengaluru, Karnataka, India

Nisha Singh School of Studies in Life Science, Pandit Ravishankar Shukla University, Raipur, India

Shakunthala Sridhara Department of Zoology, University of Agricultural Sciences, Bengaluru, Karnataka, India

S. Sushanth University of Mysore, Mysore, Karnataka, India

A. Usha Nandini Bengaluru, Karnataka, India

Juliet Vanitharani Bat Research Laboratory, Zoology Department and Research Centre, Sarah Tucker College, Tirunelveli, Tamil Nadu, India

Rakesh Verma Pineal Research Lab, Department of Zoology, Banaras Hindu University, Varanasi, India

T. N. C. Vidya Animal Behaviour and Sociogenetics Laboratory, Jawaharlal Nehru Centre for Advanced Scientific Research, Bengaluru, Karnataka, India

Chandrashekar Yashaswini Department of Biotechnology, School of Applied Sciences, REVA University, Bengaluru, Karnataka, India

Part I

General Principles Underlying Behavioural Studies in Vertebrates

Animal Behaviour and Humans: Conflict, Coexistence and Conservation in the Tropics

1

Akshay Kumar Chakravarthy

Abstract

In the present-day world, human-induced ecological and landscape changes are occurring at a pace that many species of wild animals and plants are unable to cope with. It is becoming clear that active management interventions are required to conserve wildlife. So one of the goals of the behavioural studies should be to contribute to wildlife conservation. In this chapter, examples of vertebrates are taken so that varying degrees of trends, commensalism and conservation are recorded, and accordingly, mitigation measures on the ground are executed. At the macro level, it should be recognized that human lives and the environment are linked with animals, and there are real-world consequences of losing wildlife.

Keywords

Animal behaviour · Conflict · Coexistence · Conservation

1.1 Introduction

Animal behaviour plays a pivotal role in wildlife conservation. For instance, captive breeding of endangered species requires understanding the habits, habitat, breeding and bioecology of wild animals in captivity. The introduction of cheetah (*Acinonyx jubatus*) from South Africa and Namibia to India in 2023 is a recent example. Likewise, advancements have been made to introduce mega-herbivores like elephants (*Elephas maximus*) to parts of Australia and East Africa. Conserving wildlife will yield results if aspects of behaviour are known well before animals are

A. Kumar Chakravarthy (✉)
Division of Entomology and Nematology, Indian Institute of Horticultural Research, Bengaluru, Karnataka, India

S. Sridhara et al. (eds.), *Animal Behavior in the Tropics*,
https://doi.org/10.1007/978-981-97-7324-4_1

maintained in captivity. Countries like Cambodia, Vietnam and Laos have all lost their tiger populations. To restore ecological balance in wild ecosystems, in the future, tigers may be bred in captivity in these countries. The behaviours associated with territory size, mating, food and individual-to-individual social interactions must be known prior to any initiation for captive breeding (Prater 2005). According to Breed and Moore (2022), animal behaviour studies provide scientists with the tools for measuring the levels of disturbance humans cause to wild animal populations.

Encroachment of habitats of wild animals by humans increases human–wildlife conflicts. Studies on behaviour reveal ideas to reduce conflicts and minimize adverse impacts. Amur falcon (*Falco amurensis*) and tiger (*Panthera tigris*) are instances where populations of the animals are restored. Populations of Amur falcon and tigers dwindled drastically in the northeast and central plains of India, respectively, during the last years of the twentieth century and the beginning of the twenty-first century. Currently, their populations have stabilized and are increasing.

Select species of animals are adapting to human habitations. There may be a risk of disease spread, injury or death of humans, damage to landscaping and so on. Studies on animal behaviour help us understand how to minimize these negative impacts and improve the co-existence of wild animals with humans. Bisons are coming out in herds from forests to forage in adjacent farms, in Coonoor and Kotagiri in Tamil Nadu and Kodagu in Karnataka, South India. Forest patches with contiguity with coffee estates in Kodagu bring bisons in direct conflict with humans. Humans need to have a soft approach to handling this relatively harmless species. Bisons are now sharing space with humans and giving rise to situations of peaceful co-existence with humans. Similar is the case with the wild boar (*Sus scrofa*) in South India. Wild boar upturns subsurface soil to the surface and feeds on root grubs, and wild boar is prey for top consumers in the ecosystem. Wild boar is now increasingly sighted close to humans in estates, farms and dwellings in hilly regions in South India. As human populations continue to expand and climate change impacts on wildlife intensify, understanding animal behaviour would be crucial.

The fragmentation and encroachment into habitats of wildlife and physical barriers are compelling animals to stray into villages, towns and cities. Spotting wild animals in villages/towns has become frequent news among the public in tropical countries like India. The populations of big animals are now being monitored to save them. The main goal of behavioural studies is to contribute to wildlife conservation. With this background, the current conservation status of select species is reviewed in this chapter with information on conservation efforts. Only a few animal taxa are considered, but the principles may apply to the other broader groups of animals.

1.2 Elephant (*Elephas maximus*, Linn 1758)

Elephant is the largest living animal in Asia. Since 1986, the Asian elephant has been listed as endangered. There are three species, namely, the African bush elephant, the African forest elephant and the Asian elephant. The Asian elephant, also known as the Indian elephant, is found in tropical forests and grasslands of Asia, including Thailand and Indonesia. The wild elephant population in 2023 is estimated at over 54,000. Asian elephants are crepuscular, mega-herbivores and generalist feeders, as grazers and browsers. The elephants are known to feed on 112 different species of plants. Asian elephants make sounds called growls, squeaks and snorts. Cows and calves move about together as groups, while bulls disperse from their mothers upon reaching adolescence. Bulls are solitary or form temporary 'bachelor groups'. Cow–calf units tend to be small. In some areas, humans and elephants have co-existed for hundreds of years, while in other areas, they come in conflict with each other. Factors in human–elephant conflict can be categorized into ultimate and proximate. The ultimate factor includes an increase in human populations and large-scale development projects. Proximate causes include habitat loss, expansion of agriculture and encroachment into protected areas. Poaching for ivory is quite widespread. Mukherjee (2001), Mukherjee et al. (2023) and Mukherjee (2024) extensively worked on behaviour of elephants in India.

The Asian elephant is the flagship species (Fig. 1.1), and World Elephant Day is celebrated on 12th August every year. The elephant population is threatened by loss of habitat, habitat degeneration, fragmentation and poaching. Elephant populations face worldwide challenges. Anti-poaching measures, habitat protection and restoration, community involvement, education and awareness and research and monitoring are necessary, and it is the responsibility of every country to protect. Each country's efforts to protect its elephant populations will contribute to safeguarding these magnificent creatures for future generations.

Elephants are peaceful animals. However, females are aggressive in particular situations in the wild like the presence of young ones with mothers. Elephants

Fig. 1.1 Asian elephant

engage in several forms of object play for pleasure. Elephants live in a matriarchal family group of related females called a herd. They are often led by the oldest and often the largest female in the herd. Herds may consist of 8–100 individuals. Elephants use tools. For instance, use sticks to scratch themselves in areas they cannot reach with their trunks or use large branches to shoot at flies. This behaviour is learnt from observing other elephants. Elephants use their heads, eyes, mouth, ears, tusks, feet and whole body even to communicate and signal messages to each other. Head bobbing and swaying are the two stereotypical behaviours of captive elephants.

Elephant's habitats in India are most severely fragmented. Encroachment of traditional migratory corridors is changing elephant behaviour and making the animals dangerous. Electric fences, ditches and mines block the elephant's traditional migratory routes, for instance, to Nepal, Jharkhand and Odisha from India. Elephants are forced to raid human settlements. An electric fence built on the western bank of the Medri River in Nepal not only killed many elephants but also changed their behaviour. A study by the Wildlife Protection Society of India (WPSI) revealed that over 2000 and 5000 ha of prime elephant habitats in Jharkhand and Odisha have been lost to mines.

Most elephant attacks are seen during July and August. This is due to the bushy vegetation that grows in the wild, making it difficult to spot elephants from a distance. Fodder and water sources deplete in forests due to prolonged summer, and jumbos venture out towards human habitation in search of food. Elephant is an animal in Schedule I of the Wildlife Protection Act 1972 like tiger and lion. In India, Karnataka has the highest number of elephants 6050 followed by Assam (5719) and Kerala (3054).

Elephant attacks killed 106 people between 2018–2019 and 2022–2023 in Kerala (Table 1.1). Odisha saw 542 elephant-linked deaths in this period. The available food sources for wild animals, which are fruit and vegetable crops, rubber and other herbivores, have increased the threat of elephant attacks. Some invasive varieties like lantana, a monoculture of exotic species of eucalyptus, and acacia have not been consumed by herbivores leading to a drastic impact on habitat quality. Due to the influence of humans and their habitat induced some of the animals like wild boar, and peafowl to establish in new habitats. Kerala High Court asked for efforts to be made to control human–wildlife conflict in the state (Times of India, Feb 2024a).

Table 1.1 Human deaths and damage to cattle and crops in animal attacks in Kerala

Year	Human deaths	Cattle loss, crop damage, etc. (no. of incidents)	Compensation paid (Rs. in lakh)
2019–2020	102	6063	714.2
2020–2021	97	7077	785.6
2021–2022	152	7094	1056
2022–2023	99	5800	803.1
2023–2024	105	4555	161.1
Total	**555**	**30,589**	**3519.90**

The Vantara Animal Rescue and Rehabilitation Centre in Jamnagar, Gujarat, was established as home to astonishing mammals, birds and reptiles. Vantara has rescued over 200 elephants from across India. Surgeries, endoscopies and laser operations are conducted by experts from across the world. But the elephants enjoy a simple therapy too in a fragrant forest covering 3000 acres. Then there is Ratan Tata's state-of-the-art hospital for small animals in Mumbai established by Ratan Tata (Jain 2024).

1.3 Lion (*Panthera leo,* Linn 1758)

Lion is often called the king of beasts. Today, their distribution is restricted to fragments of sub-Saharan Africa and a critically endangered population in West Africa. A small population of Asiatic lions is in India's Gir National Park. Lions are adapted to living in grasslands mixed with trees and shrubs/bushes. Compared to other cats, lions are social. A group of lions is called pride. Groups of females hunt together. Lions are carnivores and scavengers. Lions feed on buffalo, antelope, zebra, birds, hares, turtles, giraffes and young elephants. Lions keep constantly a watch for circling vultures, which indicates a dead or injured animal close by. Lions have archetypal roar, utilized in transmitting messages with conspecifics or group members, and alert intruders into their territorial boundaries. Retractable claws act like hooks and help the lion to grab and hold prey. Rough tongue aids in peeling the skin of animals away from flesh and flesh from bones. Across the belly, lions have loose skin, which allows the species to be kicked by prey with no or very little injury. About 10–40 individuals may live in a pride. Each pride has a home territory and does not allow other carnivores to hunt in that area. A territory can be as large as 250 km^2. Lion's roar is used for territorial marking and warning other individuals of a different pride. Male lions challenge the alpha of the pride to a fight to express dominance. Lions hunt in packs. This is unusual for cats. Lions do not have dens. The lioness conceals the cubs in thick bushes, gullies or rocky crops. A litter of 2–6 cubs are born, but usually 1–2 cubs survive (Wikipedia 2007). The impact of social and environmental factors on African lion have been investigated (Kotez et al. 2018).

Goswami et al. (2019) reviewed the literature on lions and explored the role of individuality (personal traits and life history) on behaviour-based welfare indices of 35 captive Asiatic lions in Sakkarbaug Zoological Garden, Junagadh, Gujarat. Results showed that intra-specific variations based on life history and personality traits consistently predict the welfare states of captive lions. Many felid populations in the wild are rapidly declining. Ex situ conservation interventions are urgently required. Animal welfare conservation of species-specific behaviour is important in modern zoos. Social behaviour was recorded most regularly with the larger pride of close relatedness.

Hunting is a cooperative activity in which the males and old lions seldom take part. Lions usually do not travel together and hunt (Packer et al. 2005). Lions prefer to hunt at night. Activities such as mining and excavation must be avoided in and around the habitats of lions.

Literature indicates that pride (10–20 lions) is the highest level of social organization. Pride may consist of groups (3–6 lions) with varying composition. These individuals will be always together, called fission–fusion. A pride has a territory, defended by 1–3 males for 2–4 years against nomadic males. The movements of the five lions were monitored by tagging radio collars. Lion social organizations may have small group sizes (3–6 individuals) because of low prey density, small body size or prey and high per capita livestock consumption (Bauer et al. 2003).

Territorial behaviour in lions buffers populations against short-term environmental disturbances. Despite changes in prey availability and vegetative cover, regional populations of Serengeti lions remained stable for 10–20-year periods and changed to new equilibria in sudden leaps. A simulation model showed that the pattern of 'saltatory equilibria' results from grouping behaviour in lions. This study proves that carnivores like lion exhibit fission–fusion social organization of African lions in the Okavango Delta, a seasonally flooded wetland. In the Okavango Delta, spatial competition arising from changing flooding regimes is a greater limiting factor for total pride sizes than food availability.

1.4 Cheetah (*Acinonyx jubatus*, Johann Christian Daniel von Schreber 1777)

Cheetahs are distributed in small, fragmented populations in northwest, East and South Africa and central Iran. The animals prefer habitats such as savannahs, arid mountain ranges in the Sahara and hilly desert terrain. Currently, cheetahs are present in South Africa, Namibia and Botswana. Among felids, cheetahs have a unique social system and camouflaging habitat. Females are solitary, while males are not. Females raise cubs on their own. Cheetah's unique body structure (narrow and lightweight), flexible spine, semi-retractable claws, long legs and tail permit the animal to catch up speed at 110 km/h. The current global population is 7100. The IUCN status for cheetahs is vulnerable.

In five case studies in North Bengal covered with fragmented forests and tea estates it was found that elephants carry the carcasses holding on to the trunks or legs to a distance before burying them. Direct human intervention was not involved in the five cases. Elephants avoided the paths where carcasses ever buried earlier. In all five cases, the legs of the carcasses were positioned upright, with the trunk and dorsal regions fully buried along the irrigation drains of the tea estates. Before burying the carcasses, members of the elephant herd issue out trumpets for 30-40 mins.

In 1947, a royal king hunted the last three cheetahs left in India. In 1952, the cheetahs were declared extinct in the country. Cheetahs require 10,000 km^2 of territory with enough prey. On 18 September 2022, eight cheetahs were released into Kuno National Park (KNP), Madhya Pradesh, from Namibia after seven decades. Only deer and blackbucks are being introduced to Kuno, and no other animal will be introduced. This is only to provide a prey base for the cheetahs (Naveen 2022). Earlier attempts to breed cheetahs in captivity in India failed due to changes in their diet, environment and other constraints in breeding. The goal of the cheetah introduction project in India is to establish a cheetah meta-population in India, perform its functional role as a top predator in the ecosystem, expand occurrence in the historic range in India and contribute to global conservation (Times of India 2023). When cheetahs are established in KNP, the grassland ecosystem will be restored, and there will be an increase in biodiversity. When cheetahs sprint, the grassland ecosystem will be restored. In November 2022, the Project Cheetah entered the second phase with the animals moving from quarantine pens into hunting enclosures prior to full release. It is difficult to identify an individual cheetah from a distance.

1.5 Leopards (*Panthera pardus fusca*, Meyer 1794)

Leopards are nocturnal animals, solitary, elusive and territorial. Leopards climb trees easily, have stealthy habits and have strong legs that make them a top consumer in forests. Leopards widely occur in Africa, Asia and Caucasia (regions between the Black Sea and the Caspian Sea include Armenia, Azerbaijan, Georgia and South Russia). In 2023, the leopard population in India was estimated at 12,172–13,535 individuals (Times of India 2024). Leopards in Nepal, Sikkim and Kashmir are not geographically separated in the Indian subcontinent. Indian leopards are not common in habitats where tigers are abundant (Qureshi et al. 2024). In tropical forests, like Nagarhole National Park, Karnataka, South India, leopards successfully co-exist with tigers without competitive exclusion or inter-species dominance hierarchies. Prey like dhole and striped hyena overlap between tiger and leopard. Hunting of Indian leopards for the illegal wildlife trade is the biggest threat to its survival. The leopard is a versatile, opportunistic hunter with a broad prey base like deer, nilgai, wild boar, common langur, Indian hare and peafowl (Mondal et al. 2012). In the Periyar Tiger Reserve in South India, primates make up a large proportion (Srivastava et al. 1996). In areas with high tiger populations such as Central India and Kanha National Park, leopards are not residents but transients (Arivazhagan et al. 2007). In India alone, over 200 leopards are killed every year (Raza et al. 2012; Wildlife Protection Society of India 2010); leopards are more likely to be killed (seven times) than tigers (Wildlife Protection Society of India 2010). In Nepal, more than 40 leopards and, in China and Tibet, nearly 130 leopards are killed every year (Banks 2004).

Encroachment into protected areas and expansion of agricultural land contribute to habitat and prey loss. As a result, leopards stray into human settlements where they prey upon pet and domestic animals like dogs, pigs, goats and cattle. At times, leopards inhabit the periphery of villages, where they come in direct conflict with humans. As urban habitats increase, the natural habitats of leopards decrease. Man-eater leopards are also found in nature.

At Jhalana, a dense forest in Rajasthan, India, has found a new tourist attraction—Leopard Safari. Sportsmen, politicians and business tycoons are visiting Rajasthan to have a glimpse of leopards in the wild. Data shows wildlife tourism in Rajasthan has grown rapidly over the past 5 years. Many forest areas in India are now prepped up for leopard safaris. Safaris have become very popular in Rajasthan.

1.6 Snow Leopards (*Panthera uncia*, Schreber 1775)

Snow leopards, called 'ghosts of the mountains', are elusive predators spread in the Himalayan mountain range and are sparsely distributed across 12 countries, including China, Tibet, Mongolia, Russia, Pakistan, Nepal and Bhutan. There are 718 snow leopards in India accounting for 1/6 to 1/9 of the global population. The first-ever survey conducted on snow leopards in India was through camera traps (Mohan 2024). The Snow Leopard Population Assessment in India (SPAI) estimates 36 individuals in Arunachal Pradesh. According to the World Wide Fund for Nature-India, the presence of snow leopards is a critical indicator of the health of the Himalayan ecosystem, including rivers like Brahmaputra (the lifeline of Assam) and the glaciers. The first step involved in evaluating the snow leopard spatial distribution was through an occupancy-based sampling approach. In the second step, abundance was determined via camera traps deployed in 1971 locations for 180,000 trap nights. A total of 241 unique snow leopards were photographed. Monitoring of glacial lakes and rivers in the Himalayan range is needed to survey snow leopards. Conservation of snow leopards warrants stronger policy push (The Setines, Guwahati, 31 Jan 2004).

1.7 Tiger (*Panthera tigris*, Linn 1758)

Tiger is an umbrella species where it conserves swathes of habitats and co-occurring species. Globally, the tiger is an endangered species due to the demand for its body parts. Tigers are territorial and solitary animals (Mellen and Sheperdson 1997). Tigers occur in Russia, North Korea, China, India, Southwest Asia and Sumatra. Tigers use visual signals, scent marks and vocalizations to communicate (Qureshi et al. 2024). According to the Tiger Census (Azad 2023), India's upper limit of population is 3925, and the lower limit is 3682 in 53 tiger reserves in 75,800 km^2 (2.3% of India's area) (Times of India, July 29, 2023). The tiger population has increased in India, Bhutan, Nepal, Thailand and Russia. But Cambodia, Vietnam and Laos lost all their tigers. Conventionally, in India, tiger numbers were estimated by Line Transect Surveys. In the 1980s, methods for capturing tigers included Radio-collaring which was standardized. Before Radio-collaring the animal was sedated. From the 1990s, camera traps were picked for occupancy sampling of tiger signs (tracks, droppings, etc.) for determining spatial distribution (Vijay 2023a, b). Habitat management, bringing connectivity between tiger habitats by creating corridors, implementing plans to reduce human–tiger conflicts and translocating tigers to suitable habitats where members are low are some of the measures. For instance, only 12 have reached carrying capacity out of 53 tiger reserves in India (Vijay 2023c). India has standardized the monitoring protocols for tigers in India. These could be applicable in other tiger areas too.

A pilot study on tigers conducted by the Indian Institute of Forest Management (IIFM) and the National Tiger Conservation Authority (NTCA) has revealed additional benefits of tiger conservation in countering climate change (Vijay 2023d; (Bashaw et al. 2007). Linking tiger conservation gains to climate adaptation plans; fostering sustainability of traditional small towns near tiger reserves, wetlands conservation and local-community-driven ecotourism with climate, fire and flood protection; and preventing prey-based depletion, encroachment and increasing vigilance in tiger areas would effectively contribute to conservation (Times of India 2023).

Land grab, transferring forest patches to tribals, shrinking tiger corridors and podu cultivation (forest trees are cut each year to clear land for farming) by tribals are deteriorating tiger habitat in Telangana and Andhra Pradesh in South India (Reddy 2024). Similipal Tiger Reserve (STR), Odisha, holds about 40 tigers, of which 16 are melanistic (Ramanath 2024).

1.8 Indian Bison (*Bos gaurus*, Smith 1827)

The estimated population of bison worldwide is 13,000–30,000 individuals. Eighty-five per cent of the bison (=gaur) population is in India. Leopards, dhole packs and mugger crocodiles attack unprotected calves or sick individuals. Humans remain the biggest enemy of Indian bison. Indian bisons are endangered animals included in Schedule I of the Wildlife Protection Act 1972 (Fig. 1.2). Bison is native to Southeast Asia and can also be found in Bangladesh, China, Cambodia, Malaysia, Nepal, Vietnam, Burma and Thailand. This bovine species prefers evergreen and moist deciduous forests, as well as dry deciduous. The Indian bison is a grazing animal, generally feeding in the early morning and the late evening. In the IUCN Red List, it is given as vulnerable and included under Schedule I of the Wildlife Protection Act, 1972. The bison population is today threatened by poaching, food scarcity, habitat loss and human–animal conflict.

According to Kumar and Kumar (2023), bison consume young leaves, tender shoots, bark and fruits. The feeding behaviour of bison varies according to season and landscape. Tigers and other carnivores prefer bisons as prey. So it is important for sustaining carnivorous predators in forests.

Bisons are most often found feeding. The human–wildlife conflicts due to bisons are increasing. Many believe that it is important to have a site- and species-specific approach to dealing with conflicts. Bisons exhibit peculiar behavioural patterns like rolling in dust and dirt, creating depressions in soil and resorting to dust baths. In

Fig. 1.2 Bison with a calf

certain forests and wild areas, bisons are nocturnal, but in other places, it is diurnal. Herds of bison are led by an old adult, but adult males are usually solitary. In the bison population, the size of the individual matters; this determines dominance. Bison is almost extinct in Sri Lanka and Bangladesh. In Nilgiris and Kodagu in South India, bisons are changing their behaviour; the animals are moving close to human habitations in roads, farms and coffee and tea estates, and moving out of grasslands and forests. This is because, in the forest floors in Nilgiris, *Lantana camara* has replaced natural grass, and bisons cannot graze in such situations. Bisons brush their horns against coffee bushes and damage them. In Nilgiris and Kodagu, bisons have overcome their shyness and freely graze in estates. Bisons are much more accepting of humans now. Bisons are now sharing space with humans and giving rise to situations of peaceful coexistence with humans.

1.9 The Bonnet Macaque (*Macaca radiata*, E. Geoffray 1812)

This species is endemic to South India and is diurnal, arboreal and terrestrial. The bonnet macaque (Fig. 1.3) feeds on vegetables, nuts, seeds, fruits, cereals and invertebrates. The bonnets are social, communicating with facial expressions. They share a linear dominance hierarchy. The alpha male is the most dominant male of the

Fig. 1.3 The bonnet macaque

troop. The beta and gamma males follow. Males dominate over females. Females remain in the same group, while the males, disperse. High-ranking males in a troop have first access to breeding females. High-ranking females attempt to kidnap lower-ranking females. Coalition formation occurs between unrelated males and females to oust another male. Males often move from one troop to another to gain a higher rank. Social grooming, food sharing, hugging, communal resting and interactive plays are common in *M. radiata* troops. Grooming is a social behaviour act in primates like *M. radiata*. Grooming promotes group cohesion and social relationships (Dileep and Jose 2014).

There is a growing presence of monkey troops in rural and urban environments, and there are hostile encounters in South India. Severe decline in populations at some urban places in cities in South India has been recorded. However, a steady increase in their populations in some other urban places has been documented. Old *Ficus* trees in urban areas are favourable haunts of these monkeys. As road construction is in progress, the continuity of habitats is being fragmented. Primates such as the bonnet monkey, *M. radiata*, are very susceptible (Anandam 2013) and the main amplifying host for the transmission of the Kyasanur forest disease virus. *Haemaphysalis spinigera*, a forest hard-bodied, obligate ectoparasitic tick also acts as a vector for disease transmission (Holbrook 2012). Humans are infected by the bite of tick nymphs. Humans get the disease only from *M. radiata* (Biswas 2022).

In urban and rural habitats, bonnets prefer large flowering and fruiting trees like banyan (*Ficus benghalensis* and *F. benjamina*), mango (*Mangifera indica*), tamarind (*Tamarindus indica*) and other trees. It has been recorded that bonnet populations dwelling in forests are shy of humans. Monkey populations around human settlements are habituated to human presence (Sinha 2021). These monkeys caused crop losses to a variety of agricultural and horticultural crops to the extent that farmers abandoned cultivation. In certain pockets of South India, the local monkey population explodes so much that individuals freely enter the houses, temples and orchards of villagers and take away fruits and other edibles. Many believe that *M. radiata* is an incarnation of God Hanuman. So it is worshipped, and people generally do not harm these animals. It is considered sacred in the Hindu religion. So people resort to providing food to these monkeys along roadsides, near temples and so on. They are of high nuisance value. Their frequent and large numbers even impacted cashew factories, beedi-manufacturing units, areca-sorting centres and agricultural operations. Threats to these monkeys include reduction and deterioration of natural habitat, human intolerance and conflict (mass poisoning of monkeys, acid attack, shooting, etc.), being locally hunted (for research and pets) and being considered as a pest on crops. In order to conserve bonnet monkey populations, the Forest Department is establishing a nursery of wild-fruit-bearing trees, distributing them to farmers and planting them in monkey-affected areas. The Forest Department is also restoring the natural food resources of monkeys in wild and forested tracts. Mewa et al. (2010) studied relationship between the lion-tailed macaque and bonnet macaques.

1.10 Hanuman Langur (*Semnopithecus entellus*, Desmarest 1822)

It is one of seven species of langur spread across Asia, and according to the IUCN, it is the least concern (LC) species. This species is named after the Hindu God Hanuman, the deity of healing and worship, and the langur is part man, part monkey. According, to Hindu scriptures, Hanuman is believed to be the son of Vayu, the wind god. His divine lineage endows him with exceptional strength and agility with monkeys. Langurs are considered sacred in Hindu culture. It is also called grey langur too, distributed from the Himalayas to Sri Lanka and from Bangladesh to Pakistan. Langurs spend half of their time on the ground and the other half in trees. They walk quadrupedally but make bipedal hops. Langurs can leap horizontally too. Nilgiri (*Semnopithecus johnii* Fisher, 1829) and Purple faced langurs ((*Semnopithecus vetulus* Erxleben, 1777) (Wilson et al. 2013; Roos et al. 2014) cluster with grey langurs. Grey langurs are diurnal, spend nights in tree canopies and prefer the highest branches for sleeping. Langurs are eaten by leopards, moles and tigers (Boggess 1976). Wolves, Jackals, Asian black bears and pythons too can prey on langurs.

Grey langurs are herbivores but do not feed on leaves and leaf buds alone. They prefer evergreen petioles, shoots, fruits, fruit buds and seeds. They also feed on coniferous needles and cones, grass, bamboo, fern rhizomes, mosses and lichens. Non-herbivorous items in their diet include termites, spiders and insect larvae (Srivastava 1991). Langurs raid agricultural crops (Starin 1978). Socially, langurs have three groups, namely, all-male, multiple males and females and one-male group, which include several females and offspring (Rajpurohit 1992; Newton 1994). Social hierarchies exist in all groups (Khan 2013). The rate of male replacement occurs depending on the group (Rajpurohit et al. 2003). Langurs use squeaks and shrieks to communicate stress (Sugiyama 1965). Grey langurs adults use whoops during displays. Alert calls on sighting a predator will be harsh barks. Grunts are made in agonistic situations (Bhaker et al. 2004; Hohmann 1989).

Countries like India have laws prohibiting the capture or killing of the langurs. But still, these animals are hunted. In India, grey langurs number about three lakhs (Mukherjee 2001). Langur populations are threatened by mining, forest fires and deforestation (Rao and Bhatnagar 2001). In few agricultural tracts, langurs are considered pests on crops (Chaudhuri et al. 2004) and are recommended for elimination from those areas (Chaudhuri et al. 2004). Langurs also steal food and raid crops and are also persecuted (Pirta et al. 1997).

1.11 Amur Falcon (*Falco amurensis*, Rudde 1863)

It is a small falcon that breeds in South-eastern Siberia and Northern China. It migrates in large flocks to India and Southern and East African coasts. During migration from October to December, the falcons pass through Nagaland, Manipur and Bangladesh in spectacular numbers before crossing the Arabian Sea and Indian Ocean into Africa. In Nagaland, a million falcons roost for a month. The town of Doxang in Manipur celebrates the Amur Falcon Festival in November, every year. Amur falcon is the longest migrating bird in India, covering 20,000 km, and is the longest travelling falcon. After the rains, East Africa produces swarms of termites, locusts, ants, beetles and dragonflies that provide ample food for the falcons, during the most arduous part of their migration route (Anderson 2009; Dixon et al. 2011).

While returning to their breeding grounds, these falcons are often found outside their normal range, such as in Italy and the United Kingdom (Rasmussen and Anderton 2005a, b; Orta 1994). Amur falcons feed mostly late in the evening or early in the morning on a wide range of insects in the air or on the ground (Orta 1994; Kopij 2009). Besides insects, falcons also feed on small birds and mammals (Alexander and Symes 2016). During migration, birds stay in open forests, grasslands or exposed perches (Orta 1994). According to the IUCN, the species is assessed as being of least concern.

During migration from breeding areas to winter quarters, birds are killed for food consumption in Northwest India and East Africa (BirdLife International 2016; Ali and Ripley 1987). The government of Nagaland in 2012 started a campaign to save the falcons. As part of the campaign, birds were fitted with 5 g satellite transmitters for tracking them (Boyes 2013).

Pangti, a village in Nagaland that became notorious for killing falcons in 2012, is now known as the falcon capital of the world. Manipur recorded 1.4 lakh birds at one roost site in 2023. In 2012, thousands of falcons had been trapped in nets and slaughtered in Nagaland. The scale of slaughter was alarming. The government of India next year, that is, in 2013, began a project on the conservation of Amur falcons in the north-east of India. Ten years gone, now Amur falcons are a cherished guest in the north-east. The Bombay Natural History Society (BNHS) launched a project to save falcons, create mass awareness and change the attitude of the local Nagaland people towards hunting. Then satellite tracking of the birds started. The conservation initiative soon became a massive community programme. The government and non-governmental organizations like Wildlife Trust of India declared 'zero hunting'. The hunters started chanting the slogans, 'We love Amur Falcons'. The engagement of local communities in falcon conservation has yielded positive results (Kalita 2023).

1.12 Vultures

Population reduction of vultures has been recorded in Pakistan, Bangladesh and Nepal (Gilbert et al. 2006). Added current declines in Egyptian vulture (*Neophron percnopterus*) and red-headed vulture (*Sarcogyps calvus*) have been recorded across India (Cuthbert et al. 2006; Elorriaga et al. 2009). Scientists are of the opinion that the causes for the reduction may be due to the use of the veterinary drug diclofenac for administering cattle (Liberatori and Penteriani 2001; Prakash 1999), and it may also be due to food shortages and habitat loss (Grubh 1974; Haenn et al. 2014).

The status of the previously mentioned six species of vultures is documented (Majumder et al. 2009; Kumar et al. 2018) and briefly summarized subsequently.

1.12.1 Cinereous Vulture (*Aegypius monachus*, Linn 1766)

The vultures inhabit scrub, grassland and forest tracts. The cinereous vulture is a carrion feeder. The birds also feed on insects, tortoises, snakes and lizards. These vultures are migratory and share a large range across Europe and Asia (Carrete and Donazar 2005).

1.12.2 Griffon Vulture (*Gyps fulvus*, Hablizl 1783)

Griffon vultures are identified with white feathers on the head and neck. The sexes morphologically appear different. The sounds produced are like that of an Indian white-backed vulture, a threatening hoarse hiss. These vultures breed up to 2500–3000 m in the lower foothills of mountains and deserts. These birds are gregarious, with habits not unlike others, and attracted to carcasses by sight. Vultures exclusively feed on carrion, also on refuse and they are migratory in habit.

1.12.3 White-Backed Vulture (*Gyps bengalensis*, Gmelin JF 1788)

The Indian white-backed vultures have a white spot on their back and upper tail coverts, which are the identification characteristics. This species is not very vocal in searching for carrion. This vulture feeds solely on carrion. These vultures are residential and mostly found in deserts and semi-arid zones. Breeding occurs in desert fringes.

1.12.4 Long-Billed Vulture (*Gyps indicus*, Scopoli 1786)

The Indian long-billed vulture is a moderate-sized bird. The neck is sparsely covered with feathers, which are white. The long-billed vulture produces various types of noise. These vultures nest and roost on forts. Specifically, the birds prefer open land and adjacent villages to towns. Vultures feed on carrion, dead livestock exclusively. These species are distributed in India, Pakistan and Nepal.

1.12.5 Himalayan Vulture (*Gyps himalayensis*, Hume 1869)

This vulture is an old-world species native to the Himalayan region and Tibetan Plateau. These vultures produce various sounds. This species prefers to inhabit terrestrial in open to slightly wooded patches. These vultures are gregarious, exclusively feeding on carrion, and are migratory. *G. himalayensis* is distributed in the Himalayas. The bird is found in China and adjacent countries, Afghanistan and Mongolia. In April 2024, 12 of these vultures were sighted in Etawah Safari Park, Uttar Pradesh, India, on a *Bombax ceiba* tree. It is listed as a near-threatened (NT) species as per the IUCN.

1.12.6 Egyptian Vulture (*Neophron percnopterus*, Linn 1758)

Egyptian vultures are small and have bills that are hooked at the tip. The whole head is naked. It is a silent bird. The birds prefer dry and open habitats. It frequents human settlements feeding on refuse. It is an opportunist scavenger.

A study jointly conducted by the National Centre for Biological Sciences (NCBS), Bengaluru; Bombay Natural History Society (BNHS), Mumbai; Department of Zoology, Cambridge University, UK; Karnataka Vulture Conservation Trust, Bengaluru; and Hume Centre for Ecology and Wildlife Biology, Kerala, employed metal barcoding technique to analyse faecal samples from four vulture species, namely, *Gyps bengalensis*, *Gyps indicus*, *Gyps tenuirostris* and *Gyps himalayensis*. Large ungulates (domestic and wild herbivores) were the main dietary components for the vultures. Harris (2013) reviewed literature on conservation of vultures in Nepal.

There should be a ban on veterinary drugs, diclofenac and other non-steroidal anti-inflammatory drugs (NSAIDs). These findings will contribute to the ground management of the natural vulture populations (Times of India 2024b).

1.13 Siberian Crane (*Leucogeranus leucogenanus*, Pallas 1773)

This majestic crane (Fig. 1.4) qualifies as critically endangered owing to rapid global population decline. This is because of the development of three big and several other smaller dams across the Yangtze River and its tributaries in China, which are the wintering grounds of the crane (BirdLife International 2024).

Fig. 1.4 Siberian crane

The world population in 2010 was 3200 birds. The eastern population breeds in northwestern Siberia and winters at Poyang Lake in the lower Yangtze River basin in China. In the western/central population, a single crane continues to winter along the south coast of the Caspian Sea in Iran covering a 5000 km migratory route. The Indian flock of Siberian cranes is now extinct. In 1974, 75 birds wintered in Bharatpur Bird Sanctuary, which declined to a pair in 2002. This is because of the mass hunting of cranes in Afghanistan every year. Stopover habitats are crucial refuelling sites for Siberian cranes. The maximum entropy model was used to identify suitable stopover habitats in Northeast China. As per the International Crane Foundation, Siberian cranes are now only found in Siberia to China, the eastern population. The western population is found on the coast of the Caspian Sea in Iran (Kanai et al. 2002a, b).

Siberian cranes not only inhabit shallow marshlands and wetlands but may also forage in deeper water than other cranes. These cranes show high fidelity in both wintering and breeding grounds. Siberian cranes are highly territorial. The cranes feed almost throughout the day. At their winter roosts, birds may feed in small and loose flocks. While feeding on submerged vegetation, they immerse their heads entirely underwater. While calling, the birds stretch their neck forward. Siberian cranes are omnivorous. In summer, they feed on a range of plants, roots, seeds, small rodents like voles and lemmings, earthworms and fish. This species breeds in two distinct regions in the arctic tundra of Russia; the western population along the

Ob and Yakutia and western Siberia. The conservation status of the Siberian crane is very serious. Of the 15 crane species, this is only the species regarded as critically endangered. The whooping crane of North America was endangered but is now well protected. Satellite telemetry is being used to track the migration route of the cranes. The cranes are sacred birds associated with the sun, spring and celestial spirits and are a symbol of long life and happiness. India and Russia have paired up to track and breed Siberian cranes; a postage stamp was issued on cranes in 1983. So, of the three regional populations of cranes in Siberia, only one survives—the one that breeds in the Yakutia region of northwestern Siberia and spends the winter in the Yangtze River basin in China (Bhattacharya 2023). Johnsgard (1983) published a book on world of cranes.

1.14 Indian Giant Flying Squirrel (*Petaurista philippensis*, Elliot 1839)

The Indian giant flying squirrel is a globally LC species, assessed as NT in South Asia. It is distributed in China, India, Lao, Myanmar, Sri Lanka, Taiwan, Thailand and Vietnam. This species occurs in tropical and subtropical dry deciduous; semi-evergreen to evergreen forests (Molur et al. 2005). India is a hot spot of flying squirrel diversity. However, the suitable ecological needs and habitats of flying squirrels are very few.

The occurrence of Indian giant flying squirrels is recorded in very few states, namely, Gujarat, Maharashtra, Rajasthan and Madhya Pradesh. It is possible that the squirrel occurs in other states but not documented. In India, 13 flying squirrel species are found. Recently, another two flying squirrel species were identified in the last decade in India, namely, *Petaurista nigra* and *Petaurista mishmiensis*. *P. philippensis* was recorded in broader distribution, except in the northeast region of the Himalayas (tropical moist deciduous forest) and the Western Ghats (tropical wet evergreen forest) (Bhatnagar et al. 2010a, b).

A survey was conducted at Sitamata Wildlife Sanctuary in tropical deciduous forests in Rajasthan, India, from the early night 1900 h to 0600 h in next day morning. A total of 268 call bouts were recorded. Call frequency was the highest in March 2009. The most vocal periods were at midnight and early night just before arriving at nest cavities (Hutton 1947). There were insignificant ($P > 0.05$) values obtained between male and female call frequencies in comparison with annual and seasonal data recorded. The 5–10-min call bout category was the highest account (40%) in the mid-canopy area of the trees. Perch height was found positively related to tree height.

The Indian flying squirrel is well-spread in the tropical regions of South Asia. The squirrel is a forest dweller, particularly in valleys and lowlands up to 4000 m. The Indian flying squirrel prefers to feed on fruits, leaves, buds and so on. However, the squirrel sustains on pith during rainy periods (Nandini 2001; Nandini and Parthasarathy 2008). The squirrel is a nocturnal, solitary animal. Fan and Jiang (2009) recorded gibbons predating on flying squirrels in China.

Recently, Molur et al. (2005) from South Asia considered it NT. Due to the nocturnal, arboreal and cryptic nature of the Indian giant flying squirrel, it is the least studied mammalian taxa in South Asia. According to Schedule II of India's Wild Life (Protection) Act 1972, it needs complete protection. The main threats to the species are habitat loss, human interruption with hunting and habitat degradation (Lee and Liao 1998). Forest fires, logging and shifting cultivation have impacted the species.

To identify the diversity behind the spatial pattern of occurrence and population pattern, the tool, species distribution models (SDMs), can be used. The tool will aid in ecological forecasting, range mapping and assessing the relationship between species occurrence and environmental characteristics. But it was difficult to collect the information due to the cryptic behaviour of flying squirrels and environmental constraints (Kuo and Lee 2012). Base estimators (combining multiple models) are the prediction process having more appropriate elements. Bagging, also known as bootstrap aggregation, is an ensemble modelling algorithm that is trained with random data subsets from the training dataset to increase model variance. Another way of accurate prediction is by using numerical prediction as a weighted average that can be used to distinguish test incidents for a specific machine learning method. Ensemble modelling is a powerful tool that uses computationally advanced algorithms to map species and forecast future habitat suitability for conservation and future work. Food habits, gliding behaviour food habits and distribution of flying squirrel have been documented (Koli et al. 2011, 2013a, b).

1.15 Crocodile (*Crocodylus palustris*, Lesson 1831)

The mugger crocodile (*Crocodylus palustris*) is a medium-sized broad-snouted crocodile, also known as mugger and marsh crocodile. It is native to freshwater habitats from southern Iran to the Indian subcontinent, where it inhabits marshes, lakes, rivers and artificial ponds. It rarely reaches a body length of 5 m (16 ft 5 in.) and is not only a powerful swimmer but also walks on land in search of suitable waterbodies during the hot season. Females dig holes in the sand as nesting sites and lay up to 46 eggs during the dry season. The sex of hatchlings depends on temperature during incubation. Both parents protect the young for up to 1 year. They feed on insects, and adults prey on fish and mammals.

With the increase in crocodile population from 96 in 1974 to 1811 in 2024, the Odisha forest department wound up in a crocodile rearing programme in Bhitarkanika National Park. Ashish (2024) highlighted in *The Times of India*, how the small habitat had big crocodiles beyond its carrying capacity. Since the crocodile by its behaviour is aggressive predators killed more than 30 people since 2013. Choudhury BC, a wildlife expert and retired scientist of the Wildlife Institute of India, Dehradun, suggested urgent measures to contain the uncontrolled growth. Locals live in fear of formidable reptiles filtrating their ponds, deterring them from venturing near the rivers and creeks. Significant concerns were initiated due to the presence of crocodiles roaming 70–100 km away from their natural habitats. 'The rear and release method adopted for the project has significantly increased the number of crocodiles'. Three species of crocodiles: saltwater crocodile (*Crocodylus porosus*), mugger (*Crocodylus palustris*) and gharial (*Gavialis gangeticus*), were part of the breeding programme that started in 1974.

In India, technology is being used to conserve wildlife in different habitats. Rotor Precision Group in collaboration with the Wildlife Institute of India (WII) is deploying drones in habitat mapping, population estimation and monitoring crocodiles in Ken Gharial Sanctuary, Madhya Pradesh (TOI, 26 Feb 2024).

1.16 Conservation: Constraints

In most of the tropical countries, there are laws/legislation to protect wildlife. But there is no implementation. The frameworks of the regulatory bodies concerned with the environment, wildlife and forests are frequently broken, and nothing consistent about consideration is implemented on the ground. For example, the great Indian bustard (*Ardeotis nigriceps*; Vigors, 1831) population in grasslands and arid regions in India is on the brink of extinction. According to IUCN, it is a critically endangered species and it is on the verge of local extinction. In the decision-making process, the local communities are ignored. Scientists/naturalists have ideas but not transmitted to the ground and implemented because conditions are imposed to implement the ideas. It also depends on the 'political will' of the government.

If the need arises to breed animals in captivity, like cheetahs in India and elephants in East African countries, then a method akin to ranching can be adopted. Captive-bred animals can be released in the open so that ecological services and balance can be restored. Wildlife conservation in the semi-urban and urban spaces can be quite demanding today. Leopards, elephants, bonnet macaques, owls and other animals are straying into the towns, outskirts of cities and around human settlements. Wild animals are also moving close to national parks, sanctuaries and other areas on the roads and adjacent areas. Farmers are mass poisoning animals that raid crops. These are difficult situations to reserve. They need to be addressed in a balanced manner and holistic way.

References

Alexander J, Symes CT (2016) Temporal and spatial dietary variation of Amur falcons (Falco amurensis) in their South African nonbreeding range. J Raptor Res. 50(3):276–288

Ali S, Ripley SD (1987) Compact handbook of the birds of India and Pakistan. 2nd Ed. Oxford University Press, Delhi, pp i–xlii, 1 l., 1–737

Anandam MV (2013) Family *Cercopithecidae*: old world monkeys. In: Mittermeier RA, Rylands Anthony B, Wilson DE (eds) Handbook of the mammals of the world, vol 3. Primates. Lynx Edicions, Barcelona, pp 238–639. ISBN 978-84-96553-89-7.

Anderson RC (2009) Do dragonflies migrate across the western Indian Ocean? J Trop Ecol 25(4):347–358

Arivazhagan C, Arumugam R, Thiyagesan K (2007) Food habits of leopard (Panthera pardus fusca), dhole (Cuon alpinus) and striped hyena (Hyaena hyaena) in a tropical dry thorn forest of southern India (PDF). J Bombay Natl Hist Soc. 104(2). Archived from the original (PDF) on 4 Mar 2009

Ashish S (2024) Crocodile breeding stopped. Times of India, 25 Feb 2024, p 13

Azad S (2023) Pegged at 3,167 in April, Indias tiger population now placed at 3,682. TNN/ updated: 30th July 2023. Times of India.

Banks D (2004) The tiger skin trail. Environmental Investigation Agency

Bashaw JM, Kelling AS, Bloomsmith MA, Maple TL (2007) Environmental effects on the behavior of zoo-housed lions and tigers, with a case study on the effects of a visual barrier on pacing. J Appl Anim Welf Sci 10:95–109. https://doi.org/10.1080/10888700701313116

Bauer H, Iough HD, Di Silvestre I (2003) Lion social behavior in the west and central African Savannah belt. Mammal Biol 68:239–243

Bhaker NR, Rajpurohit DS, Rajpurohit LS (2004) Vocalization in Hanuman langur, Semnopithecus entellus around Jodhpur, Rajasthan. Uttar Pradesh J Zool 24(3):227–233

Bhatnagar C, Koli VK, Sharma SK (2010a) Summer diet of Indian giant lying squirrel (*Petaurista philippensis* Elliot) in Sitamata wildlife sanctuary, Rajasthan, India. J Bom Nat Hist Soc 107:183–188

Bhatnagar C, Sharma SK, Koli VK (2010b) High day temperature and sleep out behaviour of Elliot's giant lying squirrel (*Petaurista philippensis*) (Elliot) in Sitamata Wildlife Sanctuary Rajasthan, India. J Bom Nat Hist Soc 107:245–246

Bhattacharya A (2023) A flight shot down in Afghan air space. Times of India, 7 Jan 2023, p 15

BirdLife International (2016) Falco amurensis. IUCN Red List of threatened species. 2016: e.T22696437A93561051

BirdLife International (2024) Species fact sheet, Leucogernus leucogernaus. https://datazone.Birdlife.org/species/factsheet/Siberian-Crane. Accessed 7 Feb 2024

Biswas BK (2022) Effect of climate change on vector-borne disease. In: Mondal S, Singh RL (eds) Emerging issues in climate smart livestock production: biological tools and techniques. Academic Press, pp 263–316, 502

Boggess JE (1976) Social behavior of the Himalayan langur (*Presbytis entellus*) in eastern Nepal. PhD dissertation, University of California, Berkeley

Boyes S (2013) Safe passage for amur falcons through India.blog.nationalgeographic.org. https://blog.nationalgeographic.org/2013/11/20/safe-passage-for-amurfalcons-through-india/. Accessed 5 June 2018

Breed MD, Moore J (2022) Animal behavior book, 3rd edn–2021. Acad Press, p 588. https://doi.org/10.1016/C2018-0-03842-6

Carrete M, Donazar JA (2005) Application of central-place foraging theory shows the importance of Mediterranean dehesas for the conservation of the Cinereous Vulture (*Aegypus monachus*). Biol Conserv 126:582–590

Chaudhuri S, Murmu A, Talukder B, Alfred JR (2004) A population survey of Hanuman langurs in the district of Purulia, West Bengal. Rec Zool Surv India. 103(3–4):47–54. https://doi.org/10.26515/rzsi/v103/i3-4/2004/159337. S2CID 251714842

Cuthbert R, Green RE, Ranade S, Saravanan S, Pain DJ, Prakash V, Cunningham AA (2006) Rapid population declines of Egyptian vulture (*Neophron percnopterus*) and red-headed vulture (*Sarcogypscalvus*) in India. Anim Conserv 9:349–354

Dileep K, Jose B (2014) Social behaviour in Bonnet macaque with reference to grooming. Int J Innov Sci Eng Technol 1(8):340–343

Dixon A, Purev-Ochir G, Batbayar N (2011) Autumn migration of an Amur Falcon *Falcon amurensis* from Mongolia to the Indian Ocean tracked by satellite. Forktail 27:81–84

Elorriaga J, Zuberogoitia I, Castilo I, Azkona A, Hidalgo S, Astorkia L, Ruiz-Moneo F, Iraeta A (2009) First documented case of long distance dispersal in the Egyptian Vulture. J Raptor Res 43:142–145

Fan P, Jiang X (2009) Predation on giant lying squirrels (*Petaurista philippensis*) by black crested gibbons (Nomascus concolor jingdongensis) at Mt. Wuliang, Yunnan, China. Primates 50:45–49. https://doi.org/10.1007/s10329-008-0110-529

Gilbert M, Watson RT, Virani MZ, Oaks JL, Ahmed S, Chaudhry MJI, Arshad M, Mahmood S, Khan AA (2006) Rapid population declines and mortality clusters in three oriental white-backed vulture Gyps bengalensis colonies in Pakistan due to diclofenac poisoning. Oryx 40:388–399

Goswami S, Tyagi PC, Mallik PC et al (2019) Effects of personality and life-history on the welfare of captive Asiatic lions. https://doi.org/10.7287/peerj.prepoints.27495v. p 48

Grubh RB (1974) The ecology and behaviour of vultures in Gir forests. PhD Thesis, University of Bombay, Mumbai

Haenn N, Schmook B, Reyes YM, Calmé S (2014) A cultural consensus regarding the king vulture? Preliminary findings and their application to Mexican conservation. Ethnobiol Conserv 3:1

Harris RJ (2013) The conservation of Accipitridae vultures of Nepal: a review. J Threat Taxa 5(2):3603–3619

Hohmann G (1989) Comparative study of vocal communication in two Asian leaf monkeys, Presbytis johnii and Presbytis entellus. Folia Primatol 52(1–2):27–57. https://doi.org/10.1159/000156380. PMID 2807093

Holbrook MR (2012) Kyasanur forest disease, Review, Antiviral Res. Decision 96(3):353–362. https://doi.org/10.1016/j.antiviral.2012.10.005

Hutton AF (1947) Nesting habits of the lying-squirrel (*Petaurista philippensis*). J Bom Nat Hist Soc 46:539–540

Mohan V (2024) Leopard numbers rise 8% in four years in India, Uttarkhand reports highest decline. Times of India March 1, p 1

Jain M (2024) Saving animals is ethical, it's evocative. And it's smart business. Times of India, 19 Mar 2024, p 12

Johnsgard P (1983) Cranes of the world (PDF). Indiana University Press, pp 129–139. ISBN 978-0-253-11255-2.

Kalita P (2023) Once massacred Amur falcons a cherished guest in North East. Times of India, 26 Nov 2023, p 11

Kanai Y, Mutsuyuki U, Germogenov N, Negandran M, Mita N, Higuchi H (2002a) Migration routes and important resting areas of Siberian cranes (*Grus leucogeranus*) between northeastern Siberian and China as revealed by satellite tracking (PDF). Biol Conserv 106(3):339–346. https://doi.org/10.1016/S0006-3207(01)00259-2

Kanai YN, Ueta M, Markin M, Rinne Y, Sorokin J, Alexander G, Hiroyoshi H, Archibald George W (2002b) Discovery of breeding grounds of a Siberian Crane (*Grus leucogeranus*) flock that winters in Iran, via satellite telemetry. Bird Conserv Int. 12(4):327–333. https://doi.org/10.1017/S0959270902002204

Khan BN (2013) Impact of captivity on social behavior of chimapanzee (Pan Troglodytes) at Lahore zoo, Pakistan. J Anim Plant Sci 23:779–785

Koli VK, Bhatnagar C, Mali D (2011) Gliding behaviour of Indian giant lying squirrel (*Petaurista philippensis* Elliot). Curr Sci 100:1563–1568

Koli VK, Bhatnagar C, Sharma SK (2013a) Food habits of Indian giant lying squirrel (*Petaurista philippensis* Elliot) in tropical deciduous forest, Rajasthan, India. Mammal Study 38:251–259. https://doi.org/10.3106/041.038.0409

Koli VK, Bhatnagar C, Sharma SK (2013b) Distribution and status of Indian giant lying squirrel (Petaurista philippensis Elliot) in Rajasthan, India. Natl Acad Sci Lett 36:27–33. https://doi.org/10.1007/s40009-012-0105-z

Kopij G (2009) Seasonal variation in the diet of the Amur kestrel (Falco amurensis) in its winter quarter in Lesotho. Afr J Ecol. https://doi.org/10.1111/j.1365-2028.2009.01130.x

Kotez R, Keith M, Winterbach CW, Winterbach HEK et al (2018) The influence of social and environmental factors on organization of African Lion (Panthera leo) prides in the Okavango Delta. J Mammol 99(4):845–858

Kumar A, Kumar S (2023) Food plants of wild gaur (*Bos gaurus* Smith, 1823) in fruiting plants and Wildlife. ISBN 978-81-955847-8-9

Kumar A, Mishra S, Yadav R, Sinha A, Kanaujia A (2018) Vultures of Pilibhit Tiger Reserve (PTR). Adv Biores 9(2):135–144. Society of Education, India

Kuo CC, Lee LL (2012) Home range and activity of the Indian giant flying squirrel (*Petaurista philippensis*) in Taiwan: Influence of diet, temperature, and rainfall. Acta Theriol 57:269–276

Lee PF, Liao CY (1998) Species richness and research trend of lying squirrel. J Taiwan Museum 51:1–20

Liberatori F, Penteriani V (2001) A long-term analysis of declining population of the Egyptian Vulture in Italian peninsula: distribution, habitat preference, productivity and conservation implications. Biol Conserv 101:381–389

Majumder A, Basu S, Sankar K, Qureshi Q (2009) Some observations on vultures in Pench Tiger Reserve, Madhya Pradesh. J Threat Taxa 1(12):627–628

Mellen JD, Sheperdson DJ (1997) Environmental enrichment for felids: an integrated approach. Int Zoo Yearbook 35:191–197. https://doi.org/10.1111/j.1748-1090.1997.tb01209.x

Mewa S, Tephillah J, Prashanth U, Werner K (2010) Male–male relationships in lion-tailed macaques (*Macaca silenus*) and bonnet macaques (*Macaca radiata*) (PDF). Int J Primatol 32:167–176. https://doi.org/10.1007/s10764-010-9448-9. S2CID 18967639. Archived from the original (PDF) on 14 Oct 2014

Mohan V (2024) Leopard numbers rise 8% in four years in India, Uttarkhand reports highest decline. Times of India March 1 .p1

Molur S, Srinivasulu C, Srinivasulu B, Walker S, Nameer PO, Ravikumar L (2005) Status of South Asian non-volant small mammals: conservation assessment and management plan (C.A.M.P.) Workshop Report. Zoo Outreach Organisation/CBSG-South Asia, Coimbatore, India, 618 pp

Mondal K, Gupta S, Bhattacharjee S, Quresh Q, Sankar K (2012) Prey selection, food habits and dietary overlap between leopard Panthera pardus (Mammalia: Carnivora) and re-introduced tiger Panthera tigris (Mammalia: Carnivora) in a semi-arid forest of Sariska Tiger Reserve, Western India. Ital J Zool 79(4):607–616. https://doi.org/10.1080/11250003.2012.687402

Mukherjee RP (2001) Status and conservation of non-human primates in India. ENVIS Bull Wildl Prot Areas 1(1):136–137

Mukherjee K (2024) Asian elephants bury dead calves, finds study. Times of India, 1 Mar 2024, pp 1 and 3

Mukherjee K, Bhattacharya KK, Ramanath R (2023) Why humans, Jumbos art at war in Bengal's border areas. Times of India, 2 Apr 2023, p 11

Nandini R (2001) The distribution and status of lying squirrels in Karnataka and Goa. Technical report. SACON, Coimbatore

Nandini R, Parthasarathy N (2008) Food habits of the Indian giant flying squirrel (*Petaurista philippensis*) in a rain forest fragment, Western Ghats. J Mammal 89(6):1550–1556. https://doi.org/10.1644/08-MAMM-A-063.1

Naveen P (2022) Seven decades after extinction cheetahs land in India's heart. TOI, Sept 18, p 1

Newton P (1994) Social stability and change among forest Hanuman langurs (*Presbytis entellus*). Primates 35(4):489–498. https://doi.org/10.1007/BF02381957. S2CID 40545787

Orta J (1994) Amur falcon. In: del Hoyo J, Elliott A, Sargatal J (eds) Handbook of birds of the world, vol 2. New world vultures to guineafowl. Lynx Edicions, Barcelona, pp 265–266

Packer C, Hilborn R, Mosser A, Kissni B et al (2005) Ecological change, group territoriality and population dynamics in Serengeti lions. Science 307:389–393

Pirta RS, Gadgil M, Kharshikar AV (1997) Management of the rhesus monkey Macaca mulatta and Hanuman langur Presbytis entellus in Himachal Pradesh, India. Biol Conserv. 79(1):97–106. https://doi.org/10.1016/0006-3207(95)00131-X

Prakash V (1999) Status of vultures in Keoladeo National Park, Bharatpur, Rajasthan, with special reference to population crash in Gyps species. J Bombay Natl Hist Soc 96(3):365–378

Prater SH (2005) The book of Indian animals, 5th edn. Bombay Natural History Society and Oxford University Press, pp 194–197

Qureshi Q, Jhala YV, Yadav SP, Tiwari VR, Garawad R, Mallick A (eds) (2024) Status of Leopards in India, 2022. National Tiger Conservation Authority, Government of India, New Delhi, and Wildlife Institute of India, Dehradun. ISBN No: 81-85496-86-2

Rajpurohit LS (1992) Origin and composition of the unisexual unit—an all-male band in Hanuman langur, (*Presbytis entellus*), around Jodhpur, India. Prim Rep 34:47–52

Rajpurohit LS, Chhangani AK, Rajpurohit RS, Mohnot SM (2003) Observation of a sudden resident-male replacement in a unimale bisexual troop of Hanuman langurs, Semnopithecus entellus, around Jodhpur (India). Folia Primatol 74(2):85–87. https://doi.org/10.1159/000070002. PMID 12778910. S2CID 46828619

Ramanath R (2024) Black tigers thrive in Odisha's Similipal. Times of India, 4 Feb 2024, p 18

Rao RJ, Bhatnagar A (2001) Primates of the Amarkantak forests, Madhya Pradesh. ENVIS Bull Wildl Prot Areas 1(1):120–123

Rasmussen PC, Anderton JC (2005a) Birds of South Asia. The ripley guide, vol 2. Smithsonian Institution/Lynx Edicions, Washington, DC/Barcelona, p 113

Rasmussen PC, Anderton JC (2005b) The birds of South Asia. The ripley guide, vol 2. Smithsonian Institution/Lynx Edicions, Washington, DC/Barcelona, p 138

Raza RH, Chauhan DS, Pasha MKS, Sinha S (2012) Illuminating the blind spot: a study on illegal trade in Leopard parts in India (2001–2010) (PDF) (Report). TRAFFIC India, WWF India, New Delhi. Archived (PDF) from the original on 24 Sept 2020. Accessed 9 Aug 2020

Reddy S (2024) Land grab in Telangana forests shrinks space for big cats. Times of India, 4 Feb 2024, p 18

Roos C, Boonratana R, Supriatna J, Fellowes JR, Groves CP, Nash SD, Rylands AB, Mittermeier RA (2014) An updated taxonomy and conservation status review of Asian primates (PDF). Asian Primates J 4(1):2–38

Sinha A (2021) The monkey in the town's commons: a natural history of the Indian Bonnet Macaque. NIAS Report. p 41

Srivastava A (1991) Insectivory and its significance to langur diets. Primates. 32(2):237–241. https://doi.org/10.1007/BF02381181. S2CID 32090680

Srivastava KK, Bhardwaj AK, Abraham CJ, Zacharias VJ (1996) Food habits of mammalian predators in Periyar Tiger Reserve, South India. Indian For 122(10):877–883. Archived from the original on 20 Jan 2015. Accessed 22 Mar 2013

Starin ED (1978) A preliminary investigation of home range use in the Gir Forest langur. Primates 19(3):551–568. https://doi.org/10.1007/BF02373316. S2CID 106751

Sugiyama Y (1965) Behavioral development and social structure in two troops of Hanuman langurs (*Presbytis entellus*). Primates. 6(2):213–247. https://doi.org/10.1007/BF01730967. S2CID 10784560

Team NSI (2023) Big cats of India, info@bigcatsindia.com Infofacts and Safari Tour

Times of India (2023) Twelve South African cheetahs to land in India tomorrow. TOI 17 Feb 2023, p 13

Times of India (2024a) Drones keeping an eye on Ken River Crocodiles, TOI 26 Feb 2024

Times of India (2024b) Vultures feed on large ungulate Carcasses: study. TOI 22 Mar 2024, p 6

Vijay P (2023a) Call of the wild: Tigress grooms mother's cub. Times of India, 26 Nov 2023, p 16

Vijay P (2023b) India has space to sustain more tigers. Times of India, 30 Jul 2023, p 6

Vijay P (2023c) Tiger populations growth rate of less than 1% a year, nothing to bring about. Times of India, 26 May 2023, p 10

Vijay P (2023d) Tiger conservation: efforts bring carbon benefits worth. Times of India, 12 Jun 2023, p 6

Wikipedia (2007) Historical and present distribution of Lion. https://en.wikipedia.org/wiki/Lion

Wildlife Protection Society of India (2010) Leopards battling for survival in India. Archived 22 Nov 2022 at the Wayback Machine. Wildlife Protection Society of India, 18 May 2010

Wilson DE, Mittermeier RA, Rylands AB (2013) Introduction. In: Handbook of the mammals of the world, vol 3. Primates, 1st edn. Lynx Edicions. ISBN 978-8496553897. Retrieved 30 Aug 2018

Is Photography Only Advantageous for Research and Ecotourism? A Reassessment to Increase Its Value in Conservation Efforts

2

Vijay Kumar Koli

Abstract

Photographing wild animals in their native habitat necessitates careful observation, patience, persistence, practice, time, and, most importantly, the right equipment. Photography is an integral part of many research projects worldwide, as well as a significant component of global ecotourism, providing economic hope for many of the world's threatened natural areas through community-based conservation. This activity is at a higher pace nowadays due to increased environmental awareness, birding activities, and the advancement of digital technology. In addition to its widespread uses, photography also has a negative impact on ecosystems and wildlife, but the majority of available literature is currently descriptive, anecdotal, and biased towards its utility. In this review, 11 types of stimuli related to four major sources, including photographers, their vehicles, cameras, and some unethical photographic activities, were identified as detrimental to wildlife and their habitats. The consequences ranged from animal disturbance to species decline. Hence, in order to achieve proper conservation outcomes, these negative effects need to be reduced. This review summarises and discusses collected data to assist workers in planning to develop technologies in response to these effects, which improves its value both in research and conservation.

Keywords

Camera trap · Unethical photography · Ecology · Negative effect · Conservation

V. K. Koli (✉)
Wildlife Research Laboratory, Department of Zoology, M L Sukhadia University, Udaipur, Rajasthan, India

S. Sridhara et al. (eds.), *Animal Behavior in the Tropics*,
https://doi.org/10.1007/978-981-97-7324-4_2

2.1 Introduction

Wildlife or nature photography is a fascinating activity and also emotionally engages people with biodiversity and nature (Hanisch et al. 2019). Photographing from micro to macro and from air to underwater, however, necessitates careful observation, patience, persistence, practice, time, and, most importantly, the proper equipment (Excell 2012). Subjects and areas vary as nature itself, and they are frequently used to convey action, ranging from a static portrait to a subject's behaviour. A wide range of cameras are now available in the market due to technical advancement, and this technology also advances in the absence of humans by combining airborne devices (e.g. drones) and remote photography technology (e.g. camera traps) (Morgan et al. 2010; Rovero et al. 2013; Tang and Shao 2015).

Photography has significantly contributed to wildlife research and ecotourism (Mossaz et al. 2015; Slater et al. 2019; Hanisch et al. 2019; Stronza et al. 2019). It has been utilised in almost every field-based wildlife research since its invention (e.g. faunal listing, behavioural documentation, population assessment and management, habitat mapping, vegetation assessments). Remote photography has improved traditional methods and allowed researchers to study many nocturnal and elusive species (Trolliet et al. 2014). Camera traps are the one of the best examples, and their use has increased even during the COVID-19 pandemic (Blount et al. 2021).

Wildlife or nature-based tourism is considered as a viable and growing economic activity that reduces human impact on wildlife habitats and promotes conservation, particularly in developing economies (Isaacs 2000; Karanth et al. 2012; Mossaz et al. 2015). Sustainability has also been accepted and adapted as a goal for wildlife tourism in both government and private sectors, necessitating management processes to demarcate negative impacts for action implementation and ongoing monitoring. Birding and wildlife photography are important components of non-consumptive wildlife tourism and income generators because they involve groups of educated, above-average-income people (Higginbottom 2004). They invest in a range of cameras and related accessories (Skrzypczak 2010). Because of improved performance and lower prices in digital single-lens reflex (DSLR) cameras and autofocus telephoto lenses, digital photography has also become an affordable option for research and recreation. Wildlife photography is the most rapidly growing activity in nature-based tourism, owing to the zeal of many photographers for adventure and their resources, providing economic (employment and entrepreneurship opportunities; Kiss 2004), infrastructure (better transportation, health care, and education), and social welfare (psychological, economic, political, and social empowerment; Karanth et al. 2012) benefits for many threatened species. Hence, photographers are considered excellent ecotourists for community-based conservation (Basnet et al. 2021).

The contribution of photographic techniques to wildlife research and management is well-established (Cutler and Swann 1999; Morgan et al. 2010; Rovero et al. 2013; Trolliet et al. 2014; Burton et al. 2015; Caravaggi et al. 2017), but rising disposable incomes, improved access of urban citizens, increased publicity, public interest in wild places and species (Karanth et al. 2012), and the boom in digital technology have greatly increased photographic activities (Excell 2012). However human presence has been identified as an important conservation issue because it disrupts animal behaviour and physiology (Weston et al. 2012). Photographers are thought to be the most likely visitors to approach birds (Klein 1993). Many times, due to a high expectation to shoot certain species, they spend more money on guides to see the species in their native habitat, where the average ecotourists are not interested (Sekercioglu 2002; Kanagavel et al. 2013). During sensitive times (e.g. nesting and breeding), these activities can have significant negative consequences. Few studies (MacArthur et al. 1979; Gabrielsen and Smith 1995) also show that when wildlife is provoked by humans rather than mechanical disturbance, the physiological defence response is much more dramatic and has a much more profound effect (Gutzwiller et al. 1994). The magnitude of the response is also affected by distance, approach movement pattern, and the animal's access to cover. Disturbance diverts time and energy away from activities that improve fitness, such as foraging, feeding, parental care, and mating displays (Frid and Dill 2002).

Thus, the rapidly growing craze (Kanagavel et al. 2013), as well as the high potential of photographic activities for wildlife research, management, and ecotourism, and their impact on wildlife must be minimised. Although modern camera traps are thought to be non-invasive (Rovero et al. 2010; Burton et al. 2015), the assumption that their presence has no effect on animal behaviour may not always be correct. As a result, this review is the first of its kind, highlighting the negative effects of various photographic activities as reported in peer-reviewed articles so that a debate in the scientific community can begin to improve its conservation value and provide a baseline for policy and camera makers. Suggestions for future research are also included.

2.2 Material and Methods

A systematic search for English articles on the use of photography and cameras in wildlife research and ecotourism was conducted using search engines of Google Scholar, Scopus, Springer, Taylor & Francis, BioOne, and Wiley-Blackwell for English articles from 1990 to 2022. Several keyword combinations were used, including 'camera', 'camera trap', 'photography', 'unethical photography', 'aerial photography', 'drone', 'wildlife research', 'ecotourism', 'animal', 'mammal', 'birds', 'nature', 'animal trade', 'animal trafficking', and 'animal disturbance'. A snowball sampling approach was used to track back references in the search for relevant literature (Handcock and Gile 2011).

I identified four major sources in photographic activities that have adverse impacts on animals and their habitats: the photographer itself, the photographer's vehicle, the camera, and other unethical photographic activities. The types of stimuli mentioned in these sources that influenced wildlife were identified and listed. To better understand these effects were also divided into two categories: short-term effects and long-term effects. Short-term effects were defined as a response that has an immediate and reversible effect and may not have an impact in the future.

Long-term effects, on the other hand, may affect the species or population in future and may not be reversed. I also recorded the International Union for Conservation of Nature (IUCN) status of each species focused on in the selected literature from the Red List (IUCN 2020) to document their global conservation status. The study locations were extracted from articles and mapped using the QGIS freeware (version 3.16; QGIS Development Team 2020).

2.3 Results

A total of 20 documents, including 19 peer-reviewed research papers and one technical report, published from 1994 to 2020 from five continents were selected to meet the objectives of this chapter (Figs. 2.1 and 2.2). Photography has been shown to be detrimental to 18 species in the literature, including one fish fauna, one reptile fauna, four avian fauna, and 12 mammalian species during the investigation. Two papers included multi-taxa during an investigation. Three of them are critically endangered (blue-crowned laughingthrush *Garrulax courtoisi*, great Indian bustard *Ardeotis nigriceps*, and hawksbill sea turtle *Eretmochelys imbricata*), three endangered (tiger, green sea turtle *Chelonia mydas*, and New Holland seahorse *Hippocampus whitei*), two vulnerable (loggerhead sea turtle *Caretta caretta* and Sri Lanka's green-billed coucal *Centropus chlororhynchos*), two near threatened (grey slender loris *Loris lydekkerianus* and Cambodian tailorbird *Orthotomus chaktomuk*), 17 least concern, one data deficient and one species was not evaluated by IUCN (Table 2.1 and Fig. 2.3).

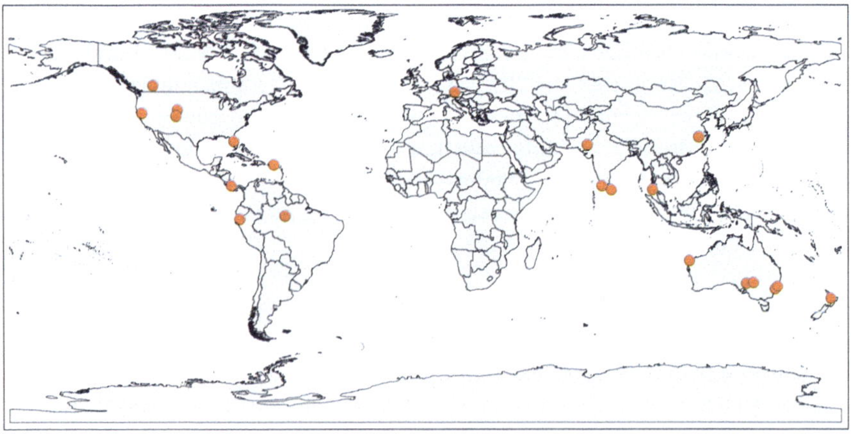

Fig. 2.1 Locations of selected studies

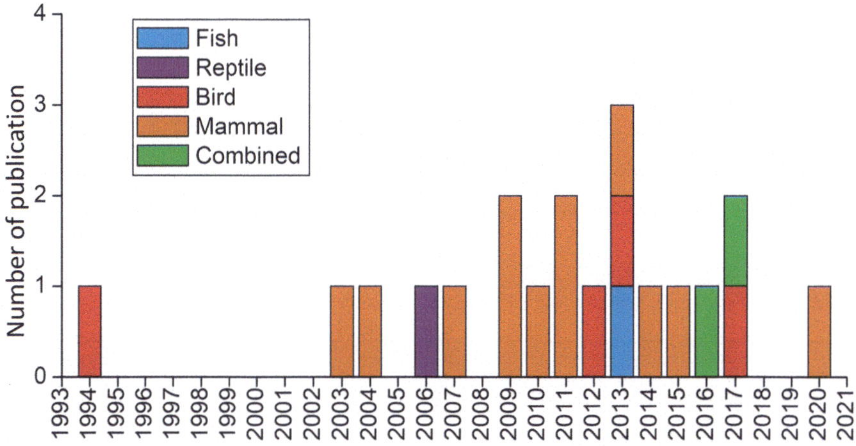

Fig. 2.2 Year-wise trend (from 1994 to 2021) of selected studies in the present chapter

2.3.1 Deleterious Effects

A total of 12 types of stimuli related to four major sources, including photographers and their vehicles, cameras and unethical photographic activities, were identified as detrimental to wildlife, their habitats, and monitoring. Summary and consequence are depicted in Table 2.1 and Fig. 2.4.

2.3.1.1 Short-Term Effects

The immediate responses of various photographic sources include animal disturbance, startle response (Gibeau and McTavish 2009), stress, and changes in natural behaviour. When disturbed, most animals flee and exhibit active fear, jumping, running, swimming, and flying rapidly, whereas some animals exhibit passive defence during a specific breeding stage (MoEF 2012). Gutzwiller et al. (1994) discovered that when humans pass through or near their territories, some subalpine birds reduce their singing activity. Animal disturbance increases vigilance time, startle response, and leaving foraging site, which ultimately reduces foraging timing. Animal distress causes immediate physiological changes, such as an increase in heart rate (Gutzwiller et al. 1994).

Although remote photography or camera traps are considered non-invasive methods of animal monitoring, avoidance of camera traps (Schipper 2007; Henrich et al. 2020) and trap shyness by animals (Wegge et al. 2004) have been reported. A study on the West Indian anole *Anolis cristatellus* found that shutter noise reduces animal display rate, but it does not change in response to flash or silence treatments (Huang et al. 2011). A nestling death has been reported while filming nesting behaviour (Cowie and Hinsley 1988).

Additional activities such as using playback add additional stress to birds. Harris and Haskell (2013) found that using playback affects the vocal behaviour of two bird species, reducing their territorial appeal and inter- and intra-specific

Table 2.1 A summary of the principal deleterious effects reported in various case studies from various photographic sources

Sl. no.	Source of stimuli	Type of stimuli	Species	Effect	Consequences	Continent	Habitat or site	Reference
1	Photographer	Approaching animal	Ruby-crowned kinglet (*Regulus calendula*), hermit thrush (*Catharus guttatus*), pine siskin (*Spinus pinus*)	Reduction in singing activity	– Increased stress response – Elevated heart rate – Reduced vigour – Energy exhaust during flight – Amendment in the body energy budget – Decreased productivity	North America	Mixed habitat forest	Gutzwiller et al. (1994)
2	Photographer and vehicle	Approaching animal	Red kangaroo (*Macropus rufus*) and euro (*M. robustus erubescens*)	Flee from site	– Increased maintenance activity – Increased energy expenditure in activity to be relaxed	Australia	Woodland and grassland	Wolf and Croft (2010)
			Great Indian Bustard (*Ardeotis nigriceps*)	Disturbed breeding activity	– Decreased breeding rate – Species decline	Asia	Grassland	MoEF (2012)
		Regular presence/ exposure of photographer	Blue-crowned laughingthrush (*Garrulax courtoisi*)	Change in nest site height	– Increased predator exposure – Increased egg/ nestling mortality – Constraining breeding success	Asia	Broad-leaved woodland	Zhang et al. (2017)

Sl. no.	Source of stimuli	Type of stimuli	Species	Effect	Consequences	Continent	Habitat or site	Reference
		Vehicle presence	Yellow-bellied marmots (*Marmota flaviventris*)	– Increased time spent in vigilance – Decreased time spent in foraging	–	North America	River valley	Li et al. (2011)
3	Camera	Presence of camera (in case of camera trap)	Coyote (*Canis latrans*)	Increased wariness	– Avoidance of camera station – Change in the use of territorial space	North America	Valley in foothills	Sequin et al. (2003)
		Shutter noise	West Indian anole (*Anolis cristatellus*)	Decreased display rate	– Disturbed animal – Change in mating and social structure	North America	Hiking trails	Huang et al. (2011)
		Camera trap sound	Mammal species	Disturbed animal	– Avoidance of camera trap site	Australia	In situ experiment	Meek et al. (2014)
		Flashlight	Kinkajous (*Potos flavus*)	Avoidance of camera trap	– Stressed animal – Preventing animals from passing the spot – Precluding effective density estimation – Precluding effective monitoring	Central America	Lowland coastal moist forests	Schipper (2007)

(continued)

Table 2.1 (continued)

Sl. no.	Source of stimuli	Type of stimuli	Species	Effect	Consequences	Continent	Habitat or site	Reference
			Red deer (*Cervus elaphus*) and Roe deer (*Capreolus capreolus*)	Avoidance of camera trap	–	Europe	Bavarian Forest National Park	Henrich et al. (2020)
			Tiger (*Panthera tigris*)	Trap shyness	– Animal changing travel route – Bias in density index measurement	Asia	Subtropical moist deciduous forest	Wegge et al. (2004)
			Grey wolf (*Canis lupus*)	Startle response	– Fleeing from site – Skewing in study result – Affecting conservation decisions	North American	Montane forest	Gibeau and McTavish (2009)
			Marine turtles (*Chelonia mydas*, *Caretta caretta*, *Eretmochelys imbricate*, *Natator depressus*) (attempting to nest)	Disturbed animal	– Turning back during emergence – Aborting the body pit – Aborting the egg chamber – Aborting covering – Disturbance during returning	Western Australia	Marine Park	Waayers et al. (2006)
		Infrared light of camera trap	Ferrets (*Mustela furo*)	Avoidance behaviour	– Depriving feed for a longer time – Body weight loss	Australia	Lab experiment	Newbold and King (2009)

Sl. no.	Source of stimuli	Type of stimuli	Species	Effect	Consequences	Continent	Habitat or site	Reference
4	Additional photographic activities	Playback	Plain-tailed Wren (*Thryothorus euophrys*) and rufous Antpittas (*Grallaria rufula*)	Affected vocal behaviour	– Reducing territorial appeal – Reducing inter- or intra-specific communication	South America	Sub-tropical area	Harris and Haskell (2013)
		Animal capturing from the wild (as photo props)	Grey slender loris (*Loris lydekkerianus*)	– Promoted trade – Increased unethical animal handling	– Species decline	Asia	Evergreen woodland forest	Kanagavel et al. (2013)
			Slow loris (*Nycticebus* sp.)	Increased animal trafficking	– Species decline	Asia (Thailand)	Patong (the main tourist resort in Phuket)	Osterberg and Nekaris (2015)
			Green anaconda and common caiman, brown-throated three-toed sloth	Increased animal trafficking	– Increased stress – Poor body condition	South America (Brazil)	State of Amazonas	D'Cruze et al. (2017)
		Nest displacement for clear view and clearance of nesting area	Cambodian tailorbird (*Orthotomus chaktomuk*) and Sri Lanka green-billed coucal	Chick exposure to predators	–	Asia (Sri Lanka)	–	Podduwage (2016)
		Animal handling	Cobra	Disturbance	–	–	–	Podduwage (2016)

(continued)

Table 2.1 (continued)

Sl. no.	Source of stimuli	Type of stimuli	Species	Effect	Consequences	Continent	Habitat or site	Reference
			New Holland seahorse (*Hippocampus whitei*)	Short-term behaviour change	–	Australia (Port Stephens estuary)	Underwater sea	Harasti and Gladstone (2013)

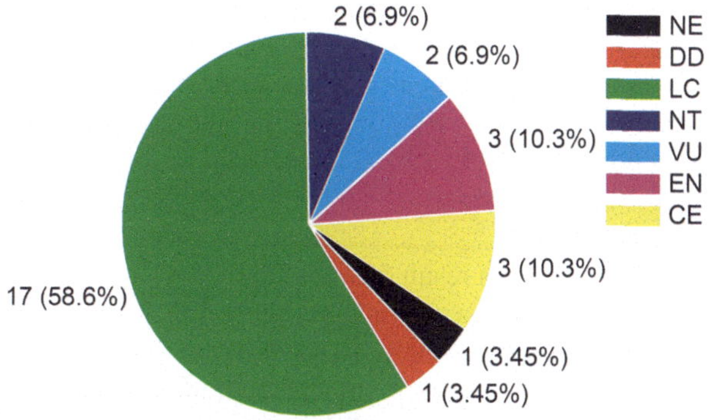

Fig. 2.3 Account of species with their IUCN status (*NE* not evaluated, *DD* data deficient, *LC* least concern, *NT* near threatened, *VU* vulnerable, *EN* endangered, *CE* critical endangered) addressed in the selected literature

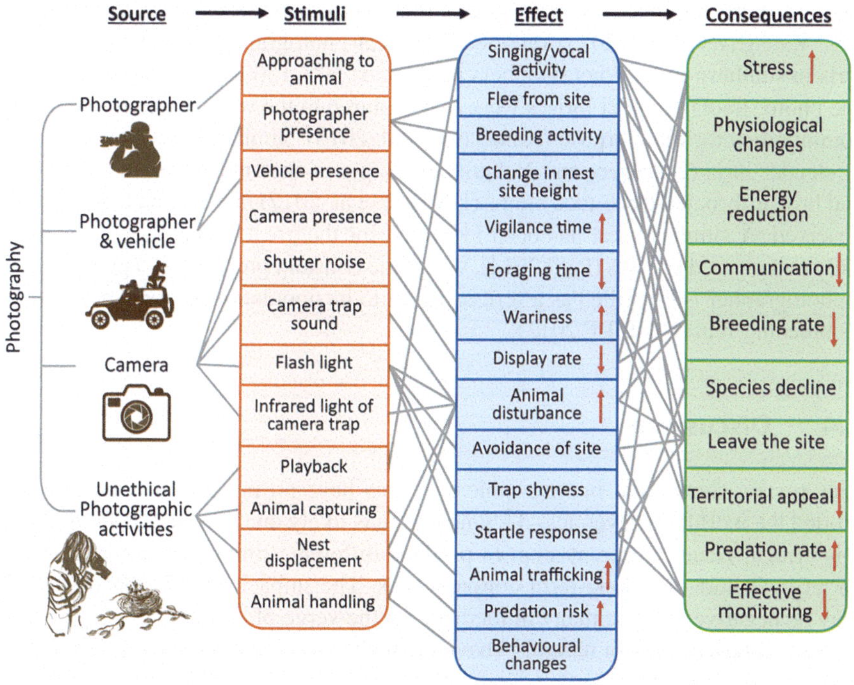

Fig. 2.4 A summary of the reported detrimental photographic activities on wildlife, natural habitats, and animal monitoring reported in the literature. Grey lines indicate the connection between two objects, and the red arrows indicate the effect. The arrows pointing up indicate increased activity, while the arrows pointing down indicate decreased activity

communication. Kanagavel et al. (2013), Osterberg and Nekaris (2015), and D'Cruze et al. (2017) have also reported animal capture for the purpose of using animals as photo props in tourist areas. Unethical practices such as nest displacement for better view and animal handling during photography disturb birds and disrupt nesting (Podduwage 2016; Harasti and Gladstone 2013).

2.3.1.2 Long-Term Effects

Long-term behavioural changes result from regular exposure to photography, photographers, and related activities, including abandonment of foraging areas and changes in food sources. Some songbirds, such as red-winged blackbirds, goldfinches, and American robins, change their behaviour after repeated human interactions and become more aggressive towards humans who visit their nest. According to research, the red-winged blackbird learned to distinguish between people who visited their nest frequently and those who were never seen before (Knight and Temple 1986a, b, c; Knight and Cole 1995).

According to some case studies, animal capture for unethical photography may have a significant impact on small populations of rare and slowly reproducing species. Kanagavel et al. (2013) reported unethical photography of endemic slender loris in southern India. Photographers paid locals 500–1500 Indian rupees to manage photo shoots. Slender lorises were also captured in the wild and housed in local communities for this purpose (Kanagavel et al. 2013). Similarly, illegal trafficking of slender loris (Osterberg and Nekaris 2015), green anacondas, common caiman, and brown-throated three-toed sloths (D'Cruze et al. 2017) in Thailand markets was observed. A similar threat has been identified for the great Indian bustard, a critically endangered species (Patil 2014), where the constant presence of a poacher and a photographer is identified as a serious threat to its population, particularly during its breeding season (MoEF 2012).

2.4 Discussion

This review shows that photographic activities have an impact on many species around the world; however, less documentation is likely due to a lack of attention to this critical issue. Different sources produce different stimuli, which can have a variety of short- and long-term consequences. The impact on threatened species may be negative because many of them are on the verge of extinction.

Stress and changes in natural behaviour, such as nest abandonment, changes in food habits, foraging, hunting, courtship, and physiological changes, including increased heart rates due to flight with increased stress hormones, are immediate responses of wildlife to various sources (Jachowski et al. 2015). Individual fitness is reduced as a result of these responses (Gabrielsen and Smith 1995). Individuals, rather than populations or communities, typically receive immediate responses (Knight and Cole 1995). The increased heart rate also reduces vigour. Energy is expended during flight, while energy acquisition is reduced due to disrupted foraging (Klein 1993). As a result, changes to the animal energy budget may reduce productivity (Knight and Cole 1995).

Pedestrian disturbance includes a wide range of on-foot photographic activities, such as approaching wildlife in the field (Knight and Cole 1995; Frid and Dill 2002; Sekercioglu 2002). Similarly, off-road and non-vehicular activities in the forest are regarded as more disruptive than vehicular traffic (Klein 1993). Animals perceive human disturbance as a threat to their survival (Walther 1969), and anti-predator behaviour has an impact on other activities (Frid and Dill 2002). Off-road photography in sensitive habitats, such as grasslands, can be extremely harmful to ground-nesting birds, plants, insects, and snakes (Dattatri and Sreenivasan 2016). Any mild sensory stimuli of approaches alert animals and cause an orienting response, which involves head and eye movements in an apparent attempt to identify it. When an animal announces the approach of a potential predator or a human, the animal uses either active or passive defence. The active defence response includes flight and fight, whereas the passive defence response includes inhibition of many activities, which results in many physiological changes. Some parameters such as behavioural activity, heart rate, metabolism, blood sugar, body temperature, respiratory rate, respiratory depth, oxygen consumption, brain blood flow, heart blood flow, and skeletal muscle blood flow rise, while others such as blood flow to the gut, gut motility, and digestive secretions decrease (Gabrielsen and Smith 1995).

Adults leave their nest or den sites temporarily due to disturbance. Reduced parental attentiveness increases the risk of young being preyed upon, disrupts feeding patterns, and exposes young to unfavourable environmental conditions. If disturbed parents are unwilling to return to the nest site, predators may visit and consume eggs and young (Verbeek 1982; Knight and Cole 1991). Aside from the breeding season, neither parent is bound to the nest site, nor are the young dependent on the parents; however, the young respond to disturbance, which either reduces energy acquisition (i.e. foraging) or increases energy expenditure (i.e. fleeing) (Knight and Cole 1991).

During incubation, birds rely on stored body reserves in the form of body fat to avoid predation and maintain a constant egg temperature. Disturbance causes repeated reheating and increases energy demand in the rerun of parents at the nest. An increase in activity of 10% per day would result in an additional 4–5 g of weight loss per day. In order to save itself, the bird would abandon incubation or give up its chicks at hatching (Gabrielsen and Smith 1995; Knight and Cole 1995; Sekercioglu 2002).

Taking photographs is a multi-modal stimulus for animals because it includes a visual stimulus from the flash, an audio stimulus from the shutter, and possibly an autofocus zoom lens (Huang et al. 2011). A study on the West Indian anole *Anolis cristatellus* (Huang et al. 2011) found that shutter noise reduces animal display rate, but it does not change in response to flash or silence treatments. As a result, lower display rates may have an impact on mating and social structure (Huang et al. 2011).

Remote photography has some effects on predators or nesting animals due to the presence of equipment, humans, human scent, and human activity (Hunt and Ogden 1991; Major and Gowing 1994; Whelan et al. 1994; Cutler and Swann 1999). These effects may disrupt nesting, attract or repel predators, and influence predator activity patterns (Whelan et al. 1994; Picman and Schriml 1994; Cutler and Swann 1999). Many mammals can detect, hear, and see the presence, sound, and illumination of a camera (Meek et al. 2014), which influences animal behaviour (Cutler and Swann 1999).

Additional activities, such as using playback, add stress to the bird and result in higher energy costs, disruption of the social system, and even pair break-up. During the breeding season, it also distracts adults from important tasks such as courtship, nest building, nest guarding, and territory defence. Using powerful flashlights at night causes animals to become temporarily blind, and chasing animals to take action photographs causes enormous stress, leaving animals with little energy for other activities or even escaping from predators. Fast driving of vehicles to see animals not only result in animals being run over, but also in those being freighted (Dattatri and Sreenivasan 2016). A photographic vehicle trip through the forest has additional consequences such as habitat destruction and air and noise pollution. When photographers congregate with their vehicles in one location, the damage to the environment becomes even more severe. Using bait on a regular basis can also lead to habituation, which can result in attacks on livestock or even humans in the case of carnivorous mammals. A GPS-tagged photograph attracts other photographers or even poachers to the same location, increasing environmental stress and poaching activity (Dattatri and Sreenivasan 2016; Neslen 2016).

Photographers' regular exposure causes long-term behavioural changes such as abandonment of foraging areas and changes in food sources (Knight and Cole 1995). Similarly, regular disturbances are known to have an impact on some populations' biodiversity, abundance, distribution, and demography. It can also have an indirect effect on population dynamics by changing individuals' investment in anti-predator behaviour (Lima 1998; Frid and Dill 2002). Long-term disturbance stimuli, according to White (1983) and Madsen (1994), can cause a species' population to decline by reducing reproductive success. On the other hand, community-level effects result from disturbance, which affects the distribution of foragers and, as a result, the density of their prey (Frid and Dill 2002). After being disturbed, an animal spends less time foraging and consumes less, which has a direct impact on the plant community structure and biomass of that area (Gill et al. 1996). Consumption of a lower proportion of available resources in a specific area may also have an effect at lower tropic levels (Chase 1998; Frid and Dill 2002).

2.5 Recommendations and Future Perspectives

According to the literature, many conservation concern species are highly sensitive to the presence of photographers, and their regular movement in the same area increases stress in the species (Sekercioglu 2002; Kanagavel et al. 2013). There are two critical periods identified for many bird and mammal species that are most affected by human disturbance: the immediate postnatal period and the breeding period (Gabrielsen and Smith 1995). As a result, photographic disturbance should be kept to a minimum during these times, and it is preferable if photography is avoided during these times or kept at a minimum approaching distance where animals feel free and safe.

Photographic quality is improved by minimising disturbance and flushing (Fernández-Juricic 2000). A low-cost digital camera combined with telephoto lenses and camouflaged clothing can assist in obtaining high-quality images from a safe distance while also minimising photographic disturbance (Ingraham 2001). The tangential slow approach should be preferred over the direct approach because birds are more tolerant of it. The approaching group should be kept small, ideally under ten people (Sekercioglu 2002). Because regular contact with wildlife causes habituation, photography in areas with some human disturbance, such as medium traffic or grime roads, can provide a closer look at wildlife. Although the vehicular approach is more tolerable (Holmes et al. 1993), it contributes to air and noise pollution in pristine areas and should thus be avoided.

Given the present state of knowledge, there are gaps in our knowledge that can be filled through well-designed studies that examine wildlife photography at different hierarchical scales. First, we must understand how photographic activities affect individuals' inclusive fitness and, ultimately, whether populations and communities are affected. If animals retreat from areas critical for feeding, reproduction, and survival, their population will decline. Similarly, if animals are disturbed while performing essential behaviours such as foraging and reproduction, the population is likely to decline. Similarly, if animals are disturbed while performing essential behaviours such as foraging and reproduction, the population is likely to decline. Animal population disruption or change can alter the dynamics of a wildlife community. A wide range of species should be studied to provide a complete picture of life history strategies. We need to gain a better understanding of how photography's multi-modal stimuli influence the predictability and perceived threat to wildlife. It is also unknown how this practice affects or changes species diversity across different habitat gradients.

Wildlife photographers make ideal ecotourists for community-based conservation. Wildlife knowledge and the expectation of photographing a variety of species in their natural habitat provide a direct link between wildlife and local income (Mossaz et al. 2015). With this high expectation, many photographers spend more money. The resulting increase in local awareness of the value of biodiversity may be critical in preserving many natural areas near human populations. Local guides who benefit directly financially from biodiversity by showing different species to photographers are more likely to conserve ecosystems that harbour unusual species (Paaby et al. 1991). Similarly, many rare species can be found outside of officially protected areas and are easy to find. Locals hosting ecotourism can thus become a source of revenue generation in their remote areas. The constant presence of photographers, as well as the restricted distribution of unusual species, increases the amount of local control as well as the profits for any given area, resulting in increased incentives for local people to protect the environment and these small patches from further destruction (Sekercioglu 2002).

Unethical photography is defined as any photographic activity that in any way harms an animal or its habitat. Photographers' high expectations are not always beneficial, and an overzealous desire to photograph certain species can have negative consequences. Displacing a nest, clearing the area around a nesting area, handling animals, using artificial elements such as spraying water to create artificial rain, refrigerating animals to slow their movement, sticking a subject to stop movement, and manipulating or artificially inserting more than two subjects are all considered unethical practices in wildlife photography (Podduwage 2016).

Effective education to vision is critical. People are more likely to support restrictions if they understand how wildlife will benefit. Photographers should feel accountable for their actions as long as they follow the rules and regulations. The importance of education in reducing the incidence of approaching animals on foot should be emphasised. Photographers, as the largest group type that disturbs the most wildlife species, should be made aware of their role in reducing access to foraging habitat and should be targeted for education programs (Klein 1993). Currently, three main guidelines on wildlife photography ethics have been published: the first is the 'Principles of Ethical Field Practices' by the North American Nature Photography Association (NANPA 1996), the second is 'The Nature Photographers' Code of Practice' by the Nature Group of the Royal Photographic Society (RPS 2007), and the third is 'An Ethical Model for the Wildlife Photography of Sri Lanka' by Podduwage (2016). Each country has its own biodiversity and landscape, necessitating a different ethical framework. At a time when most species are under severe threat of extinction, each country should form its own 'national nature photography association'. These organisations should set the standards for wildlife photography and keep an eye out for unethical behaviour. Penalties may be imposed. In this difficult time for conservation, governments could authorise these associations to grant licences to wildlife photographers who have undergone ethics and animal behaviour training. Detailing workshops and training programs on ethical practices, animal behaviour, and minimum approaching distances could be organised for people who have been sensitised to ethical photography. The alert distance, at which a species becomes aware of the presence of the observer(s), could be used as the minimum approaching distance, but keep in mind that it is a species-specific trait that varies between individuals (Blumstein et al. 2003). Recently, Møller et al. (2014) found that flight initiation distance or minimum approaching distance varies across different geographic regions due to various environmental factors, necessitating more scientific research to establish proper minimum approaching distance standards in each country or landscape separately.

References

Basnet D, Jianmei Y, Dorji T, Qianli X, Lama AK, Maowei Y, Ning W, Yantao W, Gurung K, Rujun L, Gupta N, Kanwal KA, Shaoliang Y (2021) Bird photography tourism, sustainable livelihoods, and biodiversity conservation: a case study from China. Mt Res Dev 41(2):D1–D6

Blount JD, Chynoweth MW, Green AM, Şekercioğlu ÇH (2021) COVID-19 highlights the importance of camera traps for wildlife conservation research and management. Biol Conserv 256:108984

Blumstein DT, Anthony LL, Harcourt R, Ross G (2003) Testing a key assumption of wildlife buffer zones: is flight initiation distance a species-specific trait? Biol Conserv 110:97–100

Burton AC, Neilson E, Moreira D, Ladle A, Steenweg R, Fisher JT, Bayne E, Boutin S (2015) Wildlife camera trapping: a review and recommendations for linking surveys to ecological processes. J Appl Ecol 52:675–685

Caravaggi A, Banks PB, Burton AC, Finlay CMV, Haswell PM, Hayward MW, Rowcliffe MJ, Wood MD (2017) A review of camera trapping for conservation behavior research. Remote Sens Ecol Conserv 3(3):109–122

Chase JM (1998) Central-place forager effects on food web dynamics and spatial pattern in northern California meadows. Ecology 79:1236–1245

Cowie RJ, Hinsley SA (1988) Feeding ecology of great tits (*Parus major*) and blue tits (*Parus caeruleus*), breeding in suburban gardens. J Anim Ecol 57:611–626

Cutler TL, Swann DE (1999) Using remote photography in wildlife ecology: a review. Wildl Soc Bull 27(3):571–581

Dattatri S, Sreenivasan R (2016) Stop! Don't shoot like that—a guide to ethical wildlife photography. Conservation India. http://www.conservationindia.org/resources/ethics

D'cruze N, Machado FC, Matthews N, Balaskas M, Carder G, Richardson V, Vieto R (2017) A review of wildlife ecotourism in Manaus, Brazil. Nat Conserv 22:1–16

Excell L (2012) Wildlife photography: from snapshots to great shots. Peachpit Press, Bangalore

Fernández-Juricic E (2000) Local and regional effects of pedestrians on forest birds in a fragmented landscape. Condor 102:247–255

Frid A, Dill LM (2002) Human-caused disturbance stimuli as a form of predation risk. Conserv Ecol 6(1):11

Gabrielsen GW, Smith EN (1995) Physiological responses of wildlife to disturbance. In: Knight RL, Gutzwiller KJ (eds) Wildlife and recreationists: coexistence through management and research. Island Press, Washington, DC, pp 95–107

Gibeau ML, McTavish C (2009) Not-so-candid cameras: how to prevent camera traps from skewing animal behaviour. Wildl Profess 3:35–37

Gill JA, Sutherland WJ, Watkinson AR (1996) A method to quantify the effects of human disturbance on animal populations. J Appl Ecol 33(4):786–792

Gutzwiller KJ, Wiedemann RT, Clements KL, Anderson SH (1994) Effects of human intrusion on song occurrence and singing consistency in subalpine birds. Auk 111:28–37

Handcock MS, Gile KJ (2011) Comment: on the concept of snowball sampling. Sociol Methodol 41:367–371

Hanisch E, Johnston R, Longnecker N (2019) Cameras for conservation: wildlife photography and emotional engagement with biodiversity and nature. Hum Dimens Wildl 24(3):267–284

Harasti D, Gladstone W (2013) Does underwater flash photography affect the behaviour, movement and site persistence of seahorses? J Fish Biol 83(5):1344–1353

Harris JBC, Haskell DG (2013) Simulated birdwatchers' playback affects the behavior of two tropical birds. PLoS One 8(10):e77902

Henrich M, Niederlechner S, Kröschel M, Thoma S, Dormann CF, Hartig F, Heurich M (2020) The influence of camera trap flash type on the behavioural reactions and trapping rates of red deer and roe deer. Remote Sens Ecol Conserv 6(3):399–410

Higginbottom K (2004) Wildlife tourism: impacts, management and planning. Common Ground Publishing Pty Ltd. in association with Cooperative Research Centre for Sustainable Tourism, Australia

Holmes TL, Knight RL, Stegall L, Craig G (1993) Responses of wintering grassland raptors to human disturbance. Wildl Soc Bull 21:461–468

Huang B, Lubarsky K, Teng T, Blumstein DT (2011) Take only picture, leave only...fear? The effect of photography on the West Indian anole *Anolis cristatellus*. Curr Zool 57(1):77–82

Hunt RH, Ogden JJ (1991) Selected aspects of the nesting ecology of American alligators in the Okefenokee Swamp. J Herpetol 25:448–453
Ingraham S (2001) Is there a digital camera in your birding future? Birding 33:163–165
Isaacs JC (2000) The limited potential of ecotourism to contribute to wildlife conservation. Wildl Soc Bull 28(1):61–69
IUCN (2020) The IUCN Red List of Threatened Species, https://www.iucnredlist.org
Jachowski DS, McCorquodale S, Washburn BE, Millspaugh JJ (2015) Human disturbance and the physiological response of elk in eastern Washington. Wildl Biol Pract 11(1):12–25
Kanagavel A, Sinclair C, Sekar R, Raghavan R (2013) Moolah, misfortune or spinsterhood? The plight of Slender Loris *Loris lydekkerianus* in southern India. J Threat Taxa 5(1):3585–3588
Karanth KK, DeFries R, Srivathsa A, Sankarman V (2012) Wildlife tourists in India's emerging economy: potential for a conservation constituency? Oryx 46(3):382–390
Kiss A (2004) Is community-based ecotourism a good use of biodiversity conservation funds? Trends Ecol Evol 19:232–237
Klein ML (1993) Waterbird behavioral responses to human disturbances. Wildl Soc Bull 21:31–39
Knight RL, Cole DN (1991) Effects of recreational activity on wildlife in wildlands. In: Trans 56th N Am Wildl Nat Ressour Conf, pp 238–247
Knight RL, Cole DN (1995) Factors that influence wildlife responses to recreationists. In: Knight RL, Gutzwiller KJ (eds) Wildlife and recreationists: coexistence through management and research. Island Press, Washington, DC, pp 71–79
Knight RL, Temple SA (1986a) Methodological problems in studies of avian nest defence. Anim Behav 34:561–566
Knight RL, Temple SA (1986b) Nest defense in the American goldfinch. Anim Behav 34:887–897
Knight RL, Temple SA (1986c) Why does intensity of avian nest defense increase during the nesting cycle? Auk 103:318–327
Li C, Monclus R, Maul TL, Jiang Z, Blumstein DT (2011) Quantifying human disturbance on antipredator behavior and flush initiation distance in yellow-bellied marmots. Appl Anim Behav Sci 129:146–152
Lima SL (1998) Stress and decision making under the risk of predation: recent developments from behavioral, reproductive, and ecological perspectives. Adv Study Behav 27:215–290
MacArthur RA, Johnston RH, Geist V (1979) Factors influencing heart rate in free-ranging bighorn sheep: a physiological approach to the study of wildlife harassment. Can J Zool 57:2010–2021
Madsen J (1994) Impacts of disturbance on migratory waterfowl. Ibis 137:567–574
Major RE, Gowing G (1994) An inexpensive photographic technique for identifying nest predators at active nests of birds. Wildl Res 21:657–666
Meek PD, Ballard G-A, Fleming PJS, Schaefer M, Williams W, Falzon G (2014) Camera traps be heard and seen by animals. PLoS One 9(10):e110832
MoEF (2012) Guidelines for the Great Indian Bustard recovery programme. Ministry of Environment and Forests (MoEF), Government of India, New Delhi
Møller AP, Samia DSM, Weston MA, Guay PJ, Blumstein DT (2014) American exceptionalism: population trends and flight initiation distances in birds from three continents. PLoS One 9(9):e107883
Morgan JL, Gergel SE, Coops NC (2010) Aerial photography: a rapidly evolving tool for ecological management. Bioscience 60(1):47–59
Mossaz A, Buckley RC, Castley JG (2015) Ecotourism contributions to conservation of African big cats. J Nat Conserv 28:112–118
NANPA (1996) Principles of ethical field practices. 1996.02.03 ed. North American Nature Photography Association
Neslen A (2016) Poachers using science papers to target newly discovered species. The Guardian https://www.theguardian.com/environment/2016/jan/01/poachers-using-science-papers-to-target-newly-discovered-species. Accessed 15 Mar 2021
Newbold HG, King CM (2009) Can a predator see 'invisible' light? Infrared vision in ferrets (*Mustela furo*). Wildl Res 36:309–318

Osterberg P, Nekaris KAI (2015) A case study of the slow loris *Nycticebus* spp. Traffic Bull 27(1):13

Paaby P, Clark DB, Gonzalez H (1991) Training rural residents as naturalist guides: evaluation of a pilot project in Costa Rica. Conserv Biol 5(4):542–546

Patil P (2014) Photography a serious threat to Great Indian Bustard. Conservation India. http://www.conservationindia.org/articles/photography-a-serious-threat-to-the-great-indian-bustard/. Accessed 15 Nov 2015

Picman J, Schriml LM (1994) A camera study of temporal patterns of nest predation in different habitats. Wilson Bull 106(3):456–465

Podduwage DR (2016) An ethical model for the wildlife photography of Sri Lanka. J Aesthet Fine Arts 1(1):98–129

QGIS Development Team (2020) QGIS geographic information system. Open Source Geospatial Foundation Project. http://qgis.osgeo.org

Rovero F, Tobler M, Sanderson J (2010) Camera trapping for inventorying terrestrial vertebrates. In: Eymann J, Degreef J, Hauser C, Monje JC, Samyn Y, Vandenspiegel D (eds) Manual on field recording techniques and protocols for all taxa biodiversity inventories and monitoring, vol 8. The Belgian National Focal Point to the Global Taxonomy Initiative, Abc taxa, Brussel, pp 100–128

Rovero F, Zimmermann F, Berzi D, Meek P (2013) "Which camera trap and how many do I need" a review of camera features and study designs for a range of wildlife research applications. Hystrix 24(2):148–156

RPS (2007) 'The nature photographers' code of practice. Royal Photographic Society of Great Britain

Schipper J (2007) Camera-trap avoidance by Kinkajous *Potos flavus*: rethinking the "non-invasive" paradigm. Small Carniv Conserv 36:38–41

Sekercioglu CH (2002) Impacts of birdwatching on human and avian communities. Environ Conserv 29(3):282–289

Sequin ES, Jaeger MM, Brussard PF, Barrett RH (2003) Wariness of coyotes to camera traps relative to social status and territory boundaries. Can J Zool 81:2015–2025

Skrzypczak U (2010) Wildlife photography: on safari with your DSLR: equipment, techniques, workflow. Rocky Nook, Inc., Santa Barbara, CA

Slater C, Cam G, Qi Y, Liu Y, Guay PJ, Weston MA (2019) Camera shy? Motivations, attitudes and beliefs of bird photographers and species-specific avian responses to their activities. Biol Conserv 237:327–337

Stronza AL, Hunt CA, Fitzgerald LA (2019) Ecotourism for conservation? Annu Rev Environ Resour 44(1):229–253

Tang L, Shao G (2015) Drone remote sensing for forestry research and practices. J For Res 26(4):791–797

Trolliet F, Huynen MC, Vermeulen C, Hambuckers A (2014) Use of camera traps for wildlife studies. A review. Biotechnol Agron Soc Environ 18(3):446–454

Verbeek NAM (1982) Egg predation by northwestern crows: its association with human and bald eagle activity. Auk 99:347–352

Waayers D, Newsome D, Lee D (2006) Research note observations of non-compliance behaviour by tourists to a voluntary code of conduct: a pilot study of turtle tourism in the Exmouth region, Western Australia. J Ecotourism 5(3):211–222

Walther FR (1969) Flight behaviour and avoidance of predators in Thomson's gazelle (*Gazella thomsoni:* Guenther 1884). Behaviour 34:184–221

Wegge P, Pokheral CP, Jnawali SR (2004) Effect of trapping effort and trap shyness on estimates of tiger abundance from camera trap studies. Anim Conserv 7:251–256

Weston MA, McLeod EM, Blumstein DT, Guay PJ (2012) A review of flight-initiation distances and their application to managing disturbance to Australian birds. Emu 112(4):269–286

Whelan CJ, Dilger ML, Robson D, Hallyn N, Dilger S (1994) Effects of olfactory cues on artificial-nest experiments. Auk 111:945–952

White RG (1983) Foraging patterns and their multiplier effects on productivity of northern ungulates. Oikos 40:377–384
Wolf ID, Croft DB (2010) Minimizing disturbance to wildlife by tourists approaching on foot or in a car: a study of kangaroos in the Australian rangelands. Appl Anim Behav Sci 126:75–84
Zhang W, Shi J, Huang H, Liu T (2017) The impact of disturbance from photographers on the Blue-crowned Laughingthrush (*Garrulax courtoisi*). Avian Conserv Ecol 12(1):15

Danio rerio: A Promising Tool for Neurodegenerative Dysfunctions

3

Neelakanta Sarvashiva Kiran, Chandrashekar Yashaswini, Varun Varma Sangaraju, and Senthilkumar Rajagopal

Abstract

Alzheimer's, Parkinson's, Huntington's, multiple sclerosis, and other associated brain illnesses are largely caused by neurodegeneration. Their intricate pathophysiology frequently involves mitochondrial dysfunctions, misfolded polypeptide negativity, demyelination, genetic and neurochemical deficiencies, and genetic and neurochemical deficits. Neurodegenerative illnesses often show evolutionarily conserved pathways across species albeit varied in particular underlying causes. Human brain pathology has some similarities compared to the zebrafish's central nervous system. In this chapter, we examine the role of zebrafish models in simulating the most prevalent human and animal neurodegenerative conditions. We discuss various computational models used for the neurodegenerative diseases. We also highlight how these fish models can help further understand the genetic, neurochemical, neuroanatomical, and behavioural pathogenic mechanisms.

Keywords

Brain disorders · Cellular degeneration · Demyelination · Neuroanatomy · Polymorphic · Sex determination

N. S. Kiran · C. Yashaswini · S. Rajagopal (✉)
Department of Biotechnology, School of Applied Sciences, REVA University, Bengaluru, Karnataka, India

V. V. Sangaraju
Department of Information Technology, University of Cumberlands, Williamsburg, KY, USA

S. Sridhara et al. (eds.), *Animal Behavior in the Tropics*,
https://doi.org/10.1007/978-981-97-7324-4_3

3.1 Introduction

Neurodegenerative disorders are an imminent danger to human health, particularly the rising older population. These age-related diseases, such as Parkinson's disorder, Huntington's disease, and Alzheimer's disease, have a significant effect on sufferers and their families, as well as on society. Zebrafish (*Danio rerio*), a new and effective animal model for investigating the cause of neurodegenerative disorders, have gained recognition recently. In addition to their ocular transparency, external fertilization, quick development, and adaptability to genetic manipulation, zebrafish possess many additional special features that make them desirable for research (Pitchai et al. 2019; Shen and Zuo 2020). These qualities give scientists the ability to observe and control cellular and gene expression in a mammalian model, making them an invaluable tool to combat neurodegenerative disorders (Soussi-Yanicostas 2022). Zebrafish has become a valuable model organism for scientific investigation of brain illnesses and disorders. Zebrafish have historically been utilized in developmental biology research, but they are now also being used to study a variety of neurological and neurodegenerative problems, as well as other aspects of brain diseases (Thayumanavan et al. 2022; Best and Alderton 2008).

Zebrafish are the most appropriate model for studying brain illnesses for a number of motives, including the elevated level of genetic similarity between zebrafish and humans, including several genes involved in neural function and illness (Rai et al. 2022). Scientists examine the consequences of particular genetic modifications or malfunctions in zebrafish to understand how these modifications connect with human neurological disorders. The embryos and larvae were optically transparent, ensuring the researchers could see in real-time processes related to illness, brain connection, and development (Saluja et al. 2021; Son et al. 2022). This openness makes it possible to trace the development of diseases and research; how brain circuits form and operate using non-invasive imaging techniques like fluorescence microscopy. Zebrafish are exceptionally proficient at repairing injured neural tissue, thanks to their extraordinary regenerative capabilities (Ochenkowska et al. 2022). Zebrafish, unlike mammals, regenerate neurons and repair impaired neural tissue. They are an excellent model for researching neural repair mechanisms alongside potential treatments for neurological disorders as a result of their ability to regenerate (Anand and Mondal 2020). High-throughput screening: Zebrafish may be used to test a large number of potential pharmaceutical substances or therapeutic agents because high-throughput screening methods are suitable for them (Rosa et al. 2022). This attribute is especially useful for locating compounds that might treat neurological disorders by reducing symptoms or delaying the beginning of the disease (Paduraru et al. 2023). Transgenesis, morpholino antisense oligonucleotides, mRNA infusions, morpholino antisense oligonucleotides, and CRISPR-Cas9 are just a few of the genetic engineering methods utilized to alter zebrafish (Petri et al. 2022). These techniques enable the introduction of disease-related mutations or the targeted modification of particular genes (Rothenbücher et al. 2019). Researchers better understand illness mechanisms, find possible therapeutic targets, and evaluate the effectiveness of innovative treatments using zebrafish as an illustration for neurological ailments (Lee et al. 2021). Zebrafish models are a powerful tool for analysing neurological diseases because they offer special benefits, such as optical transparency, regeneration capability, and genetic modifications, which reinforce different animal models and human studies (Verma et al. 2022).

3.2 Habitats, Life Cycle, and Diet in Natural and Laboratory Conditions

3.2.1 Natural Habitats

In particular, Southeast Asian freshwater natural habitats such as streams, rivers, and rice fields are where zebrafish (*Danio rerio*) are found. Bangladesh, India, Nepal, Myanmar, and Pakistan are among the nations where they came originally from. Ponds, ditches, as well as shallow water areas near flooded fields are all examples of slow-moving or just stagnant water areas with abundant flora that zebrafish are frequently seen throughout (Parichy and Postlethwait 2020; Aleström et al. 2020). Zebrafish are commonly discovered in warm, tropical environments, which is their natural home. As a result of their specialized adaptations, they can flourish in water between 24 °C and 30 °C (75 °F and 86 °F, respectively). Zebrafish are adaptable and withstand a variety of environmental circumstances, including decreased oxygen levels and changing water quality. Zebrafish, like to dwell in groups, commonly called as shoals or schools, depicting high socializing behaviour. To be safe and forage, they engage in social behaviour they rely entirely on their shoaling instincts. Zebrafish consume zooplankton, tiny invertebrates, and insects in their native environment (Lee et al. 2022).

Zebrafish grow quickly during the embryonic stage at 28 °C. It is simple to see and stage the developing process since the cells as well as the chorion are transparent. The chorion is what separates from an egg after fertilization, and then the transparent cytoplasm flows to the vertebral pole, which divides the blastodisc off the yolk mass (Merola et al. 2020). The blastodisc experiences meroblastic cleavage 45 min after fertilization, with synchronous cell divisions and each cell cycle lasting roughly 15 min. At the end of the fourth stage of development, some cells are unattached from the yolk, while other cells (marginal cells) keep their connection (Keenan and Currie 2019). The blastula or just blastoderm, a collection of cells upon the animal pole, is created as a result of these cleavages. The mid-blastula transition, which takes place during the tenth cell cycle, is marked by prolonged and asymmetrical divisions of cells and activation of the embryonic genetic transcription, with enhanced cell motility (Meyers 2018). A multi-nucleate primordial syncytial layer is created when the marginal cells swell and collapse. Epiboly, which starts approximately 4 h after fertilization and involves the blastula thinning and spreading to round the yolk, starts after a few more cycles. Different cellular systems coordinate epiboly, which is staged according to the proportion of the yolk that is encircled by surrounding cells (Bruce and Heisenberg 2020). Cells on the forthcoming dorsal surface of a zebrafish embryo start to involute during the epiboly process, generating the embryonic shield. Cells still involute and expand back towards the animal pole as the epiboly moves closer to the vegetable pole. The tail-bud forms close to the previous vegetal pole around 10 h after fertilization, while the head bud forms close to the initial animal pole (Williams and Solnica-Krezel 2020). At this point, the epiboly is complete.

Somite formation in the trunk mesoderm starts from anterior to posterior, with somites generating at regular intervals. Neurulation, which results in the folds of the nerve system out of the ectodermal layer, happens concurrently. By around 36 h after fertilization, the zebrafish embryos have developed their primary organ systems as a result of continued growth and specification (Schröter et al. 2008). Usually, the embryonic stage is over when the egg hatches from the chorion 48–72 h after fertilization. Between days 3 and 4, the swim bladder expands to allow upright swimming, and shortly after that, feeding activities start. It is important to remember that temperature changes can affect different developmental phases, allowing for alterations in the rate of growth (Zhang et al. 2022). Zebrafish transition into a larval stage after the embryonic stage, which lasts for around 4 weeks (Mara and Holley, 2007). Size is a more trustworthy measure for larval staging since it takes into account elements like density, temperature, and unique variations that affect larval growth. Zebrafish are categorized as juveniles until they attain a size of roughly 11 mm, which normally happens when they are approximately 4 weeks of age (Farnsworth et al. 2020). Typically, when they reach a size of above 17 mm and are between 10 and 12 weeks after fertilization, they attain sexual maturity. Zebrafish's sexual selection processes are intricate and still poorly understood. Environmental influences or a polymorphic sex determination system may have an impact in addition to the acquisition of sex chromosomes. Physical differences between male and female zebrafish, including a sleeker abdomen in males and a broader, rounder abdomen in gravid females, can help identify them (Abu-Siniyeh and Al-Zyoud 2020). The colour patterns as well as fin lengths of different zebrafish strains vary. Depending on the breed and rearing conditions, zebrafish live for an average of 3–4 years before beginning to show indications of ageing, which normally start around the age of 2 years (Di Paola et al. 2022).

Small invertebrates, such as zooplankton, pests, and their larvae, make up a large portion of the natural food of zebrafish, also called *Danio rerio* within their habitat. Zebrafish are opportunistic feeders who like to consume live food, including tiny crustaceans, worms, as well as insect larvae often located in the liquid column or on the surface of the water. They have an opportunistic feeding strategy since they also devour plant material, including algae with plant debris (Zarantoniello et al. 2020). Zebrafish exhibit selective eating of tiny, moving prey items in their natural habitat. However, it ought to be noted that zebrafish are often fed readily accessible diets that were specially developed to fulfil their nutritional needs whether kept in research settings or when kept as aquarium creatures. These diets frequently include a blend of live or frozen meals, pellets, and flakes made to mimic their preferred diets (Ma et al. 2022).

3.2.2 Laboratory Conditions

Zebrafish are frequently kept in specific aquatic environments in laboratories tailored to their demands. These systems, sometimes referred to as zebrafish habitats or zebrafish facilities, offer regulated environments for the best zebrafish colony development, reproduction, and upkeep (Tsang and Gerlai 2022). The zebrafish breeding system, which comprises several linked tanks or racks, is the main part of a zebrafish habitat facility. To preserve the stability and purity of the water in these tanks, filtering systems are installed. To replicate the circumstances found in nature for zebrafish, water factors such as temperature, pH, oxygen concentration in the water, and others are carefully monitored and managed (Shelton et al. 2020). Lighting control systems are also used in zebrafish facilities to manage the photoperiod, which simulates the day or night cycles impacting the fish's biological cycles and behaviour. To encourage breeding and synchronize experimental techniques, the lighting conditions might be changed. Zebrafish aquariums include diverse structures, including pipes made of PVC, plastic plants, and substrates, to provide ideal hiding places and enrichment (Xiao et al. 2021). These components allow the fish to engage in natural behaviour such as territorial expansion, territorial defence, and exploration. When feeding zebrafish in the lab, it is common practice to use dry pellets like flakes from commercial diets prepared to fulfil nutritional needs. Depending on the research goals and life stage, feeding schedules might change (Baratti et al. 2020). The following water conditions are suggested for zebrafish within a lab habitat: Hardness ought to be between 50 and 100 mg/L $CaCO_3$, alkalinity should be between 50 and 150 mg/L $CaCO_3$, pH levels ought to remain within 6.8 and 7.5 (although a tolerance level of 6.0–8.5 can be accepted), temperature needs to be maintained at 26–28.5 °C, unionized ammonia levels ought to be kept below 0.02 mg/L, nitrate (NO_3^-) levels should not go above 50 mg/L and nitrite (NO_2^-) levels ought to be below 0.25 mg/L. These variables help create an aquatic habitat that is appropriate and stable for zebrafish (Chaput et al. 2023). The life cycle remains the same in both natural and laboratory conditions (Dasgupta et al. 2023) (Fig. 3.1).

Zebrafish are commonly fed commercially made diets specially developed to fulfil nutritional needs in experimental settings. These diets can be found in several different formats, such as dry granulates, flakes, or granules. They are designed precisely to offer a well-balanced mix of proteins, lipids, carbs, vitamins, and minerals necessary for zebrafish development, reproduction, and general health (Ghaddar et al. 2020). The chemicals in zebrafish formulas frequently come from sources that are both animal- and plant-based. These contain several micronutrients as well as fish meal, shrimp food, soy protein, wheat flour, and cornmeal. According to the specific needs of the fish strain or the goal of the study, the diet's precise make-up may change (Farnsworth et al. 2020).

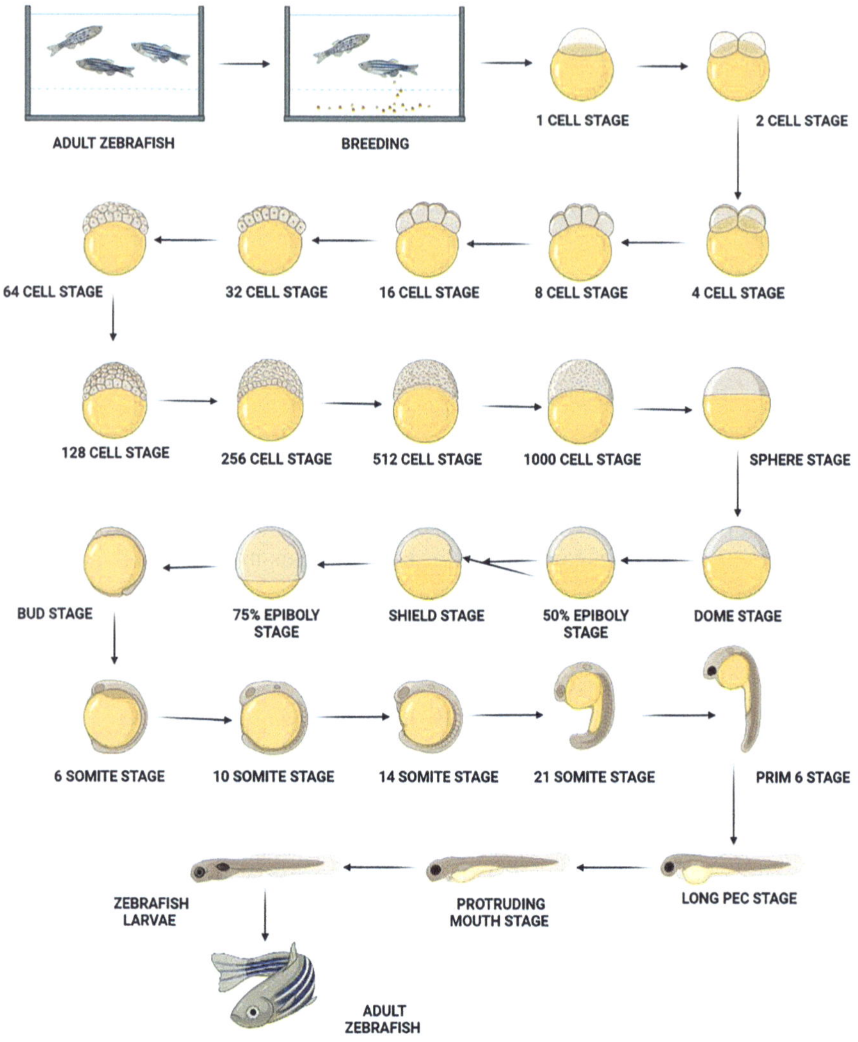

Fig. 3.1 Depiction of zebrafish life cycle

3.3 Genetics of Zebrafish

Zebrafish genetic information has been well-defined and is extensively reported in scientific literature. The zebrafish genome has 25 pairs of chromosomes and comprises around 1.7 billion base pairs, comprising 23 autosomes and 2 sexual chromosomes (XX for females and XY for males). An estimated 70% of the protein-coding genes in the zebrafish whole genome have orthologs in the human genome, according to the mapping as well as annotation in the zebrafish genome (Bradford et al. 2022). Numerous regulatory components, repetitive sequences, and non-coding

RNAs are also present in the zebrafish genome and are crucial for maintaining genome stability. The zebrafish genome sequence's accessibility has made it easier to identify and characterize the functions of zebrafish proteins involved in a variety of biological processes (Kwon et al. 2019). A comparison of human and zebrafish protein-coding genes shows intriguing results. According to Ensembl Compara, about 71.4% of a person's genes have at least one orthologous gene in zebrafish.

About 69% of the genes in zebrafish have a minimum of one ortholog in humans. Approximately 47% of the orthologous genes in humans have a 1:1 connection with their zebrafish orthologues (Sun et al. 2020). An average of 2.28 zebrafish genes are linked to every human gene in the 'one-human-to-many-zebrafish' class of orthologous genes, which is the second-largest category of orthologous genes. This discovery most likely represents the zebrafish genome's gene duplication phenomena (Ruzicka et al. 2019). Even while the receptors (lifra, lifrb, osmr, and il6r) for certain well-known human genes, including leukaemia inhibitory factor (LIF), oncostatin M (OSM), and interleukin 6 (IL-6), are present in zebrafish, these genes' actual orthologs are not always easy to identify. It is plausible that zebrafish peptides with similar functional properties to LIF, OSM, as well as IL-6 exist, but their orthologous link is obscured by severe sequence differences. Furthermore, despite the absence of a BRCA1 orthologue in the zebrafish genome, the BARD1 gene, which is linked to BRCA1, is present (Lawson et al. 2020). We compared human genes to zebrafish orthologues as well as disease-associated gene lists to evaluate the possible contribution of zebrafish to disease-related genes. We focused on 3176 genes identified by the Online Mendelian Inheritance of Man (OMIM) database as being associated with illnesses. 2601 (82%) of those disease-related genes have no less than one zebrafish orthologue (Zhu et al. 2019). We also examined the 4023 human genes involved in GWAS and found that 3075 (76%) of these genes had a relationship with a minimum of one zebrafish orthologue (Aksoy et al. 2019). Zebrafish might be a useful model organism for research into the genetic causes of illnesses. This is because a large number of human disease-related genes have corresponding equivalents in the zebrafish genome (Zhao et al. 2022; Howe et al. 2013).

3.4 Zebrafish as a Model for Biological Research

Due to the occurrence of common mutations in disease-associated genes, zebrafish have become an important model organism for researching human illnesses, including cancer, cardiovascular problems, and neurological disorders. They are ideal for simulating haematological illnesses because they resemble human red blood cells, macrophages, neutrophils, T cells, as well as B cells (Naomi et al. 2021). Studies on cardiovascular toxicity have made considerable use of zebrafish embryos, allowing for analyses of gene function and the reproduction of human disorders. Zebrafish embryos can be used in toxicological research with benefits including cost-effectiveness, high output, shortened testing times, and the possibility for positive results (Liu 2022). By combining low-cost culture tests with high-value mammalian models, this methodology has effectively detected probable adverse drug responses

Fig. 3.2 Diverse application of zebrafish model in biological research

and enhanced pre-clinical data gathering. In contrast to in vitro models, zebrafish toxicity research provides benefits over mouse models when it comes to size, progeny weight, alteration rate, target identification, target endorsement, toxicity investigations, and overall statistical analysis confidence (Bozkurt 2020). They also give unique insights into human biology. Zebrafish also share genes with humans that are linked to cancer, so their use in studies lowers the cost of rearing. Zebrafish are also useful as model organisms in medication development and toxicity research since they can absorb water-soluble substances (Dubey et al. 2022) (Fig. 3.2).

3.5 Zebrafish as a Model for Research on Neurodegenerative Disorders

3.5.1 Alzheimer's Disease

Zebrafish have distinct orthologs and co-orthologs that resemble human genes that are known to be crucial in Alzheimer's disease (AD). Zebrafish, for example, have "co-orthologs" of human APP (appa and appb), as well as orthologs of human PSEN1 (psen1), PSEN2 (psen2), and other genes. Additionally, zebrafish have orthologous genes for PSENEN (psenen), NCTN (ncstn), and APH1b (aph1b), which are all parts of the gamma-secretase complex (Bhattarai et al. 2022). Recently, zebrafish have also been shown to contain orthologs of -secretase (BACE1 and BACE2), designated as bace1 as well as bace2, respectively. Co-orthologs of the APOE gene, known as apoea and apoeb, and the microtubule-associated polypeptide tau (MAPT) gene are both present in zebrafish (Thawkar and Kaur 2021).

Due to the possibility of selectively observing loss-of-function abnormalities in certain cells the existence of duplication genes in zebrafish is advantageous for functional study. High amounts of A may encourage cerebrovascular bifurcation in the growing hindbrain of zebrafish, according to research by Kiper and Freeman (2022). They looked into the consequences of reducing A levels further and discovered that APP- and A-deficient zebrafish larvae had cerebrovascular abnormalities. The human A-42 peptide was used to treat the embryos, yet not the shorter cleavage product p3, which raises the possibility that A may play a role in preserving healthy cerebrovascular function (Nabinger et al. 2020). This discovery emphasizes the need for cautiousness when contemplating medications meant to decrease the generation of A for the treatment of Alzheimer's disease (AD). The hypothesis that amyloid is the cause of AD postulates that A aggregates are harmful, which has sparked research into aggregate-lowering substances (Wang et al. 2021). Due to the survival of Bace1 knockout mice with mild abnormalities, as opposed to the fatality seen in Psen1 null mice, BACE1 has demonstrated potential as a target (Newman et al. 2014).

3.5.2 Parkinson's Disease

Zebrafish do not have a replica of the human-synuclein gene, but they nonetheless express the three synuclein isoforms α, β-, and γ-synucleins to make up for it. Zebrafish 1-synuclein functions very similarly to human-synuclein. In zebrafish, the knockdown of either α- or β-synucleins results in motor dysfunction, which is worsened when the two synucleins are reduced (Razali et al. 2021). Zebrafish missing β- and α-synucleins show aberrant dopaminergic system development, including delayed dopaminergic neuron differentiation and decreased dopamine levels. Interestingly, an expression of human α-synuclein rescues these abnormalities. Similar to human α-synuclein, overexpression of α-synuclein in zebrafish causes neuronal aggregation and neurotoxicity (Robea et al. 2020). On the other hand, zebrafish with downregulated α-synuclein are protected against the harmful effects of the herbicide ziram, which impairs swimming behaviour and promotes dopaminergic neuron death. Apomorphine therapy or the anti-self-assembly inhibitor of amyloidogenic protein CLR01 repairs motor deficits and guards against ziram-induced neurotoxicity (Najib et al. 2020). Because of the suppression of the ubiquitin-associated proteasome process and the buildup of α-synuclein aggregates, α-synuclein is thought to be neurotoxic (Bandmann and Burton 2010). Through the reduction of α-synuclein aggregation along with neuronal cell death caused by treatment with CLR01, a protein aggregation inhibitor, cell survival is increased.

Establishing transgenic zebrafish lines is difficult because of the severe effects shown in zebrafish overexpressing human α-synuclein. Nevertheless, restricting the protein's expression to dopaminergic neurons or using a temporary expression strategy may be able to lessen the mortality brought on by α-synuclein overexpression whilst still enabling the investigation of pathogenic causes (Vaz et al. 2018).

3.5.3 Huntington's Disease

A variety of illness proteins with polyglutamine (polyQ) repeat expansions include Huntington's disease (HD), which is a particular form of neurological ailment. The age at which symptoms appear is negatively associated with the number of polyQ repeats in the protein, and these diseases are progressive and degenerative (Pal et al. 2022). Even though the genes linked to polyQ disorders are often expressed across the body, each disorder has a unique pattern of neurodegeneration that affects certain populations of neurons. In the case of HD, cortical neuronal pyramids are also impacted, but medium-sized spiny neurons (MSNs) within the striatum are more vulnerable to degeneration (Upadhayay et al. 2023). Huntington's disease (HD) is characterized by an expanded polyglutamine (polyQ) region within the N-terminus of the huntingtin protein (HTT). HTT is a large protein consisting of 3144 amino acids, and it is widely expressed in tissues. It has been proposed to act as a scaffolding protein involved in multiple cellular processes (Kumar et al. 2021). Knockout studies in mice have shown that the HTT gene is crucial for early embryonic development, indicating its essential role in development, although the specific mechanisms remain unclear (Mohamed and Ekker 2023). Histologically, HD is associated with the presence of large intracellular aggregates of polyQ proteins. These aggregates can be found in different sub-cellular regions, including the cytoplasm, perinuclear region, and nucleus. While these aggregates were initially believed causative in the disease, their exact role in HD is still not fully understood, and there is evidence suggesting that they may have a neuroprotective function (Chia et al. 2022). Through the injection of plasmid DNA or mRNA, HTT-polyQ has been overexpressed in zebrafish, which has resulted in toxicity (Angelopoulou et al. 2020). The rhodopsin promoter was used in a different zebrafish model to direct the production of mHTT-exon1 fragments just in photoreceptor cells of the eye. Retinal degeneration, however, is not an HD-specific condition. Therefore, a useful tool for further study in the area would be the creation of a zebrafish model the fact that more closely mimics certain elements of human illness (Gusella et al. 2021).

3.5.4 Multiple Sclerosis

Multiple sclerosis (MS) alongside other demyelinating illnesses, as well as other neurodegenerative diseases, can be studied using zebrafish, a potential new animal model. They have benefits including mass production of embryos, quick growth, optical transparency, and genetic manipulation, which makes them perfect for studying neurological diseases, and carrying out high-throughput drug tests, which are currently impractical using in vivo models (Burrows et al. 2019). Previous studies employed genetic cell ablation models and toxin-induced demyelination models to investigate the processes of remyelination as well as oligodendrocyte precursor cell differentiation. In these investigations, adult zebrafish were exposed to LPC to induce demyelination, which led to the optic nerve's focal demyelination, microglial infiltration, and a decrease in $Olig2^+$ cells (Bashirzade et al. 2022). Remarkably, remyelination, resolution of the microglial response, and recovery of oligodendrocytes were observed within 28 days, highlighting the rapidity of remyelination in zebrafish. Importantly, this model exhibited specific toxicity to oligodendrocytes and myelin without affecting axon numbers. Interestingly, older zebrafish (>15 months old) showed slower and incomplete recovery of myelin, even 3 months after toxin-induced demyelination, possibly due to reduced microglial response (Sanabria-Castro et al. 2020). Zebrafish models hold great potential for MS research, given their similarities in myelin and oligodendrocyte biology, as well as their rapid development and large population sizes that allow for in vivo screening of pro-myelinating therapeutics (Belo et al. 2021). Care must be taken when studying the immune system in the zebrafish model of MS, as their immune system differs from that of mammals. The future goal is to utilize zebrafish models as preclinical screening tools to identify potential therapeutic compounds before moving on to higher animal models (Liu 2023).

3.6 Computational Modelling in Neurodegenerative Diseases

The human brain consumes 20% of the body's energy and is called the house of all neurological activity (Woo et al. 2021). Computational modelling plays a key role in diagnosing various brain illnesses caused by neurodegeneration. The medical field generates huge amounts of data but is unable to analyse the data efficiently. Recent advancements in computational modelling are catching up with the data generated in multiple folds (Gulín-González et al. 2022). Using custom-made models on the existing data will help analyse some of the symptoms of these diseases and help the health-care providers assess the situation. Models are also difficult to produce due to a lack of knowledge of the neurosciences of the data scientists (Jones et al. 2022). Computational modelling is performed on various neurodegenerative diseases such as Alzheimer's, Parkinson's, and multiple sclerosis in different forms.

3.6.1 Alzheimer's Disease

Alzheimer's disease is the most prevalent neurodegenerative disease and is the most common cause of dementia (Stefanovski et al. 2021). Alzheimer's disease is caused by the disruption of mental functions and neurodegeneration of some of the brain regions (Jones et al. 2022). The disorders of the human brain and degradation of cognitive function due to Alzheimer's disease are studied and integrated to produce computational models (Jones et al. 2022). Jones et al. (2022) selected 423 patients with existing Alzheimer's conditions for their computational model development. Jones et al. (2022) reported a low-dimensional representation of possible neurodegeneration with characteristics linked to Alzheimer's disease. This model was constructed with the help of patient data representing glucose levels, mental function mapping to a manifold, external validation of manifold data with another large dataset from a multi-site study ($N = 410$), and other clinical construct validation (Jones et al. 2022). Dementia with Lewy bodies and posterior cortical atrophy exhibited abnormal activity in brain regions when modelled with external data sources (Jones et al. 2022).

The understanding of Alzheimer's disease is incomplete, and disease-modifying treatments are not present as well (Stefanovski et al. 2021). A complete understanding of high-performance computational model frameworks and biological processes of Alzheimer's disease is necessary to build meaningful computational models (Stefanovski et al. 2021). There are various types of computational models that can be incorporated into Alzheimer's disease, such as statistical disease prediction models, sub-cellular models, single neuron and neuron circuit models, and large-scale brain network models (Stefanovski et al. 2021). Tools like "The Virtual Brain" are widely used as they enable large-scale modelling of patient data, including all the underlying dynamics (Gulín-González et al. 2022; Stefanovski et al. 2021).

Research on the presence of APOE4 (Apolipoprotein E) alleles in the human brain leads to a significant risk of late-onset Alzheimer's disease (Badea et al. 2019). This was found true when tested using novel mouse models by replacing the APOE in mice with APOE3 and APOE4 in humans (Badea et al. 2019). The results also indicated that the population groups could be separated based on the lower dimensional representation of the genotypes (Badea et al. 2019). Badea et al. (2019) identified that brain loss in APOE4 mice is 6% higher than in APOE3 mice.

The computer simulations in Alzheimer's disease data showed a relationship between synaptic degradation in the hippocampus and CA3-CA1 networks with varied memory weights (Świetlik et al. 2019). Świetlik et al. (2019) aimed to create computational models of synaptic plasticity and memory in the CA3-CA1 networks hippocampal formation. This research consisted of both types of neural models, namely biophysical and connectionist, taken into consideration for brain network simulation experiments (Świetlik et al. 2019). Świetlik et al. (2019) identified that Alzheimer's disease affects different parts of the brain at various stages of the disease. The initial stage changes were identified in the mesial temporal lobe, and significant memory function degradation due to hippocampal formation (Świetlik et al. 2019).

Alzheimer's disease and vascular dementia exhibit similar neurological symptoms and are often difficult to diagnose with accuracy using MRI and clinical methods (Castellazzi et al. 2020). Castellazzi et al. (2020) applied three machine learning algorithms such as artificial neural networks (ANN), adaptive neuro-fuzzy inference system (ANFIS), and support vector machine (SVM). Castellazzi et al. (2020) used diffusion tensor imaging data and steady-state fMRI data from 60 patients (33 with Alzheimer's disease and 27 had vascular dementia) as input to the machine learning models. The predictions from machine learning models were then compared with the 3-year clinical diagnosis evidence (Castellazzi et al. 2020). Out of the three machine learning algorithms, ANFIS proved to be the best in differentiating Alzheimer's disease and vascular dementia, with an accuracy of over 84% and a small feature pattern (Castellazzi et al. 2020).

The etiopathogenesis of Alzheimer's disease is not completely explained or not vastly researched (Caligiore et al. 2020). The amyloid cascade hypothesis is one of the conceptual model frameworks generally used to study Alzheimer's disease (Caligiore et al. 2020). However, there were some clinical failures in this model due to the implementation in the later stages of Alzheimer's disease progress (Caligiore et al. 2020). Through some research and recent studies then, Caligiore et al. (2020) identified that the ventral tegmental area (VTA) was the first event in the progression of Alzheimer's disease. Caligiore et al. (2020) collected neural and behavioural data from patients for this hypothesis using computational models. Caligiore et al. (2020) showed that catecholamine release occurred due to the degeneration of VTA, and subsequent events are triggered, leading to Alzheimer's disease. Next, degeneration can occur in hippocampal and cortical areas causing Alzheimer's disease (Caligiore et al. 2020). Caligiore et al. (2020) contributed to developing a hypothesis using computational models which can formulate system-level Alzheimer's disease and support early diagnosis.

3.6.2 Parkinson's Disease

Deep Brain Simulation (DBS) is considered to be an effective method for treating Parkinson's disease's motor symptoms (Grado et al. 2018; Lu et al. 2019). However, selecting simulation parameters, such as relieving symptoms and avoiding side effects, will play a key role in the successful implementation of DBS (Grado et al. 2018). The DBS devices need inputs in the form of parameters which can be a trial-and-error method, and these inputs become more problematic to enter manually when leads and algorithms grow (Grado et al. 2018; Lu et al. 2019). Grado et al. (2018) identified the need for an automated and intelligent way of providing input parameters. Grado et al. (2018) provided a solution for this problem in the form of a Bayesian adaptive dual controller (ADC) as it is an automated parameter tuning system. Bayesian ADC uses feedback signals from patients and provides suitable parameters for the respective patient after quantifying a set of stimulation parameters (Grado et al. 2018). Bayesian ADC can improve over time, provide correct parameters to the patients, and reduce their clinical visits in the future (Grado et al. 2018). Grado et al. (2018) also used the STN-DBS algorithm in stimulation inputs, and the results showed a reduction in Parkinsonian motor signs. STN-DBS was successfully used to provide model-based indications for treatment-refractory OCD (Sangaraju and Rajagopal 2023).

The DBS methods have static stimulation parameters for whatever the patient's symptoms are, and trial-and-error input methods are causing overstimulation in some of the patients (Lu et al. 2019). Lu et al. (2019)) aimed to design a closed-loop DBS which will adjust stimulation parameters automatically from the neurophysiological signals received. Lu et al. (2019)) introduced a new closed-loop DBS system to automate stimulation parameters using a reinforcement learning method (Lu et al. 2019). The state of Parkinson's disease is called the distorted relay reliability of the thalamus, and the results of the reinforcement learning-based closed-loop DBS method showed that the distortion relay reliability of the thalamus can be restored with less DBS energy expenditure (Lu et al. 2019).

Movement disorders occur due to multiple changes to neural circuits (Humphries et al. 2018). Most movement disorders in Parkinson's, dystonia, and Huntington's disease are caused by the basal ganglia (Humphries et al. 2018). Due to these complex changes and amelioration of various symptoms, proper treatment with a success rate cannot be identified (Humphries et al. 2018). Humphries et al. (2018) used computational models to understand the causative, neural dynamics, and appropriate treatments. Humphries et al. (2018) reviewed the models related to the effects of losing dopamine on basal ganglia for causative mechanisms. Similarly, Humphries et al. (2018) discussed beta band oscillations in Parkinsonian basal ganglia for understanding neural dynamics. Lastly, Humphries et al. (2018) tried to understand deep brain stimulation for treatments using computational models.

Information communication in the brain is performed through the collective firing of the neurons (Pavlides et al. 2015). This communication process is disrupted in the brain neurons when a patient is suffering from Parkinson's disease (Pavlides et al. 2015). In Parkinson's disease, neurons produce larger oscillations in their activity due to over-synchronization (Pavlides et al. 2015). The oscillation frequency of the initial Parkinson's disease diagnosis range is 13–30 Hz, which is known as the beta band (Pavlides et al. 2015). Pavlides et al. (2015) aimed to understand these beta band oscillations to improve the treatments for Parkinson's disease. Pavlides et al. (2015) hypothesized two different reasons for the cause of the beta band oscillations. The first hypothesis is that the beta band oscillations could emerge from the motor cortex and resonate with the cortical input through the basal ganglia (Pavlides et al. 2015). The other hypothesis is that the return of feedback from the basal ganglia to the motor cortex could be one of the strong reasons for continued beta oscillations (Pavlides et al. 2015). Pavlides et al. (2015) built two computational models based on the hypotheses using the STN-GPe (sub thalamic nucleus—globus pallidus) circuits.

In Parkinson's disease, the normal behaviour of the brain's basal ganglia-thalamus-cortex (BG-T-C) neuronal circuit is linked to the death of dopaminergic neurons (Ranieri et al. 2021). Although the molecular pathogenesis of the ailment has been quite established, yet no cure has been identified (Ranieri et al. 2021). The research took the direction of analysing the computational models of similar neurobiological aspects identified in the animals' models to produce treatments for Parkinson's disease (Ranieri et al. 2021). In this concept, Ranieri et al. (2021) proposed a biophysical computational model of brain regions affected by Parkinson's disease based on the data collected from the brain of marmoset monkeys. Ranieri et al. (2021) performed a data-driven approach in their computational model where they used differential evolution on biological parameters. This computational model is the first experimental model that explored seven different brain regions of marmoset monkeys (Ranieri et al. 2021) (Table 3.1).

3.6.3 Multiple Sclerosis

Multiple sclerosis is a complex heterogeneous disease that is caused due to environmental and genetic factors involving bad immune responses (Pappalardo et al. 2020). In this autoimmune disorder, the immune response is primarily directed towards myelin components, and neurodegenerative processes have occurred lately, leading to neurological damage (Pappalardo et al. 2020). There were around 20 disease-modifying drugs approved for treating multiple sclerosis, and they were categorized into high-efficacy and moderate efficacy (Pappalardo et al. 2020). Doctors prescribe any of these drugs based on the patient's condition, clinical diagnosis, MRI data, and experience (Pappalardo et al. 2020). To avoid misjudgments and experimental medicines, Pappalardo et al. (2020) aimed to create a modelling infrastructure to predict the progress of multiple sclerosis in patients and prescribe appropriate treatments. Pappalardo et al. (2020) built a computational model called Universal Immune System Simulator (UISS), which simulates the patient's immune system activities and symptoms. UISS was later extended to respond to all other multiple sclerosis pathogenesis interactions with the immune system (Pappalardo et al. 2020).

Even though there is a lot of research performed, the aetiology of multiple sclerosis seems to be not completely known (Pernice et al. 2020). The immune response of various individuals can be greatly influenced by different endogenous and

Table 3.1 Computational models used in neurodegenerative diseases

Neurodegenerative disease	Computational model or technique used	Contribution to neurodegenerative disease	Source
Alzheimer's	Three machine learning models (ANN, SVM, ANFIS) and MRI	Differentiation between Alzheimer's and Vascular dementia; early prediction of two diseases	Castellazzi et al. (2020)
Alzheimer's	Biophysical and connectionist neural models	Synaptic plasticity and memory in the CA3-CA1 networks hippocampal formation	Świetlik et al. (2019)
Parkinson's	Bayesian adaptive dual controller and STN-DBS	Provides suitable parameters for the respective patient after quantifying a set of stimulation parameters	Grado et al. (2018)
Parkinson's	Closed-loop DBS system using Reinforcement learning	Adjusts stimulation parameters automatically from the neurophysiological signals received	Lu et al. (2019)
Multiple sclerosis	Universal Immune System Simulator	Simulates the patient's immune system activities and symptoms	Pappalardo et al. (2020)

exogenous factors, which makes understanding the disease complex (Pernice et al. 2020). Pernice et al. (2020) proposed a model that studies the immune response in relapsing-remitting multiple sclerosis (RRMS). Alternate episodes of symptom exacerbation are called relapses, while periods of disease stability are known as remission, and both combine to be the most common type of multiple sclerosis (Pernice et al. 2020). This computational model represents both peripheral lymph nodes and the central nervous system (Pernice et al. 2020). Simulation results of this computational model proved to reproduce the immune T cell balance and additionally confirmed the timely intervention of the disease course when applied to patient data (Pernice et al. 2020).

One of the main causes of multiple sclerosis is due to synaptic and neuronal loss contributing to disability and network dysfunction (Huiskamp et al. 2022). However, the effect of excitatory and inhibitory synapses is not completely known, and their impact on network functioning loss is not measurable (Huiskamp et al. 2022). To overcome these problems, Huiskamp et al. (2022) aimed to quantify excitatory and inhibitory synapses and their losses on network functioning through computational models. Huiskamp et al. (2022) compared densities of glutamatergic and GABAergic synapses and the neurons between post-mortem multiple sclerosis and non-neurological control cases using immunofluorescent staining and confocal microscopy. Huiskamp et al. (2022) concluded that cognitive impairment occurs due to network disinhibition, and inhibitory synapsis is more relevant to multiple sclerosis.

Decision-making, mood and motivational symptoms, and neurocognitive impairments are some of the commonly associated terms with multiple sclerosis (Fernández et al. 2022). Fernández et al. (2022) had an objective to study decision-making impairments in multiple sclerosis and their interaction with neuropsychiatric and neurocognitive measures based on computational models. Iowa Gambling Task (IGT) was performed on 29 multiple sclerosis patients and 26 matched control subjects to evaluate their neurocognitive skills (Fernández et al. 2022). Fernández et al. (2022) applied hierarchical Bayesian analysis to understand the three computational models used on the patient data. Fernández et al. (2022) concluded that neurocognitive impairments, latent computational processes, and mood/motivational symptoms played a crucial role in the decision-making deficits of multiple sclerosis patients.

Acknowledgement The authors sincerely acknowledge REVA University and the University of the Cumberlands for providing an opportunity for research.

References

Abu-Siniyeh A, Al-Zyoud W (2020) Highlights on selected microscopy techniques to study zebrafish developmental biology. Lab Anim Res 36(1):12

Aksoy YA, Nguyen DT, Chow S, Chung RS, Guillemin GJ, Cole NJ, Hesselson D (2019) Chemical reprogramming enhances homology-directed genome editing in zebrafish embryos. Commun Biol 2(1):198

Aleström P, D'Angelo L, Midtlyng PJ, Schorderet DF, Schulte-Merker S, Sohm F, Warner S (2020) Zebrafish: housing and husbandry recommendations. Lab Anim 54(3):213–224
Anand SK, Mondal AC (2020) Neuroanatomical distribution and functions of brain-derived neurotrophic factor in zebrafish (Danio rerio) brain. J Neurosci Res 98(5):754–763
Angelopoulou E, Paudel YN, Piperi C (2020) Exploring the role of high-mobility group box 1 (HMGB1) protein in the pathogenesis of Huntington's disease. J Mol Med 98:325–334
Badea A, Wu W, Shuff J, Wang M, Anderson RJ, Qi Y, Johnson GA, Wilson JG, Koudoro S, Garyfallidis E, Colton CA (2019) Identifying vulnerable brain networks in mouse models of genetic risk factors for late onset Alzheimer's disease. Front Neuroinform 13:72
Bandmann O, Burton EA (2010) Genetic zebrafish models of neurodegenerative diseases. Neurobiol Dis 40(1):58–65
Baratti G, Potrich D, Sovrano VA (2020) The environmental geometry in spatial learning by zebrafish (Danio rerio). Zebrafish. 17(2):131–138
Bashirzade AA, Zabegalov KN, Volgin AD, Belova AS, Demin KA, de Abreu MS, Babchenko VY, Bashirzade KA, Yenkoyan KB, Tikhonova MA, Amstislavskaya TG (2022) Modeling neurodegenerative disorders in zebrafish. Neurosci Biobehav Rev 138:104679
Belo MA, Oliveira MF, Oliveira SL, Aracati MF, Rodrigues LF, Costa CC, Conde G, Gomes JM, Prata MN, Barra A, Valverde TM (2021) Zebrafish as a model to study inflammation: a tool for drug discovery. Biomed Pharmacother 144:112310
Best JD, Alderton WK (2008) Zebrafish: an in vivo model for the study of neurological diseases. Neuropsychiatr Dis Treat 4(3):567–576
Bhattarai P, Turgutalp B, Kizil C (2022) Zebrafish as an experimental and preclinical model for Alzheimer's disease. ACS Chem Neurosci 13(20):2939–2941
Bozkurt Y (2020) Introductory chapter: importance of zebrafish (Danio rerio) as model organism in biomedical research. In: Zebrafish in biomedical research. IntechOpen
Bradford YM, Van Slyke CE, Ruzicka L, Singer A, Eagle A, Fashena D, Howe DG, Frazer K, Martin R, Paddock H, Pich C (2022) Zebrafish information network, the knowledgebase for Danio rerio research. Genetics 220(4):iyac016
Bruce AE, Heisenberg CP (2020) Mechanisms of zebrafish epiboly: a current view. Curr Top Dev Biol 136:319–341
Burrows DJ, McGown A, Jain SA, De Felice M, Ramesh TM, Sharrack B, Majid A (2019) Animal models of multiple sclerosis: from rodents to zebrafish. Mult Scler J 25(3):306–324
Caligiore D, Silvetti M, D'Amelio M, Puglisi-Allegra S, Baldassarre G (2020) Computational modeling of catecholamines dysfunction in Alzheimer's disease at pre-plaque stage. J Alzheimers Dis 77(1):275–290
Castellazzi G, Cuzzoni MG, Cotta Ramusino M, Martinelli D, Denaro F, Ricciardi A, Vitali P, Anzalone N, Bernini S, Palesi F, Sinforiani E (2020) A machine learning approach for the differential diagnosis of Alzheimer and vascular dementia fed by MRI selected features. Front Neuroinform 14:25
Chaput SL, Burggren WW, Hurd PL, Hamilton TJ (2023) Zebrafish (Danio Rerio) shoaling in light and dark conditions involves a complex interplay between vision and lateral line. Behav Brain Res 439:114228
Chia K, Klingseisen A, Sieger D, Priller J (2022) Zebrafish as a model organism for neurodegenerative disease. Front Mol Neurosci 15:940484
Dasgupta S, LaDu JK, Garcia GR, Li S, Tomono-Duval K, Rericha Y, Huang L, Tanguay RL (2023) A CRISPR-Cas9 mutation in sox9b long intergenic noncoding RNA (slincR) affects zebrafish development, behavior, and regeneration. Toxicol Sci 194(2):153–166
Di Paola D, Capparucci F, Abbate JM, Cordaro M, Crupi R, Siracusa R, D'Amico R, Fusco R, Genovese T, Impellizzeri D, Cuzzocrea S (2022) Environmental risk assessment of oxaliplatin exposure on early life stages of zebrafish (Danio rerio). Toxics. 10(2):81
Dubey A, Ghosh NS, Singh R (2022) Zebrafish as an emerging model: an important testing platform for biomedical science. J Pharm Negat Results 13(3):2
Farnsworth DR, Saunders LM, Miller AC (2020) A single-cell transcriptome atlas for zebrafish development. Dev Biol 459(2):100–108

Fernández RS, Crivelli L, Pedreira ME, Allegri RF, Correale J (2022) Computational basis of decision-making impairment in multiple sclerosis. Mult Scler J 28(8):1267–1276

Ghaddar B, Veeren B, Rondeau P, Bringart M, Lefebvre d'Hellencourt C, Meilhac O, Bascands JL, Diotel N (2020) Impaired brain homeostasis and neurogenesis in diet-induced overweight zebrafish: a preventive role from A. borbonica extract. Sci Rep 10(1):14496

Grado LL, Johnson MD, Netoff TI (2018) Bayesian adaptive dual control of deep brain stimulation in a computational model of Parkinson's disease. PLoS Comput Biol 14(12):e1006606

Gulín-González J, Bringas-Vega ML, Martínez-Montes E, Ritter P, Solodkin A, Valdes-Sosa MJ, Valdes-Sosa PA (2022) Brain modeling of neurogenerative disorders. Front Neuroinform 16:937790

Gusella JF, Lee JM, MacDonald ME (2021) Huntington's disease: nearly four decades of human molecular genetics. Hum Mol Genet 30(R2):R254–R263

Howe K, Clark MD, Torroja CF, Torrance J, Berthelot C, Muffato M, Collins JE, Humphray S, McLaren K, Matthews L, McLaren S (2013) The zebrafish reference genome sequence and its relationship to the human genome. Nature 496(7446):498–503

Huiskamp M, Kiljan S, Kulik S, Witte ME, Jonkman LE, Bol GJM, Schenk GJ, Hulst HE, Tewarie P, Schoonheim MM, Geurts JJ (2022) Inhibitory synaptic loss drives network changes in multiple sclerosis: an ex vivo to in silico translational study. Mult Scler J 28(13):2010–2019

Humphries MD, Obeso JA, Dreyer JK (2018) Insights into Parkinson's disease from computational models of the basal ganglia. J Neurol Neurosurg Psychiatry 89(11):1181–1188

Jones D, Lowe V, Graff-Radford J, Botha H, Barnard L, Wiepert D, Murphy MC, Murray M, Senjem M, Gunter J, Wiste H (2022) A computational model of neurodegeneration in Alzheimer's disease. Nat Commun 13(1):1643

Keenan SR, Currie PD (2019) The developmental phases of zebrafish myogenesis. J Dev Biol 7(2):12

Kiper K, Freeman JL (2022) Use of zebrafish genetic models to study etiology of the amyloid-beta and neurofibrillary tangle pathways in Alzheimer's disease. Curr Neuropharmacol 20(3):524

Kumar V, Singh C, Singh A (2021) Zebrafish an experimental model of Huntington's disease: molecular aspects, therapeutic targets and current challenges. Mol Biol Rep 48(12):8181–8194

Kwon RY, Watson CJ, Karasik D (2019) Using zebrafish to study skeletal genomics. Bone. 126:37–50

Lawson ND, Li R, Shin M, Grosse A, Yukselen O, Stone OA, Kucukural A, Zhu L (2020) An improved zebrafish transcriptome annotation for sensitive and comprehensive detection of cell type-specific genes. Elife. 9:e55792

Lee JG, Cho HJ, Jeong YM, Lee JS (2021) Genetic approaches using zebrafish to study the microbiota–gut–brain axis in neurological disorders. Cells. 10(3):566

Lee CJ, Paull GC, Tyler CR (2022) Improving zebrafish laboratory welfare and scientific research through understanding their natural history. Biol Rev 97(3):1038–1056

Liu S (2022) Utilization of zebrafish as a model system in medical research. BIO Integr 3(4):188–192

Liu Y (2023) Zebrafish as a model organism for studying pathologic mechanisms of neurodegenerative diseases and other neural disorders. Cell Mol Neurobiol 43(6):2603–2620

Lu M, Wei X, Che Y, Wang J, Loparo KA (2019) Application of reinforcement learning to deep brain stimulation in a computational model of Parkinson's disease. IEEE Trans Neural Syst Rehabil Eng. 28(1):339–349

Ma K, Chen S, Wu Y, Ma Y, Qiao H, Fan J, Wu H (2022) Dietary supplementation with microalgae enhances the zebrafish growth performance by modulating immune status and gut microbiota. Appl Microbiol Biotechnol 106(2):773–788

Mara A, Holley SA (2007) Oscillators and the emergence of tissue organization during zebrafish somitogenesis. Trends Cell Biol 17(12):593–599

Merola C, Perugini M, Conte A, Angelozzi G, Bozzelli M, Amorena M (2020) Embryotoxicity of methylparaben to zebrafish (Danio rerio) early-life stages. Comp Biochem Physiol C Toxicol Pharmacol 236:108792

Meyers JR (2018) Zebrafish: development of a vertebrate model organism. Curr Protoc Essent Lab Tech 16(1):e19

Mohamed WM, Ekker M (2023) Zebrafish as a translational neuroscience model: today's science and tomorrow's success. Front Physiol 14:1202198

Nabinger DD, Altenhofen S, Bonan CD (2020) Zebrafish models: Gaining insight into purinergic signaling and neurological disorders. Prog Neuropsychopharmacol Biol Psychiatry 98:109770

Najib NH, Nies YH, Abd Halim SA, Yahaya MF, Das S, Lim WL, Teoh SL (2020) Modeling Parkinson's disease in zebrafish. CNS Neurol Disord Drug Targets 19(5):386–399

Naomi R, Bahari H, Yazid MD, Embong H, Othman F (2021) Zebrafish as a model system to study the mechanism of cutaneous wound healing and drug discovery: advantages and challenges. Pharmaceuticals. 14(10):1058

Newman M, Ebrahimie E, Lardelli M (2014) Using the zebrafish model for Alzheimer's disease research. Front Genet 5:189

Ochenkowska K, Herold A, Samarut É (2022) Zebrafish is a powerful tool for precision medicine approaches to neurological disorders. Front Mol Neurosci 15:944693

Paduraru E, Iacob D, Rarinca V, Plavan G, Ureche D, Jijie R, Nicoara M (2023) Zebrafish as a potential model for neurodegenerative diseases: a focus on toxic metals implications. Int J Mol Sci 24(4):3428

Pal N, Jain A, Nathiya D, Sharma RK, Patel A, Bareth H, Goswami S, Chaturvedi B (2022) A review on role of zebrafish in Huntington's disease. J Pharm Res Int. 34:9–14

Pappalardo F, Russo G, Pennisi M, Parasiliti Palumbo GA, Sgroi G, Motta S, Maimone D (2020) The potential of computational modeling to predict disease course and treatment response in patients with relapsing multiple sclerosis. Cells. 9(3):586

Parichy DM, Postlethwait JH (2020) The biotic and abiotic environment of zebrafish. In: Behavioral and neural genetics of zebrafish. Academic Press, pp 3–16

Pavlides A, Hogan SJ, Bogacz R (2015) Computational models describing possible mechanisms for generation of excessive beta oscillations in Parkinson's disease. PLoS Comput Biol 11(12):e1004609

Pernice S, Follia L, Maglione A, Pennisi M, Pappalardo F, Novelli F, Clerico M, Beccuti M, Cordero F, Rolla S (2020) Computational modeling of the immune response in multiple sclerosis using epimod framework. BMC Bioinformatics. 21:1–20

Petri K, Zhang W, Ma J, Schmidts A, Lee H, Horng JE, Kim DY, Kurt IC, Clement K, Hsu JY, Pinello L (2022) CRISPR prime editing with ribonucleoprotein complexes in zebrafish and primary human cells. Nat Biotechnol 40(2):189–193

Pitchai A, Rajaretinam RK, Freeman JL (2019) Zebrafish as an emerging model for bioassay-guided natural product drug discovery for neurological disorders. Medicines. 6(2):61

Rai AR, Joy T, Rashmi KS, Rai R, Vinodini NA, Jiji PJ (2022) Zebrafish as an experimental model for the simulation of neurological and craniofacial disorders. Vet World. 15(1):22

Ranieri CM, Pimentel JM, Romano MR, Elias LA, Romero RA, Lones MA, Araujo MF, Vargas PA, Moioli RC (2021) A data-driven biophysical computational model of Parkinson's disease based on marmoset monkeys. IEEE Access 9:122548–122567

Razali K, Othman N, Mohd Nasir MH, Doolaanea AA, Kumar J, Ibrahim WN, Mohamed Ibrahim N, Mohamed WM (2021) The promise of the zebrafish model for Parkinson's disease: today's science and tomorrow's treatment. Front Genet 12:655550

Robea MA, Balmus IM, Ciobica A, Strungaru S, Plavan G, Gorgan LD, Savuca A, Nicoara M (2020) Parkinson's disease-induced zebrafish models: focussing on oxidative stress implications and sleep processes. Oxid Med Cell Longev 2020:1370837

Rosa JG, Lima C, Lopes-Ferreira M (2022) Zebrafish larvae behavior models as a tool for drug screenings and pre-clinical trials: a review. Int J Mol Sci 23(12):6647

Rothenbücher TS, Ledin J, Gibbs D, Engqvist H, Persson C, Hulsart-Billström G (2019) Zebrafish embryo as a replacement model for initial biocompatibility studies of biomaterials and drug delivery systems. Acta Biomater 100:235–243

Ruzicka L, Howe DG, Ramachandran S, Toro S, Van Slyke CE, Bradford YM, Eagle A, Fashena D, Frazer K, Kalita P, Mani P (2019) The Zebrafish Information Network: new support for

non-coding genes, richer Gene Ontology annotations and the Alliance of Genome Resources. Nucleic Acids Res 47(D1):D867–D873
Saluja D, Jhanji R, Kaushal S, Verma B, Sharma N, Singh R, Agrawal S, Yadav M, Kumar A, Singh C, Singh A (2021) Importance of zebrafish as an efficient research model for the screening of novel therapeutics in neurological disorders. CNS Neurol Disord Drug Targets 20(2):145–157
Sanabria-Castro A, Flores-Díaz M, Alape-Girón A (2020) Biological models in multiple sclerosis. J Neurosci Res 98(3):491–508
Sangaraju VV, Rajagopal S (2023) Application of computational modeling in obsessive compulsive disorder. In: Rajagopal S, Essa MM (eds) Nutrition and obsessive-compulsive disorder: the interplay. Routledge/Taylor & Francis Group
Schröter C, Herrgen L, Cardona A, Brouhard GJ, Feldman B, Oates AC (2008) Dynamics of zebrafish somitogenesis. Dev Dyn 237(3):545–553
Shelton DS, Shelton SG, Daniel DK, Raja M, Bhat A, Tanguay RL, Higgs DM, Martins EP (2020) Collective behavior in wild zebrafish. Zebrafish. 17(4):243–252
Shen C, Zuo Z (2020) Zebrafish (Danio rerio) as an excellent vertebrate model for the development, reproductive, cardiovascular, and neural and ocular development toxicity study of hazardous chemicals. Environ Sci Pollut Res 27(35):43599–43614
Son M, Kim DY, Kim CH (2022) Disease modeling of rare neurological disorders in zebrafish. Int J Mol Sci 23(7):3946
Soussi-Yanicostas N (2022) Zebrafish as a model for neurological disorders. Int J Mol Sci 23(8):4321
Stefanovski L, Meier JM, Pai RK, Triebkorn P, Lett T, Martin L, Bülau K, Hofmann-Apitius M, Solodkin A, McIntosh AR, Ritter P (2021) Bridging scales in Alzheimer's disease: biological framework for brain simulation with the virtual brain. Front Neuroinform 15:630172
Sun Y, Zhang B, Luo L, Shi DL, Wang H, Cui Z, Huang H, Cao Y, Shu X, Zhang W, Zhou J (2020) Systematic genome editing of the genes on zebrafish Chromosome 1 by CRISPR/Cas9. Genome Res 30(1):118–126
Świetlik D, Białowąs J, Moryś J, Klejbor I, Kusiak A (2019) Computer modeling of Alzheimer's disease—simulations of synaptic plasticity and memory in the CA3-CA1 hippocampal formation microcircuit. Molecules 24(10):1909
Thawkar BS, Kaur G (2021) Zebrafish as a promising tool for modeling neurotoxin-induced Alzheimer's disease. Neurotox Res 39:949–965
Thayumanavan G, Jeyabalan S, Fuloria S, Sekar M, Ravi M, Selvaraj LK, Bala L, Chidambaram K, Gan SH, Rani NN, Begum MY (2022) Silibinin and naringenin against bisphenol A-induced neurotoxicity in zebrafish model—potential flavonoid molecules for new drug design, development, and therapy for neurological disorders. Molecules. 27(8):2572
Tsang B, Gerlai RT (2022) Common aquarium plants as an enrichment strategy in zebrafish facilities. Zebrafish 19(6):218–223
Upadhayay S, Jamwal S, Kumar P (2023) Animal models of Huntington's disease and their applicability to novel drug discovery and development. Expert Opin Drug Discov 18(5):527–538
Vaz RL, Outeiro TF, Ferreira JJ (2018) Zebrafish as an animal model for drug discovery in Parkinson's disease and other movement disorders: a systematic review. Front Neurol 9:347
Verma R, Choudhary PR, Nirmal NK, Syed F, Verma R (2022) Neurotransmitter systems in zebrafish model as a target for neurobehavioural studies. Mater Today Proc 69:1565–1580
Wang X, Zhang JB, He KJ, Wang F, Liu CF (2021) Advances of zebrafish in neurodegenerative disease: from models to drug discovery. Front Pharmacol 12:713963
Williams ML, Solnica-Krezel L (2020) Cellular and molecular mechanisms of convergence and extension in zebrafish. Curr Top Dev Biol 136:377–407
Woo J, Cho H, Seol Y, Kim SH, Park C, Yousefian-Jazi A, Hyeon SJ, Lee J, Ryu H (2021) Power failure of mitochondria and oxidative stress in neurodegeneration and its computational models. Antioxidants. 10(2):229
Xiao F, Zhu W, Yu Y, He Z, Wu B, Wang C, Shu L, Li X, Yin H, Wang J, Juneau P (2021) Host development overwhelms environmental dispersal in governing the ecological succession of zebrafish gut microbiota. NPJ Biofilms Microbiomes 7(1):5

Zarantoniello M, Randazzo B, Gioacchini G, Truzzi C, Giorgini E, Riolo P, Gioia G, Bertolucci C, Osimani A, Cardinaletti G, Lucon-Xiccato T (2020) Zebrafish (Danio rerio) physiological and behavioural responses to insect-based diets: a multidisciplinary approach. Sci Rep 10(1):10648
Zhang W, Scerbo P, Delagrange M, Candat V, Mayr V, Vriz S, Distel M, Ducos B, Bensimon D (2022) Fgf8 dynamics and critical slowing down may account for the temperature independence of somitogenesis. Commun Biol 5(1):113
Zhao C, Zhai Y, Geng R, Wu K, Song W, Ai N, Ge W (2022) Genetic analysis of activin/inhibin β subunits in zebrafish development and reproduction. PLoS Genet 18(12):e1010523
Zhu W, Xu X, Wang X, Liu J (2019) Reprogramming histone modification patterns to coordinate gene expression in early zebrafish embryos. BMC Genomics 20(1):1–3

Chronobiology: Circadian Rhythm and Cancer

4

Richa Pathak, Nisha Singh, and Arti Parganiha

Abstract

Recent studies have proved that in mammals, molecular oscillators are present in every cell of the body. They comprise a group of core circadian genes forming a transcription–translation feedback loop to confer circadian programming in physiology, metabolism, and behavior. Mutations of the gene and physical destruction of the suprachiasmatic nucleus (SCN) proved that the master clock is responsible for the circadian rhythm. Multiple reports explain the tissue-specific altered circadian clock gene expression in various malignancies, including head and neck, breast, colon, and lung cancers. Extensive lines of emerging evidence collected for over two decades convincingly demonstrated that proper functioning of circadian rhythms is crucial for well-being and survival. There is highly credible evidence that circadian rhythm dysregulation results in a multitude of pathologies, including cancer. The circadian rhythms of melatonin, cortisol, TSH, growth hormone, core temperature, and rest–activity and sleep–wake cycle are disrupted due to shift work, jet lag, and light at night in cancer patients. These rhythm alterations are studied with the help of parameters, such as 24-h average or mesor, amplitude, phase advance or delay, and so on. This chapter addresses the prospects of applying the principles of chronobiology to treating cancer malignancies. Cancer cells exhibit a distinct circadian rhythm compared to normal healthy cells, necessitating a more nuanced treatment approach. Cancer chronotherapy, a promising field in clinical oncology, underscores the importance of administering anticancer medica-

R. Pathak · N. Singh
School of Studies in Life Science, Pandit Ravishankar Shukla University, Raipur, India

A. Parganiha (✉)
School of Studies in Life Science, Pandit Ravishankar Shukla University, Raipur, India

Center for Translational Chronobiology, Pandit Ravishankar Shukla University, Raipur, India

S. Sridhara et al. (eds.), *Animal Behavior in the Tropics*,
https://doi.org/10.1007/978-981-97-7324-4_4

tions in alignment with the patient's circadian rhythms to optimize therapeutic efficacy.

Keywords

Circadian rhythm · Cancer · Chronotherapy · Clock genes · Rest–activity cycle

4.1 Introduction

Earth rotates on its axis approximately every 24 h, resulting in a periodic pattern of important geophysical events: day and night. Various other periodic events, namely, seasons, lunar, tidal, and temperature cycles, are also present in nature. The frequency of oscillations of numerous biological processes (biochemical, molecular, and physiological) coincides with the day and night cycle, one of the dominant cycles. These oscillations (rhythms) are omnipresent and are exhibited by all eukaryotes and at least one prokaryote (cyanobacteria) (Pati 2001). Studies of these rhythmic aspects of biology fall in the domain of Chronobiology.

Chronobiology is an interdisciplinary branch of science that deals with the study of biological rhythms. A circadian rhythm is any biological rhythm that follows an approximately 24-h cycle. This internal time-keeping system is regulated by the suprachiasmatic nuclei (SCN); the master clock is located in the hypothalamus of the brain. Light is considered one of the most potent zeitgebers (time cue) with other non-photic factors that entrain the circadian rhythm. Examples of circadian rhythms in humans include the sleep–wake cycle, rest–activity rhythm, immune system—related secretions (such as cytokines and antigens), core body temperature, and hormone secretion. This temporal system aids in predicting the rhythmic aspects of human physiology, such as growth, replication, apoptosis, metabolism, the cell cycle, and proliferation.

4.2 Background

Jean-Jacques d'Ortous de Mairan, a French astronomer, in 1729 observed the biological rhythm in a plant, *Mimosa pudica*. He observed that a rhythmic movement of opening and closing of petals continues even in the absence of external stimuli. In 1832, De Candolle validated Mairan's experiments by demonstrating that under constant conditions, the leaf movements in *Mimosa pudica* continue showing a period of 22–23 h. In 1914, Wihem Pfeffer documented the presence of circadian rhythm in *Flemingia congesta*. Sir Jagadish Chandra Bose, an Indian scientist with his sophisticated experiments on *Mimosa pudica* (Bose 1913), *Cassia alata* (1927), *Water lily*, *Nymphaea*, *Jhinga*, and *Luffa acutangula* (1927) demonstrated the entrainment of rhythms with day and night changes (Patten 1927; Pati 2001). The 1960s and early 1970s witnessed the publication of numerous significant books (Bunning 1964; Halberg 1969; Conroy and Mills 1970; Mills 1973) that left a notable influence on the study of circadian rhythms. The field of chronobiology emerged

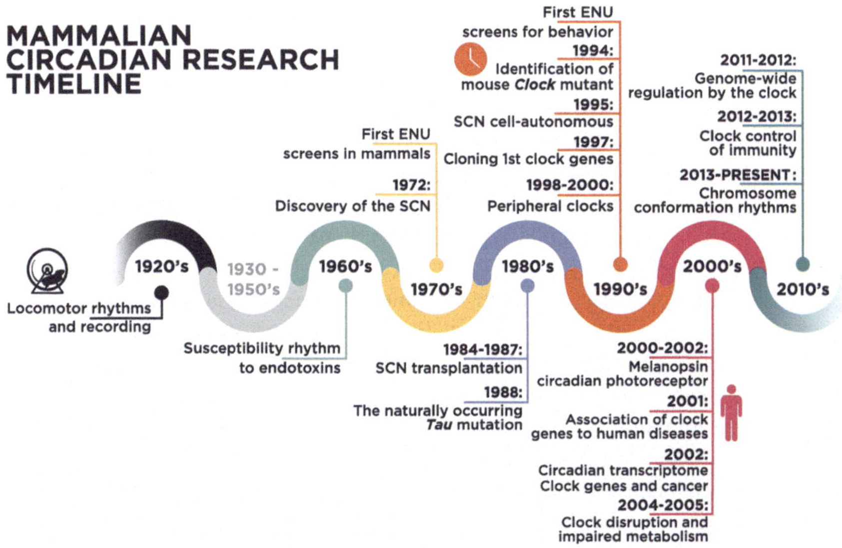

Fig. 4.1 Overview of major discoveries in the field of mammalian circadian clock research. (Permission taken from Rijo-Ferreira and Takahashi 2019)

following the discovery of circadian rhythms in various biological systems, ranging from single-cell activities to basic human behaviors. This advancement led to a comprehensive understanding of biological rhythms and their regulation.

The first clock mutations in *Drosophila melanogaster* (fruit fly) (Konopka and Benzer 1971) and *Mesocricetus auratus* (Syrian or golden hamster) (Ralph and Menaker 1988) are the initial experiments that started the new era of chronobiology. This discovery became a breakthrough and marked the beginning of an era of mammalian circadian research (Fig. 4.1). In 1984, Jeffrey C. Hall, Michael Rosbash, and Michael W. Young successfully isolated the "*period*" gene (Reddy et al. 1984). Hall and Rosbash discovered that the PER protein, generated by the *period* gene, accumulates at nighttime and degrades during the daytime (Hardin et al. 1990; Liu et al. 1992). Consequently, the levels of PER protein exhibit a 24-h cycle that aligns with the circadian rhythm. In 2017, J.C. Hall, M. Rosbash, and M.W. Young were awarded the Nobel Prize in medicine and physiology for the breakthrough discovery of the molecular aspects of circadian rhythm in *Drosophila*.

4.3 Physiology

Physiologically, SCN has since been identified as the major circadian pacemaker in mammals, situated in the anterior hypothalamus above the optic chiasm. Two groups of scientists, Bob Moore's group and Irving Zucker's group, independently demonstrated that the SCN is responsible for regulating the adrenal cortical and behavioral drinking rhythms in mice through lesioning experiments. In some species, the retina and pineal gland were also shown to have circadian clocks (Moore and Eichler 1972; Takahashi and Menaker 1984). Welsh et al. (1995) proved the existence of electrical activity in isolated single SCN cells exhibiting circadian rhythms, indicating that each cell can control its rhythms. Approximately 10,000–15,000 neurons form the SCN, which synchronizes the circadian rhythms and entrains them through the retinohypothalamic pathway. The SCN uses humoral and neurological systems to coordinate the step-by-step oscillation mechanism in cells, tissues, and organs in humans and many other animal species (Reppert and Weaver 2001). Further, the SCN through humoral and neural mechanisms controls the tissue-specific activities of peripheral clocks like the liver, brain, lung, heart, and kidney (Kramer-Hämmerle et al. 2005; Haus 2009; Logan and McClung 2019) (Fig. 4.2a). Peripheral tissues contain the peripheral clocks. Food restrictions appear to impact peripheral rhythms without compromising the SCN's core pacemaker activity. As a result, the master clock directs the peripheral clocks to generate their own circadian rhythm (Shearman et al. 2000; Schibler and Sassone-Corsi 2002; Hastings et al. 2003). The SCN generates a signal that is synchronized and maintained throughout all its neurons or cells, while every peripheral cell has its rhythm irrespective of the neighboring cell. The constrained feeding schedules and the ganglion cells in the retina are responsible for the entrainment of the

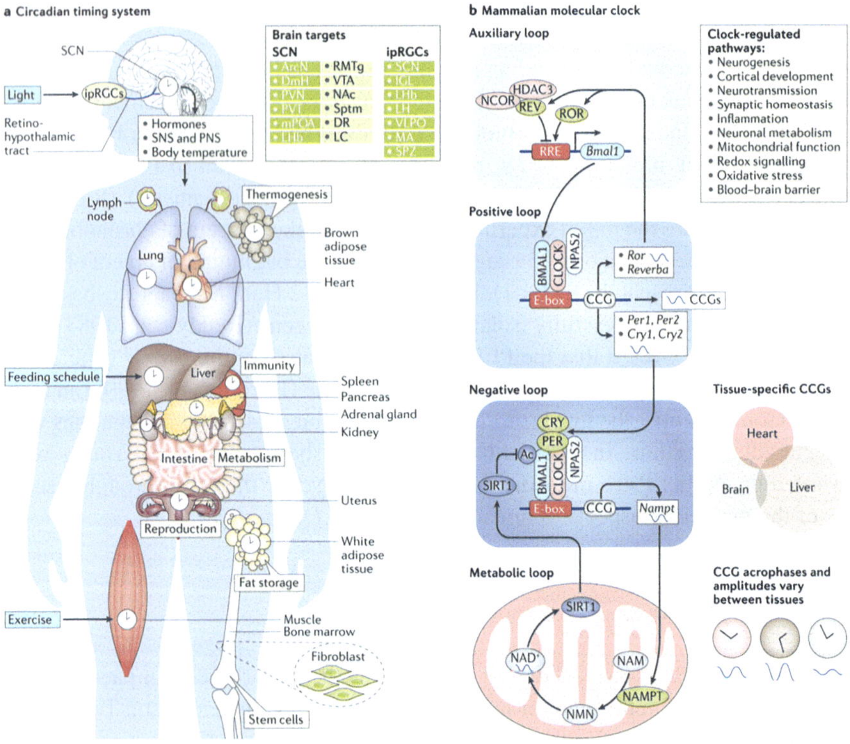

Fig. 4.2 (**a**) The circadian timing system and (**b**) the transcriptional and translational feedback loops oscillate for nearly 24 h in mammals. (Permission taken from Logan and McClung 2019)

clock by the day and night cycle or the artificial light. The circadian clock, an intrinsic time-keeping mechanism known as the circadian timing system (CTS), consists of central and peripheral clocks (Lowrey and Takahashi 2011).

4.4 Types of Rhythm

Biological rhythms are categorized into the following three major types based on the periods (τ) they display.

- **Circadian Rhythm**
- Biological variations that occur once every 24 ± 4 h or those that persist under constant conditions of light, temperature, and other geophysical factors for approximately about a day are referred to as circadian rhythms. The human sleep–wake cycle is a significant and extensively studied example of a circadian rhythm. Franz Halberg first introduced the term "circadian" in 1959 (Halberg 1959). The word "circadian" comes from the Latin words "circa," meaning "around," and "dies," meaning "day."
- **Ultradian Rhythm**
- Biological rhythms having a period (τ) < 20 h are termed ultradian rhythms. Heart rates, blood circulation, hormone secretions, thermoregulation, cellular activities, respiration, sleep stages, and food patterns are examples of ultradian rhythm.
- **Infradian Rhythm**
- Biological rhythms that exhibit a period (τ) > 28 h are termed infradian rhythms. It includes circannual rhythms (1 year), circatidal rhythms (12.44 days), and circalunar rhythms (29.5 days). Menstrual cycle, bird migration, fiddler crab feeding habits, and hibernation are examples of infradian rhythm.

All biological rhythms are studied with the help of four basic characteristics, such as period (τ), time series average (mesor), rhythm-dependent variation limits (amplitude), and peak time (acrophase) (Gupta and Pati 1992; Pati 2001; Pati et al. 2006, 2007; Parganiha et al. 2014). Mammals exhibit circadian rhythms in most of the biological mechanisms, such as subjective drowsiness (Gupta and Pati 1994), tumor necrosis factor (Liebmann et al. 1998), rest–activity cycle (Mormont et al. 2000), body temperature (Cajochen et al. 2005), hormone secretion (Oster et al. 2006), sleep–wake cycle (Wulff et al. 2009), immune response (Keller et al. 2009), blood pressure (Agarwal 2010), and gene expression (Lowrey and Takahashi 2004, 2011). Among various physiological and behavioral parameters, rest–activity rhythm is considered a noninvasive and robust marker for assessing the state of the circadian timing system (CTS) in humans (Mormont et al. 2000). This rhythm synchronizes with the natural light–dark cycle, exhibiting a circadian period (Zavada et al. 2005). Rest–activity rhythm is already utilized as a marker of endogenous circadian function in various clinical and nonclinical studies, including isolation studies (Aschoff 1965, 1994), psychological studies (Teicher et al. 1993), and cancer studies (Mormont et al. 2000; Ancoli-Israel et al. 2006; Taj et al. 2013; Lévi et al. 2014; Ortiz-Tudela et al. 2014; Parganiha et al. 2014; Sultan et al. 2014, 2017a, b, c, 2018, 2020).

4.5 Molecular Chronobiology

During the last two decades, gene-based behavioral studies conducted on many model systems (including mostly eukaryotes and a few prokaryotes) have revealed the molecular mechanism underlying circadian clocks (Lowrey and Takahashi 2011). Table 4.1 summarizes the key circadian genes and their function in lower organisms. The molecular circadian oscillators comprise a group of core circadian genes that establish a transcriptional–translational feedback loop to provide living forms with circadian programming in physiology, metabolism, and behavior. The efficient functioning of these oscillators is vital for the well-being and survival of

Table 4.1 The circadian clock genes in various lower model organisms

Organism	Circadian clock genes	Mechanism of circadian clock	References
Bacteria (*Cyanobacteria*)	*KaiA* *KaiB* *KaiC*	• The *KaiA* gene controls its own expression and produces the KaiA protein, which positively regulates the *KaiB* and *KaiC* gene transcription • The KaiC protein inhibits its transcription, establishing a negative feedback loop	Liu et al. (1995), Ishiura et al. (1998), Iwasaki et al. (1999)
Fungi (*Neurospora crassa*)	*wc1* *wc2* *frq*	• The dimerization of WC1 and WC2 proteins (translated from *wc1* and *wc2* genes, respectively) occurs followed by phosphorylation of WC1 due to exposure to light • The dimer of WC1 and WC2 enters the nucleus and positively regulates the transcription of the *frq* gene • FRQ protein translated from *frq* gene inhibits its own transcription	McClung et al. (1989), Ballario and Macino (1997), Garceau et al. (1997), Linden and Macino (1997), Liu et al. (1997), Merrow et al. (1999)

(continued)

Table 4.1 (continued)

Organism	Circadian clock genes	Mechanism of circadian clock	References
Insects (*Drosophila melanogaster and Antheraea pernyi*)	*bmal1* *dclock* *(jerk)* *dcry/cry* *dbt* *per* *tim*	• *Clock* and *Bmal1* genes encode for CLOCK and BMAL1 protein. These two proteins form a heterodimer and positively regulate the *per* and *tim* genes transcription • The heterodimer of PER and TIM moves to the nucleus and inhibits its own transcription • *Dbt* gene encodes Casein kinase 1 that helps in the degradation of the *per* monomers by ubiquitin pathway (DBT-dependent phosphorylation) • The heterodimer of proteins CYCLE and CLOCK binds to the promoter of *per* and *tim* to activate their transcription • *Cry* gene inhibits the PER-TIM dimer as it is structurally similar to photolyases (an activated DNA repair enzyme)	Konopka and Benzer (1971), Sehgal et al. (1994), Sauman and Reppert (1996), Allada et al. (1998), Darlington et al. (1998), Price et al. (1998), Stanewsky et al. (1998), Ishikawa et al. (1999), Selby and Sancar (1999)

these organisms and is found in almost every cell of the body (Balsalobre et al. 1998; Lowrey and Takahashi 2004; Welsh et al. 2004; Yoo et al. 2004). In mammals, *Clock*, *Bmal1*, *Per1*, *Per2*, *Cry1*, and *Cry2* are some of the important circadian clock genes.

4.6 Molecular Mechanism of the Mammalian Circadian Clock

In 1988, Ralph and Menaker identified a spontaneously occurring mutation known as *tau* that affected the biological clock in the golden hamster (Ralph and Menaker 1988). Subsequent research determined that *tau* is a semidominant autosomal mutation responsible for generating a circadian rhythm over a short period. The discovery of a second mutant, named circadian locomotor output cycles kaput (*clock*), was achieved through *N*-ethyl-*N*-nitrosourea (ENU) mutagenesis. This mutation led to the complete disruption of circadian locomotor rhythm under constant darkness conditions (Vitaterna et al. 1994). Later, the mammalian *period* gene was identified, and its transcript was found to exhibit periodicity in the suprachiasmatic nucleus (SCN), with a slow rise before dawn and a peak in the afternoon, while its expression in the retina lagged 4 h behind (Tei et al. 1997). Additional genes involved in the core clock mechanism, such as *timeless* (*tim*) (Sangoram et al. 1998), *Bmal1* (Gekakis et al. 1998), and *cryptochrome* (*Cry1* and *Cry2*) (van der Horst et al. 1999), have also been cloned in mammals.

The key findings in the circadian timing system (autoregulatory feedback) in mammals as per current knowledge have been summarized and described in Fig. 4.3. It starts with the dimerization of the CLOCK and BMAL1 (activates transcription), leading to the positive regulation of the expression levels of *period* and *cryptochrome* genes. The protein products (PER and CRY) accumulate in the cytoplasm and subsequently form a heterodimer. This PER-CRY heterodimer moves back into the nucleus to suppress its own transcription (Fig. 4.3). This creates a loop that runs for approximately 24 h, and the turnover rate of PER and CRY protein is governed by SCF E3 ubiquitin ligases, posttranscriptionally. Additional feedback loops such as *Rev-erb* and *Rorα* are connected to the core PER–CRY loop. This loop results in the antiphase oscillation of *BMAL1* transcription (a little effect on *CLOCK* also). CLOCK:BMAL1 also controls a large number of target genes known as *clock-controlled genes* (*CCGs*). The PER and CRY proteins are phosphorylated by Casein

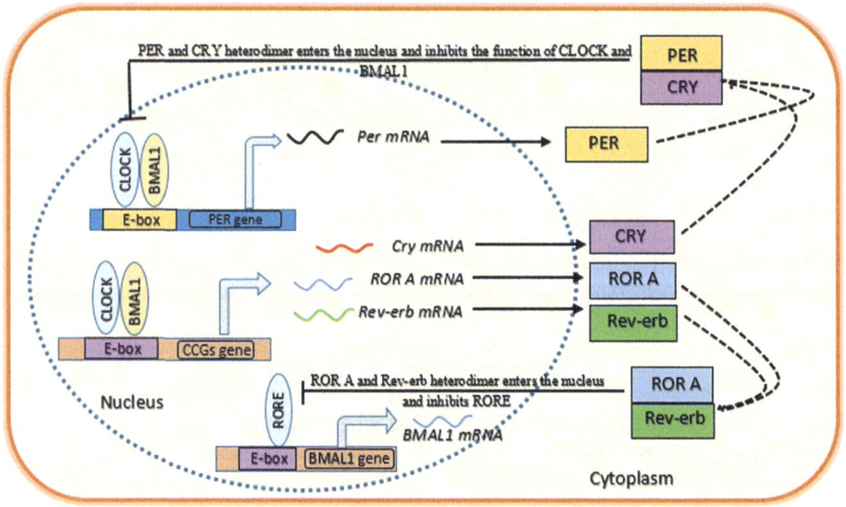

Fig. 4.3 Oversimplified circadian core clock mechanism in mammals: also known as a transcriptional–translational feedback loop that governs the circadian rhythm. *CRY* cryptochrome, *CCG* clock-controlled gene, *BMAL1* brain and muscular aryl-hydrocarbon receptor nuclear translocator-like 1, *CLOCK* circadian locomotor output cycles kaput, *PER* period, *Rev-erb* nuclear receptor subfamilies, *ROR α* retinoidrelated orphan receptors α. (Modified, Permission taken from Sultan 2019)

kinase 1 (CK1) and AMP kinase (AMPK), respectively, to increase the polyubiquitination (E3 ubiquitin ligases tag the PER and CRY proteins for degradation by the 26S proteasome complex) (Lowrey and Takahashi 2004, 2011; Takahashi et al. 2008; Mohawk et al. 2012). DEC1, DEC2, and other proteins are transcriptional targets of CLOCK–BMAL1 (Lowrey and Takahashi 2004; Takahashi et al. 2008; Mohawk et al. 2012; Partch et al. 2014; Sultan et al. 2017c; Sultan 2019).

The above transcriptional–translational feedback loop is seen in both SCN as well as other peripheral tissues (Panda et al. 2002; Buhr and Takahashi 2013; Sultan 2019) (Figs. 4.2b and 4.3). Besides clock genes, hundreds of genes have been identified that have a circadian disposition. There are a considerable number of differences between the rhythmicity of various genes residing in different organs (Panda et al. 2002; Storch et al. 2002). Thus, clock gene expression seems time-specific to achieve the best possible function of a particular organ (Buhr and Takahashi 2013). The molecules (oscillators) involved in the circadian timing system might show heterogeneity across species, but there is substantial conservation in the molecular mechanisms throughout the various levels of organisms (Zordan et al. 2002). The advancement in the tools and techniques of molecular biology has shown a dramatic rise in the molecular evidence of such biological phenomena.

4.7 Cancer Chronobiology

Tumor cell development exhibits a predictable rhythmicity, equivalent to the circadian rhythm that drives the cell proliferation process in normal tissues. The circadian clock controls the expression level of cell-cycle and tumor-suppressor genes at both the mRNA and protein levels. Circadian clock proteins regulate DNA synthesis and repair at the molecular level during normal S-phase advancement. The genes involved in the circadian clock have a molecular and functional relationship with the cell cycle and DNA damage checkpoints that are crucial for maintaining genome stability and integrity (Collis and Boulton 2007). The molecular components of the circadian clockwork control important genes of cell proliferation (e.g., *wee1*) in rejuvenating normal tissues (Matsuo et al. 2003). The circadian clock genes affect a cell's genotoxic stress vulnerability. This could be brought by the time-based alterations in carcinogenesis caused by genotoxic chemicals.

Lévi et al. (2007) suggested the presence of a relationship between clock and cell cycle proteins, highlighting a connection between circadian rhythm disturbances and an increased risk of cancers, including endometrial, breast, colon, and prostate cancers, observed in humans and most animal models (Lévi et al. 2007). The circadian rhythm in cell proliferation offers the potential to improve the efficacy and tolerance of anticancer therapies by synchronizing treatment with specific phases of the host's or tumor's circadian system.

It was demonstrated through experimental modification of the circadian cycle in an animal model that dysregulation in CTS resulted in tumor growth (Filipski et al. 2002). The tumors can follow the body's circadian rhythm or have their tumor-specific rhythm pattern. In well-differentiated, slow-growing cancers, host-tumor synchronization is most common, while ultradian rhythms are more frequently observed and dominate in fast-growing or advanced-stage malignancies (Filipski et al. 2002). Following radiation therapy, a transition from a mainly ultradian rhythm to a predominantly circadian rhythm in tumor DNA replication was observed, suggesting a potential alteration in the tumor's biological behavior (Haus 2002).

Cancer patients experience changes in the circadian parameters of several variables, including melatonin, cortisol, lymphocytes, and rest–activity (Mormont et al. 2002; Pati et al. 2006, 2007; Parganiha et al. 2014). The circadian timing system (CTS) was found to be disrupted in metastatic colorectal cancer patients and accelerates the progression of cancer in rodents with tumors (Lévi 1997, 2001; Mormont et al. 2000, 2002; Mormont and Lévi 2003; Innominato et al. 2009; Lévi et al. 2010). Moreover, diurnal rhythms in cortisol were observed flattened or disrupted in metastatic breast cancer (Sephton et al. 2000).

4.8 Rest–activity Rhythm

Rest–activity is a reflector rhythm of the biological clock functioning in humans and several animal species (Moore and Eichler 1972). In humans, rest–activity is one of the important biomarkers of the endogenous circadian clock's function. At the gene level, a mutation in the genes responsible for the circadian regulation induced disruption in rest–activity rhythm in *Drosophila*, hamsters, or mice (Ralph et al. 1990; Vitaterna et al. 1994; King et al. 1997).

Mormont and Lévi (2003) documented that the action of certain anticancer drugs (chronopharmacology) showed a phase relationship comparable to that of the rest–activity cycles in both humans and mice. Rodents showed a significant improvement in their therapeutic index when the anticancer drugs were administered at specific stages of the rest–activity cycle (Lévi 1997).

4.9 Monitoring of Rest–activity Circadian Rhythms

The evaluation of circadian rhythms in rest–activity using actigraphy has become widely accepted as a biomarker in various clinical fields. Actigraphy is a noninvasive technique for measuring rest–activity by converting wrist movements into activity counts, and it has been proven to be a valid and reliable instrument. "Activity count, a device-specific arbitrary unit, represents the frequency and amplitude of acceleration events over a user-defined measurement epoch" (Parganiha et al. 2014). In the 1970s, Colburn et al. (1970), McPartland et al. (1976), and Kripke et al. (1978) released the first credible actigraphy data. They demonstrated that wrist actigraphy provided cost-effective and precise estimations of sleep time comparable to other methods. They also compared wrist activity movement with EEG, EOG, and EMG data, showing that the accuracy of wrist actigraphy was approximately

90% of the values obtained by polysomnography (Louis et al. 1997). Subsequently, numerous studies have been conducted using actigraphy in healthy individuals, cancer patients, shift workers, and patients with cardiovascular disease and sleep disorders (Sadeh and Acebo 2002; Pati et al. 2006, 2007; Soni et al. 2008; Innominato et al. 2010; Paudel et al. 2011; Taj et al. 2013; Parganiha et al. 2014; Sultan et al. 2014, 2017a, b, 2018, 2020; Jeon et al. 2020; Roh et al. 2022). Currently, this technique is increasingly employed in clinical, sleep, and chronotherapeutic research, offering an advantage over conventional sleep recording methods (Ancoli-Israel et al. 2002). Various actigraphy devices such as Motionlogger (Ambulatory Monitoring, Inc., Ardsley, NY); Actical (Mini Mitter Co., Respironics, Philips, USA); Actiwatch 2 (AW2), Actiwatch Spectrum (AW-S), and Actiwatch Spectrum Plus (AWSP) (Philips Respironics, USA); Research Activity Monitor RT3 (Stayhealthy Inc.); ActiGraph GT3X, GT1M (ActiGraph, LLC, Pensacola, FL, USA); and MotionWatch 8 (CamNtech, USA) are used for the assessment of rest–activity rhythm in humans.

The regulation of the sleep–wake cycle and various pathological processes is influenced by rest–activity circadian rhythms (RARs) (Mormont et al. 2000; Paudel et al. 2010). As individuals age, these rhythms typically deteriorate, showing reduced amplitude, rhythm loss (Kripke et al. 2005), delayed phase (Chaudhury et al. 2005), and diminished adaptation to phase-resetting signals (Hofman and Swaab 2006). Research has demonstrated that changes in rest–activity rhythms (RARs) are significantly correlated with a deteriorating health index in elderly individuals (Paudel et al. 2010). Furthermore, abnormal RARs have been identified in several clinical conditions, including cancer and dementia (Mormont et al. 2000; Gehrman et al. 2004; Tranah et al. 2010, 2011; Montaruli et al. 2012).

4.10 Rest–activity Rhythm in Cancer Patients

Three common modifications in rest–activity rhythms in cancer patients were identified by Touitou et al. (1995), Mormont and Lévi (1997), and Pati et al. (2006, 2007): (1) amplitude and 24-h average reduction, (2) phase shift advance/delay, and (3) repression of one or more circadian outputs. Mormont et al. (2000) demonstrated that in metastatic colorectal cancer, the rest–activity rhythm positively correlated with tumor response and survival of patients. Changes in rest–activity rhythm might be due to large tumors, an increased number of metastatic sites, metastasis in the liver, poor performance index, and a staged tumor (Mormont and Lévi 1997; Mormont et al. 2000; Pati et al. 2006, 2007). Figure 4.4a–d provides an illustrative example of the rest–activity profiles for cancer inpatients (Fig. 4.4a, c) and healthy individuals (Fig. 4.4b, d) (Pati et al. 2007). The figure clearly demonstrates that healthy subjects, irrespective of gender, displayed an apparent and consistent day–night pattern in rest–activity, with higher activity levels during the day (Fig. 4.4b, d). In contrast, this distinction and regularity were absent in cancer inpatients, regardless of gender (Fig. 4.4a, c), with activity occurring both day and night.

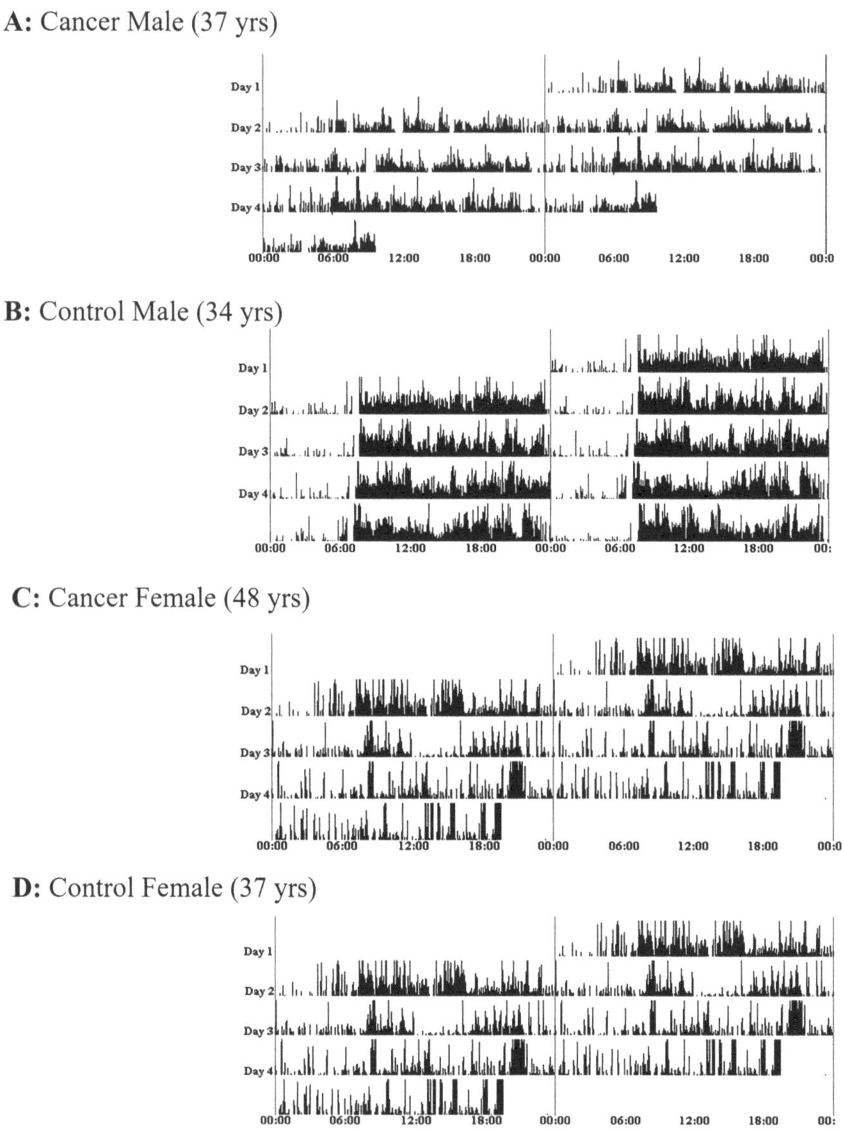

Fig. 4.4 Examples are shown of double-plotted rest–activity profiles (actograms) for cancer inpatients (**a** and **c**) and healthy subjects (**b** and **d**). The *x*-axis represents the clock hour, with each row spanning two 24-h periods (1 day and the next). The activity levels (measured in relative units) in each row are indicated by the black bars at that time on the *x*-axis. Notable activity was observed in cancer patients during the nighttime. (Permission taken from Pati et al. 2007)

Additionally, cancer patients demonstrated a lower overall activity level compared to healthy controls, with noticeable nighttime activity (Pati et al. 2007).

Different cancer forms are linked to a dysregulated rest–activity rhythm, with breast cancer being one of them (Ancoli-Israel et al. 2006; Zhu et al. 2008). Our laboratory's findings revealed deterioration in the circadian rest–activity rhythm in newly diagnosed patients with breast cancer in Chhattisgarh (Sultan et al. 2017a, b, 2018, 2020). The disruption in rest–activity rhythm could be linked to the effects of the disease or its treatment, such as chemotherapy (Sultan et al. 2018, 2020). Savard et al. (2009) documented that the circadian rhythm parameters, such as rhythmicity, amplitude, and mesor (up-mesor or down-mesor), were severely disturbed during the first week of chemotherapy cycles. Sultan et al. (2018) observed similar outcomes, indicating a gradual decline in the characteristics of the rest–activity rhythm and performance among both breast cancer inpatients and outpatients as they progressed through chemotherapy cycles. Roveda et al. (2019) demonstrated that the parameters of the rest–activity rhythm were altered in breast cancer survivors 5 years after diagnosis. Several theories propose that the expression of circadian genes is influenced by the rest–activity pattern (Merrow et al. 2005; Rijo-Ferreira and Takahashi 2019). Moreover, the habitat type (rural versus urban) may impact gene expression, including circadian patterns (Amoah et al. 2014; Carvalho et al. 2014; Bickler et al. 2020).

4.11 Circadian Clock Gene Expression in Cancer Patients

The identification of X-chromosome-linked *period* (*Per*) mutations in *Drosophila* by Konopka and Benzer (1971) laid the groundwork for molecular chronobiology. The significance of the *Per* gene in *Drosophila* circadian rhythms sparked interest in studying circadian clock genes in cyanobacteria, green algae, fungi, higher plants, and animals. The molecular foundation of circadian clocks was unraveled by Devlin and Kay (2001), Xu and Johnson (2001), Loros and Dunlap (2001), Reppert and Weaver (2001), and Williams and Sehgal (2001). The molecular organization of the circadian rhythm consists of an intricate network of clock components that work together to maintain the body's physiological processes. Disruptions in these clocks can lead to various illnesses, including cancer and metabolic disorders. The operation of the clock machinery involves the heterodimerization of the *Clock* and *Bmal1* genes, which stimulates the transcription of other clock genes like *Cry* and *Per*. Additionally, the *Clock/Bmal1* complex regulates "*clock-controlled genes*" (*CCGs*). Fourteen core clock genes have been identified, which regulate transcriptional/translational feedback loops (Ko and Takahashi 2006).

Initial research focused on single-clock genes. For example, Gery et al. (2006) investigated the role of *Per1* in cell development and DNA damage management. Disruptions in the functioning of clock-controlled circadian genes are associated with carcinogenesis. These genes typically act as tumor suppressors, but when their expression patterns are disrupted, it can accelerate the progression and development of cancer (Fu et al. 2002; Fu and Lee 2003; Zhu et al. 2005, 2007a, b, 2008, 2009a, b; Hoffman et al. 2009; Fu and Kettner 2013). Luo et al. (2012) investigated a similar set of genes in human gliomas to better understand their functional relationships. A systematic investigation of clock genes across various cancer types has highlighted the importance of individual genes in specific cancers (Ye et al. 2018).

Numerous experimental and epidemiological studies have established a link between altered circadian clock genes, their polymorphisms, and cell cycle genes, resulting in an increased risk of breast cancer (Sahar and Sassone-Corsi 2007; Fu et al. 2011; Monsees et al. 2012; Zienolddiny et al. 2013). The *Per* family genes (*Per1, Per2,* and *Per3*) are thought to function as tumor suppressors (Li et al. 2013). *Per1* expression is connected to cell cycle phase transitions (Gery et al. 2006). In response to DNA damage, *Per1* interacts with *Chk2* and *Atm*, triggering the *Atm* checkpoint pathway and subsequently activating *Chk2*, which aids in preventing cellular transformation into a cancerous state (Gery et al. 2006).

The altered expression of clock genes has been associated with poor prognosis and differentiation in breast cancer (Cadenas et al. 2014). Their research found that decreased gene expression significantly linked with a higher likelihood of metastasis. The study highlighted a strong relationship between *Cry2* and *Per3* in slow-growing, well-differentiated tumors, which was absent in fast-growing tumors. This lack of association between *Cry2* and *Per3* was also linked to decreased estrogen receptor expression, *Her2* oncogene amplification, and higher histological grade. About 96% of human breast cancer cells exhibited abnormalities in the promoter regions of *Per* genes (Gery et al. 2006). Mutations in the *Per* genes, particularly *Per2*, have been associated with an increased risk of breast cancer in women (Hoffman et al. 2010). *Dec1* and *Dec2* are crucial factors that regulate cells' circadian cycles, proliferation, tumorigenesis, differentiation, and apoptosis (Liu et al. 2010). Tumor growth factor (TGF) can induce an increase in *Dec1* expression (Ehata et al. 2007). *Dec1* expression is closely associated with tumor invasiveness and malignancy, rising with tumor stage and hypoxia, suggesting a role in tumor suppression and potentially apoptosis (Chakrabarti et al. 2004). In MCF-7 and MDA-MB-231 cancer cell lines, lower *Dec1* expression is linked to higher claudin-1 expression, which is thought to suppress tumor invasion (Li et al. 2013). Additionally, claudin-1 dysregulation may enhance the metastatic potential of breast cancer cells (Liu et al. 2013). Despite belonging to the same gene family, *Dec1* and *Dec2* have distinct roles in the cell cycle process. *Dec1* promotes cell proliferation and metastatic progression, whereas *Dec2* inhibits the latter (Li et al. 2013).

Table 4.2 Alteration in cellular function attributed to clock gene deregulation in various cancer types

Cancer type	Circadian genes deregulated	Alteration in cellular function	Reference(s)
Abdominal malignant tumors	*Clock, Bmal1, Per1, Per2* and *Per3, Cry1* and *Cry2*	• Uncontrolled cell growth • Malignant transformation	Yang et al. (2017)
Acute lymphocytic leukemia (ALL)	*Bmal1* and *Clock*	• Cell cycle control • Chromatin remodeling	Taniguchi et al. (2009)
Acute myeloid leukemia (AML)	*Bmal1, Per1, Per2,* and *Per3*	• Cell cycle control • Chromatin remodeling	Puram et al. (2016)
Breast cancer	*Bmal1, Clock, Cry1, Cry2, Per1, Per2, Per3,* and *Npas2*	• Apoptosis • Cell cycle control • Chromatin remodeling • DNA damage repair • Telomere length	Chen et al. (2005), Blask et al. (2011), Erdem et al. (2017), Li et al. (2024)
Chronic lymphocytic leukemia (CLL)	*Cry1* and *Per2*	• Cell cycle control • Chromatin remodeling • DNA damage repair	Lewintre et al. (2009), Eisele et al. (2009), Hernández-Rosas et al. (2020)
Chronic myeloid leukemia (CML)	*Bmal1, Cry1, Cry2, Per1, Per2,* and *Per3*	• Apoptosis • Cell cycle control • Chromatin remodeling	Sun et al. (2010), Wang et al. (2020)
Colorectal cancer	*Clock, Per1, Per2, Per3, Bmal1* and *Cry1*	• Chromatin remodeling • DNA damage repair	Garufi et al. (2016), Momma et al. (2017)
Endometrial cancer	*Cry1, Per1, Per2,* and *Per3*	• Chromatin remodeling	Shih et al. (2006), Viswanathan et al. (2007), Viswanathan and Schernhammer (2009)
Glioma	*Cry1, Cry2, Per1, Per2, Per3* and *Arntl2*	• Apoptosis • Cell cycle control	Fujioka et al. (2006), Xia et al. (2010), Luo et al. (2012), Armstrong et al. (2017), Arafa and Emara (2020)
Head and neck squamous cell carcinoma	*Bmal1, CKIε, Cry1, Cry2, Per1, Per2, Per3,* and *Tim*	• Cell cycle control • Cell proliferation	Hsu et al. (2012), Papagerakis et al. (2014), Rahman et al. (2019)

(continued)

Table 4.2 (continued)

Cancer type	Circadian genes deregulated	Alteration in cellular function	Reference(s)
Hepatocellular carcinoma	*Cry2, Per1, Per2, Per3, Tim* and *Rev-Erbα*	• Cell cycle control • Chromatin remodeling	Lin et al. (2008), Hassan et al. (2020)
Lung cancer	*Clock, Per1, Per2,* and *Per3*	• Cell cycle control • Chromatin remodeling • DNA damage repair	Gery et al. (2007), Kim et al. (2012)
Malignant pleural mesothelioma	*Bmal1, Cry2, Per1, Per3, Npas2, Rev-Erbα, Rev-Erbβ,* and *Tim*	• Apoptosis • Cell cycle control • Chromatin remodeling • Deregulated chemotherapy drug response	Elshazley et al. (2012)
Non-Hodgkin's lymphoma (NHL)	*Npas2*	• Cell cycle control • DNA damage repair • Immune deficiency	Lahti et al. (2008), Hoffman et al. (2008)
Diffuse large B-cell lymphoma	*Bmal1* and *Clock*	• Chromatin remodeling	Taniguchi et al. (2009)
Oral squamous cell carcinoma	*Per2*	• Cell proliferation • Apoptosis • Cell cycle progression	Wang et al. (2016), Su et al. (2017), Nirvani et al. (2018)
Osteosarcoma	*CK1ε* and *Per2*	• Apoptosis • Cell cycle control	Kong et al. (2022)
Ovarian cancer	*Bmal1, CK1ε, Clock, Cry1, Cry2, Per1, Per2,* and *Per3*	• Apoptosis • Cell cycle control • Deregulated chemotherapy drug response	Tokunaga et al. (2008)
Pancreatic cancer	*Bmal1, CK1ε, Clock, Cry1, Cry2, Dec1, Per1, Per2, Per3, Tim,* and *Tipin*	• Apoptosis • Cell cycle control • Chromatin remodeling	Suzuki et al. (2008), Oda et al. (2009), Relles et al. (2013)

(continued)

Table 4.2 (continued)

Cancer type	Circadian genes deregulated	Alteration in cellular function	Reference(s)
Prostate cancer	*Bmal1, CK1ε, Clock, Cry1, Cry2, Npas2, Per1, Per2,* and *Per3*	• Apoptosis • Cell cycle control • DNA damage repair • Trans-activation of androgen receptor	Cao et al. (2009), Kiss and Ghosh (2016)

Overall, it is evident that disrupting clock gene function through various mechanisms promotes carcinoma growth, progression, and metastasis. Disruptions and polymorphisms in circadian genes are more prevalent and are associated with an increased risk of cancers such as prostate, breast, and non-Hodgkin's lymphoma (Stevens and Rea 2001; Hsu et al. 2007; Hoffman et al. 2009, 2010; Cerliani and Richard 2013). Table 4.2 depicts the alteration in cellular parameters attributed to the clock gene deregulation in various cancer types.

4.12 Cancer Chronotherapy

Circadian timing system (CTS) consists of oscillators that result in 24-h rhythms in metabolism, cell cycle, cell proliferation, immune system, angiogenesis, apoptosis, DNA replication, DNA repair mechanism, and many other cellular pathways. Cancer cells are proven to have a different circadian rhythm compared to normal healthy cells. Cancer chronotherapy is one of the most promising fields for clinical oncologists, which emphasizes the fact that the administration of anticancer medications should be done in synchronization with the patient's circadian rhythms to anticancer action to enhance the quality of life and patient's life (Lévi et al. 2020; Ozturk et al. 2017; Kuo and Ladurner 2020; Qian et al. 2021; Yang et al. 2021).

Lévi and his group are actively engaged in cancer chronotherapeutics. Chronotherapies increase the effectiveness and tolerability of anticancer medications (Lévi 1996, 1997, 2001, 2006; Lévi et al. 1988, 1997; Mormont and Lévi 2003; Lévi and Schibler 2007; Ortiz-Tudela et al. 2014). In 1966, Mauricio Garcia-Sainz and Franz Halberg were pioneer clinicians and researchers who started time-dependent medical treatment in human mammary and basal cancer cells through radiotherapy. This time-dependent cancer therapeutic approach was way ahead of time as the molecular mechanisms of circadian rhythmicity were discovered in the 1980s by Michael W. Young, Michael Rosbash, and Jeffrey C. Hall (Kuo and Ladurner 2020). Haus et al. (1972) and Lévi et al. (1988) discovered that the host toxicity to certain drugs was different at different time points. The time-dependent administration of about 50 anticancer drugs that include anthracyclines, platinum analogs, alkylating agents, antimetabolites, and plant alkaloids influences the extent of host toxicity (Pati 2001; Lévi et al. 2010; Dallmann et al. 2016). The toxicity of drugs markedly differs by their cell cycle stages in normal and tumor tissues, respectively. This forms the basis of the time-dependent chronotherapeutic approach in oncological studies (Mormont and Lévi 2003). Numerous cancer drugs have been investigated for their chronopharmacodynamics and chronotoxicity (Smolensky and Peppas 2007). The group of Lévi has experimented with 5-FU, oxaliplatin, and 5-FUDR on colorectal cancer patients and found that the proper timing of administration of oxaliplatin increases the success rate of clinical therapy and patient survival suffering from colon cancer (Lévi et al. 1997, 2007; Mormont and Lévi 2003; Altinok et al. 2007).

Exposure to morning light or taking melatonin before bedtime are two strategies for synchronizing daily rhythms and benefiting from chronotherapy. The link between cancer and melatonin, known as the sleep hormone, released by the pineal gland, has been extensively studied for over 40 years (Reiter et al. 2017). In cancer chronotherapy, melatonin has been used as an adjuvant drug to help re-entrain the sleep–wake cycle in various cancer types. Li et al. (2017) identified a connection between melatonin and cancer, suggesting it may have an oncostatic effect on tumors such as gastric, prostate, breast, and colorectal cancers. Melatonin is regarded as a proapoptotic, antiproliferative, anti-inflammatory, antiangiogenic, and powerful antitumor agent (Bonmati-Carrion and Tomas-Loba 2021; Amiama-Roig et al. 2022). It serves as a biomarker rhythm alongside rest–activity and cortisol rhythms in cancer patients (Bartsch et al. 1981; Carlson et al. 2007; Reiter et al. 2017; Zaki et al. 2021). Önder et al. (2022) demonstrated that melatonin, when used in combination with treatment, enhances the in vitro cell death of breast cancer cells. Additionally, Li et al. (2017) suggested that melatonin could be used as an adjuvant drug to increase tumor cell sensitivity to both chemotherapy and radiotherapy.

Karaboué et al. (2024) recently suggested that the increasing understanding of the time-dependent antitumor regulation by the immune system can also be a great help in establishing individualized clinical trials. Kuo and Ladurner (2020) explained that the most effective approach would be to assess each patient's biological clock at the molecular level to reduce the cytotoxicity of anticancer drugs. They summarize the challenges of chronotherapy as follows: (a) designing clinical trials to study the interaction between anticancer drugs and biological clocks, (b) investigating how factors such as age, sex, and comorbidities affect chronotherapy, and (c) encouraging pharmaceutical companies to use chronotherapy criteria to assess the efficacy and toxicity of their products. Ozturk et al. (2017), Kuo and Ladurner (2020), and Yang et al. (2021) also suggested that the optimal chronomodulated drug delivery schedules could be developed by understanding the links between pharmacokinetics and chronobiology.

4.13 Conclusion

Cancer therapy has evolved beyond a purely biological field, with numerous interdisciplinary researchers contributing to a vast amount of knowledge. Mathematical models have gained attention in cancer research by considering environmental factors affecting tumor growth, known as "control theory" (Lecca 2021). The increasing data from tumors enables the development of mathematical models to predict optimal parameters for therapy delivery, tumor progression control, and response time to specific treatments (Li et al. 2011; Ledzewicz et al. 2013). Despite ongoing challenges, integrating molecular and clinical findings into a more sophisticated systemic design could significantly advance cancer treatment.

Acknowledgments The authors are thankful to the University Grants Commission, New Delhi, through the DRS-SAP scheme sanctioned in the thrust area—Chronobiology to the School of Studies in Life Sciences, Pt. Ravishankar Shukla University, Raipur; Chhattisgarh Council of Science & Technology (CCOST), Raipur; and Center for Translation Chronobiology, Pt. Ravishankar Shukla University, Raipur, India, for financial support. The authors are grateful to Dr. Joseph S. Takahashi, Dr. Colleen A. McClung, Dr. Armiya Sultan, and Dr. Atanu Kumar Pati for permitting us to use Figs. 4.1, 4.2, 4.3, and 4.4, respectively.

References

Agarwal R (2010) Regulation of circadian blood pressure: from mice to astronauts. Curr Opin Nephrol Hypertens 19(1):51–58

Allada R, White NE, So WV, Hall JC, Rosbash M (1998) A mutant Drosophila homolog of mammalian Clock disrupts circadian rhythms and transcription of period and timeless. Cell 93(5):791–804

Altinok A, Lévi F, Goldbeter A (2007) A cell cycle automaton model for probing circadian patterns of anticancer drug delivery. Adv Drug Deliv Rev 59(9–10):1036–1053

Amiama-Roig A, Verdugo-Sivianes EM, Carnero A, Blanco JR (2022) Chronotherapy: circadian rhythms and their influence in cancer therapy. Cancers 14(20):5071

Amoah AS, Obeng BB, May L, Kruize YC, Larbi IA, Kabesch M, Wilson MD, Hartgers FC, Boakye DA, Yazdanbakhsh M (2014) Urban–rural differences in the gene expression profiles of Ghanaian children. Genes Immun 15(5):313–319

Ancoli-Israel S, Martin JL, Kripke DF, Marler M, Klauber MR (2002) Effect of light treatment on sleep and circadian rhythms in demented nursing home patients. J Am Geriatr Soc 50(2):282–289

Ancoli-Israel S, Liu L, Marler MR, Parker BA, Jones V, Sadler GR, Dimsdale J, Cohen-Zion M, Fiorentino L (2006) Fatigue, sleep, and circadian rhythms prior to chemotherapy for breast cancer. Support Care Cancer 14(3):201–209

Arafa K, Emara M (2020) Insights about circadian clock and molecular pathogenesis in gliomas. Front Oncol 10:199

Armstrong TS, Shade MY, Breton G, Gilbert MR, Mahajan A, Scheurer ME, Vera E, Berger AM (2017) Sleep-wake disturbance in patients with brain tumors. Neuro-Oncology 19(3):323–335

Aschoff J (1965) Circadian rhythms in man: a self-sustained oscillator with an inherent frequency underlies human 24-hour periodicity. Science 148(3676):1427–1432

Aschoff J (1994) Naps as integral parts of the wake time within the human sleep-wake cycle. J Biol Rhythm 9(2):145–155

Ballario P, Macino G (1997) White collar proteins: passing the light signal in *Neurospora crassa*. Trends Microbiol 5(11):458–462
Balsalobre A, Damiola F, Schibler U (1998) A serum shock induces circadian gene expression in mammalian tissue culture cells. Cell 93(6):929–937
Bartsch C, Bartsch H, Jain AK, Laumas KR, Wetterberg L (1981) Urinary melatonin levels in human breast cancer patients. J Neural Transm 52(4):281–294
Bickler G, Morton S, Menne B (2020) Health and sustainable development: an analysis of 20 European voluntary national reviews. Public Health 180:180–184
Blask DE, Hill SM, Dauchy RT, Xiang S, Yuan L, Duplessis T, Mao L, Dauchy E, Sauer LA (2011) Circadian regulation of molecular, dietary, and metabolic signaling mechanisms of human breast cancer growth by the nocturnal melatonin signal and the consequences of its disruption by light at night. J Pineal Res 51(3):259–269
Bonmati-Carrion MA, Tomas-Loba A (2021) Melatonin and cancer: a polyhedral network where the source matters. Antioxidants 10(2):210
Bose JC (1913) On diurnal variation of moto-excitability in Mimosa. Ann Bot 27(108):759–779
Buhr ED, Takahashi JS (2013) Molecular components of the mammalian circadian clock. Handb Exp Pharmacol 217:3–27
Bunning E (1964) Circadian leaf movements in bean plants: earlier reports. Science 146(3643):551–551
Cadenas C, van de Sandt L, Edlund K, Lohr M, Hellwig B, Marchan R, Schmidt M, Rahnenfúhrer Oster H, Hengstler JG (2014) Loss of circadian clock gene expression is associated with tumor progression in breast cancer. Cell Cycle 13(20):3282–3291
Cajochen C, Munch M, Kobialka S, Krauchi K, Steiner R, Oelhafen P, Orgul S, Wirz-Justice A (2005) High sensitivity of human melatonin, alertness, thermoregulation, and heart rate to short wavelength light. J Clin Endocrinol Metabol 90(3):1311–1316
Cao Q, Gery S, Dashti A, Yin D, Zhou Y, Gu J, Koeffler HP (2009) A role for the clock gene per1 in prostate cancer. Cancer Res 69(19):7619–7625
Carlson LE, Campbell TS, Garland SN, Grossman P (2007) Associations among salivary cortisol, melatonin, catecholamines, sleep quality and stress in women with breast cancer and healthy controls. J Behav Med 30(1):45–58
Carvalho FG, Hidalgo MP, Levandovski R (2014) Differences in circadian patterns between rural and urban populations: an epidemiological study in countryside. Chronobiol Int 31(3):442–449
Cerliani MB, Richard SM (2013) Polymorphism analysis of genes involved in xenobiotic metabolism and circadian rhythm in human breast cancer. Int J Med Sci 1(4):57–64
Chakrabarti J, Turley H, Campo L, Han C, Harris AL, Gatter KC, Fox SB (2004) The transcription factor DEC1 (stra13, SHARP2) is associated with the hypoxic response and high tumour grade in human breast cancers. Br J Cancer 91(5):954–958
Chaudhury D, Wang LM, Colwell CS (2005) Circadian regulation of hippocampal long-term potentiation. J Biol Rhythm 20(3):225–236
Chen ST, Choo KB, Hou MF, Yeh KT, Kuo SJ, Chang JG (2005) Deregulated expression of the PER1, PER2 and PER3 genes in breast cancers. Carcinogenesis 26(7):1241–1246
Colburn TR, Smith BM, Guarini JJ, Simmons NN (1970) An ambulatory activity monitor with solid state memory. Biomed Sci Instrum 12:117–122
Collis SJ, Boulton SJ (2007) Emerging links between the biological clock and the DNA damage response. Chromosoma 116(4):331–339
Conroy RTWL, Mills JN (1970) Human circadian rhythms. Churchill, London, ix, pp 236
Dallmann R, Okyar A, Lévi F (2016) Dosing-time makes the poison: circadian regulation and pharmacotherapy. Trends Mol Med 22:430–445
Darlington TK, Wager-Smith K, Ceriani MF, Staknis D, Gekakis N, Steeves TD, Weitz CJ, Takahashi JS, Kay SA (1998) Closing the circadian loop: CLOCK-induced transcription of its own inhibitors per and tim. Science 280(5369):1599–1603
Devlin PF, Kay SA (2001) Circadian photoperception. Annu Rev Physiol 63(1):677–694
Ehata S, Hanyu A, Hayashi M, Aburatani H, Kato Y, Fujime M, Saitoh M, Miyazawa K, Imamura T, Miyazono K (2007) Transforming growth factor-β promotes survival of mammary car-

cinoma cells through induction of antiapoptotic transcription factor DEC1. Cancer Res 67(20):9694–9703
Eisele L, Prinz R, Klein-Hitpass L, Nückel H, Lowinski K, Thomale J, Moeller LC, Dührsen U, Dürig J (2009) Combined PER2 and CRY1 expression predicts outcome in chronic lymphocytic leukemia. Eur J Haematol 83(4):320–327
Elshazley M, Sato M, Hase T, Yamashita R, Yoshida K, Toyokuni S, Ishiguro F, Osada H, Sekido Y, Yokoi K, Usami N (2012) The circadian clock gene BMAL1 is a novel therapeutic target for malignant pleural mesothelioma. Int J Cancer 131(12):2820–2831
Erdem SJ, Notø HØ, Skare Ø, Lie JAS, Petersen-Øverleir M, Reszka E, Peplońska B, Zienolddiny S (2017) Mechanisms of breast cancer risk in shift workers: association of telomere shortening with the duration and intensity of night work. Cancer Med 6(8):1988–1997
Filipski E, King VM, Li X, Granda TG, Mormont MC, Liu X, Claustrat B, Lévi F (2002) Host circadian clock as a control point in tumor progression. J Natl Cancer Inst 94(9):690–697
Fu L, Kettner NM (2013) The circadian clock in cancer development and therapy. Prog Mol Biol Transl Sci 119:221–282
Fu L, Lee CC (2003) The circadian clock: pacemaker and tumour suppressor. Nat Rev Cancer 3(5):350–361
Fu LW, Zhang YM, Liang YJ, Yang XP, Pan QC (2002) The multidrug resistance of tumour cells was reversed by tetrandrine in vitro and in xenografts derived from human breast adenocarcinoma MCF-7/adr cells. Eur J Cancer 38(3):418–426
Fu Y, Sun Y, Li Y, Li J, Rao X, Chen C, Xu A (2011) Differential genome-wide profiling of tandem 3′ UTRs among human breast cancer and normal cells by high-throughput sequencing. Genome Res 21(5):741–747
Fujioka A, Takashima N, Shigeyoshi Y (2006) Circadian rhythm generation in a glioma cell line. Biochem Biophys Res Commun 346(1):169–174
Garceau NY, Liu Y, Loros JJ, Dunlap JC (1997) Alternative initiation of translation and time-specific phosphorylation yield multiple forms of the essential clock protein FREQUENCY. Cell 89(3):469–476
Garufi C, Giacomini E, Torsello A, Sperduti I, Melucci E, Mottolese M, Zeuli M, Ettorre GM, Ricciardi T, Cognetti F, Magnani M (2016) Gender effects of single nucleotide polymorphisms and miRNAs targeting clock-genes in metastatic colorectal cancer patients (mCRC). Sci Rep 6:34006
Gehrman P, Marler M, Martin JL, Shochat T, Corey-Bloom J, Ancoli-Israel S (2004) The timing of activity rhythms in patients with dementia is related to survival. J Gerontol A Biol Sci Med Sci 59(10):1050–1055
Gekakis N, Staknis D, Nguyen HB, Davis FC, Wilsbacher LD, King DP, Takahashi JS, Weitz CJ (1998) Role of the CLOCK protein in the mammalian circadian mechanism. Science 280(5369):1564–1569
Gery S, Komatsu N, Baldjyan L, Yu A, Koo D, Koeffler HP (2006) The circadian gene per1 plays an important role in cell growth and DNA damage control in human cancer cells. Mol Cell 22(3):375–382
Gery S, Komatsu N, Kawamata N, Miller CW, Desmond J, Virk RK, Marchevsky A, Mckenna R, Taguchi H, Koeffler HP (2007) Epigenetic silencing of the candidate tumor suppressor gene Per1 in non-small cell lung cancer. Clin Cancer Res 13(5):1399–1404
Gupta S, Pati AK (1992) Data analysis methodology in chronobiological studies. J Parasitol Appl Anim Biol 1:151–163
Gupta S, Pati AK (1994) Characteristics of circadian rhythm in six variables of morning active and evening active healthy human subjects. Indian J Physiol Pharmacol 38:101–107
Halberg F (1959) Physiologic 24-hour periodicity; general and procedural considerations with reference to the adrenal cycle. Int Z Vitaminforsch Beih 10:225–296
Halberg F (1969) Chronobiology. Annu Rev Physiol 31(1):675–726
Hardin PE, Hall JC, Rosbash M (1990) Feedback of the Drosophila period gene product on circadian cycling of its messenger RNA levels. Nature 343(6258):536–540

Hassan SA, Schmithals C, Von Harten M, Piiper A, Korf HW, von Gall C (2020) Time-dependent changes in proliferation, DNA damage and clock gene expression in hepatocellular carcinoma and healthy liver of a transgenic mouse model. Int J Cancer 148(1):226–237
Hastings MH, Reddy AB, Maywood ES (2003) A clockwork web: circadian timing in brain and periphery, in health and disease. Nat Rev Neurosci 4(8):649–661
Haus E (2002) Chronobiology of the mammalian response to ionizing radiation potential applications in oncology. Chronobiol Int 19(1):77–100
Haus E (2009) Chronobiology in oncology. Int J Radiat Oncol Biol Phys 73(1):3–5
Haus E, Halberg F, Scheving LE, Pauly JE, Cardoso S, Kühl JF, Hwang DS (1972) Increased tolerance of leukemic mice to arabinosyl cytosine with schedule adjusted to circadian system. Science 177(4043):80–82
Hernández-Rosas F, López-Rosas CA, Saavedra-Vélez MV (2020) Disruption of the molecular circadian clock and cancer: an epigenetic link. Biochem Genet 58(1):189–209
Hoffman AE, Zheng T, Ba Y, Zhu Y (2008) The circadian gene NPAS2, a putative tumor suppressor, is involved in DNA damage response. Mol Cancer Res 6(9):1461–1468
Hoffman AE, Zheng T, Yi C, Leaderer D, Weidhaas J, Slack F, Zhu Y (2009) microRNA miR-196a-2 and breast cancer: a genetic and epigenetic association study and functional analysis. Cancer Res 69(14):5970–5977
Hoffman AE, Zheng T, Yi CH, Stevens RG, Ba Y, Zhang Y, Leaderer D, Holford T, Hansen J, Zhu Y (2010) The core circadian gene cryptochrome 2 influences breast cancer risk, possibly by mediating hormone signaling CRY2 is a novel circadian biomarker for breast cancer. Cancer Prev Res 3(4):539–548
Hofman MA, Swaab DF (2006) Living by the clock: the circadian pacemaker in older people. Ageing Res Rev 5(1):33–51
Hsu HM, Wang HC, Chen ST, Hsu GC, Shen CY, Yu JC (2007) Breast cancer risk is associated with the genes encoding the DNA double-strand break repair Mre11/Rad50/Nbs1 complex. Cancer Epidemiol Biomarkers Prev 16(10):2024–2032
Hsu CM, Lin SF, Lu CT, Lin PM, Yang MY (2012) Altered expression of circadian clock genes in head and neck squamous cell carcinoma. Tumour Biol 33(1):149–155
Innominato PF, Focan C, Gorlia T, Moreau T, Garufi C, Waterhouse J, Giacchetti S, Coudert B, Iacobelli S, Genet D, Tampellini M, Chollet P, Lentz MA, Mormont MC, Lévi F, Bjarnason GA (2009) Circadian rhythm in rest and activity: a biological correlate of quality of life and a predictor of survival in patients with metastatic colorectal cancer. Cancer Res 69(11):4700–4707
Innominato PF, Lévi FA, Bjarnason GA (2010) Chronotherapy and the molecular clock: clinical implications in oncology. Adv Drug Deliv Rev 62(9–10):979–1001
Ishikawa T, Matsumoto A, Kato T Jr, Togashi S, Ryo H, Ikenaga M, Tanimura T (1999) DCRY is a Drosophila photoreceptor protein implicated in light entrainment of circadian rhythm. Genes Cells 4(1):57–65
Ishiura M, Kutsuna S, Aoki S, Iwasaki H, Andersson CR, Tanabe A, Golden SS, Johnson TH, Kondo T (1998) Expression of a gene cluster kaiABC as a circadian feedback process in cyanobacteria. Science 281(5382):1519–1523
Iwasaki H, Taniguchi Y, Ishiura M, Kondo T (1999) Physical interactions among circadian clock proteins KaiA, KaiB and KaiC in cyanobacteria. EMBO J 18(5):1137–1145
Jeon S, Conley S, Redeker NS (2020) Rest-activity rhythms, daytime symptoms, and functional performance among people with heart failure. Chronobiol Int 37(8):1223–1234
Karaboué A, Innominato PF, Wreglesworth NI, Duchemann B, Adam R, Lévi FA (2024) Why does circadian timing of administration matter for immune checkpoint inhibitors' efficacy? Br J Cancer. https://doi.org/10.1038/s41416-024-02704-9
Keller M, Mazuch J, Abraham U, Eom GD, Herzog ED, Volk HD, Kramer A, Maier B (2009) A circadian clock in macrophages controls inflammatory immune responses. Proc Natl Acad Sci U S A 106(50):21407–21412
Kim KS, Kim YC, Oh IJ, Kim SS, Choi JY, Ahn RS (2012) Association of worse prognosis with an aberrant diurnal cortisol rhythm in patients with advanced lung cancer. Chronobiol Int 29(8):1109–1120

King DP, Zhao Y, Sangoram AM, Wilsbacher LD, Tanaka M, Antoch MP, Steeves TD, Vitaterna MH, Kornhauser JM, Lowrey PL, Turek FW (1997) Positional cloning of the mouse circadian clock gene. Cell 89(4):641–653

Kiss Z, Ghosh PM (2016) Women in cancer thematic review: circadian rhythmicity and the influence of 'clock' genes on prostate cancer. Endocr Relat Cancer 23(11):T123–T134

Ko CH, Takahashi JS (2006) Molecular components of the mammalian circadian clock. Hum Mol Genet 15(2):R271–R277

Kong D, Liu Y, Zhang M (2022) Expression of the circadian clock gene ARNTL associated with DNA repair gene and prognosis of patient with osteosarcoma. Mutat Res 825:111801

Konopka RJ, Benzer S (1971) Clock mutants of *Drosophila melanogaster*. Proc Natl Acad Sci U S A 68(9):2112–2116

Kramer-Hämmerle S, Rothenaigner I, Wolff H, Bell JE, Brack-Werner R (2005) Cells of the central nervous system as targets and reservoirs of the human immunodeficiency virus. Virus Res 111(2):194–213

Kripke DF, Mullaney DJ, Messin S, Wyborney VG (1978) Wrist actigraphic measures of sleep and rhythms. Electroencephalogr Clin Neurophysiol 44(5):674–676

Kripke DF, Youngstedt SD, Elliott JA, Tuunainen A, Rex KM, Hauger RL, Marler MR (2005) Circadian phase in adults of contrasting ages. Chronobiol Int 22(4):695–709

Kuo TT, Ladurner AG (2020) Corrigendum: exploiting the circadian clock for improved cancer therapy: perspective from a cell biologist. Front Genet 11:601

Lahti TA, Partonen T, Kyyrönen P, Kauppinen T, Pukkala E (2008) Night-time work predisposes to non-Hodgkin lymphoma. Int J Cancer 123(9):2148–2151

Lecca P (2021) Control theory and cancer chemotherapy: how they interact. Front Bioeng Biotechnol 8:621269

Ledzewicz U, Schättler H, Gahrooi MR, Dehkordi SM (2013) On the MTD paradigm and optimal control for multi-drug cancer chemotherapy. Math Biosci Eng 10(3):803

Lévi F (1996) Chronopharmacology and chronotherapy of cancers. Pathol Biol 44(7):631–644

Lévi F (1997) Chronopharmacology of anticancer agents. In: Redfern PH, Lemmer B (eds) Physiology and pharmacology of biological rhythms. Springer, Heidelberg, pp 299–331

Lévi F (2001) Circadian chronotherapy for human cancers. Lancet Oncol 2(5):307–315

Lévi F (2006) Chronotherapeutics: the relevance of timing in cancer therapy. Cancer Cause Control 17:611–621

Lévi F, Schibler U (2007) Circadian rhythms: mechanisms and therapeutic implications. Annu Rev Pharmacol Toxicol 47:593–628

Lévi F, Blazsek I, Ferle-Vidovic A (1988) Circadian and seasonal rhythms in murine bone marrow colony-forming cells affect tolerance for the anticancer agent 4′-O-tetrahydropyranyladriamycin (THP). Exp Hematol 16(8):696–701

Lévi F, Zidani R, Misset JL (1997) Randomized multicentre trial of chronotherapy with oxaliplatin, fluorouracil, and folinic acid in metastatic colorectal cancer. Lancet Oncol 350:681–686

Lévi F, Filipski E, Iurisci I, Li XM, Innominato P (2007) Cross-talks between circadian timing system and cell division cycle determine cancer biology and therapeutics. Cold Spring Harb Symp Quant Biol 72(1):465–475

Lévi F, Okyar A, Dulong S, Innominato PF, Clairambault J (2010) Circadian timing in cancer treatments. Annu Rev Pharmacol 50:377–421

Lévi F, Dugué PA, Innominato P, Karaboué A, Dispersyn G, Parganiha A, Giacchetti S, Moreau T, Focan C, Waterhouse J, Spiegel D, Chronotherapy Group (2014) Wrist actimetry circadian rhythm as a robust predictor of colorectal cancer patient's survival. Chronobiol Int 31(8):891–900

Lévi F, Komarzynski S, Huang Q, Young T, Ang Y, Fuller C, Bolborea M, Brettschneider J, Fursse J, Finkenstädt B, White DP (2020) Tele-monitoring of cancer patients rhythms during daily life identifies actionable determinants of circadian and sleep disruption. Cancer 12(7):1938

Lewintre EJ, Martín CR, Ballesteros CG, Montaner D, Rivera RF, Mayans JR, García-Conde J (2009) Cryptochrome-1 expression: a new prognostic marker in B-cell chronic lymphocytic leukemia. Haematologica 94(2):280–284

Li D, Xu W, Guo Y, Xu Y (2011) Fluctuations induced extinction and stochastic resonance effect in a model of tumor growth with periodic treatment. Phys Lett A 375(5):886–890

Li L, Yuan L, Luo J, Gao J, Guo J, Xie X (2013) MiR-34a inhibits proliferation and migration of breast cancer through down-regulation of Bcl-2 and SIRT1. Clin Exp Med 13(2):109–117

Li Y, Li S, Zhou Y, Meng X, Zhang JJ, Xu DP, Li HB (2017) Melatonin for the prevention and treatment of cancer. Oncotarget 8(24):39896

Li SY, Hammarlund JA, Wu G, Lian JW, Howell SJ, Clarke RB, Adamson AD, Gonçalves CF, Hogenesch JB, Anafi RC, Meng QJ (2024) Tumor circadian clock strength influences metastatic potential and predicts patient prognosis in luminal A breast cancer. Proc Natl Acad Sci U S A 121(7):e2311854121

Liebmann PM, Reibnegger G, Lehofer M, Moser M, Pürstner P, Mangge H, Schauenstein K (1998) Circadian rhythm of the soluble p75 tumor necrosis factor (sTNF-R75) receptor in humans—a possible explanation for the circadian kinetics of TNR-alpha effects. Int Immunol 10(9):1393–1396

Lin YM, Chang JH, Yeh KT, Yang MY, Liu TC, Lin SF, Chang JG (2008) Disturbance of circadian gene expression in hepatocellular carcinoma. Mol Carcinog 47(12):925–933

Linden H, Macino G (1997) White collar 2, a partner in blue-light signal transduction, controlling expression of light–regulated genes in *Neurospora crassa*. EMBO J 16(1):98–109

Liu X, Zwiebel LJ, Hinton D, Benzer S, Hall JC, Rosbash M (1992) The period gene encodes a predominantly nuclear protein in adult Drosophila. J Neurosci 12(7):2735–2744

Liu Y, Tsinoremas NF, Johnson CH, Lebedeva NV, Golden SS, Ishiura M, Kondo T (1995) Circadian orchestration of gene expression in cyanobacteria. Genes Dev 9(12):1469–1478

Liu Y, Garceau NY, Loros JJ, Dunlap JC (1997) Thermally regulated translational control of FRQ mediates aspects of temperature responses in the Neurospora circadian clock. Cell 89(3):477–486

Liu Y, Chen N, Cui X, Zheng X, Deng L, Price S, Karantza V, Minden A (2010) The protein kinase Pak4 disrupts mammary acinar architecture and promotes mammary tumorigenesis. Oncogene 29(44):5883–5894

Liu CY, Xu JY, Shi XY, Huang W, Ruan TY, Xie P, Ding JL (2013) M2-polarized tumor-associated macrophages promoted epithelial–mesenchymal transition in pancreatic cancer cells, partially through TLR4/IL-10 signaling pathway. Lab Investig 93(7):844–854

Logan RW, McClung CA (2019) Rhythms of life: circadian disruption and brain disorders across the lifespan. Nat Rev Neurosci 20(1):49–65

Loros JJ, Dunlap JC (2001) Genetic and molecular analysis of circadian rhythms in Neurospora. Annu Rev Physiol 63:757

Louis J, Cannard C, Bastuji H, Challamel MJ (1997) Sleep ontogenesis revisited: a longitudinal 24-hour home polygraphic study on 15 normal infants during the first two years of life. Sleep 20(5):323–333

Lowrey PL, Takahashi JS (2004) Mammalian circadian biology: elucidating genome-wide levels of temporal organization. Annu Rev Genomics Hum Genet 5:407–441

Lowrey PL, Takahashi JS (2011) Genetics of circadian rhythms in mammalian model organisms. Adv Genet 74:175–230

Luo Y, Wang F, Chen LA, Chen XW, Chen ZJ, Liu PF, Li FF, Li CY, Liang W (2012) Deregulated expression of cry1 and cry2 in human gliomas. Asian Pac J Cancer Prev 13(11):5725–5728

Matsuo T, Yamaguchi S, Mitsui S, Emi A, Shimoda F, Okamura H (2003) Control mechanism of the circadian clock for timing of cell division in vivo. Science 302(5643):255–259

McClung CR, Fox BA, Dunlap JC (1989) The Neurospora clock gene frequency shares a sequence element with the Drosophila clock gene period. Nature 339(6225):558–562

McPartland RJ, Kupfer DJ, Gordon Foster F (1976) The movement-activated recording monitor: a third-generation motor-activity monitoring system. Behav Res Meth Instrum 8(4):357–360

Merrow M, Brunner M, Roenneberg T (1999) Assignment of circadian function for the Neurospora clock gene frequency. Nature 399(6736):584–586

Merrow M, Spoelstra K, Roenneberg T (2005) The circadian cycle: daily rhythms from behaviour to genes. EMBO Rep 6(10):930–935

Mills JN (1973) Transmission processes between clock and manifestations. In: Mills JN (ed) Biological aspects of circadian rhythms. Springer, Boston, MA, pp 27–84
Mohawk JA, Green CB, Takahashi JS (2012) Central and peripheral circadian clocks in mammals. Annu Rev Neurosci 35:445
Momma T, Okayama H, Saitou M, Sugeno H, Yoshimoto N, Takebayashi Y, Ohki S, Takenoshita S (2017) Expression of circadian clock genes in human colorectal adenoma and carcinoma. Oncol Lett 14(5):5319–5325
Monsees GM, Kraft P, Hankinson SE, Hunter DJ, Schernhammer ES (2012) Circadian genes and breast cancer susceptibility in rotating shift workers. Int J Cancer 131(11):2547–2552
Montaruli A, Patrini P, Roveda E, Carandente F (2012) Physical activity and breast cancer. Sport Sci Health 8(1):1–13
Moore RY, Eichler VB (1972) Loss of a circadian adrenal corticosterone rhythm following suprachiasmatic lesions in the rat. Brain Res 42(1):201–206
Mormont MC, Lévi F (1997) Circadian-system alterations during cancer processes: a review. Int J Cancer 70(2):241–247
Mormont MC, Lévi F (2003) Cancer chronotherapy: principles, applications, and perspectives. Cancer 97(1):155–169
Mormont MC, Waterhouse J, Bleuzen P, Giacchetti S, Jami A, Bogdan A, Lellouch J, Misset JL, Touitou Y, Lévi F (2000) Marked 24-h rest/activity rhythms are associated with better quality of life, better response, and longer survival in patients with metastatic colorectal cancer and good performance status. Clin Cancer Res 6(8):3038–3045
Mormont MC, Langouët AM, Claustrat B, Bogdan A, Marion S, Waterhouse J, Touitou Y, Lévi F (2002) Marker rhythms of circadian system function: a study of patients with metastatic colorectal cancer and good performance status. Chronobiol Int 19(1):141–155
Nirvani M, Khuu C, Utheim TP, Sand LP, Sehic A (2018) Circadian clock and oral cancer. Mol Clin Oncol 8(2):219–226
Oda A, Katayose Y, Yabuuchi S, Yamamoto K, Mizuma M, Shirasou S, Rikiyama T (2009) Clock gene mouse period2 overexpression inhibits growth of human pancreatic cancer cells and has synergistic effect with cisplatin. Anticancer Res 29(4):1201–1209
Önder GÖ, Sezer G, Özdamar S, Yay A (2022) Melatonin has an inhibitory effect on MCF-7 and MDA-MB-231 human breast cancer cell lines by inducing autophagy and apoptosis. Fundam Clin Pharmacol 36(6):1038–1056
Ortiz-Tudela E, Iurisci I, Beau J, Karaboue A, Moreau T, Rol MA, Madrid JA, Lévi F, Innominato PF (2014) The circadian rest-activity rhythm, a potential safety pharmacology endpoint of cancer chemotherapy. Int J Cancer 134(11):2717–2725
Oster H, Damerow S, Hut RA, Eichele G (2006) Transcriptional profiling in the adrenal gland reveals circadian regulation of hormone biosynthesis genes and nucleosome assembly genes. J Biol Rhythm 21(5):350–361
Ozturk N, Ozturk D, Halil Kavakli I, Okyar A (2017) Molecular aspects of circadian pharmacology and relevance for cancer chronotherapy. Int J Mol Sci 18(10):2168
Panda S, Antoch MP, Miller BH, Su AI, Schook AB, Straume M, Hogenesch JB (2002) Coordinated transcription of key pathways in the mouse by the circadian clock. Cell 109(3):307–320
Papagerakis S, Zheng L, Schnell S, Sartor MA, Somers E, Marder W, McAlpin B, Kim D, McHugh J, Papagerakis P (2014) The circadian clock in oral health and diseases. J Dent Res 93(1):27–35
Parganiha A, Taj S, Chandel P, Sultan A, Choudhary V (2014) Effect of hospitalization on rest-activity rhythm and quality of life of cancer patients. Indian J Exp Biol 52(5):549–558
Partch CL, Green CB, Takahashi JS (2014) Molecular architecture of the mammalian circadian clock. Trends Cell Biol 24(2):90–99
Pati AK (2001) Chronobiology: the dimension of time in biology and medicine. Proc Indian Natl Sci Acad 67(6):323–372
Pati AK, Parganiha A, Kar A, Soni R, Roy S, Choudhary V (2006) Implications of the study of rest-activity circadian rhythm in head and neck cancer patients. Biol Rhythm Res 37(6):497–505
Pati AK, Parganiha A, Kar A, Soni R, Roy S, Choudhary V (2007) Alterations of the characteristics of the circadian rest-activity rhythm of cancer in-patients. Chronobiol Int 24(6):1179–1197

Patten CJ (1927) Plant autographs and their revelations. Nature 119(3008):919–920
Paudel KS, Milewski M, Swadley CL, Brogden NK, Ghosh P, Stinchcomb AL (2010) Challenges and opportunities in dermal/transdermal delivery. Ther Deliv 1(1):109–131
Paudel ML, Taylor BC, Ancoli-Israel S, Stone KL, Tranah G, Redline S, Barrett-Connor E, Stefanick ML, Ensrud KE (2011) Rest/activity rhythms and cardiovascular disease in older men. Chronobiol Int 28(3):258–266
Price JL, Blau J, Rothenfluh A, Abodeely M, Kloss B, Young MW (1998) Double-time is a novel Drosophila clock gene that regulates PERIOD protein accumulation. Cell 94(1):83–95
Puram RV, Kowalczyk MS, de Boer CG, Schneider RK, Miller PG, McConkey M, Chen MC (2016) Core circadian clock genes regulate leukemia stem cells in AML. Cell 165(2):303–316
Qian DC, Kleber T, Brammer B, Xu KM, Switchenko JM, Janopaul-Naylor JR, Zhong J, Yushak ML, Harvey RD, Paulos CM, Lawson DH (2021) Effect of immunotherapy time-of-day infusion on overall survival among patients with advanced melanoma in the USA (MEMOIR): a propensity score-matched analysis of a single-centre, longitudinal study. Lancet Oncol 22(12):1777–1786
Rahman S, Kraljević Pavelić S, Markova-Car E (2019) Circadian (De)regulation in head and neck squamous cell carcinoma. Int J Mol Sci 20(11):2662
Ralph MR, Menaker M (1988) A mutation of the circadian system in golden hamsters. Science 241(4870):1225–1227
Ralph MR, Foster RG, Davis FC, Menaker M (1990) Transplanted suprachiasmatic nucleus determines circadian period. Science 247(4945):975–978
Reddy P, Zehring WA, Wheeler DA, Pirrotta V, Hadfield C, Hall JC, Rosbash M (1984) Molecular analysis of the period locus in *Drosophila melanogaster* and identification of a transcript involved in biological rhythms. Cell 38(3):701–710
Reiter RJ, Rosales-Corral S, Tan DX, Jou MJ, Galano A, Xu B (2017) Melatonin as a mitochondria-targeted antioxidant: one of evolution's best ideas. Cell Mol Life Sci 74(21):3863–3881
Relles D, Sendecki J, Chipitsyna G, Hyslop T, Yeo CJ, Arafat HA (2013) Circadian gene expression and clinicopathologic correlates in pancreatic cancer. J Gastrointest Surg 17(3):443–450
Reppert SM, Weaver DR (2001) Molecular analysis of mammalian circadian rhythms. Annu Rev Physiol 63(1):647–676
Rijo-Ferreira F, Takahashi JS (2019) Genomics of circadian rhythms in health and disease. Genome Med 11(1):1–16
Roh HW, Choi SJ, Jo H, Kim D, Choi J-G, Son SJ, Joo EY (2022) Associations of actigraphy derived rest activity patterns and circadian phase with clinical symptoms and polysomnographic parameters in chronic insomnia disorders. Sci Rep 12(1):1–9
Roveda E, Bruno E, Galasso L, Mulè A, Castelli L, Villarini A, Caumo A, Esposito F, Montaruli A, Pasanisi P (2019) Rest-activity circadian rhythm in breast cancer survivors at 5 years after the primary diagnosis. Chronobiol Int 36(8):1156–1165
Sadeh A, Acebo C (2002) The role of actigraphy in sleep medicine. Sleep Med Rev 6(2):113–124
Sahar S, Sassone-Corsi P (2007) Circadian clock and breast cancer: a molecular link. Cell Cycle 6(11):1329–1331
Sangoram AM, Saez L, Antoch MP, Gekakis N, Staknis D, Whiteley A, Fruechte EM, Vitaterna MH, Shimomura K, King DP, Young MW, Takahashi JS (1998) Mammalian circadian autoregulatory loop: a timeless ortholog and mPer1 interact and negatively regulate CLOCK-BMAL1-induced transcription. Neuron 21(5):1101–1113
Sauman I, Reppert SM (1996) Circadian clock neurons in the silkmoth *Antheraea pernyi*: novel mechanisms of Period protein regulation. Neuron 17(5):889–900
Savard J, Liu L, Natarajan L, Rissling MB, Neikrug AB, He F, Dimsdale JE, Mills PJ, Parker BA, Sadler GR, Ancoli-Israel S (2009) Breast cancer patients have progressively impaired sleep-wake activity rhythms during chemotherapy. Sleep 32(9):1155–1160
Schibler U, Sassone-Corsi P (2002) A web of circadian pacemakers. Cell 111(7):919–922
Sehgal A, Price JL, Man B, Young MW (1994) Loss of circadian behavioral rhythms and per RNA oscillations in the Drosophila mutant timeless. Science 263(5153):1603–1606

Selby CP, Sancar A (1999) A third member of the photolyase/blue-light photoreceptor family in Drosophila: a putative circadian photoreceptor. Photochem Photobiol 69(1):105–107

Sephton SE, Sapolsky RM, Kraemer HC, Spiegel D (2000) Diurnal cortisol rhythm as a predictor of breast cancer survival. J Natl Cancer Inst 92(12):994–1000

Shearman LP, Sriram S, Weaver DR, Maywood ES, Chaves I, Zheng B, Kume K, Lee CC, van der Horst GT, Hastings MH (2000) Interacting molecular loops in the mammalian circadian clock. Science 288(5468):1013–1019

Shih MC, Yeh KT, Tang KP, Chen JC, Chang JG (2006) Promoter methylation in circadian genes of endometrial cancers detected by methylation-specific PCR. Mol Carcinog 45(10):732–740

Smolensky MH, Peppas NA (2007) Chronobiology, drug delivery, and chronotherapeutics. Adv Drug Deliv Rev 59(9–10):828–851

Soni R, Dubey P, Kar A, Parganiha A, Pradhan RK, Pati AK (2008) Permanent night work alters characteristics of circadian rhythm of rest-activity in human subjects. Biol Rhythm Res 39:481–492

Stanewsky R, Kaneko M, Emery P, Beretta B, Wager-Smith K, Kay SA, Rosbash M, Hall JC (1998) The cryb mutation identifies cryptochrome as a circadian photoreceptor in Drosophila. Cell 95(5):681–692

Stevens RG, Rea MS (2001) Light in the built environment: potential role of circadian disruption in endocrine disruption and breast cancer. Cancer Causes Control 12(3):279–287

Storch KF, Lipan O, Leykin I, Viswanathan N, Davis FC, Wong WH, Weitz CJ (2002) Extensive and divergent circadian gene expression in liver and heart. Nature 417(6884):78–83

Su X, Chen D, Yang K, Zhao Q, Zhao D, Lv X, Ao Y (2017) The circadian clock gene PER2 plays an important role in tumor suppression through regulating tumor-associated genes in human oral squamous cell carcinoma. Oncol Rep 38(1):472–480

Sultan A (2019) Identification and development of clock-modulating small molecules—an emerging approach to fine-tune the disrupted circadian clocks. Biol Rhythm Res 50(5):769–786

Sultan A, Choudhary V, Parganiha A (2014) Characteristics of circadian rhythms in rest-activity and energy expenditure in cancer in-patients. S Asian J Exp Biol 4(6):327–335

Sultan A, Choudhary V, Parganiha A (2017a) Worsening of rest-activity circadian rhythm and quality of life in female breast cancer patients along progression of chemotherapy cycles. Chronobiol Int 34(5):609–623

Sultan A, Choudhary V, Parganiha A (2017b) Monitoring of rest-activity rhythm in cancer patients paves the way for the adoption of patient-specific chronotherapeutic approach. Biol Rhythm Res 48(2):189–205

Sultan A, Parganiha A, Sultan T, Choudhary V, Pati AK (2017c) Circadian clock, cell cycle, and breast cancer: an updated review. Biol Rhythm Res 48(3):353–369

Sultan A, Pati AK, Choudhary V, Parganiha A (2018) Hospitalization induced exacerbation of the ill effects of chemotherapy on rest-activity rhythm and quality of life of breast cancer patients: a prospective and comparative cross-sectional follow-up study. Chronobiol Int 35(11):1513–1532

Sultan A, Pati AK, Choudhary V, Parganiha A (2020) Repeated chemotherapy cycles produced progressively worse and enduring impairments in the sleep-wake profile of hospitalized breast cancer patients. Biol Rhythm Res 51(8):1166–1181

Sun CM, Huang SF, Zeng JM, Liu DB, Xiao Q, Tian WJ, Feng WL (2010) Per2 inhibits k562 leukemia cell growth in vitro and in vivo through cell cycle arrest and apoptosis induction. Pathol Oncol Res 16(3):403–411

Suzuki T, Sato F, Kondo J, Liu Y, Kusumi T, Fujimoto K, Kijima H (2008) Period is involved in the proliferation of human pancreatic MIA-PaCa2 cancer cells by TNF-α. Biomed Res J 29(2):99–103

Taj S, Choudhary V, Parganiha A (2013) Temporal profiles of physical activity and energy expenditure in cancer in-patients. Biol Rhythm Res 44(2):219–235

Takahashi JS, Menaker M (1984) Multiple redundant circadian oscillators within the isolated avian pineal gland. J Comp Physiol 154(3):435–440

Takahashi JS, Hong HK, Ko CH, McDearmon EL (2008) The genetics of mammalian circadian order and disorder: implications for physiology and disease. Nat Rev Genet 9(10):764–775
Taniguchi H, Fernández AF, Setién F, Ropero S, Ballestar E, Villanueva A, Esteller M (2009) Epigenetic inactivation of the circadian clock gene BMAL1 in hematologic malignancies. Cancer Res 69(21):8447–8454
Tei H, Okamura H, Shigeyoshi Y, Fukuhara C, Ozawa R, Hirose M, Sakaki Y (1997) Circadian oscillation of a mammalian homologue of the Drosophila period gene. Nature 389(6650):512–516
Teicher MH, Glod CA, Harper D, Magnus E, Brasher C, Wren F, Pahlavan K (1993) Locomotor activity in depressed children and adolescents: I. Circadian dysregulation. J Am Acad Child Adolesc Psychiatry 32(4):760–769
Tokunaga H, Takebayashi Y, Utsunomiya H, Akahira JI, Higashimoto M, Mashiko M, Ito K, Niikura H, Takenoshita SI, Yaegashi N (2008) Clinicopathological significance of circadian rhythm-related gene expression levels in patients with epithelial ovarian cancer. Acta Obstet Gynecol Scand 87(10):1060–1070
Touitou Y, Lévi F, Bogdan A, Benavides M, Bailleul F, Misset JL (1995) Rhythm alteration in patients with metastatic breast cancer and poor prognostic factors. J Cancer Res Clin Oncol 121(3):181–188
Tranah GJ, Parimi N, Blackwell T, Ancoli-Israel S, Ensrud KE, Cauley JA, Redline S, Lane N, Paudel ML, Hillier TA, Yaffe K (2010) Postmenopausal hormones and sleep quality in the elderly: a population based study. BMC Womens Health 10(1):1–8
Tranah GJ, Blackwell T, Stone KL, Ancoli-Israel S, Paudel ML, Ensrud KE, Cauley JA, Redline S, Hillier TA, Cummings SR, Yaffe K (2011) Circadian activity rhythms and risk of incident dementia and mild cognitive impairment in older women. Ann Neurol 70(5):722–732
van der Horst GT, Muijtjens M, Kobayashi K, Takano R, Kanno SI, Takao M, Wit J, Verkerk A, Eker AP, Leenen DV, Buijs R, Bootsma D, Hoeijmakers JH, Yasui A (1999) Mammalian Cry1 and Cry2 are essential for maintenance of circadian rhythms. Nature 398(6728):627–630
Viswanathan AN, Schernhammer ES (2009) Circulating melatonin and the risk of breast and endometrial cancer in women. Cancer Lett 281(1):1–7
Viswanathan AN, Hankinson SE, Schernhammer ES (2007) Night shift work and the risk of endometrial cancer. Cancer Res 67(21):10618–10622
Vitaterna MH, King DP, Chang AM, Kornhauser JM, Lowrey PL, McDonald JD, Dove WF, Pinto LH, Turek FW, Takahashi JS (1994) Mutagenesis and mapping of a mouse gene, Clock, essential for circadian behavior. Science 264(5159):719–725
Wang Q, Ao Y, Yang K, Tang H, Chen D (2016) Circadian clock gene Per2 plays an important role in cell proliferation, apoptosis and cell cycle progression in human oral squamous cell carcinoma. Oncol Rep 35(6):3387–3394
Wang N, Mi M, Wei X, Sun C (2020) Circadian clock gene Period2 suppresses human chronic myeloid leukemia cell proliferation. Exp Ther Med 20(6):147
Welsh DK, Logothetis DE, Meister M, Reppert SM (1995) Individual neurons dissociated from rat suprachiasmatic nucleus express independently phased circadian firing rhythms. Neuron 14(4):697–706
Welsh DK, Yoo SH, Liu AC, Takahashi JS, Kay SA (2004) Bioluminescence imaging of individual fibroblasts reveals persistent, independently phased circadian rhythms of clock gene expression. Curr Biol 14(24):2289–2295
Williams JA, Sehgal A (2001) Molecular components of the circadian system in Drosophila. Annu Rev Physiol 63(1):729–755
Wulff K, Porcheret K, Cussans E, Foster RG (2009) Sleep and circadian rhythm disturbances: multiple genes and multiple phenotypes. Curr Opin Genet Dev 19(3):237–246
Xia HC, Niu ZF, Ma H, Cao SZ, Hao SC, Liu ZT, Wang F (2010) Deregulated expression of the Per1 and Per2 in human gliomas. Can J Neurol Sci 37(3):365–370
Xu Y, Johnson CH (2001) A clock-and light-regulated gene that links the circadian oscillator to LHCB gene expression. Plant Cell 13(6):1411–1426
Yang SL, Ren QG, Wen L, Hu JL, Wang HY (2017) Research progress on circadian clock genes in common abdominal malignant tumors. Oncol Lett 14(5):5091–5098

Yang Y, Lindsey-Boltz LA, Vaughn CM, Selby CP, Cao X, Liu Z, Hsu DS, Sancar A (2021) Circadian clock, carcinogenesis, chronochemotherapy connections. J Biol Chem 297(3):101068
Ye Y, Xiang Y, Ozguc FM, Kim Y, Liu CJ, Park PK, Guo AY (2018) The genomic landscape and pharmacogenomic interactions of clock genes in cancer chronotherapy. Cell Syst 6(3):314–328
Yoo SH, Yamazaki S, Lowrey PL, Shimomura K, Ko CH, Buhr ED, Siepka SM, Hong HK, Oh WJ, Yoo OJ, Menaker M (2004) PERIOD2::LUCIFERASE real-time reporting of circadian dynamics reveals persistent circadian oscillations in mouse peripheral tissues. PNAS 101(15):5339–5346
Zaki NF, Sabri YM, Farouk O, Abdelfatah A, Spence DW, Bahammam AS, Pandi-Perumal SR (2021) Depressive symptoms, sleep profiles and serum melatonin levels in a sample of breast cancer patients. Nat Sci Sleep 12:135
Zavada A, Gordijn MC, Beersma DG, Daan S, Roenneberg T (2005) Comparison of the Munich Chronotype Questionnaire with the Horne-Östberg's morningness-eveningness score. Chronobiol Int 22(2):267–278
Zhu Y, Brown HN, Zhang Y, Stevens RG, Zheng T (2005) Period3 structural variation: a circadian biomarker associated with breast cancer in young women. Cancer Epidemiol Biomarkers Prev 14(1):268–270
Zhu Q, Tannenbaum S, Kurtzman SH (2007a) Optical tomography with ultrasound localization for breast cancer diagnosis and treatment monitoring. Surg Oncol Clin N Am 16(2):307–321
Zhu Y, Leaderer D, Guss C, Brown HN, Zhang Y, Boyle P, Stevens RG, Hoffman A, Qin Q, Han X, Zheng T (2007b) Ala394Thr polymorphism in the clock gene NPAS2: a circadian modifier for the risk of non-Hodgkin's lymphoma. Int J Cancer 120(2):432–435
Zhu Y, Stevens RG, Leaderer D, Hoffman A, Holford T, Zhang Y, Brown HN, Zheng T (2008) Non-synonymous polymorphisms in the circadian gene NPAS2 and breast cancer risk. Breast Cancer Res Treat 107(3):421–425
Zhu W, Qin W, Atasoy U, Sauter ER (2009a) Circulating microRNAs in breast cancer and healthy subjects. BMC Res Notes 2(1):89
Zhu Y, Stevens RG, Hoffman AE, FitzGerald LM, Kwon EM, Ostrander EA, Davis S, Zheng T, Stanford JL (2009b) Testing the circadian gene hypothesis in prostate cancer: a population-based case-control study. Cancer Res 69(24):9315–9322
Zienolddiny S, Haugen A, Lie JAS, Kjuus H, Anmarkrud KH, Kjaerheim K (2013) Analysis of polymorphisms in the circadian-related genes and breast cancer risk in Norwegian nurses working night shifts. Breast Cancer Res 15(4):53
Zordan M, Costa R, Fukuhara C, Tosini G (2002) The genetics and molecular biology of circadian clocks. In: Kumar V (ed) Biological rhythms. Springer, Berlin, pp 46–59

Statistical Methods for Animal Behavior Inference 5

Akshay Kumar Chakravarthy

Abstract

This contribution is on statistical methods used by ethologists to analyze data on animal behavior. Methods applicable in ethology are not often covered by general statistical books. Workers in animal behavior now have several statistical methods and approaches to select for data analysis. Methods like capture–recapture, survey sampling, and trapping may not be used frequently under field conditions nowadays. Ethologists can find references to methods in journals like *Animal Behavior*, *Behavioral Ecology and Sociobiology*, *Ecology*, *Indian Journal of Ecology*, *Journal of Wildlife Management*, and *American Journal of Primatology*. Animal behavior presents some unique situations for data analysis that need to be addressed. Software programs are designed with versions for behavioral studies. Before implementing the study on behavior, the investigator ought to know the design of the experiment, observations, and what patterns of observations the data will contain, including the method of analysis.

Keywords

Statistical method · Animal behavior · Sampling · Data analysis

5.1 Statistical Considerations

A number of publications are available on "Animal behavior statistics." The publications that are selected include: Bavry (1996), Cohen (1988), Taylor and Gerodette (1993), Zar (1996), Hanks et al. (2015), Langrock et al. (2014), Snedecor and Cochran (1980), Weis and Sol (2019), Agostinelli and Agostinelli (2001), Agostinelli

A. Kumar Chakravarthy (✉)
Division of Entomology and Nematology, Indian Institute of Horticultural Research, Bengaluru, Karnataka, India

S. Sridhara et al. (eds.), *Animal Behavior in the Tropics*,
https://doi.org/10.1007/978-981-97-7324-4_5

and Lund (2020), and Gadagkar (2022). Some of the considerations that merit attention are summarized subsequently.

Statistical inferences do not generally apply beyond the population sampled. It is also necessary that the population and study area are well-defined. The timing and season of recording observations on animals are also important. Sampling the entire population often is difficult practically. However, sampling along borders or length and breadth of a definite size and period of time (peak activity of animals) at night or day may be preferred. The investigator must try to sample large sample sizes. Previous knowledge about the habitats, landscapes, and the target animal is desirable before initiating sampling. However, under field conditions, several factors like weather conditions, migrants, and predation frequency and rates are not under the control of the workers. In such cases, it is assumed that the population sampled is representative and that the statistical inferences drawn are applicable. The researcher should also determine the standard error and/or standard deviation at a 5% level of significance. In behavior ecology, if one wants to extend the findings of a study at one location to other locations, species, and times, then the study needs replicates with larger samples. Biologists/ethologists use the following statistical parameters to quantify or qualify the animal behavior: mean, mode, proportion, correlation, regression, similarity index, diversity index, standard error, and standard deviation. Several studies in animal behavior use the population unit as an animal or plant or troupe; location as a transect, block, zone, or habitat; and time as an hour, 30 min, or 10 min. The time intervals are repeated several times in a day. Sampling error or bias is said to be properties of the estimator. Statistical bias arises as a result of procedures, data analysis, and statistical assumptions.

5.2 Randomization Techniques

Data on animal behavior are often randomly distributed and are not amenable to analysis by parametric tests. Ethologists use rank-transformation tests. Animal ethologists use nonparametric tools when datasets do not follow a normal distribution (Martin and Bateson 1986). Most of these methods involve ranking the data sequentially and performing statistics on the ranks (Sokal and Rohlf 1995). The non-normal datasets assume that the distributions so compared are symmetrical (Day and Quinn 1989). Alternative tests are available based on resampling methods.

Biologists adopt the most widely known of the resampling methods: bootstrap, which generates a confidence interval for a given statistic by taking N samples (of size N) with replacement from a dataset and calculates an estimate of the statistic.

The variations between treatments using Kruskal–Wallis ($H = 49.12$, $P < 0.001$) were found. The frequency distribution of possible ss_t from the randomization test ranged from 0.25 to 648.59. The raw data ss_t was 648.59, larger than 4999 of the randomly calculated ss_t. These results demonstrate that the differences between treatments in the original data were not likely to have occurred by chance (Adams and Antony 1996). Results from Kruskal–Wallis test and randomization tests were compared for two behavioral datasets. Randomization tests proved more powerful than Kruskal–Wallis and detected smaller effect sizes in the data. The variance was determined around the *P* value at eight levels of replication ranging from 500 to 10,000, to find the optimal number of replications for the randomization test (Dean Adams and Antony 1996).

5.3 Nonparametric Tests

The first step is to set up a hypothesis and decide on the level of significance. Then calculate and compare the test statistic to the decision role. Nonparametric tests are used when very little is known about the data and do not assume that data are derived from a given probability distribution. Nonparametric tests are used when sample sizes are very small and the population distribution is unknown. These tests are used when datasets are categorized/ranked and fewer assumptions about data are made. But nonparametric tests may be imparted by assumptions of normal distribution and homogeneity of variance.

Depending on the situation under field conditions and the target animal, the Mann–Whitney U test, Spearman correlation test, the sign test, Friedman's ANOVA test, the Wilcoxon signed-rank test, and the Kruskat–Wallis tests are used to analyze the datasets. Mann–Whitney U test is the most widely used because it compares two unpaired groups; the data may originate from two different populations of unknown origin. It assumes that the individual dataset collected is sampled independently. The chi-square test is a nonparametric statistic used when all variables are normal. The z-test is a parametric test to find if the means of two datasets differ and is based on the presumption that the sample mean is normal. The t-test is used based on the student's t distribution.

Cramer's V test is used to determine the association between two categorical variables when there is more than a 2 × 2 contingency.

5.4 Association Behavior

Researchers need to evaluate associations among individual animals in a group. For example, song dialects between two individuals in a group, social interactions among individuals in a group, arrangement of positions in a flock of birds, and feeding and foraging patterns of individuals in a group need to be evaluated. Since association values between pairs of individuals are mathematically related statistical tests used for finding significant differences become inappropriate. Tests that consider the presence of interdependence must be chosen (Schnell et al. 1985). Mantel (1967) developed a general method of matrix comparisons used in animal behavior studies.

5.5 Movement Behavior

The study of motion and trajectories through space and time is important for every scientific discipline, especially in ecology where observations of animal movement represent one of the primary opportunities for learning about species relationships with their environment. With the advent of portable, remotely sensed devices that can transmit time stamped measurements of position to distant retrivers. It has become possible to track animal movements in remote locations with increasing frequency, duration, and precision (Anthony 1995; Arnold et al. 2018).

For example, researchers interested in learning how seals, often in the open ocean for several days or weeks at a time during foraging excursions, divide their time between searching and feeding behaviors (SHJ17, MB19) or scientists may seek to understand what parts of the landscape facilitate/hinder the movement of mountain lions utilizing kill sites in varied mountain terrain [HHA15, HSM19]. Insights into the behavior of movement help in formulating movement data that can inform national and international policies responsible for management and conservation.

Movement data have historically defining characteristic, and it has been telemetry. Observations of trajectories are generally made using remote-sensing technology, wherein a transmitter is affixed to a moving individual, and a receiver records the location of the transmitter at some combination of predefined and opportunistic times. Some commonly used techniques include very/ultrahigh-frequency (V/UHF), radio telemetry, which requires that the receiver and transmitter be connected by line-of-sight, and satellite-based telemetry, in which transmitters are tracked through hard-to-access parts of the globe such as polar regions and the open ocean.

It remains a fundamental challenge to transform raw observations of animals' positions, usually contaminated with non-ignorable measurement error, into real scientific learning. Several nonstatistical methods exist for interpolating observations by utilizing statistical models for trajectories over simpler, quicker computational tools. There are three primary benefits; statistical approaches are fundamentally probabilistic and so provide a natural way to quantify uncertainty about inferences made from data about the interactions between an individual and its environment driven by biological and ecological mechanisms that could; for example, be important to understand the purposes of management and conservation. Statistical approaches offer methods for formally coupling these mechanisms to observations through the use of parametric models in which the values of the parameters inform scientific understanding of animal behavior. The statistical models used in the analysis of animal movement are generative of a model and validate the theory underpinning the mechanisms of interest. In practice, many commonly used statistical validation methods such as k-fold cross validation are impractical for highly parameterized hierarchical models for movement.

5.6 Parametric Models

In most cases, the goals of researchers studying animal movement include inferring important biological features of behavior for individuals and populations. Statistical approaches provide one way to achieve these goals through the specification of parametric models. A parametric model is a collection, or family, of probability distributions defined by a finite number of parameters. Simple examples of parametric models are location-scale families, such as the family of univariate Gaussian distributions with unknown mean (location) and variance (scale).

Useful parametric models pose the following qualities:

1. The parameters represent interpretable quantities such that specific values have meaningful implications for the application.
2. They are flexible enough that at least for certain combinations of parameter values, the observed data could conceivably represent a realization from the corresponding probability distribution (i.e., the model has a "good fit").
3. The family of distributions is sufficiently constrained that the observed data contain information about which collections of parameters are likely to have generated the data (i.e., the parameters are "identifiable"). It is the role of the analyst to specify and "fit" a useful statistical model to data and then to interpret the estimated parameters. One large class of estimation, or model fitting, techniques is based on the *likelihood* function of the parameters, θ, given the observed data, $\mathbf{y}$, defined as $L(\theta|\mathbf{y}) = p(\mathbf{y}|\theta)$,(1), where $p(\mathbf{y}|\theta)$ denotes the probability density function for the data evaluated at $\mathbf{y}$, conditioned on parameter values θ. The functional form of $(\mathbf{y}|\theta)$ is specified by the analyst as part of the model-building process.

5.7 Phenomenological Models

"Phenomenological" models provide one approach for estimating the true position process with uncertainty from observations trajectory estimation from behavioral inference in mathematically and computationally convenient ways. Observations of animal movement data are almost always gathered remotely through the use of transmitters and receivers that introduce measurement errors (Berger-Tal and Bar-David 2015). The scale of measurement error can vary dramatically depending on the measurement device.

5.8 Foraging Behavior

Generalized linear mixed models that allowed for zero inflation and dispersion parameters were constructed with the package of glmm TMB (Brooks et al. 2017) in the statistical software R (R core Team 2020). This model design comprised three parts, and details of variables are given in the study of Loren et al. (2022). The species encountered were possum, brown bandicoot, black rat, and brown rat, and the study area was in the suburban district of New South Wales, Australia. Loren et al. (2022) discuss at length the foraging behavior of rodents including squirrels and small animals like foxes and other species.

5.9 Feeding Behavior

Grivet monkeys (*Chlorocebus aethiops*) in Ethiopia are restricted to forest patches when space and food are limited. Focal and scan sampling techniques having 15-min sessions were used to record observations. The activity patterns showed no significant seasonal variations. The most frequent activity was feeding. Statistical packages for social sciences (SPSS version 2.0) and Microsoft Excel were used to analyze data at a 95% confidence level. The chi-square test was used to examine the significance of seasonal differences in feeding and activity trends. A number of publications are available on data analysis of feeding behavior (Serio-Silva et al. 2007; Hohmann et al. 2006; Mekonnen et al. 2010; Mammides et al. 2009; Altmann 1974; Silver et al. 1998; Linkie et al. 2007; Stanishaus 2006; Segal 2008). Two-sample Watson test (Zar 1976) was used to compare the homogeneity of the two fitted distributions. Data on a number of other variables on feeding was recorded and analyzed (Ossi et al. 2020; Alelign et al. 2023).

BECOMING REARER!

References

Adams DC, Antony CD (1996) Using randomization techniques to analyse behavioural data. Anim Behav 51:733–738

Agostinelli C, Agostinelli M (2001) S-plus original by Ulric Lund and R port by Claudio Agostinelli (2018). CircStats: circular statistics, from “topics in circular statistics” (2001) 2018. R package version 0.2-6. https://cran.r-project.org/web/packages/CircStats/CircStats.pdf. Accessed 12 Oct 2020

Agostinelli C, Lund U (2020) R package ‘circular’: circular statistics (version 0.4-93). 2013. https://r-forge.r-project.org/projects/circular/. Accessed 12 Oct 2020

Alelign A, Yonas M, Moges E (2023) Feeding habits and activity patterns of grivet monkey (Chlorocebus aethiops L.) in Batiero Church Forest, northern Ethiopia. Int J Zool 2023:5828576. https://doi.org/10.1155/2023/5828576

Altmann J (1974) Observational study of behaviour, sampling methods. Behaviour 49(3–4):227–266

Anthony CD (1995) Competitive interactions within and between two species of Plethodon in the Ouachita Mountains: effects of territoriality and parasitism. Ph.D. thesis, University of Southwestern Louisiana, Lafayette

Arnold JM, Gerhardt P, Steyaert SMJG, Hochbichler E, Hackländer K (2018) Diversionary feeding can reduce red deer habitat selection pressure on vulnerable forest stands, but is not a panacea for red deer damage. For Ecol Manag 407:166–173

Bavry JL (1996) Statistical Design Analysis Software (Version 3): User’s Guide. QEI Systems, Portland, Oregon

Berger-Tal O, Bar-David S (2015) Recursive movement patterns: review and synthesis across species. Ecosphere 6:1–12

Brooks ME, Kasper K, van Benthem Koen J, Arni M, Berg Casper W, Anders N, Skaug Hans J, Martin M, Bolker Benjamin M (2017) glmmTMB balances speed and flexibility among packages for zero-inflated generalized linear mixed modeling contributed research article, pp 380–400

Cohen J (1988) Statistical power analysis for the behavioral sciences, 2nd edn. Lawrence Erlbaum, Hillsdale, New Jersey

Day RW, Quinn GP (1989) Comparisons of treatments after an analysis of variance in ecology. Ecol Monogr 59:433–463

Gadagkar R (2022) Experiments in animal behaviour: cutting edge research at trifling cost. Indian Academic of Sciences, CV Raman Avenue, Bangalore, p 328

Hanks EM, Hooten MB, Alidredge MW (2015) Continuous-time discrete-space models for animal movement. Ann Appl Stat 9(1):145–165. https://doi.org/10.1214/14-AOAS803. MR3341111

Hohmann G, Robbins MM, Boesch C (2006) Feeding ecology in apes and other primates ecological, physical and behavioural aspects. Cambridge University Press, Cambridge

Langrock R, Hopcraft JGC, Blackwell PG, Goodall V, Niu RKM, Patterson TA, Pedersen MW, Skarin A, Schick RS (2014) Modelling group dynamic animal movement. Methods Ecol Evol 5(2):190–199. Available at 1308.5850

Linkie M, Dinata Y, Nofrianto A, Leader-Willams N (2007) Patterns and perceptions of wildlife crop raiding in and around Kerinci Seblat National Park, Sumatra. Anim Conserv 10(1):175–181

Loren LF, Catherine EMN, Chris RP, Dickman CR (2022) Small prey animal habitat use in landscapes of fear: effects of predator presence and human activity along an urban disturbance gradient. Front Ecol Evol 10:1–24

Mammides C, Cords M, Peters MK (2009) Effects of habitat disturbance and food supply on population densities of three primate species in the Kakamega Forest, Kenya. Afr J Ecol 47(1):87–96

Mantel N (1967) The detection of disease clustering and a generalized regression approach. Cancer Res 27:209–220

Martin P, Bateson P (1986) Measuring behavior. Cambridge University Press, Cambridge

Mekonnen A, Bekele A, Fashing PJ, Hemson G, Atickem A (2010) Diet, activity patterns, and ranging ecology of the Bale monkey (*Chlorocebus djamdjamensis*) in Odobullu Forest, Ethiopia. Int J Primatol 31(3):339–363

Ossi F, Ranc N, Moorcroft P, Bonanni P, Cagnacci F (2020) Ecological and behavioral drivers of supplemental feeding use by roe deer *Capreolus capreolus* in a peri-urban context. Animals 2020(10):2088. https://doi.org/10.3390/ani10112088

Schnell GD, Watt DJ, Douglas ME (1985) Statistical comparison of proximity matrices: applications. Anim Behav 33:239–253

Segal C (2008) Foraging behaviour and diet in chacma baboons in Suikerbosrand Nature Reserve, M.Sc. Thesis. Witwatersand University, Johanneshburg, South Africa

Serio-Silva JCS, Alfaro JL, Salazar ITH (2007) Ecology and behavior of tropical primates. Trop Biol Conserv Manag 8:1–7

Silver SC, Ostro LET, Yeager CP, Horwich R (1998) Feeding ecology of the black howler monkey (Alouatta caraya) in northern Belize. Am J Primatol 43(3):263–279

Snedecor GW, Cochran WG (1980) Statistical methods, 8th edn. Wiley Black Well Publishing, IOWA, p 503

Sokal RR, Rohlf FJ (1995) Biometry, 3rd edn. W. H. Freeman, New York

Stanishaus MK (2006) Feeding ecology and diurinal activity pattern of the Grevy's zebra (Equus grevyi) in Samburu Community Lands, Kenya. Master's thesis. School of Graduate Studies, Addis Ababa University, Ethiopia

Taylor BL, Gerodette T (1993) The uses of statistical power in conservation biology; the Vaquita and Northern spotted owl. Conserv Biol 7:489–500

Weis JS, Sol D (2019) Biological invasions and animal behaviour. Cambridge University Press, p 365

Zar JH (1976) Watson's non parametric two-sample test. Behav Res Methods Instrum 8:513

Zar JH (1996) Biostatistical analysis, 3rd edn. Prentice-Hall, London

Part II

Exploring Behaviour in Vertebrates

6 Personality Is the Key to Success: An Amphibian Perspective

Deyatima Ghosh and N. A. Aravind

Abstract

Amphibians have attracted researchers in view of unique ecological traits, particularly their biphasic development characterised by a metamorphic phase involving drastic behavioural and physiological transformations. However, what renders them especially intriguing is their personality, which is intricately linked to various life history aspects that facilitate their survival and reproductive success. Being anmniotes, amphibians exhibit a parallel evolution to the higher taxa following their divergence approximately 365 million years ago. While their brain retains a primitive structure, it also has the standard major brain areas and basic connectional patterns of the forebrain found in all amniotes, that is, reptiles, birds and mammals. This connection underscores the importance of amphibians as a model for understanding the evolutionary origins and adaptations of cognitive skills and complex behaviours in higher vertebrates. Considering the unique life history and the evolutionary significance of amphibians, this chapter explores the impact of personality on essential aspects, such as disease resistance, dispersal and mating which eventually impacts their survival.

D. Ghosh (✉)
Laboratory of Animal Behaviour and Conservation, College of Life Sciences, Nanjing Forestry University, Nanjing, Jiangsu, People's Republic of China

SM Sehgal Foundation Centre for Biodiversity and Conservation, Ashoka Trust for Research in Ecology and the Environment (ATREE), Bangalore, India

School of Life Sciences, University of Lincoln, Lincoln, UK
e-mail: deyatima@atree.org

N. A. Aravind
SM Sehgal Foundation Centre for Biodiversity and Conservation, Ashoka Trust for Research in Ecology and the Environment (ATREE), Bangalore, India

Yenepoya Research Centre, Yenepoya (Deemed to be University), Mangalore, India

S. Sridhara et al. (eds.), *Animal Behavior in the Tropics*,
https://doi.org/10.1007/978-981-97-7324-4_6

The chapter also encourages the integration of animal personality and behaviour into conservation strategies. By focusing on the specific needs and behaviours of amphibians, rather than adopting an anthropocentric approach, conservation efforts can become more tailored and effective, ultimately enhancing the survival of these vulnerable species.

Keywords
Amphibians · Behavioural syndrome · Conservation · Conservation behaviour Life history · Tropics

6.1 Introduction

6.1.1 Why Study Amphibians?

Class Amphibia consists of the anurans (the frogs and toads), the urodeles (the tailed amphibians, including salamanders and newts) and the apodans or gymnophiones (caecilians). This class consisting of more than 8000 species (Iyer and Briggman 2021) is diverse like mammals, inhabits diverse landscapes and occurs across all continents except Antarctica. Amphibians diverged from ancestral tetrapods around 365 million years ago, carving out a unique evolutionary path distinct from amniotes—reptiles, birds and mammals. This long evolutionary journey has allowed amphibians to develop along a parallel branch, resulting in a fascinating divergence of life forms and adaptations. One of the most striking outcomes of this evolutionary path is seen in their brains. Despite retaining a primitive structure, amphibian brains boast the major brain areas and fundamental connectional patterns of the forebrain found in all amniotes. This remarkable blend of primitive and advanced features provides a unique window into the evolutionary history of vertebrate brains, showcasing both the conservation of ancient traits and the innovation of new adaptations. The study of amphibians, therefore, not only illuminates the evolutionary processes that shaped their own lineage but also offers exciting insights into the origins and development of complex behaviours and neurological functions in higher vertebrates.

There is evolutionary specialisation in the expanded visual and auditory parts of the midbrain in amphibians, which is different from other vertebrates (Wilczynski 2009). Additionally, this anamniote group links aquatic and terrestrial life. Considering all these features, amphibians provide a fantastic model system for exploring behavioural plasticity across several ecological contexts. Further, amphibians provide an ideal study system for the mechanistic bases for the coupling and uncoupling of behavioural syndromes and life history characteristics during ontogeny (Wilson and Krause 2012). Even though amphibians are abundant and diverse, many aspects of behavioural biology remain understudied, undiscovered, unacknowledged and misunderstood. This chapter synthesises information on amphibian personality and how it influences several aspects of their lives.

6.2 Animal Personality and Behavioural Syndrome

Animal personality is the temporally repeatable, among-individual behavioural differences consistent and predictable through time and/or contexts. With advances in research on animal personality, behavioural syndrome is the key for the animal to show optimal behaviour in varying environments. Some individual personalities (or genotypes) might be generally more aggressive, and these differences remain consistent over time. Aggressiveness, for example, can be correlated to individuals being more exploratory or bold. Timid individuals on the other hand are less exploratory and shy (Sih et al. 2004). Thus, individuals in a population have a personality type or behaviour type (for instance, aggressive type vs. shy type). The behaviours associated with personality types are termed personality traits (e.g., aggression and boldness vs. docile and shyness). Five personality traits are generally characterised in vertebrates: active/passive, aggressive/docile, bold/shy, exploratory/stationary and sociable/non-sociable (Réale et al. 2007).

Behavioural syndrome, or behavioural profile, is the suite of correlated behavioural traits. Behavioural syndrome (Sih et al. 2004) are related to survival, foraging, predator–prey interactions (Carlson and Langkilde 2013), reproductive potential (Cole and Quinn 2014), disease risk (Koprivnikar et al. 2012), habitat selection (Bégué et al. 2023), dispersal abilities and even cognitive styles (Cote et al. 2010; Fig. 6.1). The fitness consequence of an animal's personality is extremely complex. and is currently attracting significant attention from researchers (Réale et al. 2010).

6.3 Maintaining Personality and Behavioural Syndrome Through Evolution

Personality traits are genetically correlated (Dingemanse and Réale 2005) and are a reflection of genetics, environment-dependent phenotypic plasticity or a combination of both factors and are 10–40% heritable across generations (Réale et al. 2007). Behavioural responses impact evolution. When individuals are shielded from novel selection pressures behavioural responses can slow evolution while speeding up evolution as individuals experience novel selection pressures. Thus, selection pressure on behaviour traits leads to evolutionary changes. Therefore, behaviour is crucial for affecting a population's stability, persistence and evolution over a longer period (Wolf and Weissing 2012). Additionally, the process of natural selection selecting a personality and evolution maintaining and modifying it depends on the condition that variation is maintained through frequency-dependent selection, and no rare mutant is favoured.

For instance, in a population that harbours both bold and shy individuals, selection might favour bolder individuals who are aggressive and reproduce more than shy or docile individuals. Thus, boldness traits will be selected, and bolder individuals will increase in proportion. However, over time an increase in bold individuals in a population may lead to heightened aggression and conflicts, resulting in physical injuries. This hyper-aggressiveness can diminish reproductive activities and

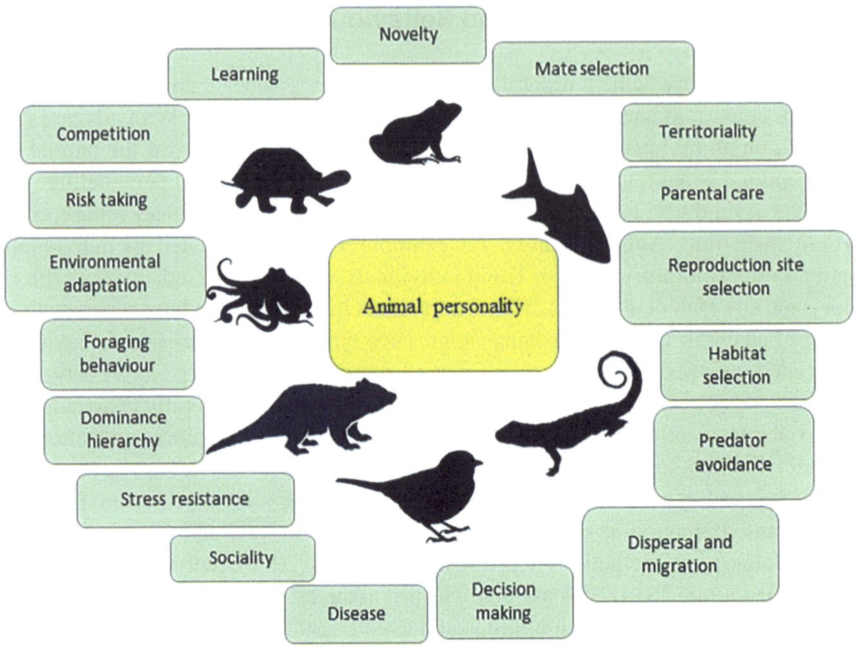

Fig. 6.1 The various essential survival traits in animals influenced by personality across taxa, including vertebrates and invertebrates

overall fitness, further endangering the population. In this situation, a shyer strategy would be beneficial for the population, and shyer individuals would increase in proportion. Also, boldness and shyness are negatively correlated traits. Therefore, directional selection for one trait imposes an evolutionary constraint, hindering both traits from evolving to their independent optima while simultaneously preserving genetic variation in both behavioural traits.

For a behavioural syndrome to endure, it must either confer a distinct advantage or be upheld by shared constraints that prevent its divergence, even if it occasionally leads to suboptimal responses. This persistence underscores the complex balance of benefits and limitations shaping evolutionary trajectories (Bell 2005; Dingemanse et al. 2010). For instance, bold individuals are more exploratory and proactive. They excel at quickly learning new tasks but tend to retain less information over time, are less flexible, and exhibit more mistakes in avoidance or reversal learning tests (Sih and Giudice 2012). Shy individuals are slower to explore, less reactive and more sensitive to environmental factors but store information more effectively and excel in avoidance and reversal learning tasks (Carere and Locurto 2011). In environments where boldness and quick learning maximise rewards and shyness and precision enhance safety, personality and cognitive style would co-vary within a population. This suggests that different strategies can thrive under varying circumstances. However, the evolutionary origins and maintenance of such phenotypic variation in animal personality remain largely unexplored and require further investigation. Understanding these mechanisms could unveil new insights into the adaptive significance of personality traits across species.

6.4 Maintaining Personality Through Metamorphosis

Amphibian metamorphosis requires more extreme and abrupt shifts in habitat and ecological niche than any other animal group. So amphibians offer fundamental insights into how personality traits are related to developmental changes in physiology, morphology and ecology (Wilson and Krause 2012). This drastic shift from aquatic to terrestrial life affects a wide range of behaviours, including mating, diet, foraging, predator evasion, movement or dispersal and even social grouping. A study by Wilson and Krause (2012) revealed that lake frogs (*Rana ridibunda*) maintain personality traits such as activity level and exploration in novel environments through metamorphosis and that bold individuals spend more time in risky open areas of familiar environments pre- and post-metamorphosis. Being bold and exploratory is advantageous as these individuals turn out more efficient while foraging in the presence of predators.

6.5 Personality and Disease Resistance

In the wood frog (*Lithobates sylvaticus*), green frog (*L. clamitans*) and the American bullfrog (*L. catesbeianus*), individual persoanlity has been correlated with parasite load (Koprivnikar et al. 2012; Szuroczki and Richardson 2012). More activity and exploration have been related to lower parasite loads due to anti-parasite behaviours such as parasite evasion and removal by shaking off the not-yet-attached cercariae stage of the parasite (Koprivnikar et al. 2012). However, the lower parasitic load may not be just linked to behavioural evasion. Aggressive and non-aggressive individuals have different coping mechanisms for stressors in a proactive–reactive behavioural syndrome. There is an existing positive link between immunity and the proactive–reactive axis where proactive animals may have more activity and exploration levels and have superior body conditions. These individuals are also better able to resist/tolerate parasite infection through non-behavioural mechanisms (Jm 2008). Aggressive individuals proactively prevent or manipulate any stressors associated with high sympathetic reactivity, whereas non-aggressive individuals accept or react to stressors (Jm 2008).

6.6 Personality and Dispersal

Different personality types in animals aid the various stages of dispersal which are departure from the current patch, movement between patches (transience) and settlement in a new patch (Bowler and Benton 2005; Stamps 2001). Investigation on the neotropical poison frog (*Allobates femoralis*) shows that the bolder frogs always select complex habitats that offer more diversity in microhabitats for breeding. Additionally, bolder frogs were also found to move more between different habitat structures (Bégué et al. 2023). Brodin et al. (2013) compared the personality of the European common frog (*Rana temporaria*) between a source and an isolated population. Observations showed that both tadpoles and froglets from isolated populations were bolder and more exploratory compared to the less isolated mainland populations. Therefore, successful range extension in amphibians could be linked to the personality traits of the migrating subset of a population that is bolder than the others.

Interestingly, successful colonisation can lead to invasion. Bolder cane toads (*Rhinella marina*) are asocial, readily feed from unoccupied feeding sites and are more exploratory (González-Bernal et al. 2014). On the contrary, shy individuals depend on social cues, select occupied foraging grounds and get outcompeted in feeding competition with bolder individuals (González-Bernal et al. 2014). Another study by Gruber et al. (2017) examined that cane toads at the range front are more bold and exploratory allowing them to explore new foraging grounds where competition is less compared to those individuals in the range core who experience a lot more competition due to familiarity of the environment and hence have less opportunities to take risk or explore unknown territories.

Evidently, personality traits are correlated to sociality in amphibians contributing to dispersal. Therefore, population density can also drive dispersal as the sociality of the individuals depends on their attraction towards conspecifics (Cote et al. 2010). A social individual has fitness at low densities and remains away from other individuals and disperse when densities are at peak, exceeding their threshold for social tolerance, potentially offering more conspecific interactions (Cote et al. 2010) On the contrary social animals seek conspecifics as their fitness increase at high density and might disperse when density is too low.

Aggressiveness has also been shown to enable settling where aggressive individuals replace and remove a competitor (Duckworth and Badyaev 2007). On the contrary, submissive individuals benefit from settling in a new population by increasing their acceptance by the conspecifics. Such situations have been documented in birds, mammals and reptiles (Aragón et al. 2006; Brandt 1992; Duckworth and Badyaev 2007) and may be impacting amphibian dispersal.

Several ecological factors can also determine which type of personality is better for the three stages of dispersal one of which is predation. While bold and aggressive individuals are good at initiating the dispersal to new territories, this decision can be driven by predation pressure and predation risk.

6.7 Personality in Mating

Investigations revealed that personality types and mating success in amphibians are correlated (Chaloupka et al. 2022; Réale et al. 2007; Souza et al. 2021). Due to high diversity and the transition from aquatic to terrestrial life, amphibians have evolved a wide range of mating and nesting behaviour and parental care that takes different forms, from actively guarding nests and spawns to transporting the tadpoles to building mud enclosures. Neotropical poison frogs (*Allobates femoralis*) have been extensively studied for their territoriality and parental care. Poison frogs show prominent territorial aggressiveness necessary to claim and maintain a territory eventually enhancing reproductive success (Peignier et al. 2023). Peignier et al. (2022) also investigated the relationship between boldness and territory occupation in wild poison frogs and revealed that more aggressive males respond more fiercely to territory intruders. This heightened aggression, however, comes at a significant cost, as these individuals more frequently engage in energy-intensive and potentially harmful conflicts (Peignier et al. 2022).

Boldness is also evident in the advertisement calls of males. Those exhibiting higher risk-taking behaviours call at higher rates and amplitudes, over longer durations, from more exposed locations, and move or turn more frequently while calling (Peignier et al. 2022), which positively impacts reproductive success (Souza et al. 2021). Males with high aggressiveness can fend off intruders better and defend their territory, but the increase in calling amplitude and calling activity also results in greater energy expenditure, increasing the risk of injury or predation—the fundamental conundrum. Hence, males exhibiting lower aggression levels might enjoy increased survival prospects until the next season. Even in females, bolder individuals are known to be choosier while selecting mates (Feagles and Höbel 2023).

Another effect of personality can be seen in the poison frog males who rely on spatial memory to select wetlands to distribute tadpoles to reduce the risks of desiccation or predation (Pašukonis et al. 2016). Exploratory males are better suited for prospecting novel sites for tadpole deposition or a better territory to establish and can risk themselves as well as the tadpoles as longer pathways require longer exposure to threats. Therefore, which personality gets selected over time is context-dependent.

6.8 What Drives Individual Personality?

Corticosterone regulates foraging behaviour, locomotion and overall body condition (Dufty et al. 2002; Spencer and Verhulst 2007). Carere et al. (2003) have shown that in the great tit (*Parus major*), bold individuals display a lower rise in corticosteroid levels after social stress than shy ones. A very interesting development in animal behaviour shows that maternal stress hormone levels influence offspring behaviour (Meylan et al. 2002). High maternal stress levels cause lower body size and poor body condition, reducing the growth of juveniles (Meylan et al. 2002). Even dispersal is influenced by maternal stress levels. High maternal stress reduces dispersal, whereas lower maternal stress increases the chances of dispersal in lizards (Meylan and Clobert 2005). Cote et al. (2010) proposed two separate pathways that establish a link between maternal, genetic and early developmental conditions and the postnatal environment to drive dispersal in animals. Maternal conditions and genes, along with environmental factors, may predispose dispersal inclination and the personality of an offspring, which when acted upon by postnatal environment can drive dispersal decision. On the contrary, maternal conditions and environment along with genes determine personality in offspring, and postnatal environment finally impacts personality that develops dispersal inclination in individuals. Thus, personality-driven ecological decision-making is a complex phenomenon that includes environmental stressors as proximate causes, whereas genetics and prenatal environment as distal causes.

6.9 How Is Personality Important for Conservation?

Amphibians are the most diverse and most threatened compared to other vertebrate taxa (Quaranta et al. 2009), with one-third of currently known species being endangered with extinction (IUCN 2023) and the majority experiencing population declines (Moor et al. 2022). Recent extinctions have been linked to the fungal disease chytridiomycosis and global warming (Becker et al. 2007). In addition to these critical stressors, habitat loss (Becker et al. 2007), overharvesting for food (Warkentin et al. 2009), pet trade (Hughes et al. 2021) and increased ultraviolet radiation (Kiesecker et al. 2001) are contributing significantly to population declines (Beebee and Griffiths 2005). These multifaceted threats underscore the urgent need for comprehensive conservation efforts to mitigate further biodiversity loss. Conservation initiatives, including habitat management or captive breeding, for introducing, reintroducing and reinforcing amphibian populations can benefit from animal personality research. For decades, conservationists and behavioural ecologists have been working independently. However, in recent times, there has been an increase in the integration of both fields, but the extent of implication of this integration is low (Angeloni et al. 2008). This current integrated emerging field is referred to as conservation behaviour, in which conservation actions draw reference from animal behaviour.

Bold individuals explore more, prospect novel environments and food, forage more and reproduce better than shy individuals (de Azevedo and Young 2021). Furthermore, exploration of the environment may benefit individuals in constructing a cognitive map of the area; identifying feeding areas, shelters and mates, and establishing memory of locations. There are existing evidence of Blanding's turtles (*Emydoidea blandingii*) and Tasmanian devils (*Sarcophilus harrisii*), where bold and exploratory individuals survived more and comparatively for a longer time than shy individuals (Allard et al. 2019; Sinn et al. 2014). A meta-analysis by Smith and Blumstein (2008) showed risky behavioural type to have a 6% higher chance of survival in the wild. However, a contrasting case also exists where bold individuals have been seen to die sooner than shy individuals (Bremner-Harrison et al. 2004). Additionally, personality traits such as boldness increase mating chances and fertilise more eggs and, therefore, reproduce more. Further, bold and aggressive individuals are better at resisting diseases through immunological behavioural responses (Fig. 6.2). Therefore, given that bolder individuals are more exploratory, reproduce more effectively and demonstrate better disease resilience, a pressing question arises: Could utilising bolder animals to bolster populations be a more effective strategy for conserving species endangered by anthropogenic stress? Or can the reintroduction of bolder individuals help establish new territory rapidly to replenish

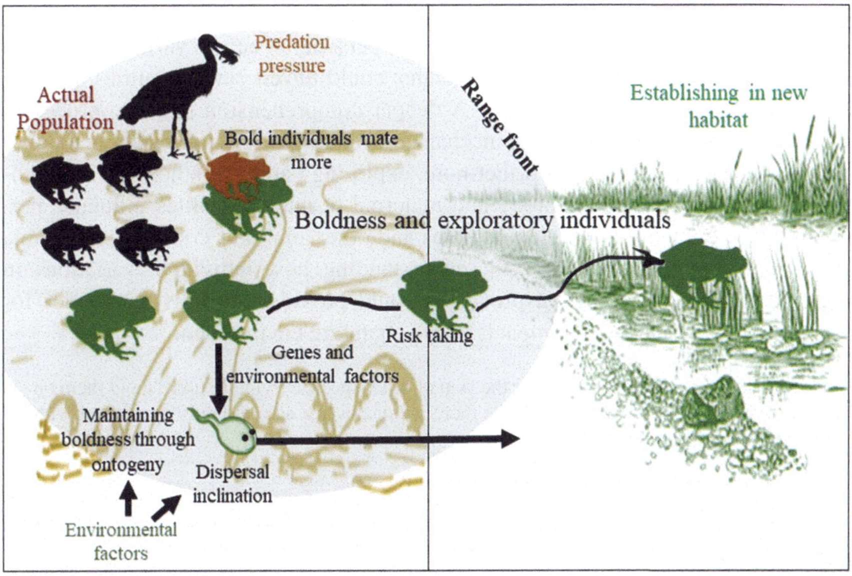

Fig. 6.2 Summarising the ideas that can inform future conservation actions. The figure uses boldness as a personality trait that can be used to drive conservation actions. It displays bold individuals being more prone to risk-taking in evading predations that increases reproductive success, and this trait is predisposed in offspring, eventually allowing dispersal and establishment in novel habitats. Green frogs indicate boldness and black frogs indicate shy individuals

the declining population? de Azevedo and Young (2021) mention that plans for conservation can include different proportions of bold:shy individuals varying, with a range of 100:0, 80:20, 50:50, 20:80 and 0:100 under various environmental conditions, such as the presence and absence of predators, to evaluate how different frequencies of personality traits affect survival.

In the future, we can expect to develop scientifically informed strategies which can improve the outcomes of animal translocation for successful conservation. At the same time, understanding personality and the link with species dispersal can help prevent phenomena such as biological invasions. The relation between personality and conservation behaviour presents a new and potential avenue that awaits proper exploration and implications for safeguarding species amidst an incessant increase in anthropogenic intervention.

6.10 Future Directions

This discipline is relatively new and still evolving and has scope for exploration. Investigating the interplay between behavioural ecology and amphibian personality has the potential to revolutionise conservation. To encourage greater engagement in this realm, we need to understand how environmental factors, social interactions and ecological pressures shape individual personalities. Moreover, delving into the neurobiological foundations of behaviour and personality across various taxa, such as reptiles, mammals, birds and amphibians, could unveil commonalities and distinctions in underlying mechanisms. A deeper comprehension of these processes might emerge by comparing brain mechanisms that drive cognition and personality traits in these diverse groups. Furthermore, exploring parallels in how environmental elements, stressors, physiology and maternal hormonal profiles influence personality development between anamniotes and other amniotes could offer valuable cross-species perspectives. Overall, understanding how individual variations in behaviour influence ecological processes could provide innovative strategies for managing and mitigating the impacts of the changing environment.

> We (humans) are trying to get under the skin of other species, trying to understand them on their terms. And more we succeed, the more we discover a natural landscape dotted with magic wells—Frans de Wall.

References

Allard S, Fuller G, Torgerson-White L, Starking MD, Yoder-Nowak T (2019) Personality in zoo-hatched Blanding's turtles affects behavior and survival after reintroduction into the wild. Front Psychol 10:2324

Angeloni L, Schlaepfer MA, Lawler JJ, Crooks KR (2008) A reassessment of the interface between conservation and behaviour. Anim Behav 75(2):731–737. https://doi.org/10.1016/j.anbehav.2007.08.007

Aragón P, Clobert J, Massot M (2006) Individual dispersal status influences space use of conspecific residents in the common lizard, Lacerta vivipara. Behav Ecol Sociobiol 60(3):430–438. https://doi.org/10.1007/s00265-006-0183-3

Becker CG, Fonseca CR, Haddad CFB, Batista RF, Prado PI (2007) Habitat split and the global decline of amphibians. Science 318(5857):1775–1777. https://doi.org/10.1126/science.1149374

Beebee TJC, Griffiths RA (2005) The amphibian decline crisis: a watershed for conservation biology? Biol Conserv 125(3):271–285. https://doi.org/10.1016/j.biocon.2005.04.009

Bégué L, Peignier M, Ringler E (2023) The link between animal personality and habitat selection in males of the neotropical poison frog *Allobates femoralis*. Behaviour 160(3–4):257–274. https://doi.org/10.1163/1568539X-bja10202

Bell A (2005) Behavioural differences between individuals and two populations of stickleback (*Gasterosteus aculeatus*). J Evol Biol 18:464–473. https://doi.org/10.1111/j.1420-9101.2004.00817.x

Bowler DE, Benton TG (2005) Causes and consequences of animal dispersal strategies: relating individual behaviour to spatial dynamics. Biol Rev Camb Philos Soc 80(2):205–225. https://doi.org/10.1017/s1464793104006645

Brandt CA (1992) Social factors in immigration and emigration. In: Stenseth NC, Lidicker WZ (eds) Animal dispersal: small mammals as a model. Springer, Dordrecht, pp 96–141

Bremner-Harrison S, Prodohl PA, Elwood RW (2004) Behavioural trait assessment as a release criterion: boldness predicts early death in a reintroduction programme of captive-bred swift fox (*Vulpes velox*). Anim Conserv 7(3):313–320. https://doi.org/10.1017/S1367943004001490

Brodin T, Lind M, Johansson F (2013) Personality trait differences between mainland and Island populations in the common frog (Rana temporaria). Behav Ecol Sociobiol 67:135–143. https://doi.org/10.1007/s00265-012-1433-1

Carere C, Locurto C (2011) Interaction between animal personality and animal cognition. Curr Zool 57:491–498. https://doi.org/10.1093/czoolo/57.4.491

Carere C, Groothuis TGG, Möstl E, Daan S, Koolhaas JM (2003) Fecal corticosteroids in a territorial bird selected for different personalities: daily rhythm and the response to social stress. Horm Behav 43(5):540–548. https://doi.org/10.1016/S0018-506X(03)00065-5

Carlson BE, Langkilde T (2013) Personality traits are expressed in bullfrog tadpoles during open-field trials. J Herpetol 47(2):378–383

Chaloupka S, Peignier M, Stückler S, Araya-Ajoy Y, Walsh P, Ringler M, Ringler E (2022) Repeatable territorial aggression in a neotropical poison frog. Front Ecol Evol 10:881387

Cole EF, Quinn JL (2014) Shy birds play it safe: personality in captivity predicts risk responsiveness during reproduction in the wild. Biol Lett 10(5):20140178. https://doi.org/10.1098/rsbl.2014.0178

Cote J, Fogarty S, Weinersmith K, Brodin T, Sih A (2010) Personality traits and dispersal tendency in the invasive mosquitofish (*Gambusia affinis*). Proc R Soc B Biol Sci 277(1687):1571–1579. https://doi.org/10.1098/rspb.2009.2128

de Azevedo CS, Young RJ (2021) Animal personality and conservation: basics for inspiring new research. Animals (Basel) 11(4):1019. https://doi.org/10.3390/ani11041019

Dingemanse NJ, Réale D (2005) Natural selection and animal personality. Behaviour 142(9–10):1159–1184. https://doi.org/10.1163/156853905774539445

Dingemanse NJ, Dochtermann N, Wright J (2010) A method for exploring the structure of behavioural syndromes to allow formal comparison within and between data sets. Anim Behav 79:439–450. https://doi.org/10.1016/j.anbehav.2009.11.024

Duckworth RA, Badyaev AV (2007) Coupling of dispersal and aggression facilitates the rapid range expansion of a passerine bird. Proc Natl Acad Sci U S A 104(38):15017–15022. https://doi.org/10.1073/pnas.0706174104

Dufty AM, Clobert J, Møller AP (2002) Hormones, developmental plasticity and adaptation. Trends Ecol Evol 17(4):190–196. https://doi.org/10.1016/S0169-5347(02)02498-9

Feagles OS, Höbel G (2023) Personality affects female mate choice: frogs displaying more consistent bold behaviors are choosier. Behav Ecol 35(1):1–8. https://doi.org/10.1093/beheco/arad101

González-Bernal E, Brown GP, Shine R (2014) Invasive cane toads: social facilitation depends upon an individual's personality. PLoS One 9(7):e102880. https://doi.org/10.1371/journal.pone.0102880

Gruber J, Brown G, Whiting M, Shine R (2017) Geographic divergence in dispersal-related behaviour in cane toads from range-front versus range-core populations in Australia. Behav Ecol Sociobiol 71:38. https://doi.org/10.1007/s00265-017-2266-8

Hughes AC, Marshall BM, Strine CT (2021) Gaps in global wildlife trade monitoring leave amphibians vulnerable. eLife 10:e70086. https://doi.org/10.7554/eLife.70086

IUCN (2023) The IUCN Red List of threatened species 2023-1. https://www.iucnredlist.org/search

Iyer AA, Briggman KL (2021) Amphibian behavioral diversity offers insights into evolutionary neurobiology. Curr Opin Neurobiol 71:19–28. https://doi.org/10.1016/j.conb.2021.07.015

Jm K (2008) Coping style and immunity in animals: making sense of individual variation. Brain Behav Immun 22(5):662. https://doi.org/10.1016/j.bbi.2007.11.006

Kiesecker JM, Blaustein AR, Belden LK (2001) Complex causes of amphibian population declines. Nature 410(6829):681–684. https://doi.org/10.1038/35070552

Koprivnikar J, Gibson CH, Redfern JC (2012) Infectious personalities: behavioural syndromes and disease risk in larval amphibians. Proc R Soc B Biol Sci 279:1544. https://doi.org/10.1098/rspb.2011.2156

Meylan S, Clobert J (2005) Is corticosterone-mediated phenotype development adaptive? Maternal corticosterone treatment enhances survival in male lizards. Horm Behav 48(1):44–52. https://doi.org/10.1016/j.yhbeh.2004.11.022

Meylan S, Belliure J, Clobert J, de Fraipont M (2002) Stress and body condition as prenatal and postnatal determinants of dispersal in the common lizard (Lacerta vivipara). Horm Behav 42(3):319–326. https://doi.org/10.1006/hbeh.2002.1827

Moor H, Bergamini A, Vorburger C, Holderegger R, Bühler C, Egger S, Schmidt BR (2022) Bending the curve: simple but massive conservation action leads to landscape-scale recovery of amphibians. Proc Natl Acad Sci U S A 119(42):e2123070119. https://doi.org/10.1073/pnas.2123070119

Pašukonis A, Trenkwalder K, Ringler M, Ringler E, Mangione R, Steininger J, Warrington I, Hödl W (2016) The significance of spatial memory for water finding in a tadpole-transporting frog. Anim Behav 116:89–98. https://doi.org/10.1016/j.anbehav.2016.02.023

Peignier M, Araya-Ajoy YG, Bégué L, Chaloupka S, Dellefont K, Leeb C, Walsh P, Ringler M, Ringler E (2022) Exploring links between personality traits and their social and non-social environments in wild poison frogs. Behav Ecol Sociobiol 76(7):93. https://doi.org/10.1007/s00265-022-03202-9

Peignier M, Bégué L, Ringler M, Szabo B, Ringler E (2023) Regardless of personality, males show similar levels of plasticity in territory defense in a neotropical poison frog. Sci Rep 13:3435. https://doi.org/10.1038/s41598-023-30546-7

Quaranta A, Bellantuono V, Cassano G, Lippe C (2009) Why amphibians are more sensitive than mammals to xenobiotics. PLoS One 4(11):e7699. https://doi.org/10.1371/journal.pone.0007699

Réale D, Reader SM, Sol D, McDougall PT, Dingemanse NJ (2007) Integrating animal temperament within ecology and evolution. Biol Rev 82(2):291–318. https://doi.org/10.1111/j.1469-185X.2007.00010.x

Réale D, Garant D, Humphries MM, Bergeron P, Careau V, Montiglio P-O (2010) Personality and the emergence of the pace-of-life syndrome concept at the population level. Philos Trans R Soc B Biol Sci 365(1560):4051–4063. https://doi.org/10.1098/rstb.2010.0208

Sih A, Giudice MD (2012) Linking behavioural syndromes and cognition: a behavioural ecology perspective. Philos Trans R Soc B Biol Sci 367:2762–2772. https://doi.org/10.1098/rstb.2012.0216

Sih A, Bell AM, Johnson JC, Ziemba RE (2004) Behavioral syndromes: an integrative overview. Q Rev Biol 79(3):241–277. https://doi.org/10.1086/422893

Sinn DL, Cawthen L, Jones SM, Pukk C, Jones ME (2014) Boldness towards novelty and translocation success in captive-raised, orphaned Tasmanian devils. Zoo Biol 33(1):36–48. https://doi.org/10.1002/zoo.21108

Smith BR, Blumstein DT (2008) Fitness consequences of personality: a meta-analysis. Behav Ecol 19(2):448–455. https://doi.org/10.1093/beheco/arm144

Souza JRD, Kaefer IL, Lima AP (2021) Calling activity determines male mating success in a territorial frog with parental care. Ethology 127(4):359–365. https://doi.org/10.1111/eth.13135

Spencer KA, Verhulst S (2007) Delayed behavioral effects of postnatal exposure to corticosterone in the zebra finch (*Taeniopygia guttata*). Horm Behav 51(2):273–280. https://doi.org/10.1016/j.yhbeh.2006.11.001

Stamps J (2001) Habitat selection by dispersers: integrating proximate and ultimate approaches. In Dispersal (eds J. Clobert, E. Danchin, A. A. Dhondt & J. D. Nichols), pp. 110-122. New York, NY: Oxford University Press

Szuroczki D, Richardson JML (2012) The behavioral response of larval amphibians (Ranidae) to threats from predators and parasites. PLoS One 7(11):e49592. https://doi.org/10.1371/journal.pone.0049592

Warkentin IG, Bickford D, Sodhi NS, Bradshaw CJA (2009) Eating frogs to extinction. Conserv Biol 23(4):1056–1059. https://doi.org/10.1111/j.1523-1739.2008.01165.x

Wilczynski W (2009) Evolution of the brain in amphibians. In: Binder MD, Hirokawa N, Windhorst U (eds) Encyclopedia of neuroscience. Springer, Berlin, pp 1301–1305

Wilson ADM, Krause J (2012) Metamorphosis and animal personality: a neglected opportunity. Trends Ecol Evol 27(10):529–531. https://doi.org/10.1016/j.tree.2012.07.003

Wolf M, Weissing FJ (2012) Animal personalities: consequences for ecology and evolution. Trends Ecol Evol 27(8):452–461. https://doi.org/10.1016/j.tree.2012.05.001

A Global Synthesis of the Impacts of Urbanisation on Amphibians

7

Madhushri Mudke and N. A. Aravind

Abstract

Land-use changes due to urbanisation remain one of the most ubiquitous causes of ecological disturbance. Urbanisation affects amphibians more than any other tetrapod. This chapter documents scientific evidence of urban stressors like roads, noise, light, and alien invasive species that disturb amphibian habitats and impact the amphibian population and summarises the potential changes in amphibian behaviours in response to urban stressors. This chapter concludes with recommendations for conservation and insights on future studies that must be carried out in the Global South, including rapidly developing nations like India.

Keywords

Amphibians · Urbanisation · Impact assessment · Conservation

M. Mudke
SM Sehgal Foundation Centre for Biodiversity and Conservation, Ashoka Trust for Research in Ecology and the Environment (ATREE), Bangalore, India

Manipal Academy of Higher Education (MAHE), Manipal, India

N. A. Aravind (✉)
SM Sehgal Foundation Centre for Biodiversity and Conservation, Ashoka Trust for Research in Ecology and the Environment (ATREE), Bangalore, India

Yenepoya Research Centre, Yenepoya (Deemed to be University), Mangalore, India
e-mail: arvind@atree.org

S. Sridhara et al. (eds.), *Animal Behavior in the Tropics*,
https://doi.org/10.1007/978-981-97-7324-4_7

7.1 Introduction

All tetrapods or animals that walk on all four legs and are non-amniotes are clustered under the class 'Amphibia'. All modern amphibians are included in the subclass or clade 'Lissamphibia' and have evolved from a common ancestor (Sullivan 2008). Amphibians are classified into three orders—anurans (e.g. tailless tetrapods, frogs and toads), Caudata or Urodela (e.g. the salamanders), and Gymnophiona or Apoda (limbless worm-like creatures and the caecilians). Most amphibians need freshwater for reproduction and have complex life histories (Sullivan 2008). Amphibians are either explosive breeders or breed for prolonged periods of time at any given site (Sullivan 2008). Amphibians, unlike amniotes, undergo metamorphoses and become adults with developed lungs from a larval stage with gills (Catenazzi and von May 2021).

Amphibians are some of the most sensitive animals living on planet Earth (Sullivan 2008). Characteristics like their permeable skin, external fertilisation and small distribution ranges make them susceptible to changes (Sullivan 2008; Alroy 2015). However, not all amphibian species can be called 'sensitive' even when they share some or all of these features. In wetlands and aquatic ecosystems, amphibians serve as biological indicators (Sullivan 2008). An increase or decrease in any species of amphibians can give us information about the health of a given ecosystem. For example, an increase in the number of toads or bullfrogs have been attributed to the favorable changes in habitats. For instance, more generalist species is often seen along with anthropogenised and human-occupied areas (Sullivan 2008; Alroy 2015). Along with this, the role of amphibians in the ecosystem is well-established, and they play a role in all four categories of the Millennium Ecosystem Assessment (MEA) framework (Hocking and Babbitt 2014). They offer provisioning services where amphibians are a food source to some people and serve as models for medical and pharmaceutical research. Cultural services to people is by increasing quality of life through recreation, religion, spirituality, and aesthetics. Supporting services (e.g., soil burrowing, aquatic bioturbation, decomposition and nutrient cycling all are critical for a proper functioning of any ecosystem) and Regulating services (e.g., potential control of other pests and through predation of insect pollinators) (Hocking and Babbitt 2014) are also significant.

However, despite their significance in the ecosystem and their spectacular life histories, urbanisation currently threatens more than one-third of the world's known amphibian species (Hamer and McDonnell 2008).

7.2 Urbanisation and Amphibian Decline

The process of urbanisation involves continuous changes over a patch of land that 'homogenises the physical environment', which has effects larger than that of traditional farming, logging and many other land uses (Marzluff and Ewing 2008). Urbanisation increases human interferences, causing hydrological, chemical, physical, ecological and changes that alter the existing flora and fauna (McKinney 2002).

There is justification that urbanisation is one of the major hurdles to amphibian populations, leading to complete wiping out and regional, localised extinctions (Cushman 2006; Windmiller and Calhoun 2007; Hamer and McDonnell 2008; Measey and Tolley 2011; Mokhatla et al. 2012). Amphibians are at the peak of the biodiversity extinction crisis (IPBES et al. 2019). The global amphibian assessment report showed that 1893 (32%) out of 5881 amphibian species assessed were threatened with extinction and about 80% of amphibian populations have declined far more significantly than birds and mammals (Baillie et al. 2008). A total of 21% of amphibians are categorised as 'critically endangered' or 'endangered', while for mammals and birds, the rates are 5.4% and 10.5%, respectively (Collins and Storfer 2003).

One of the reasons why amphibians are more affected by habitat changes and associated urban changes compared to other terrestrial vertebrates is because they have very poor dispersal abilities (Semlitsch and Bodie 2003). Although some anurans move relatively long distances (up to 500 m; Lemckert and Mahony 2004), most generally tend to have poor dispersal abilities (Gibbs and Reed 2007; Semlitsch and Bodie 2003; Pillsbury and Miller 2008). As urban sprawl increases, dispersal corridors are disrupted, and wetlands are degraded, reducing amphibian richness through reduced dispersal (Delis et al. 1996). This restricts species' migration routes and limits their recolonisation to wetlands or habitats from which they have been extirpated (Pillsbury and Miller 2008). Urbanisation is widely studied in the Global North, but scientific evidence from the Global South remains scarce (Shackleton et al. 2021). Despite having higher biodiversity in the Global Southern countries, scientific studies on the impacts of unplanned, unregulated urbanisation and associated human population densities on amphibian diversity are rare (Ficetola and De Bernardi 2004; Hamer and McDonnell 2008) and urbanisation on behavioural responses are practically nil.

7.3 Impacts of Roads and Traffic

A busy road network is often a direct result of urbanisation (McKinney 2002). Amphibians have the highest median road kill rate compared to other taxa like reptiles, mammals or birds (Pinto et al. 2020a). Almost 70–88% of all road kills among terrestrial vertebrates are amphibians (Glista et al. 2008). However, the 'anthropause' explained as shutdowns of all human-related activities, including notable reductions of traffic levels in cities due to the COVID-19 pandemic, had a reverse effect on road mortalities of animals (LeClair et al. 2021). In the United States alone, during the time of disruption of human activities in the spring of 2020, an almost 50% decline was detected in all amphibian road mortalities (LeClair et al. 2021). Earlier studies from countries like Canada, Netherlands, England, Bulgaria, Denmark, Poland and Serbia have shown that amphibian species that are generally found in urban areas have suffered considerable mortalities affecting populations with increasing traffic vehicles (Fahrig et al. 1995; Cooke and Sparks 2004; Hels and Buchwald 2001; Elzanowski et al. 2009; Anđelković and Bogdanović 2022). Usually, generalist species that breed in the same, permanent, urban water body show less road mortalities than species that move slower or migrate from breeding areas to roosting areas (Carr and Fahrig 2001; Gibbs and Shriver 2005).

The impacts of roads on amphibians include not just road mortalities but also skeletal malformations from injuries/deformities due to collision with vehicles and genetic isolation (Pinto et al. 2020a, b; Langevelde et al. 2009; Anđelković and Bogdanović 2022) There is a high correlation between the proximity of roads and the increased number of frog abnormalities (Reeves et al. 2008). Roads act as barriers causing an increase in genetic differentiation and reduced gene flow, thereby decreasing genetic diversity among Moor frogs (*Rana arvalis*), the midwife toad (*Alytes obstetricans*), the palmate newt (*Lissotriton helveticus*) and salamanders such as red-backed salamander (*Plethodon cinereus*), Shenandoah salamander (*Plethodon shenandoah*), otter salamander (*Plethodon hubrichti*) and cheat mountain salamander *Plethodon nettingi* (Herrington and Larsen 1985; Marsh et al. 2005; Holderegger and Di Giulio 2010). Studies suggest that the construction of passages for amphibians across roads will help mitigate the potential fragmentation of populations and reduce the barrier effects of roads, especially in areas where roads are constructed within the range of existing endangered species (Jaarsma 1997; Jaarsma and Willems 2002; Marsh et al. 2005; Garcia-Gonzalez et al. 2012).

7.4 Impacts of Anthropogenic Noise

Military operations; industry; modes of transport, especially roads and traffic; loudspeakers; human voices and infrastructure construction lead to the production of 'anthropogenic noise' (Blickley and Patricelli 2010; Herrera-Montes and Aide 2011). Anthropogenic noise has a frequency range of 5 to 7 kHz (concentrated below 2 kHz) and is known to overlap and mask animal acoustic communication (Brumm and Slabbekoorn 2005; Cunnington and Fahrig 2010; Herrera-Montes and Aide 2011; Simmons and Narins 2018). Amphibians use acoustic communication in the range of 1–5 kHz for mating, territorial defence, advertising and signalling (Gerhardt and Huber 2003; Simmons and Narins 2018; Zaffaroni-Caorsi et al. 2023). Anthropogenic noise affects the acoustic behaviour of amphibians by altering the pitch and frequency of the call (Gerhardt and Huber 2003). Frogs like common mist frogs (*Litoria rheocola*), the southern brown tree frog (*Litoria ewingii*) and the common eastern froglet (*Crinia signifera*) call at a higher pitch and more frequently in the presence of noise (Gerhardt and Huber 2003; Parris et al. 2009). A large number of anuran families like Bufonidae, Dendrobatidae, Hylidae, Hyperoliidae, Myobatrachidae and Ranidae show an increase in the dominant frequency of the advertisement calls in order to cope with anthropogenic noises (Zaffaroni-Caorsi et al. 2023). Also, most species that increase their dominant frequency also increase their call rates and amplitude (Parris et al. 2009; Cunnington and Fahrig 2010; Zaffaroni-Caorsi et al. 2023). Significant alterations in the call duration, call rates and call lengths of frogs like *Boana bischoffi*, *Boana leptolineata*, *Microhyla butleri*, *Rana nigrovittata*, *Kaloula pulchra*, *Hyla arborea*, *H. versicolor*, *L. clamitans*, *L. pipiens* and *R. taipehensis* were noted when exposed to aeroplane or traffic sounds (Bee and Swanson 2007; Cunnington and Fahrig 2013; Zaffaroni-Caorsi et al. 2023). Frogs that call in chorus, like *Rana taipehensis*, reduce the number of hours of calling per night or the number of nights spent calling in chorus when exposed to anthropogenic noise (Parris et al. 2009) (Table 7.1).

Acoustic signalling among amphibians conserves a lot of energy, and the prolonged calling exhausts energy reserves in male frogs, reducing overall mating success (Ryan 1986; Murphy 1999; Wollerman and Wiley 2002; Bee and Swanson 2007; Lengagne 2008; Parris et al. 2009; Zaffaroni-Caorsi et al. 2023). Some female frogs, like grey treefrog (*Hyla chrysoscelis*), wood frog (*Lithobates sylvaticus*) and Ezo brown frog (*Rana pirica*), show decreased orientation towards male advertisement calls in the presence of road traffic noises (Bee and Swanson 2007; Tennessen et al. 2014; Senzaki et al. 2018). Species such as the marsh frog (*Pelophylax ridibundus*), the Leopard frog (*Rana pipiens*) and the spadefoot toad (*Scaphiopus couchii*) freeze or remain motionless due to sudden, loud noises like single honks, higher than 70 dB SPL (Grace and Noss 2018). Some of the other lesser-known effects of continuous traffic noise exposure are increased corticosterone levels (stress hormones) and reduced sperm count (Tennessen et al. 2014; Kaiser et al. 2015). The presence of a flowing stream, dense vegetation alongside roads and the use of regulation about quieter hours in densely populated urban areas mask traffic noises and have beneficial effects on amphibian communities (Goosem et al. 2008). An elaborate review of the impacts of anthropogenic noise and relevant conservation strategies is discussed in detail by Caorsi et al.

7.5 Impacts of Artificial Lights

Light pollution is the result of excessive light emitted through common anthropogenic sources such as street lighting and vehicle lights and lights emitted through various other human activities due to urbanisation (Cinzano et al. 2001). Artificial light at night (ALAN) has adverse impacts on wildlife and ecosystems (Longcore and Rich 2004; Wise 2007; Gaston et al. 2013; Hebbar et al. 2019). ALAN impacts the reproductive and breeding behaviours of amphibians like green frogs (*Rana clamitans melanota*) and common toads (*Bufo bufo*), including time and duration of amplexus, mate choices, egg laying, metamorphosis and tadpoles (Gerhardt and Huber 2003; Baker and Richardson 2006; Touzot et al. 2019). With higher intensities of light, fertilisation rates decrease, and male body mass increases (Lengagne 2008; Touzot et al. 2019). Prolonged exposure in a laboratory set-up shows down-regulation of 62% of genes at night and 61.5% of genes during the day related to the innate immune system and lipid metabolism (Touzot et al. 2019).

ALAN negatively impacts seasonal and diel calling in anurans (Longcore and Rich 2004). Some species like *L. gracilis*, *Odontophrynus* sp., *Physalaemus* spp., *P. minuta,* and *S. squalirostri* show longer and stronger vocalisations in the absence of artificial light. ALAN can cause a cessation of amphibian calling (Dias et al. 2019). However, *Dendropsophus* species, generalist species or species that call during the day are known to exhibit less sensitivity towards artificial lights

Table 7.1 Summary of the impact of anthropogenic stressors on amphibian behaviour

Sr. no.	Anthropogenic stressors	Impact on amphibian behaviour	Direct or indirect impact	Kind of behaviour impacted	References
1	Roads	Imbalance between sex ratios of males and females in a community due to road mortalities and genetic isolation	Indirect	Movement, migration, feeding, reproduction	Carr and Fahrig (2001), Gibbs and Shriver (2005), Hels and Buchwald (2001), Elzanowski et al. (2009), Anđelković and Bogdanović (2022)
2	Vehicular traffic	Freezing and immobility on encounter, malformations, deformities and population decline due to accidents giving rise to skewed amphibian communities and increased breeding pressure	Direct	Movement, migration, feeding, reproduction	Mazerolle et al. (2005)
3	City and vehicular lights	Freezing and immobility during migration, thus reduced overall breeding success and reduced acoustic behaviour	Direct	Movement, migration, reproduction	Grace and Noss (2018)
4	Traffic noises like motor cycle and other vehicular sounds	Freezing and immobility on sudden exposure, less occupancy in louder areas, decreased orientation of female frogs towards male advertisement calls, increase in corticosterone levels (stress hormones) and a reduction in sperm count, leading to breeding pressures and skewed male–female ratios	Both	Reproduction (acoustic)	Ryan (1986), Bee and Swanson (2007), Parris et al. (2009). Tennessen et al. (2014), Kaiser et al. (2015), Senzaki et al. (2018)

(continued)

Table 7.1 (continued)

Sr. no.	Anthropogenic stressors	Impact on amphibian behaviour	Direct or indirect impact	Kind of behaviour impacted	References
5	Loud honks	Freezing and motionlessness along with abandoning roosting burrows	Direct	Reproduction (acoustic)	Brumm and Slabbekoorn (2005), Grace and Noss (2018)
6	Aeroplane noises	Increased calling rate, reduction in the number of hours of calling per night and reduced number of nights spent calling in chorus, thereby causing exhaustion and reduced breeding success	Both	Reproduction (acoustic)	Simmons and Narins (2018), Parris et al. (2009), Cunnington and Fahrig (2013)
7	Street lights or other urban lighting	Weaker and shorter vocalisations, sometimes complete cessation of calling and freezing response leading to reduced breeding success and size-selective predation where smaller-sized prey (e.g., tadpoles) are preferred by predators influencing community structure and reduced foraging	Indirect	Movement, migration, feeding, reproduction	Gerhardt and Huber (2003), Longcore and Rich (2004), Baker and Richardson (2006), Wise (2007), Touzot et al. (2019), Dias et al. (2019), Shidemantle et al. (2022)
8	Habitat degradation	Restriction in the dispersal and migration of amphibians, changes in community structure with some amphibians breeding explosively, leading to wiping out of others	Direct	Movement, migration, feeding, reproduction	Marsh (2001), McKinney (2002), Cushman (2006), Hamer and McDonnell (2008)

(continued)

Table 7.1 (continued)

Sr. no.	Anthropogenic stressors	Impact on amphibian behaviour	Direct or indirect impact	Kind of behaviour impacted	References
9	Salts	Reduction in egg mass densities, suppresses larval growth, thus making breeding habitats unsuitable and increasing breeding pressure	Direct	Feeding, reproduction	Karraker et al. (2008), Collins and Russell (2009), Georgiev et al. (2009), Hintz et al. (2021)
10	Chemical pollutants (nitrates)	Irreversible physiological changes causing decrease in survival and impact on reproduction	Indirect	Feeding, reproduction	Hecnar (1995), Baker and Waights (1994)
11	Diseases	Spreading of Bd fungus and ranavirus, leading to skewed male–female ratios and malformations impacting reproductive success	Direct	Reproduction	Carey and Alexander (2003)
12	Alien invasive species	Alteration in predation pressure, inability to locate the predator and resource-sharing leading to limited resources for native species causing reduced breeding and amplexus	Direct	Feeding, reproduction	Kats and Ferrer (2003), Bucciarelli et al. (2014), Nunes et al. (2019)

(Buchanan 2006; Wise 2007). Exposure to artificial lights from vehicles causes some amphibian species, especially the ones that migrate, to freeze (Buchanan 2006; Mazerolle et al. 2005). About six commonly occurring species, namely, spotted salamanders (*Ambystoma maculatum*), blue-spotted salamanders (*Ambystoma laterale*), green frogs (*Rana clamitans*), American toads (*Anaxyrus americanus*), spring peepers (*Pseudacris crucifer*) and wood frogs (*Rana sylvatica*), remain immobile with approaching headlights from vehicles exposing them to road traffic mortalities (Mazerolle et al. 2005).

Urban-lit habitats have a beneficial effect on amphibian species, especially toads (*Bufo bufo*) who benefit from the availability of more insects attracted by streetlights (González-Bernal et al. 2016). The effects of ALAN are highest in peri-urban and rural areas where toads (including alien invasive toads) benefit from consuming increased amounts of food (Komine et al. 2020). ALAN also impacts size-selective predation where smaller-sized prey (e.g. tadpoles) are preferred by predator newts and larger-sized tadpoles of American toads and wood frogs are left behind, influencing community structure (Shidemantle et al. 2022). ALAN reduces the feeding behaviour of red-backed salamander (*Plethodon cinereus*) on the moist forested floor, where they usually forage (Buchanan 2006; Wise 2007). ALAN also alters skin pigmentation in the tadpoles of spring peeper (*Pseudacris crucifer*) (Shidemantle et al. 2022). It is unclear whether the changes in skin pigmentation are due to the alteration of melatonin in the frog's skin (Sugden et al. 2004; Touzot et al. 2019; Shidemantle et al. 2022). Few studies show that lights at night are a potential stressor impacting the biological functioning of amphibians and contribute to existing population decline (Buchanan 2006; Wise 2007).

7.6 Habitat Degradation and Destruction Due to Urbanisation

One of the most well-known causes of amphibian population decline is habitat degradation and destruction, including, but not limited to, a complete disappearance and/or conversion of existing habitats into newer ones (Scheffers et al. 2012; Bugnot et al. 2019). Due to urbanisation, amphibian habitats that undergo changes restrict the dispersal and migration of amphibians, further restricting their colonisation leading to the decline of native species and localised extinction (Marsh 2001; Cushman 2006; Hamer and McDonnell 2008). As part of urban planning and architecture, artificial water bodies are created that change the structure of amphibian communities and allow some amphibians to thrive while wiping out others (McKinney 2002; Hamer and McDonnell 2008). Terrestrial and aquatic habitats both play a major role in sustaining amphibian species in a given landscape (Trenham and Shaffer 2005). Studies from several urbanised regions across the world show that the number of amphibian species within urban cores is less compared to rural and suburban areas (Hamer and McDonnell 2008).

Widespread conversion of forests into urban land in North Carolina resulted in a significant decline in the populations of the southern two-lined salamander (*Eurycea cirrigera*) and the northern dusky salamander (*Desmognathus fuscus*) (Price et al. 2007). In Plovdiv and Bulgaria, only four species were seen in the city compared to ten in the rural areas (Kambourova et al. 2012). About one-third of amphibian species from Pune city in India were locally extinct in a span of just 25 years due to habitat destruction and industrial growth of the city (Padhye et al. 2002). In Australia's Greater Melbourne region, species like the common eastern froglet (*Crinia signifera*) and the southern brown tree frog (*Litoria ewingii*) were likely to adapt to newer urban areas, but the Victorian smooth froglet (*Geocrinia victoriana*) was hardly detected in urban environments (Hamer and Parris 2011). Reduced genetic diversity, reduced gene flow and smaller fragmented populations of the eastern golden frog (*Pelophylax plancyi*) were reported when urban changes occurred within a 2 km radius of existing populations of the frog's habitat (Zhang et al. 2016; Wei et al. 2021). The population of spadefoot toads (*Pelobates fuscus*) from the Cultural Capital of Romania, Iaşi, was reduced to only 40 individuals due to the destruction of feeding and breeding habitats (Nicoara and Nicoara 2008). The population estimates of two salamander species *Eurycea cirrigera* and *Desmognathus fuscus* have declined by 44% and 30%, respectively, in three decades, due to the widespread conversion of forest land into urban land (Price et al. 2007). Naturally occurring temporary ponds and streams supported the breeding population of ten amphibian species, while farmland ponds and artificially created water supported only seven species (Ferreira and Beja 2013).

Urban watersheds have significantly fewer amphibian species than agro-watersheds and evergreen watersheds (Barrett and Guyer 2008). Several anthropogenic influences like isolated deep water bodies, small size of water bodies, vertical pond walls, shoreline length and development ratio, absence of vegetation and terrestrial areas around the water bodies, pollution and discharge from households or industries disturb the existence of amphibians within urban areas (Ficetola and De Bernardi 2004; Rubbo and Kiesecker 2004; Hamer and McDonnell 2008; Pillsbury and Miller 2008; Hamer and Parris 2011; Smallbone et al. 2011; Nowakowski et al. 2017). The composition of amphibian assemblages is influenced by changes in the physical conditions of wetlands, such as the alteration of hydrologic regimes (Vickers et al. 1985). Excessive groundwater withdrawal also affects native anurans (Guzy et al. 2006; Bunnell and Ciraolo 2010).

With increasing urbanisation, the amount of industrial effluents or contaminants like nitrates and salts (NaCl) reduce the quality of freshwater habitats, thereby reducing the number and diversity of amphibians in urban areas (Georgiev et al. 2015). Common salts are used as de-icing agents on roads in urban areas in countries that receive heavy snowfall to reduce >78% of road accidents (Hintz et al. 2021). De-icing agents increase salinity within water bodies and impact amphibians across their life cycles (Collins and Russell 2009). Salts reduced egg mass densities of spotted salamanders (*Ambystoma maculatum*) and wood frogs (*Rana sylvatica*) to almost half when compared to egg mass densities within unaffected forested areas (Karraker et al. 2008). Salts also impact the larval survival of amphibians like the wood frogs (*Lithobates sylvatica*) and spotted salamanders (*Ambystoma maculatum*) (Karraker et al. 2008). Salts (beyond 1000 mg/L) and other pollutants cause physical deformities (Collins and Russell 2009). This, combined with other urban stressors, shows a 21% higher likelihood of amphibian abnormalities and deformities (Eigenbrod et al. 2008). The presence of salts within urban water bodies reduces amphibian occupancy and leads to local extinction (Collins and Russell 2009).

On the other hand, nitrates occur in urban water bodies due to slaughterhouses, animal feedlots, untreated industrial effluents, domestic cleaning agents, lawns, urban run-off and chemical fertilisers. Concentrations of nitrates because of these human activities exceed approximately 100 mg/L (Lucey and Goolsby 1993; Rouse et al. 1999). Nitrate levels of more than 1 mg/L are harmful to amphibians, and those exceeding 10 mg/L show chronic and often irreversible physiological changes. More than 13 mg/L is lethal for amphibians (Rouse et al. 1999). Several species, like the Western chorus frog (*Pseudacris triseriata*), northern leopard frog (*Rana pipiens*), green frogs (*Rana clamitans*), common toads (*Bufo bufo*), American toad (*B. americanus*) and White's tree frog (*Litoria caerulea*) are highly susceptible to nitrate exposure of 3 mg/L for 96 h (Hecnar 1995; Baker and Waights 1994; Waights and Baker 1993). Vegetation, woods or grassy patches of approximately 19–24 m can drastically reduce nitrates (from 12 to 1 mg/L) from entering the water bodies (Haycock and Burt 1993).

A few generalist and resilient species like pickerel frogs (*R. palustris*), American toad (*Bufo americanus*) and blue ballon flower (*B. woodhouseii fowleri*), spring peepers (*Pseudacris crucifer*), grey treefrogs (*Hyla versicolor*), green frogs (*R. clamitans*) or bullfrogs (*R. catesbeiana*) tend to occupy urban spaces more than sensitive species (Padhye et al. 2002; Rubbo and Kiesecker 2004; Hamer and McDonnell 2008; Georgiev et al. 2009). For amphibian populations to sustain within urban areas, habitat quality and microhabitats like vegetation cover, shallow waters and terrestrial habitats around the wetlands must be maintained as per the needs of native amphibians (McKinney 2002; Hamer and McDonnell 2008; Smallbone et al. 2011; Wei et al. 2021). It is also important to allow corridors that connect 2–3 wetlands, reducing impervious surfaces between wetlands within urban areas (Hamer 2022a, b). Suitable habitat for some amphibians may still be available within urban areas, but the diversity is reasonably low.

7.7 Urban Changes Spread Diseases Among Amphibians

Increased human density can be linked to changes in light and thermal environments, water chemistry and hydrology (Hamer and McDonnell 2008; Georgiev et al. 2009). These characteristic shifts provide a changing backdrop for the infection of amphibian hosts by macroparasites such as trematodes and ranavirus (Skelly et al. 2006). The virulence and density of the pathogens intensify due to urban and suburban changes in degraded, disturbed habitats (Johnson et al. 2001; Carey and Alexander 2003; Skelly et al. 2006; Hamer and McDonnell 2008). Urban changes can potentially hinder the transmission of parasites to frogs by limiting access to other vertebrates that live within wetlands (King et al. 2007).

Chytridiomycosis, a fungal infection caused by the pathogen *Batrachochytrium dendrobatidis* (Bd), is identified as one of the major threats to amphibian populations (Hof et al. 2011). Bd fungus does not directly spread with urbanisation. However, the transport of biotic elements, trade and movements is more in urban areas than in rural or less developed areas (Daszak et al. 2004; Turner et al. 2021). Human contact through the release of pet species and the introduction of amphibians to new regions like permanently constructed water bodies inadvertently transfer Bd pathogens into urban landscapes (Carey and Alexander 2003; Kiesecker et al. 2001). Bullfrogs are resilient towards Bd pathogens but act as host species to spread the virus among native amphibians, especially in urban areas (Kiesecker et al. 2001; Daszak et al. 2004).

7.8 Impacts of Alien Invasive Species in Urban Spaces

Construction of permanent water bodies is commonly seen around urban and suburban areas, and most of these water bodies are stocked with non-native, exotic fish and other introduced species that impact the diversity of amphibians (Kentula et al. 2004; Riley et al. 2005; Paul and Meyer 2001; Pilliod et al. 2014). An estimated 16% of the total amphibians on the planet are threatened by introduced species alone (Nunes et al. 2019). Introduced species affect amphibians through predation, competition and altering native habitat and biota structure (Kats and Ferrer 2003; Bucciarelli et al. 2014; Capinha et al. 2017; Nunes et al. 2019).

Several species of exotic fish like trouts, bass (*Micropterus* spp.), bluegill (*Lepomis macrochirus*), and mosquitofish (*Gambusia affinis*) reduce the abundance of frogs (Knapp and Matthews 2000; Riley et al. 2005). Crayfish *Procambarus clarkii*, alien turtle species (e.g. the red-eared slider [*Trachemys scripta elegans*]), the Argentine ant (*Linepithema humile*) and invasive floating plant (*Azolla filiculoides*) reduce amphibian abundance, sometimes causing local extinctions (Riley et al. 2005; Pinero-Rodríguez et al. 2021; Melotto et al. 2021).

Invasive, introduced amphibian species like the American bullfrog (*Lithobates catesbeianus*) and the African clawed frog (*Xenopus laevis*) often benefit from urban areas and deeper pits seen during road construction or within artificial water bodies (Kiesecker et al. 2001; Hamer and McDonnell 2008; Measey et al. 2012, 2017; Mora et al. 2019). These alien amphibians prey on native species and have a potential to spread two amphibian diseases, chytridiomycosis and ranavirus, thus reducing native species diversity and abundance (Wu et al. 2005; Bucciarelli et al. 2014; Nunes et al. 2019). Habitat restoration projects in urban and suburban areas must focus on the removal of exotics (Nunes et al. 2019). Wetlands must focus on connectivity that facilitates the dispersal of native species and retention of shallow, ephemeral water bodies that support more biodiversity than deeper, permanent water bodies (Hamer 2022a, b).

7.9 Research Gaps and Future Directions

Urbanisation will progress with the times. To evolve conservation strategies specific for urbanised and/or rapidly urbanising landscapes, especially for understudied regions like most tropical countries and developing countries like India, efforts are immediately required. Most tropical countries have high human population densities (Tan et al. 2022). Natural processes and biodiversity, when integrated into the place where people live and work, improve the quality of life and make urban environments more liveable for amphibians (Miller and Hobbs 2002; Boyer and Polasky 2004; Pillsbury and Miller 2008).

New studies must focus on understanding the habitat use and occupancy by amphibians from an urban-to-rural gradient perspective. Regional, long-term studies should explore whether amphibian assemblages sustain themselves in urban and peri-urban areas. Telemetry studies at local scales can address urban-habitat-use patterns at meso- and micro-scales and can help answer questions like 'What constitutes preferred habitat?' From the data available, roads appear to be a major mortality factor for adult frogs. While population models can help predict rates of population decline, focused studies on road mortality that address various population parameters are needed. Research must also examine how habitat quality (including roads) allows gene flow across urbanising landscapes. Pesticides and endocrine-disrupting chemicals continue to loom as the least-studied factors that may be of greatest importance.

There is also a need to assess the level of ecosystem services provided by amphibians in urban and peri-urban areas. Giving value to the functions provided by the amphibians will help in influencing policymakers and managers to put management action in place, even in developed and developing cities. Caecilians are the most understudied amphibian group. There are no long-term population studies that are carried out because caecilians are rarely seen and hard to locate (San Mauro et al. 2004). Long-term quantitative data on population sizes and other population estimating parameters for amphibians are rare from across the globe. New studies are also needed to address mechanisms of amphibian movement, as studies from China (Wei et al. 2021) show the significance of urban corridor networks. Demographic data can help reveal how life-stage-specific survival and natality rates affect population resilience to disturbance events (i.e., types and degree of urbanisation).

In most urbanised landscapes, there are two major strategies towards conservation–restoration of the built environment or preservation of undeveloped land to promote biodiversity conservation (McKinney 2002; Ettinger et al. 2021). Both these means of conservation are costly due to the increasing costs of real estate. However, a few key recommendations that are in line with other studies like Goosem et al. (2008), Pillsbury and Miller (2008), Hamer and McDonnell (2008), Parris et al. (2009) and Scheffers et al. (2012) that can be of immediate effect are as follows:

1. Recognition and conservation of existing amphibian microhabitats
2. Making urban water bodies amphibian friendly, allowing native vegetation to grow and terrestrial land and shallow waters to thrive, limiting pollution and run-offs within urban water bodies, and also periodically checking pollution levels and maintaining water quality
3. Restoring connectivity in areas of hyper-diversity of amphibians to maintain gene flow, species diversity and richness
4. Traffic noise amelioration, at least in areas and waterbodies with active, breeding amphibian populations
5. Canopy cover maintenance and revegetation with native vegetation wherever required
6. Control on artificial lights, especially bright streetlights and commercial lights (e.g., banners), that must have time limits to promote effective natural darkness and implementing time limits around urban water bodies and parks
7. Organising educational events in schools/colleges and awareness campaigns to inform the public about the role of amphibians and conservation in urban areas

and encouraging local communities, schools and stakeholders in citizen science projects and habitat restoration to participate by conducting periodic 'amphibian walks'

8. Limiting the use of pesticides, fertilisers and other harmful chemicals in urban areas that can contaminate water bodies and affect amphibian populations, stopping the domestic sewage and industrial effluents from entering water bodies such as urban lakes and ponds, and encouraging organic and environmentally friendly practices in gardening and landscaping

References

Alroy J (2015) Current extinction rates of reptiles and amphibians. Proc Natl Acad Sci U S A 112(42):13003–13008. https://doi.org/10.1073/pnas.1508681112

Anđelković M, Bogdanović N (2022) Amphibian and reptile road mortality in special nature reserve Obedska bara, Serbia. Animals 12(5):561. https://doi.org/10.3390/ani12050561

Baillie JE, Collen B, Amin R, Akcakaya HR, Butchart SH, Brummitt N et al (2008) Toward monitoring global biodiversity. Conserv Lett 1(1):18–26

Baker BJ, Richardson JML (2006) The effect of artificial light on male breeding-season behaviour in green frogs, *Rana clamitans melanota*. Can J Zool 84(10):1528–1532

Baker JMR, Waights V (1994) The effects of nitrate on tadpoles of the tree frog (*Litoria caerulea*). Herpetol J 4(3):106–108

Barrett K, Guyer C (2008) Differential responses of amphibians and reptiles in riparian and stream habitats to land use disturbances in western Georgia, USA. Biol Conserv 141(9):2290–2300

Bee MA, Swanson EM (2007) Auditory masking of anuran advertisement calls by road traffic noise. Anim Behav 74(6):1765–1776

Blickley JL, Patricelli GL (2010) Impacts of anthropogenic noise on wildlife: research priorities for the development of standards and mitigation. J Int Wildl Law Policy 13(4):274–292

Boyer T, Polasky S (2004) Valuing urban wetlands: a review of non-market valuation studies. Wetlands 24(4):744–755

Brumm H, Slabbekoorn H (2005) Acoustic communication in noise. Adv Stud Behav 35:151–209

Bucciarelli GM, Blaustein AR, Garcia TS, Kats LB (2014) Invasion complexities: the diverse impacts of non-native species on amphibians. Copeia 2014(4):611–632

Buchanan BW (2006) Observed and potential effects of artificial night lighting on anuran amphibians. In: Ecological consequences of artificial night lighting, pp 192–220

Bugnot AB, Hose GC, Walsh CJ, Floerl O, French K, Dafforn KA et al (2019) Urban impacts across realms: making the case for inter-realm monitoring and management. Sci Total Environ 648:711–719

Bunnell JF, Ciraolo JL (2010) The potential impact of simulated ground-water withdrawals on the oviposition, larval development, and metamorphosis of pond-breeding frogs. Wetl Ecol Manag 18:495–509

Capinha C, Seebens H, Cassey P, García-Díaz P, Lenzner B, Mang T et al (2017) Diversity, biogeography and the global flows of alien amphibians and reptiles. Divers Distrib 23(11):1313–1322

Carey C, Alexander MA (2003) Climate change and amphibian declines: is there a link? Divers Distrib 9(2):111–121

Carr LW, Fahrig L (2001) Effect of road traffic on two amphibian species of differing vagility. Conserv Biol 15(4):1071–1078

Catenazzi A, von May R (2021) Systematics and conservation of Neotropical amphibians and reptiles. Diversity 2021(13):45. https://doi.org/10.3390/d13020045

Cinzano P, Falchi F, Elvidge CD (2001) The first world atlas of the artificial night sky brightness. Mon Not R Astron Soc 328(3):689–707

Collins SJ, Russell RW (2009) Toxicity of road salt to Nova Scotia amphibians. Environ Pollut 157(1):320–324

Collins JP, Storfer A (2003) Global amphibian declines: sorting the hypotheses. Divers Distrib 9:89–98

Cooke AS, Sparks TH (2004) Population declines of common toads (Bufo bufo): the contribution of road traffic and monitoring value of casualty counts. Herpetol Bull 88:13–26

Cunnington GM, Fahrig L (2010) Plasticity in the vocalizations of anurans in response to traffic noise. Acta Oecol 36(5):463–470

Cunnington GM, Fahrig L (2013) Mate attraction by male anurans in the presence of traffic noise. Anim Conserv 16(3):275–285

Cushman SA (2006) Effects of habitat loss and fragmentation on amphibians: a review and prospectus. Biol Conserv 128(2):231–240

Daszak P, Strieby A, Cunningham AA, Longcore JE, Brown CC, Porter D (2004) Experimental evidence that the bullfrog (Rana catesbeiana) is a potential carrier of chytridiomycosis, an emerging fungal disease of amphibians. Herpetol J 14:201–208

Delis PR, Mushinsky HR, McCoy ED (1996) Decline of some west-central Florida anuran populations in response to habitat degradation. Biodivers Conserv 5:1579–1595

Dias KS, Dosso ES, Hall AS, Schuch AP, Tozetti AM (2019) Ecological light pollution affects anuran calling season, daily calling period, and sensitivity to light in natural Brazilian wetlands. Sci Nat 106:1–10

Eigenbrod F, Hecna SJ, Fahrig L (2008) The relative effects of road traffic and forest cover on anuran populations. Biol Conserv 141(1):35–46

Elzanowski A, Ciesiołkiewicz J, Kaczor M, Radwańska J, Urban R (2009) Amphibian road mortality in Europe: a meta-analysis with new data from Poland. Eur J Wildl Res 55:33–43

Ettinger AK, Buhle ER, Feist BE, Howe E, Spromberg JA, Scholz NL, Levin PS (2021) Prioritizing conservation actions in urbanizing landscapes. Sci Rep 11(1):818

Fahrig L, Pedlar JH, Pope SE, Taylor PD, Wegner JF (1995) Effect of road traffic on amphibian density. Biol Conserv 73(3):177–182

Ferreira M, Beja P (2013) Mediterranean amphibians and the loss of temporary ponds: are there alternative breeding habitats? Biol Conserv 165:179–186

Ficetola GF, De Bernardi F (2004) Amphibians in a human-dominated landscape: the community structure is related to habitat features and isolation. Biol Conserv 119(2):219–230

Garcia-Gonzalez C, Campo D, Pola IG, Garcia-Vazquez E (2012) Rural road networks as barriers to gene flow for amphibians: species-dependent mitigation by traffic calming. Landsc Urban Plan 104(2):171–180

Gaston KJ, Bennie J, Davie TW, Hopkins J (2013) The ecological impacts of nighttime light pollution: a mechanistic appraisal. Biol Rev 88(4):912–927

Georgiev MI, Abrashev R, Krumova E, Demirevska K, Ilieva M, Angelova M (2009) Rosmarinic acid and antioxidant enzyme activities in Lavandula vera MM cell suspension culture: a comparative study. Appl Biochem Biotechnol 159:415–425

Georgiev D, Petrova S, Gecheva G, Velcheva I, Tsekov A, Yancheva V et al (2015) Freshwater habitats in Plovdiv town and its surroundings and their importance for the biodiversity. J Biosci Biotechnol 4(2):139–148

Gerhardt HC, Huber F (2003) Acoustic communication in insects and anurans: common problems and diverse solutions. Bibliovault OAI Repository, the University of Chicago Press, p 114. https://doi.org/10.1121/1.1591773

Gibbs J, Reed J (2007) Population and genetic linkages of vernal pool-associated amphibians. In: Science and conservation of Vernal Pools in Northeastern North America, pp 149–167. Research gate. https://doi.org/10.1201/9781420005394.ch8

Gibbs JP, Shriver WG (2005) Can road mortality limit populations of pool-breeding amphibians? Wetl Ecol Manag 13(3):281–289. https://doi.org/10.1007/s11273-004-7522-9

Glista DJ, DeVault TL, DeWoody JA (2008) Vertebrate road mortality predominantly impacts amphibians. Herpetol Conserv Biol 3(1):77–87

González-Bernal E, Greenlees MJ, Brown GP, Shine R (2016) Toads in the backyard: why do invasive cane toads (*Rhinella marina*) prefer buildings to bushland? Popul Ecol 58:293–302

Goosem M, Searle L, Pohlman C (2008) Assessing the habitat quality of remnant and riparian vegetation. In: Marine and tropical sciences research facility annual conference, pp 36–40

Grace MK, Noss RF (2018) Evidence for selective avoidance of traffic noise by anuran amphibians. Anim Conserv 21:343–351. https://doi.org/10.1111/acv.12400

Guzy JC, Campbell TS, Campbell KR (2006) Effects of hydrological alterations on frog and toad populations at Morris Bridge Wellfield, Hillsborough County, Florida. Florida Sci 69(4):276–287

Hamer AJ (2022a) Exotic predatory fish reduce amphibian reproduction at wetlands in an urbanising landscape. Hydrobiologia 849:121–139. https://doi.org/10.1007/s10750-021-04716-5

Hamer AJ (2022b) A multi-scale, multi-species approach highlights the importance of urban greenspace and pond design for amphibian communities. Urban Ecosyst 25:393–409. https://doi.org/10.1007/s11252-021-01162-y

Hamer AJ, McDonnell MJ (2008) Amphibian ecology and conservation in the urbanising world: a review. Biol Conserv 141(10):2432–2449. https://doi.org/10.1016/j.biocon.2008.07.020

Hamer AJ, Parris KM (2011) Local and landscape determinants of amphibian communities in urban ponds. Ecol Appl 21(2):378–390

Haycock NE, Burt TP (1993) The sensitivity of rivers to nitrate leaching: the effectiveness of near-stream land as a nutrient retention zone. pp 261–272

Hebbar P, Ravikanth G, Aravind NA (2019) A review on the conservation genetic studies of Indian amphibians and their implications on developing strategies for conservation. J Genet 98(5):114. https://doi.org/10.1007/s12041-019-1159-1

Hecnar (1995) Acute and chronic toxicity of ammonium nitrate fertilizer to amphibians from southern Ontario. https://setac.onlinelibrary.wiley.com/doi/abs/10.1002/etc.5620141217

Hels T, Buchwald E (2001) The effect of road kills on amphibian populations. Biol Conserv 99(3):331–340. https://doi.org/10.1016/S0006-3207(00)00215-9

Herrera-Montes MI, Aide TM (2011) Impacts of traffic noise on anuran and bird communities. Urban Ecosyst 14(3):415–427. https://doi.org/10.1007/s11252-011-0158-7

Herrington RE, Larsen JH (1985) Current status, habitat requirements and management of the Larch Mountain salamander *Plethodon larselli* burns. Biol Conserv 34(2):169–179. https://doi.org/10.1016/0006-3207(85)90107-7

Hintz W, Fay L, Relyea R (2021) Road salts, human safety, and the rising salinity of our fresh waters. Front Ecol Environ 20:22. https://doi.org/10.1002/fee.2433

Hocking D, Babbitt K (2014) Amphibian contributions to ecosystem services. Herpetol Conserv Biol 9:1–17

Hof C, Araújo M, Jetz W, Rahbek C (2011) Additive threats from pathogens, climate and land-use change for global amphibian diversity. Nature 19(480):516–519. https://doi.org/10.1038/nature10650

Holderegger R, Di Giulio M (2010) The genetic effects of roads: a review of empirical evidence. Basic Appl Ecol 11(6):522–531. https://doi.org/10.1016/j.baae.2010.06.006

IPBES (2019) Global assessment report on biodiversity and ecosystem services of the Intergovernmental Science-Policy Platform on Biodiversity and Ecosystem Services. In: Brondizio ES, Settele J, Díaz S, Ngo HT (eds). IPBES secretariat. Bonn, Germany. 1148p. https://doi.org/10.5281/zenodo.3831673

Jaarsma R (1997) Approaches for the planning of rural road networks according to sustainable land use planning. Landsc Urban Plan 39(12):47–54. https://doi.org/10.1016/S0169-2046(97)00067-4

Jaarsma C, Willems G (2002) Reducing habitat fragmentation by minor rural roads through traffic calming. Landsc Urban Plan 58:125–135. https://doi.org/10.1016/S0169-2046(01)00215-8

Johnson PTJ, Lunde KB, Ritchie EG, Reaser JK, Launer AE (2001) Morphological abnormality patterns in a California amphibian community. Herpetologica 57(3):336–352

Kaiser K, Devito J, Jones CG, Marentes A, Perez R, Umeh L, Weickum RM, McGovern KE, Wilson EH, Saltzman W (2015) Effects of anthropogenic noise on endocrine and reproductive function in White's treefrog, Litoria caerulea. Conserv Physiol 3(1):61. https://doi.org/10.1093/conphys/cou061. PMID: 27293682; PMCID: PMC4778486

Kambourova N, Koshev Y, Popgeorgiev G, Ragiov D, Pavlova M, Mollov I, Nedyalkov N (2012) Effect of traffic on mortality of amphibians, reptiles, birds and mammals on two types of roads between Pazardzhik and Plovdiv region (Bulgaria)—preliminary results. Acta Zool Bulg 1(01):57–67

Karraker NE, Gibbs JP, Vonesh JR (2008) Impacts of road deicing salt on the demography of vernal pool-breeding amphibians. Ecol Appl 18(3):724–734

Kats LB, Ferrer RP (2003) Alien predators and amphibian declines: review of two decades of science and the transition to conservation. Divers Distrib 9(2):99–110. https://doi.org/10.1046/j.1472-4642.2003.00013.x

Kentula ME, Gwin SE, Pierson SM (2004) Tracking changes in wetlands with urbanization: sixteen years of experience in Portland, Oregon, USA. Wetlands 24(4):734–743

Kiesecker JM, Blaustein AR, Belden LK (2001) Complex causes of amphibian population declines. Nature 410(6829):681–684

King KC, McLaughlin JD, Gendron AD, Pauli BD, Giroux I, Rondeau B et al (2007) Impacts of agriculture on the parasite communities of northern leopard frogs (Rana pipiens) in southern Quebec, Canada. Parasitology 134(14):2063–2080

Knapp RA, Matthews KR (2000) Non-native fish introductions and the decline of the mountain yellow-legged frog from within protected areas. Conserv Biol 14(2):428–438

Komine H, Fukasawa K, Akasaka M, Watari Y, Iwai N, Kaji K (2020) Rapid behavioural responses of native frogs caused by past predation pressure from invasive mongooses. J Zool 310(2):126–134

Langevelde F, Dooremalen C, Jaarsma CF (2009) Traffic mortality and the role of minor roads. J Environ Manag 90(1):660–667

LeClair G, Chatfield M, Frederick CA (2021) Influence of the COVID-19 pandemic on amphibian road mortality. Conserv Sci Pract 3(11):e535

Lemckert F, Mahony M (2004) Two hundred and ten years looking for the giant burrowing frog. Aust Zool 32(4):597–604. https://doi.org/10.7882/AZ.2004.005

Lengagne T (2008) Traffic noise affects communication behaviour in a breeding anuran, Hyla arborea. Biol Conserv 141(8):2023–2031

Longcore T, Rich C (2004) Ecological light pollution. Front Ecol Environ 2(4):191–198

Lucey KJ, Goolsby DA (1993) Effects of climatic variations over 11 years on nitrate-nitrogen concentrations in the Raccoon River, Iowa. J Environ Qual 22(1):38–46

Marsh DM (2001) Fluctuations in amphibian populations: a meta-analysis. Biol Conserv 101(3):327–335. https://doi.org/10.1016/S0006-3207(01)00076-3

Marsh DM, Milam GS, Gorham NP, Beckman NG (2005) Forest roads as partial barriers to terrestrial salamander movement. Conserv Biol 19:2004–2008. https://doi.org/10.1111/j.1523-1739.2005.00238.x

Marzluff JM, Ewing K (2008) Restoration of fragmented landscapes for the conservation of birds: a general framework and specific recommendations for urbanizing landscapes. In: Urban ecology: an international perspective on the interaction between humans and nature, pp 739–755

Mazerolle MJ, Huot M, Gravel M (2005) Behavior of amphibians on the road in response to car traffic. Herpetologica 61(4):380–388

McKinney ML (2002) Urbanization, biodiversity, and conservation: the impacts of urbanization on native species are poorly studied, but educating a highly urbanized human population about these impacts can greatly improve species conservation in all ecosystems. Bioscience 52(10):883–890

Measey GJ, Tolley KA (2011) Investigating the cause of the disjunct distribution of Amietophrynus pantherinus, the endangered South African western leopard toad. Conserv Genet 12:61–70

Measey GJ, Rödder D, Green SL, Kobayashi R, Lillo F, Lobos G et al (2012) Ongoing invasions of the African clawed frog, Xenopus laevis: a global review. Biol Invasions 14:2255–2270

Measey J, Davies SJ, Vimercati G, Rebelo A, Schmidt W, Turner A (2017) Invasive amphibians in southern Africa: a review of invasion pathways. Bothalia 47(2):1–12

Melotto A, Ficetola GF, Alari E, Romagnoli S, Manenti R (2021) Visual recognition and coevolutionary history drive responses of amphibians to an invasive predator. Behav Ecol 32(6):1352–1362

Miller JR, Hobbs RJ (2002) Conservation where people live and work. Conserv Biol 16(2):330–337

Mokhatla MM, Measey GJ, Chimimba CT, Van Rensburg BJ (2012) A biogeographical assessment of anthropogenic threats to areas where different frog breeding groups occur in South Africa: implications for anuran conservation. Divers Distrib 18(5):470–480

Mora M, Pons DJ, Peñafiel-Ricaurte A, Alvarado-Rybak M, Lebuy S, Soto-Azat C (2019) High abundance of invasive African clawed frog *Xenopus laevis* in Chile: challenges for their control and updated invasive distribution. Manag Biol Invasion 10(2):377–388. https://doi.org/10.3391/mbi.2019.10.2.11

Murphy CG (1999) Nightly timing of chorusing by male barking treefrogs (Hyla gratiosa): the influence of female arrival and energy. Copeia 1999:333–347

Nicoara A, Nicoara M (2008) Surveying an isolated population of Pelobates fuscus (Anura, Amphibia) in the urban area of the town of Iasi (Romania). Oceanol Hydrobiol Stud 37(1):61–69

Nowakowski AJ, Thompson ME, Donnelly MA, Todd BD (2017) Amphibian sensitivity to habitat modification is associated with population trends and species traits. Glob Ecol Biogeogr 26(6):700–712. https://doi.org/10.1111/geb.12571

Nunes AL, Fill JM, Davies SJ, Louw M, Rebelo AD, Thorp CJ, Measey J (2019) A global meta-analysis of the ecological impacts of alien species on native amphibians. Proc R Soc B 286(1897):20182528

Padhye AD, Mahabaleshwarkar M, Ghate HV (2002) An overview of amphibian fauna of Pune District with special reference to their status in and around Pune City. Zoos' Print J 17(4):757–763

Parris KM, Velik-Lord M, North JM (2009) Frogs call at a higher pitch in traffic noise. Ecol Soc 14(1):25

Paul MJ, Meyer JL (2001) Streams in the urban landscape. Annu Rev Ecol Syst 32(1):333–365

Pilliod DS, Goldberg CS, Arkle RS, Waits LP (2014) Factors influencing detection of eDNA from a stream-dwelling amphibian. Mol Ecol Resour 14(1):109–116

Pillsbury FC, Miller JR (2008) Habitat and landscape characteristics underlying anuran community structure along an urban–rural gradient. Ecol Appl 18(5):1107–1118

Pinero-Rodríguez MJ, Fernández-Zamudio R, Arribas R, Gomez-Mestre I, Díaz-Paniagua C (2021) The invasive aquatic fern *Azolla filiculoides* negatively impacts water quality, aquatic vegetation and amphibian larvae in Mediterranean environments. Biol Invasions 23:755–769

Pinto FA, Clevenger AP, Grilo C (2020a) Effects of roads on terrestrial vertebrate species in Latin America. Environ Impact Assess Rev 81:106337

Pinto F, Clevenger A, Grilo C (2020b) Effects of roads on terrestrial vertebrate species. Environmental Impact Assessment Review. Environmental Science. Published 1 March 2020. https://doi.org/10.1016/j.eiar.2019.106337. Corpus ID: 210718286

Price SJ, Howe RW, Hanowski JM, Regal RR, Niemi GJ, Smith CR (2007) Are anurans of Great Lakes coastal wetlands reliable indicators of ecological condition? J Great Lakes Res 33:211–223

Reeves MK, Dolph CL, Zimmer H, Tjeerdema RS, Trust KA (2008) Road proximity increases risk of skeletal abnormalities in wood frogs from National Wildlife Refuges in Alaska. Environ Health Perspect 116(8):1009–1014

Riley SP, Busteed GT, Kats LB, Vandergon TL, Lee LF, Dagit RG et al (2005) Effects of urbanization on the distribution and abundance of amphibians and invasive species in southern California streams. Conserv Biol 19(6):1894–1907

Rouse JD, Bishop CA, Struger J (1999) Nitrogen pollution: an assessment of its threat to amphibian survival. Environ Health Perspect 107(10):799–803

Rubbo M, Kiesecker J (2004) Leaf litter composition and community structure: translating regional species changes into local dynamics. Ecology 85(9):2519–2525. https://doi.org/10.1890/03-0653

Ryan MJ (1986) Factors influencing the evolution of acoustic communication: biological constraints. Brain Behav Evol 28(1–3):70–82

San Mauro D, Gower DJ, Oommen OV, Wilkinson M, Zardoya R (2004) Phylogeny of caecilian amphibians (Gymnophiona) based on complete mitochondrial genomes and nuclear RAG1. Mol Phylogenet Evol 33(2):413–427

Scheffers BR, Joppa LN, Pimm SL, Laurance WF (2012) What we know and don't know about Earth's missing biodiversity. Trends Ecol Evol 27(9):501–510

Semlitsch RD, Bodie JR (2003) Biological criteria for buffer zones around wetlands and riparian habitats for amphibians and reptiles. Conserv Biol 17(5):1219–1228

Senzaki M, Kadoya T, Francis CD, Ishiyama N, Nakamura F (2018) Suffering in receivers: negative effects of noise persist regardless of experience in female anurans. Funct Ecol 32(8):2054–2064

Shackleton CM, Cilliers SS, du Toit MJ, Davoren E (2021) The need for an urban ecology of the Global South. Springer, Cham, pp 1–26

Shidemantle G, Blackwood J, Horn K, Velasquez I, Ronan E, Reinke B, Hua J (2022) The morphological effects of artificial light at night on amphibian predators and prey are masked at the community level. Environ Pollut 30(8):119604

Simmons AM, Narins PM (2018) Effects of anthropogenic noise on amphibians and reptiles. In: Effects of anthropogenic noise on animals, pp 179–208

Skelly DK, Bolden SR, Holland MP, Freidenburg LK, Freidenfelds NA, Malcolm TR et al (2006) Urbanization and disease in amphibians. In: Disease ecology: community structure and pathogen dynamics, pp 153–167

Smallbone LT, Luck GW, Wassens S (2011) Anuran species in urban landscapes: relationships with biophysical, built environment and socio-economic factors. Landsc Urban Plan 101(1):43–51

Sugden D, Davidson K, Hough KA, Teh MT (2004) Melatonin, melatonin receptors and melanophores: a moving story. Pigment Cell Res 17(5):454–460

Sullivan B (2008) The ecology and behavior of amphibians, Kentwood D. Wells. University of Chicago Press, Chicago, p 1148

Tan YL, Chen JE, Yiew TH, Habibullah MS (2022) Habitat change and biodiversity loss in South and Southeast Asian countries. Environ Sci Pollut Res 29(42):63260–63276

Tennessen JB, Parks SE, Langkilde T (2014) Traffic noise causes physiological stress and impairs breeding migration behaviour in frogs. Conserv Physiol 2(1):32–39

Touzot M, Teulier L, Lengagne T, Secondi J, Théry M, Libourel PA et al (2019) Artificial light at night disturbs the activity and energy allocation of the common toad during the breeding period. Conserv Physiol 7(1):02

Trenham PC, Shaffer HB (2005) Amphibian upland habitat use and its consequences for population viability. Ecol Appl 15(4):1158–1168

Turner A, Wassens S, Heard G, Peters A (2021) Temperature as a driver of the pathogenicity and virulence of amphibian chytrid fungus *Batrachochytrium dendrobatidis*: a systematic review. J Wildl Dis 57(3):477–494

Vickers CR, Harris LD, Swindel BF (1985) Changes in herpetofauna resulting from ditching of cypress ponds in coastal plains flatwoods. For Ecol Manag 11(1–2):17–29

Waights V, Baker J (1993) The effect of sodium nitrate on the growth and survival of toad tadpoles (Bufo bufo) in the laboratory. Herpetol J 3(4):147–148

Wei X, Huang M, Yue Q, Ma S, Li B, Mu Z et al (2021) Long-term urbanization impacts the eastern golden frog (*Pelophylax plancyi*) in Shanghai City: demographic history, genetic structure, and implications for amphibian conservation in intensively urbanizing environments. Evol Appl 14(1):117–135

Windmiller B, Calhoun A (2007) 12 Conserving vernal pool wildlife in urbanizing landscapes. https://doi.org/10.1201/9781420005394.ch12

Wise S (2007) Studying the ecological impacts of light pollution on wildlife: amphibians as models. In: Cipriano M, Jafar J (eds) Starlight: a common heritage. Cambridge University Press, Cambridge, pp 209–218

Wollerman L, Wiley RH (2002) Background noise from a natural chorus alters female discrimination of male calls in a neotropical frog. Anim Behav 63(1):15–22

Wu Z, Li Y, Wang Y, Adams MJ (2005) Diet of introduced bullfrogs (*Rana catesbeiana*): predation on and diet overlap with native frogs on Daishan Island, China. J Herpetol 39(4):668–674

Zaffaroni-Caorsi V, Both C, Márquez R, Llusia D, Narins P, Debon M, Borges-Martins M (2023) Effects of anthropogenic noise on anuran amphibians. Bioacoustics 32(1):90–120

Zhang W, Li B, Shu X, Pei E, Yuan X, Sun Y et al (2016) Responses of anuran communities to rapid urban growth in Shanghai, China. Urban For Urban Green 20:365–374

8 Behavioral Responses of Olive Ridley Sea Turtles (*Lepidochelys olivacea*) to Beach Cleaning and Conservation Measures Off the West Coast of India

Akshay Kumar Chakravarthy and Ashutosh Dey

Abstract

This chapter reports numerical responses of olive ridley sea turtles (*Lepidochelys olivacea*) to the 2 km long stretch of cleaned beaches, Kodi and M Kodi, Kundapura, Udupi, Karnataka, on the west coast of India. Annually, very few (4–10 turtles at few locations) turtles were found along the 320 km stretch of the west coast of India (12.87°N, 74.88°E, 22 m amsl). As part of the Swachhata Action Plan of the Government of India, a cleaning and awareness campaign by the Environmental Management and Policy Research Institute (EMPRI) adopted five beaches for 5 months (November 2020–March 2021) and collected 65 tons of garbage. As a result, for the first time, 3600 olive ridley's eggs were counted without disturbing the nest and eggs at Kodi and M Kodi beaches, implying a greater number of eggs present than the one estimated above because the eggs were also underneath the sand. A dozen, nylon fishing net cages (2 m^3) were erected on the beaches to protect turtle eggs from predators. Hatchlings of turtles were found crawling, orienting, and moving to seawaters from January 2021, indicating that beaches are preferred nesting sites. Observations showed that olive ridley's active foraging in coastal waters confirms that Kodi and M Kodi beaches served as nesting habitats for the turtles. Even small nesting grounds are critical for species' survival because large nesting congregations are rare and have become rarer. Turtles play a unique role in the marine food web and ecosystem; populations are declining and are categorized as vulnerable by the

A. Kumar Chakravarthy (✉)
Division of Entomology and Nematology, Indian Institute of Horticultural Research, Bengaluru, Karnataka, India

A. Dey
Society for Science and Technology Applications (SSTA), Bengaluru, Karnataka, India

Environmental Management and Policy Research Institute, Bengaluru, Karnataka, India

S. Sridhara et al. (eds.), *Animal Behavior in the Tropics*,
https://doi.org/10.1007/978-981-97-7324-4_8

International Union for Conservation of Nature and Natural Resources (IUCN) Red List. However, olive ridley and the remaining four species of turtles are categorized as endangered because of their few remaining nesting sites in the Indian subcontinent. Extending the cleaning campaign to more beaches will help in long-term conservation of turtles and marine biodiversity. The tools and techniques deployed to gain insight into behavior and ecology of sea turtles needs refinement.

Keywords

Olive ridley turtle · West coast of India · Cleaning and conservation campaign · Engendered species · Marine biodiversity

8.1 Introduction

India has more than 8000 km of coastline, and five species of sea turtles are found to occur in the coastal waters. The species are olive ridley sea turtle (*Lepidochelys olivacea*), green sea turtle (*Chelonia mydas*), hawksbill sea turtle (*Eretmochelys imbricata*), loggerhead sea turtle (*Caretta caretta*), and the leatherback (*Dermochelys coriacea*). Of these, green sea turtles according to the International Union for Conservation of Nature and Natural Resources (IUCN) Red List are endangered. The hawksbill sea turtle is critically endangered, and the other three are vulnerable. Except for the loggerhead, the other four species nest along the Indian west coast. However, all the species of turtles found in India are endangered. All are protected by law under Schedule I of the Indian Wildlife Protection Act 1972. The Convention of International Trade in Engendered Species of Wild Fauna and Flora (CITES) and the Convection on the Conservation Migratory Species of Wild Animals (CMS) also prohibit trade in turtle products, placing turtles in Appendix I and Appendices I and II of the conventions, respectively. Sea turtles have a low survival rate, often slaughtered in thousands for meat; eggs are illegally harvested and are abandoned as bycatch; and mortality of turtles is generally high due to coastal fisheries that do not use turtle excluder devices (TEDs) in their nets. Turtle populations worldwide are rapidly declining. Human activities and increasing carbon emissions are having serious impacts on turtle survival. Turtles are finding it hard to cope with sea acidification and rising temperatures. Tropical and subtropical coastal areas are considered prime areas for tourism but are also important sea turtle nesting grounds worldwide. So efforts to protect turtles' major beaches are vital. Turtles are critically important for marine ecosystems as scavengers, herbivores, and carnivores and contribute significant biomass to the ecosystems. Turtles break down the energy of plant materials and convert them into proteins. Turtles feed on seaweeds and sea grasses and regulate their impacts on marine ecosystems.

8.1.1 Olive Ridley Turtle

Olive ridley gets its name from the olive green color of its heart-shaped shell, which shares morphological similarities only with the smallest Kemp's ridley (*Lepidochelys kempii*) turtle. Olive ridleys are widely known for "arribadas" or man nesting events that occur only in a few locals in Mexico, Costa Rica, and India (Pritchand 2007). However, there is little knowledge of their distribution and abundance in the Western Pacific and Southeast Asia (Bernando and Plotkin 2007). The olive ridley turtles are the most abundant of all turtles found across the tropical regions of the Antarctica, Pacific, and Indian Oceans spread across 80 countries (Farias et al. 2022). It is distributed along the coast of the North Pacific Ocean, North Atlantic Ocean, Indian Ocean, South Atlantic Ocean, and South Pacific Ocean (Fig. 8.1). This species exhibits one of the most spectacular natural phenomena with thousands of females coming out of the water to nest in a highly synchronized event on sandy beaches. Olive ridley is the most cosmopolitan of the seven sea turtle species. Worldwide olive ridleys have experienced a 30–50% decline, putting the population at risk and being considered endangered by IUCN. Sea turtles are important components of healthy oceans. Sea turtles play multiple roles as prey, consumers, hunters, competitors, as well as carriers or hosts for other organisms (Farias et al. 2022). Olive ridleys were earlier thought to overexploitation because of their abundance. This was not true. In the period from the 1960s to the 1980s, large-scale industrial extraction caused huge population crashes, as seen, for example, in Mexico (Chamboult et al. 2017). As per the Government of India's year-long Swacchtha Action Plan, the beach cleaning activity was initiated by Environmental Management and Policy Research Institute (EMPRI), Bangalore, with the regional directors, Udupi and Karwar of Aranya Jeevi matthu paristhithi Matthu Parisara under Karnataka Forest

Fig. 8.1 Garbage dumps at Kodi and M Kodi beaches

Department (KFD), Government of Karnataka, Bangalore. Two nongovernment organizations, namely, Clean kundapura and Iddya Beach Friends, also participated in the campaign. The activity began in November 2020 and concluded in March 2021. Following beaches along the Arabian Sea were adopted for cleaning: Dakshina Kannada district, Bengre and Iddya beaches; Udupi district, Kundapura taluk—Kodi; and M Kodi, Uttarakannada district-Gokarna beach. The beaches were selected on the basis of whether they come under the blue flag list, not leased by the district administration and potential beaches with more footfall. On average, 16 beach workers participated in the campaign daily and were provided with dress materials, gloves, caps, shoes, goggles, cleaning rakes, and bags to collect the garbage. Daily cleaning activity extended from 8:00 am to 6:00 pm. On some days, the cleaning activity continued until 9:00 pm. All the garbage bags were sent to the taluk municipality office for recycling (Fig. 8.1). The garbage collected was weighed and recorded daily at each beach. Observations were recorded on tourists and other animals visiting the beaches. EMPRI staff and regional directors coordinated and monitored the entire cleaning and awareness program besides recording observations on the activity of turtles on the beaches and in seawaters at edges.

From 22 January 2021, onwards, olive ridley Sea turtles were sighted at Kodi and M Kodi beaches (Fig. 8.2). The turtles were identified by their small head, rusty green olive carapace, and bony without ridges. Olive ridley looks remarkably like Kemp's ridley but has a deeper body and slightly upturned edges to its carapace. The adult turtles were sighted out of the seawaters at 8:00 pm onwards to the beaches. The trend of turtles' numbers visiting the beach is indicated in Fig. 8.3. Only the female turtles visit the beaches to lay eggs at night and return to see the same night. The egg-laying activity of turtles at Kodi and M Kodi beaches continued up to 3 March 2021 (Fig. 8.4). An estimated 35 adult turtles were sighted between 22 January and 3 March 2021. Pandav and Choudhury (2000) measured the population of olive ridley sea turtles along the Orissa coast. Long-term monitoring is essential for the detection of population trends. Seminoff (2004), Tripathy (2005) and Tripathy and Rajasekhar (2009) studied the ecology and conservation of olive ridley sea turtles in 2004 and 2005, respectively. The turtle eggs at the above two beaches were estimated at 3600 without disturbing the nests and eggs. To protect

Fig. 8.2 Adults reaching the seawater

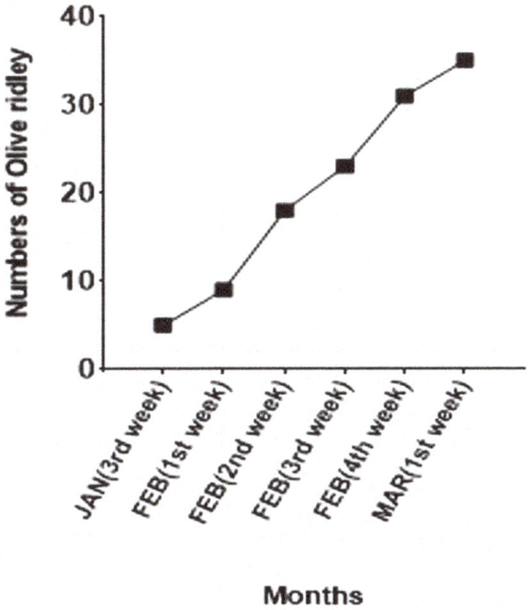

Fig. 8.3 Visiting pattern of olive ridley adults at Kodi and M Kodi beaches

Fig. 8.4 Eggs of olive ridley turtles

turtle eggs from predation and destruction at 2 km coastal stretch, a dozen nylon fishnet cages (2 m^3) were erected leaving space at the bottom (Figs. 8.5 and 8.6). This is because, at Kodi and M Kodi, predators—Brahminy kites (*Haliastur indus* and Jungle crows (*Corvus macrorhynchos*)—were recorded. The juveniles of turtles were observed returning to the seawaters from 23 March 2021 onwards. These turtles are known to nest in mass synchronized nests called arribadas (meaning in Spanish mass arrival). The hatchlings walk the last stretch themselves (Fig. 8.7) as

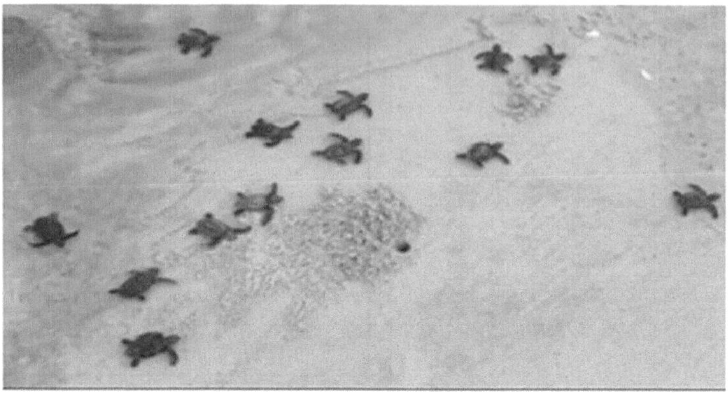

Fig. 8.5 Hatchlings returning to seawater

Fig. 8.6 Pattern of olive ridley hatchling to seawater at Kodi and M Kodi beaches

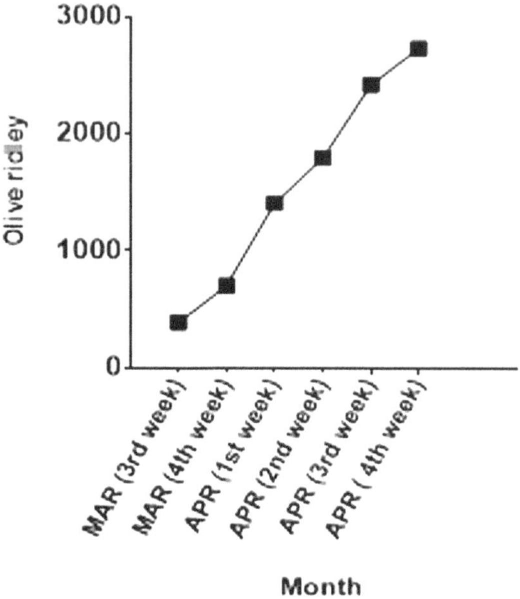

it is in these steps they geomagnetically imprint the sand on the belly, which helps them return to the same beach. The trend in the numbers of hatchlings returning to sea is shown in Fig. 8.5. Following cues of lunar cycles, tides, and weather conditions, the females began to arrive at beaches as per the officials of the Forest Department Government of Karnataka. This is the first time the turtles in large populations were found in Kundapura beaches. In view of the cleaning of beaches, the waters and beach habitat of the coast became favorable for the turtles to breed and nest. An average clutch size is over 110 eggs requiring an incubation period of 52–58 days. White eggs are laid in conical nests, about 0.5 m deep, dug using their

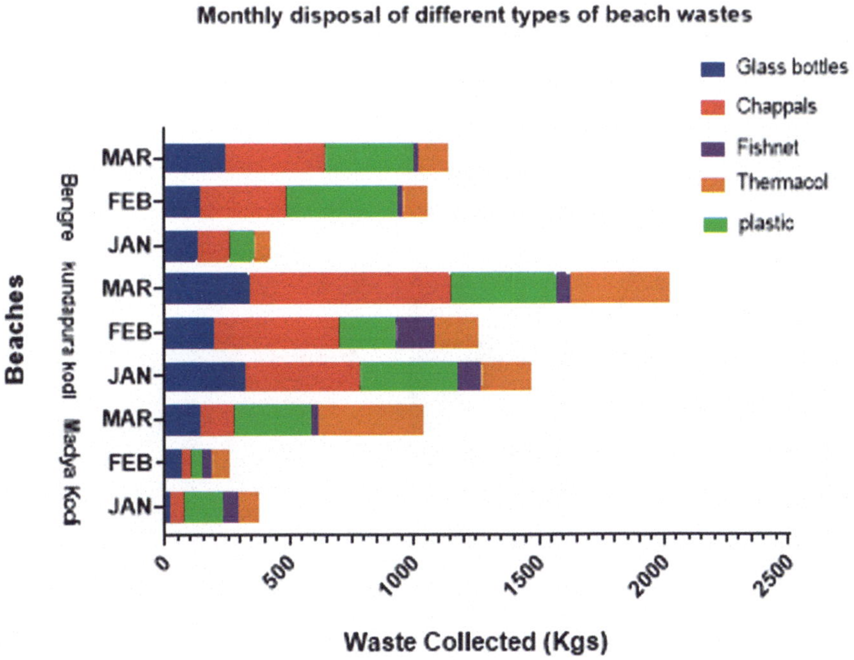

Fig. 8.7 Monthly disposal of different types of beach wastes

hind flippers, scooping out the sand. Instinct guides the adults to emerge from the sea and after laying eggs return to the sea.

Sasithitlu is an old and first nesting site near Mangalore sighted. There are 15 nesting sites of olive ridley along 320 km of coastline. Eggs relocated to avoid high tides and fungal infections. There was 78% success in hatching rate (Sanjeev 2024).

Garbage of 65 MT in 5 months comprising plastics, papers, metals, glass, textiles, rubber, wood, and plant waste was collected from five beaches and sent to the municipality office (Fig. 8.8). The tourist numbers before and during the cleaning initiative to the two beaches were very less (on an average, 5–10/day, $n = 25$). So the beach situation at Kodi and M Kodi after cleaning turned favorable for the turtles to nest. Although turtles forage in deep waters, they were sighted foraging at the surface waters near the beach indicating an abundant presence of food. The diet of olive ridley's comprises crustaceans—shrimps, crabs, mollusks, tunicates, and fish.

Adult olive ridleys are generally sighted with other three species of turtles, namely, green sea turtle, leatherback turtle, and hawksbill turtle along the west coast of India earlier at Apsarakonda beach at Kasargod, Mavinakurve beach at Honnavara and at Laval of Kumta taluk but in negligible numbers (2–4 adults/beach/year). Nesting of olive ridley's has been sighted only in a season (November to April) off the west coast of India in contrast to two seasons in a year in the Pacific and Atlantic Oceans.

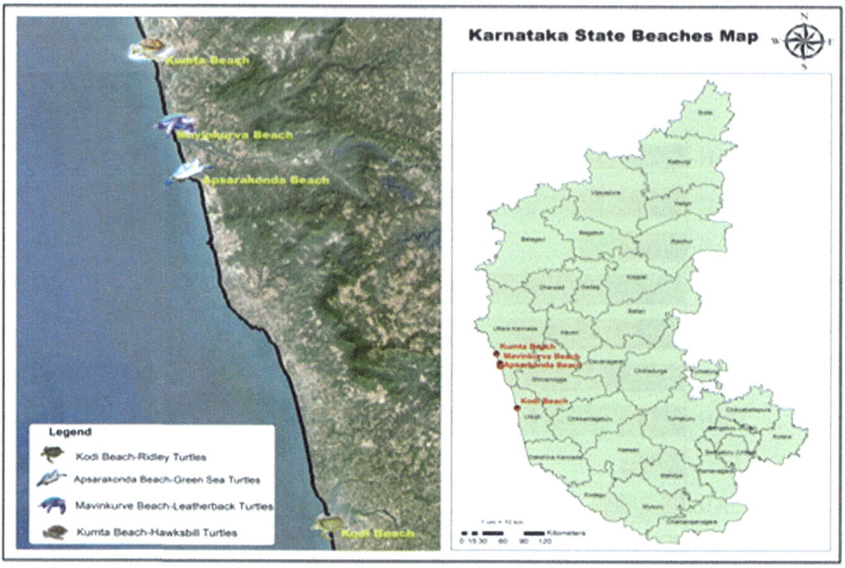

Fig. 8.8 Species of turtles cited along the west coast of India

8.2 Threats and Conservation

Olive ridleys are found along the coast of the North Pacific Ocean, North Atlantic Ocean, Indian Ocean, South Atlantic Ocean, and South Pacific Ocean.

The main threats facing turtles today are habitat deterioration and loss, wildlife trade, incidental captures, climate change, and pollution. Important conservation measures required for turtles are cleaning and protecting beaches, monitoring migration patterns of sea turtles, addressing impacts of climate change, sea turtle conservation needed to be integrated into coastal management policies, and promoting smart fishing activities. Tripathy and Rajasekhar (2009), studied nesting habitats of olive ridley sea turtles on the east coast of India. The workers made threat assessments and listed sand mining, beach armoring, artificial illumination, highways marine drives, exotic plantations, ports harbors, and jetties, offshore water threats that include pollution, fisheries, aquaculture, and tourism. The recent advances in conservation and management practices enable us to understand marine turtles' population boundaries and connectivity (Maxwell et al. 2019). Tagging of turtles regarding migration also suggests behavioral status and ecology (Guzman et al. 2019). Olive ridleys often forage via passive drifting rather than active swimming that shows they eat in same food web for many days. The trophic position of the olive ridley may be different than other large vertebrates. Despite these, olive ridley's open ocean trophic ecology has not been sufficiently investigated. One instance of action is worth mentioning. The Defence Research and Development Organization (DRDO) paused missile testing at Wheeler Island off the coast of Odisha during the mass nesting season of olive ridley sea turtles from January to March 2024 to ensure the endangered species is protected. Missile testing, noisy mechanized boats, and movement of people adversely impact the mass nesting and breeding of turtles at Wheeler Island, Odisha. About 5 lakh olive ridleys nested at Wheeler Island in 2023. The Indian Army and coast guard patrol the coast to keep trawlers and fishing boats from venturing close to the narrow strip of sand near bays and estuaries where turtles lay eggs, (Times of India 2023).

8.3 Tools and Techniques

The delineation of the inter-nesting habitat for gravid sea turtles is a crucial factor for their survival. Aggregation of turtles in large ground during the inter-nesting period exposes them to increased threats to their well-being. This is the time when the beaches on the west coast are used by tourists and fishermen. Olive ridleys have the capacity to adjust behaviorally to conditions of wetland they have the most generalist diet. However, the turbid water poses a threat to the turtle from the viewpoint of visibility and food. Chamboult et al. (2017) tracked individuals of olive ridleys by satellite tags recording the diving parameters and location of the feeding area. The use of satellite tracking to locate turtle movements and their habitat is crucial during the breeding season due to aggregation of turtles close to the beaches. Nesting makes turtles vulnerable in view of the high energy costs of reproduction and nesting activities (Scott et al. 2017; Pendoley et al. 2014). An evaluation of the home range of turtles is an essential tool for the survival of the endangered turtle species because it establishes the core area for activities of these sea animals. The satellite trackings has been successfully deployed for supporting conservation initiatives for leatherback, Kemp's ridley, loggershead, green hawkbill, and olive ridley turtles globally (Chamboult et al. 2017). A study by Pikesley et al. (2013) revealed that nine olive ridleys exhibited strong behavioral plasticity in dispersal and diving. Based on the geographical area and conditions in the sea, olive ridleys adopt either a capital breeding strategy in which they accumulate fat reserves at the foraging zone and then cease to feed during the breeding/nesting period or turtles continue to feed through the reproductive period (Colman et al. 2019). Chamboult et al. (2017) used satellite tracking to unravel information on habitat use, behavioral shifts, and conditions in the seas for olive ridley turtles. Chamboult et al. (2016) investigated the inter-nesting behavioral adjustments of Green Turtles (Chelomia mydas) in estuarine habitat in French Guyana. Satellite telemetry was used to track home range and diving behavior. Data showed that the turtles spent the maximum time resting in surface waters. The turtles intermittently exhibited short and shallow dives. The turtle species showed the ability to tolerate extreme environments. To restrict the loss of energy during the inter-nesting period, sea turtles have developed strategies. Gravid females rest near to nesting beaches, females rest for extended periods on beaches, or females rest on the seabed during 'U'-shaped dives. As listed, nesting has been observed in Caribbean hawksbill turtles (Chamboult et al. 2016). Stable carbon and nitrogen isotopic compositions have been used extensively to gain insight into foraging strategy, particularly for consumers whose behavior and dietary preferences are difficult to observe directly. In practice, measuring $\delta^{13}N$ values of tissues from multiple individuals within a population can provide insights into local nitrogen sources, trophic level, and niche space. The method consists of the following steps in order to execute bulk tissue analysis: (i) mass spectrometry and (ii) exploring δ^{13}NShift with size, gender, and isotopic niche area.

Acknowledgments The authors thank MoEF and CC Government of India for financial help, the director general of EMPRI and director of research of EMPRI for providing facilities for conducting the research and cooperation extended by regional directors, Udupi and Karwar of Karnataka Forest Department (KFD), and Clean Kundapura and Iddya Beach NGOs and volunteers.

References

Bernando J, Plotkin P (2007) An evolutionary perspective on the arribada phenomenon and reproductive behavioral polymorphism of olive ridley sea turtles (L. olivacea). In: Plotkin P (ed) Biology and conservation of ridley sea turtles. John Hopkins Univ. Press, Baltimare, pp 59–88

Lucie CPG, Benoît DT, Marc B, Laurent K, Virginie DR, Fabian B, Yvan LM, Damien C (2017) Habitat use and diving behaviour of gravid olive ridley sea turtles under riverine conditions in French Guiana. J Mar Syst 165:115–123. https://doi.org/10.1016/j.jmarsys.2016.10.005

Chamboult P, de Benolt T, Laurent K et al (2016) Inter-resting behavioral adjustments of green turtles to an estuarine habitat in French Guiana. Mar Ecol Prog Ser 555

Chamboult P, Lucie G, de Benolt T, Marc B et al (2017) Habitat use and diving behavior of gravid olive ridley Sea turtles under riverine conditions in French Guiana. J Mar Syst 165:115–123. https://doi.org/10.1016/j.jmarsys.2016.10.005

Colman LP, Thomé JCA, Almeida AP, Baptistotte C, Barata PCR, Broderick AC, Ribeiro FA, Vila-Verde L, Godley BJ (2019) Thirty years of leatherback turtle (Dermochelys coriacea) nesting in Espírito Santo, Brazil, 1988–2017: reproductive biology and conservation. Endanger Species Res 39:147–158

Farias LC, Resendiz E, Espinoza J (2022) Threats and vulnerabilities for globally distributed olive ridley (*Lepidochelys olivacea*) sea turtle; a historical and current status evaluation. Animals 12:1837. https://doi.org/10.3390/ani12141837

Guzman HM, Rogers G, Gomez CG (2019) Behavioral states related to environmental conditions and fisheries during olive ridley turtle migration from Pacific Panama. Front Mar Sci 6:770

Maxwell SM, Broderick AC, Dutton PH, Fossette-Halot S, Fuentes MM, Reina RD (2019) Advances in the biology and conservation of marine turtles. Front Mar Sci 6:9

Pandav B, Choudhury BC (2000) Conservation and management of olive ridley sea turtle (Lepidochelys olivacea) in Orissa. Final report. Wildlife Institute of India

Pendoley KL, Bell CD, McCracken R, Ball KR, Sherborne J, Oates JE, Becker P, Vitenbergs A, Whittock PA (2014) Reproductive biology of the flatback turtle Natator depressus in Western Australia. Endanger Species Res 23:115–123

Pikesley SK, Maxwell SM, Pendoley K, Costa DP and others (2013) On the front line:integrated habitat mapping for olive ridley sea turtles in the southeast Atlantic. Divers Distrib 19: 1518–1530

Pritchand PCH (2007) Arribadas I have known. In: Plotkin P (ed) Biology and conservation of ridley sea turtles. Johns Hopkins Univ. Press, Baltimore, pp 7–21

Sanjeev D (2024) 88 Olive ridley hatchlings enter sea from Mangalore nesting sites. Times of India p 4

Scott R, Biastoch A, Agamboue PD, Bayer T, Boussamba FL, Formia A, Godley BJ, Mabert BDK, Manfoumbi JC, Schwarzkopf FU, Sounguet GP, Wagner P, Witt MJ (2017) Spatio-temporal variation in ocean current-driven hatchling dispersion: Implications for the world's largest leatherback sea turtle nesting region. Divers Distrib 23:604–614

Seminoff JA (2004) Sea turtles, red listing, and the need for regional assessments. Mar Turt Newsl 106:4–6

Times of India (2023) Missile testing passed in Odisha to save sea turtle. TOI, dt.9/12/23. p 6

Tripathy B (2005) A study on the ecology and conservation of olive ridley sea turtle Lepidochelys olivacea at the Rushikulya rookery of Orissa coast, India. Unpublished PhD Thesis. Andhra University, Visakhapatnam, India

Tripathy B, Rajasekhar PS (2009) Natural and anthropogenic threats to olive ridley sea turtles (Lepidochelys olivacea) at the rushikulya rookery of Orissa coast. India Indian J Mar Sci 38(4):439–443

Reptile Love: Are We Treating Them Humanely?

9

Deyatima Ghosh and N. A. Aravind

Abstract

Reptiles are fascinating creatures that for a long time were believed to have limited mental capacities compared to birds and mammals. However, recent developments in the field of animal sentience show that reptiles can feel a wide range of states, including stress, pain, suffering, and pleasure. Given this new understanding, how well we treat captive reptiles is an open question. Reptiles form one of the important groups of animals in research, pet trade, and zoos globally. However, given the vast diversity in their behaviour, physiology, and species specificity and therefore their very specific needs, there is still no definitive protocol for maintaining them in captivity. Owners are unaware of the high sentience their reptilian pets possess, and this gives rise to poor animal welfare that is physically and mentally harmful to the animals. The reason for no scientific standards in reptile welfare stems from both the absence of enough studies and the lack of uptake of information by the owners and authorities. Intending to produce a good grasp of the needs of these cryptic yet magnificent creatures, this chapter details information on abnormal behaviours in captive reptiles and rec-

D. Ghosh (✉)
Laboratory of Animal Behaviour and Conservation, College of Life Sciences, Nanjing Forestry University, Nanjing, Jiangsu, People's Republic of China

SM Sehgal Foundation Centre for Biodiversity and Conservation, Ashoka Trust for Research in Ecology and the Environment (ATREE), Royal Enclave, Bangalore, India

School of Life Sciences, University of Lincoln, Lincoln, UK
e-mail: deyatima@atree.org

N. A. Aravind
SM Sehgal Foundation Centre for Biodiversity and Conservation, Ashoka Trust for Research in Ecology and the Environment (ATREE), Royal Enclave, Bangalore, India

Yenepoya Research Centre, Yenepoya (Deemed to be University), Mangalore, India

S. Sridhara et al. (eds.), *Animal Behavior in the Tropics*,
https://doi.org/10.1007/978-981-97-7324-4_9

ommends husbandry practices, which are essentially 'controlled deprivation' of the environment that reduce abnormal behaviour and increase positive responses. This book chapter also suggests research aspects that can aid in filling the existing gap to treat reptiles better, physically and mentally.

Keywords
Animal welfare · Captivity · Emotion · Environmental enrichment · Reptile · Sentience

Sentience

(i) Whether they (animals) are aware of what is happening around them; (ii) whether they are capable of cognitive processing; and (iii) whether they can have feelings such as pain— (Broom 2014). 'Behaviourally simple' reptiles are capable of all of these.

Recent research revealed that reptiles possess mental capacities on par with higher taxa, such as birds and mammals. Reptiles are just ahead of being sluggish, dumb, and dim-witted. They are capable of emotions and states such as anxiety, stress, distress, excitement, fear, frustration, suffering, pain, and pleasure (Lambert et al. 2019). In short, they possess sentience. However, there is an immense lack of concern for reptiles' mental well-being, as observed in a review by Melfi (2009). Reptiles form a major part of commercial trade, most of which are for pets. Additionally, almost 445 different species of reptiles are housed globally in zoos and aquariums, representing approximately 11% of the total terrestrial vertebrate species (Conde et al. 2013). Reptiles also form a major model species for laboratory experiments. Captivity in zoos, pet shops, and laboratories presents many challenging situations that adversely affect the welfare of the reptiles irrespective of whether they are wild-caught or captive-bred.

9.1 Stress

Stress is defined as "a normal biologically adaptive response of an individual to internal or external stimuli representing a threat to homeostasis" (Martínez Silvestre 2014), and animals can exhibit four types of reactions to stress situations: (1) behavioural changes, (2) alterations in the functioning of the autonomic nervous system, (3) neuroendocrine responses, and (4) immune responses (Warwick et al. 2013). Stress arising from captivity takes several forms, such as increased abnormal behaviour, behavioural inhibition, vigilance, hiding, mock striking, fearfulness and frequent startled responses, aggression, freezing behaviour, decreased exploratory and reproductive behaviour, and other behavioural complexities. Chronic and progressive stress leads to a predisposition to obesity and hepatic lipidosis and high levels of corticosterone resulting in muscle degeneration and growth suppression (Benn et al. 2019). However, acutely or intensely stressed reptile species are predisposed to emaciation, immune depression, and reproductive difficulty (Benn et al. 2019). On the contrary, the behaviour exhibited by a relaxed reptile includes species-specific behaviours like foraging, thermoregulation, exploration, and behavioural diversity, including unhurried body movements, regular resting habits, relaxed eating, space use in the enclosure for locomotion, and other natural responses (Bacon 2018).

To avoid behavioural abnormalities in captive animals and sustain good quality of their physical and mental state, animal welfare has evolved into three approaches over the decades (Hewson 2006): the first approach is the naturalist approach that allows an animal to perform its full behavioural repertoire, which is almost equal to living in the wild, i.e., the captive environment is similar to the animal's *umwelt* (Sueur and Pelé 2019); the second approach is the adaptive or behavioural approach based on the principle that animals adapt to the changes in their captive environment, provided the change is not drastic; the third approach is the mental approach, which states that the animal's captive environment must be designed to avoid prolonged negative emotions such as fear, pain, and frustration. Alongside enclosures being stressful, handling reptiles is very common and often mandatory in research and zoos and one of the commonest acts by pet owners that cause stress in reptiles. Additionally, frequent changes made to the enclosure, including lighting, substrate moving, adding conspecifics, and so on in enclosures, are other sources of severe stress in reptiles (Warwick et al. 1995).

9.2 Measuring Stress

Along with understanding stress response through behaviour in captive animals, animal husbandry needs to adopt methods to measure the changes in stress due to different welfare approaches. Higher levels of plasma corticosterone in the wild enhance fitness by facilitating access to energy stores and thus enhance the animal's ability to escape rapidly from a predator and fight with conspecifics (Lance et al. 2001). Therefore, concentrations of glucocorticoid hormones, including cortisol and corticosterone (CORT), are used as indicators of stress levels and are predictors of individual fitness in captive animals as well (Bonier et al. 2009). Schuett et al. (2004) showed that wild-caught captive adult male western diamond-backed rattlesnakes (*Crotalus atrox*) show a rapid and significant CORT response to mild handling stress within 30 min post handling. In another study, Moore et al. (1991) subjected tree lizards (*Urosaurus ornatus*) to acute and chronic stress by restraining lizards in bags for over 4 h and maintaining lizards in individual cages in the laboratory for up to 3 weeks. This caused an increase in corticosterone levels and a decrease in testosterone. However, the hormonal response was much higher in acute stress conditions. As a standardisation to properly measure the stress response, Tylan et al. (2020) suggest a "3-min" rule after stress to obtain corticosterone samples in reptiles based on species. Several behavioural exhibits also indicate stress, such as behavioural or emotional fever, where reptiles voluntarily seek higher temperature after stress, increased interactions with transparent boundaries, maxillary abscess, leg biting, even lithophagy (eating stones reported in captive tortoises), chronic friction lesion, rostral lesion on maxillae, and mandibles. A non-exhaustive list of such behavioural anomalies in reptiles that indicate stress of varied kinds, the probable causes for stress, and potential solutions is provided in Table 9.1. As evident from the literature and the table, very rarely would there be a single behavioural display to a cause and more often combination of several behavioural exhibits. There the first step to good animal welfare requires the owner to be well-equipped with the species' natural behavioural responses to identify the most likely causes of the discomfort and take necessary measures.

9.3 How Do We Enrich Our Reptiles?

Given the range of physiological needs, the sensory systems of reptiles, and the great variation among reptilian taxa, more research is needed to fully understand reptile husbandry. One of the solutions to reducing stress is resource provisioning in the form of environmental enrichment. Environmental enrichment considers modifying the captive environment to have physiological, psychological, and behavioural requirements of inhabitants, thus improving their welfare (Mellen and Sevenich MacPhee 2001; Hill and Broom 2009). This can be achieved in a variety of ways by providing social stimuli, for example, housing individuals in pairs or groups; occupational stimuli, that is, provisioning stimuli that keep animals occupied, such as hunting prey and cognitive tasks; or physical stimuli, that is, structural changes in the enclosure to keep animals motivated (Fig. 9.1). It is also the most common form of enrichment explored in reptile husbandry to date. Enrichment also enhances sensory and nutritional requirements (Bloomsmith et al. 1991; Azevedo et al. 2007) and allows animals to fully manifest their natural behaviour.

Corn snakes (*Elaphe obsoleta*) have been shown to prefer enriched habitats over standard habitats following exposure to both kinds of habitat (Hoehfurtner et al. 2021). Spain et al. (2020) in a recent study showed that Madagascar snakes (*Leioheterodon madagascariensis*) tend to spend more time in complex environments that are large enclosures with complexity and display more diversity in their behaviours. Excessive hiding/concealment is not a normal behaviour of caged reptiles. Therefore, increased exposure indicates welfare in exploratory snake species. A similar result was reported by Rose et al. (2014) in corn snakes (*E. obsoleta*) where physically enriched enclosures with plants and complex furnishings allowed climbing, hence increasing interactions with components within the enclosure thereby increasing activity and wellbeing of the reptile. A similar response towards an enriched naturalistic environment was reported in box turtles (*Terrapene carolina*) irrespective of the environment they grew up in (Case et al. 2005).

Cognitive and behavioural response to novelty as an indicator of exploration has been shown as an efficient way to understand stress response and good or poor

Table 9.1 List of abnormal reptile behaviours in captivity owing to stress response, indicators of such behaviour, the reason for the stress, and probable solutions that can improve the welfare of reptiles in zoos, those used for research, and pet animals

Behaviour	Description	Solution
Sentience: Fear		
Hiding or seeking shelter	Reptiles hide most of their time in shelters, substrates, or other secluded areas	Provide ample hiding spots and shelters in the enclosure, ensuring they are appropriately sized and located to offer security for the reptile. Minimise disturbances and allow the reptile time to acclimate to its environment
Freezing or remaining motionless	Reptiles remain still or freeze in response to perceived threats	Evaluate potential environmental stressors, such as loud noises, sudden movements, or visual stimuli. Provide hiding spots and visual barriers to reduce stress and promote safety. Minimise handling and disturbances
Aggression or defensive postures	Reptiles display defensive behaviours, such as hissing, biting, or puffing up	Approach the reptile calmly and avoid sudden movements or loud noises. Provide visual barriers and hiding spots to reduce feelings of vulnerability. Minimise handling and disturbances to promote a sense of safety
Attempted escape	Reptiles attempt to flee or escape from perceived threats or stressful situations	Evaluate enclosure design and security to prevent escape attempts. Ensure enclosure furnishings and equipment are secure and free from hazards. Provide hiding spots and visual barriers to reduce stress and promote feelings of safety
Excessive basking or soaking	Reptiles spend prolonged periods under heat lamps or in water, a potential coping mechanism for stress	Monitor temperature gradients and ensure they are appropriate for the species. Provide a variety of hiding spots and visual barriers to reduce stress. Evaluate environmental conditions and adjust as needed to promote comfort and well-being
Increased agitation or restlessness	Reptiles display increased agitation or restlessness, pacing along enclosure walls or exhibiting erratic movements	Evaluate potential environmental stressors, such as changes in routine, handling, or enclosure conditions. Provide hiding spots and visual barriers to reduce stress and promote feelings of security. Minimise disturbances and allow the reptile time to acclimate

(continued)

Table 9.1 (continued)

Behaviour	Description	Solution
Excessive vocalisation	Reptiles vocalise excessively such as whines and squeaks, displaying signs of distress or agitation	Identify and address potential sources of stress or discomfort in the environment. Provide a quiet and calm environment to minimise stress. Minimise handling and disturbances to promote a sense of security. Seek veterinary attention if necessary
Flight response	Reptiles exhibit a strong flight response, attempting to flee or escape perceived threats	Minimise handling and disturbances to avoid triggering a flight response. Provide a secure and comfortable environment with ample hiding spots and visual barriers. Avoid sudden movements
Reduced appetite or food avoidance	Reptiles show decreased interest in food or refuse to eat	Monitor for signs of stress or anxiety and evaluate potential environmental stressors. Provide a quiet and calm feeding area free from disturbances. Offer a varied diet to keep motivation and monitor feeding behaviour closely
Increased defecation or digestive issues	Reptiles exhibit increased faecal discharge or signs of digestive disorder	Evaluate environmental conditions and husbandry practices to minimise stress. Provide a quiet and secure environment for feeding. Offer a varied diet with appropriate food items and monitor digestive health closely
Reduced activity or lethargy	Reptiles exhibit decreased movement or lethargy	Evaluate environmental conditions and enrichment. Seek veterinary attention if illness is suspected
Glass surfing or pacing	Reptiles repeatedly move along enclosure walls	Review enclosure conditions and provide distractions. Monitor for stressors
Repetitive behaviours	Reptiles engage in repetitive movements or displays	Provide environmental enrichment and space. Monitor for stressors
Abnormal shedding	Reptiles experience difficulty shedding or incomplete sheds	Ensure proper humidity and hydration. Provide hiding spots. Monitor for retained shed
Increased hiding or withdrawal	Reptiles become increasingly withdrawn or hide more than usual	Assess enclosure conditions. Provide secure hiding spots and minimise disturbances
Sentience: Pain		
Decreased activity	Reptiles exhibit reduced movement or become lethargic	Monitor for signs of illness or injury. Provide a comfortable environment for rest. Seek veterinary attention for diagnosis
Altered gait or limping	Reptiles display abnormal gait or favour one limb while moving	Check for signs of injury. Provide padded substrate. Seek veterinary assistance for diagnosis and treatment

(continued)

Table 9.1 (continued)

Behaviour	Description	Solution
Guarding or protecting a body part	Reptiles exhibit defensive postures or guard a specific area of their body	Examine the guarded area for signs of injury. Provide a stress-free environment. Minimise handling
Reduced appetite	Reptiles refuse to eat or show decreased interest in food	Assess for signs of illness or stress. Offer different food items. Seek veterinary advice if loss of appetite persists
Aggression or agitation	Reptiles display aggressive behaviour towards conspecifics or handlers	Evaluate potential stressors. Provide hiding spots and barriers. Minimise handling and disturbances. Seek veterinary consultation if aggression continues
Vocalisation	Reptiles vocalise excessively or exhibit signs of distress	Identify and address potential sources of discomfort. Provide a calm environment. Monitor for illness or injury. Seek veterinary attention if necessary
Changes in posture	Reptiles assume abnormal resting postures	Evaluate for signs of injury or discomfort. Provide supportive bedding. Seek veterinary assistance for diagnosis and treatment
Increased hiding or withdrawal	Reptiles spend more time hiding or avoiding interaction	Monitor for stress, illness, or injury. Provide a secure hiding spot. Minimise disturbances. Allow time for rest and recovery
Excessive soaking	Reptiles spend prolonged periods soaking in water	Check for signs of dehydration or discomfort. Provide access to clean water. Monitor for skin issues. Seek veterinary advice if soaking persists excessively
Increased respiratory rate	Reptiles display rapid or difficult breathing	Monitor for respiratory issues. Ensure proper enclosure ventilation. Seek veterinary attention for persistent respiratory symptoms
Sentience: Boredom		
Repetitive behaviours	Reptiles exhibit repetitive movements or displays, such as pacing, head bobbing, or tail wagging	Provide environmental enrichment, such as climbing structures, hiding spots, and interactive toys. Rotate enrichment regularly. Offer novel food items or foraging opportunities
Excessive basking or soaking	Reptiles spend prolonged periods under heat lamps or in water	Ensure proper temperature gradients and humidity levels. Provide various basking spots. Offer opportunities for natural behaviours like climbing or exploring
Aggression or agitation	Reptiles show aggressive behaviour towards conspecifics or handlers	Evaluate enclosure size and compatibility of tank mates. Provide visual barriers and hiding spots. Minimise handling and disturbances

(continued)

Table 9.1 (continued)

Behaviour	Description	Solution
Excessive vocalisation	Reptiles vocalise excessively, indicating distress or boredom	Provide environmental enrichment to stimulate mentally and physically. Offer hiding spots, climbing structures, and exploration opportunities. Ensure a quiet and calm environment
Pacing or restlessness	Reptiles display restless behaviour, pacing along enclosure walls or glass surfaces	Evaluate enclosure design and provide various textures and surfaces for exploration. Offer environmental enrichment like branches, rocks, and plants. Provide opportunities for natural behaviours
Decreased activity or exploration	Reptiles display reduced movement or lethargy	Review the enclosure layout and provide space for natural exploration and movement. Offer various environmental enrichment items. Monitor for signs of illness or stress and seek veterinary attention if necessary
Excessive faecal smearing	Reptiles smear faeces around the enclosure	Ensure proper hygiene and cleanliness. Offer various environmental enrichment items. Monitor for signs of stress or discomfort and address underlying causes
Increased restlessness	Reptiles show increased restlessness or agitation	Provide environmental enrichment to encourage natural behaviours. Offer hiding spots, climbing structures, and foraging opportunities. Ensure a consistent daily routine
Reduced appetite	Reptiles show decreased interest in food or refuse to eat	Offer a varied diet with different food items and textures. Provide opportunities for natural hunting or foraging behaviours. Ensure proper husbandry conditions
Excessive handling or interaction seeking	Reptiles seek excessive handling or interaction	Minimise handling and disturbances. Provide environmental enrichment to promote mental stimulation and natural behaviours. Ensure a quiet and stress-free environment
Disease/infection/injury		
Limping or favouring a limb	Reptiles exhibit a reluctance to use a limb or favour one side while moving	Inspect for visible signs of injury. Provide a safe substrate. Seek veterinary attention for diagnosis
Visible wounds or lacerations	Reptiles display open wounds, cuts, or lesions on its body	Assess the cause of injury. Maintain a clean and hygienic environment. Seek veterinary assistance for treatment
Loss of tail or limb	Reptiles are missing part of their tail or limb	Determine the cause. Ensure proper husbandry. Seek veterinary evaluation for wound care and rehabilitation

(continued)

Table 9.1 (continued)

Behaviour	Description	Solution
Bleeding	Reptiles exhibit signs of bleeding	Apply gentle pressure to control bleeding. Seek immediate veterinary attention for evaluation
Guarding or protecting a body part	Reptiles display defensive behaviours or guard a specific area of its body	Examine the guarded area. Provide a stress-free environment. Seek veterinary guidance for treatment
Reduced mobility or activity	Reptiles exhibit decreased movement or reluctance to explore	Assess for signs of injury. Provide supportive care. Seek veterinary consultation for diagnosis
Abnormal posture	Reptiles assume an unusual or awkward posture	Check for spinal injury or musculoskeletal issues. Seek veterinary assistance for imaging and treatment
Hiding or avoiding interaction	Reptiles spend excessive time hiding or avoiding interaction	Provide a secure hiding spot. Minimise disturbances. Seek veterinary evaluation for underlying issues
Difficulty eating or drinking	Reptiles show reluctance to eat or drink, possibly due to oral injuries or pain	Examine the mouth for injuries. Offer softer foods. Seek veterinary assistance for oral examination
Behavioural changes	Reptiles exhibit altered behaviour patterns, such as increased aggression or withdrawal	Consider underlying pain or discomfort. Monitor closely. Seek veterinary consultation for assessment
Defence		
Hissing	Reptiles emit hissing sounds as a warning signal	Respect the reptile's boundaries and avoid further provocation. Provide a calm and quiet environment. Minimise handling and disturbances
Biting	Reptiles bite or attempt to bite when approached or handled	Use proper handling techniques and avoid sudden movements. Approach calmly and confidently. Provide positive reinforcement for calm behaviour
Striking	Reptiles strike out with its head or body when feeling threatened	Respect the reptile's space and avoid unnecessary handling. Provide hiding spots and visual barriers. Use handling tools or protective equipment when necessary
Tail-wagging	Reptiles wag its tail as a warning signal or defensive display	Respect the reptile's boundaries and avoid further provocation. Provide environmental enrichment and hiding spots. Monitor for signs of aggression and adjust handling techniques
Puffing-up	Reptiles inflate their body or display an enlarged posture	Approach calmly and avoid sudden movements. Provide hiding spots and visual barriers. Minimise handling and disturbances

(continued)

Table 9.1 (continued)

Behaviour	Description	Solution
Defensive postures	Reptiles adopt defensive postures, such as arching their back or raising their head	Being respectful and avoid further provocation. Provide hiding spots and visual barriers. Monitor for signs of aggression and adjust handling techniques
Struggling or writhing	Reptiles struggle or writhe when approached or handled	Use proper handling techniques and avoid excessive restraint. Approach calmly and confidently. Provide environmental enrichment and hiding spots
Defecation	Reptiles defecate as a defensive response	Respect the reptile's boundaries and avoid unnecessary handling. Provide a calm and quiet environment. Minimise handling and disturbances
Vocalisation	Reptiles vocalise, such as hissing or growling, as a warning or defensive signal	Respect the reptile's space and avoid further provocation. Provide hiding spots and visual barriers. Minimise handling and disturbances
Retreating or hiding	Reptiles retreat or hide in response to perceived threats	Respect the reptile's boundaries and avoid excessive handling or disturbances. Provide ample hiding spots and visual barriers. Allow time for acclimation and feeling secure
Escape		
Repeatedly rubbing or pressing against enclosure walls	Reptiles repeatedly rub their body against the walls of the enclosure in an attempt to escape	Assess the enclosure for potential stressors, such as inadequate space, poor environmental conditions, or lack of hiding spots. Ensure the enclosure is properly secured and free from gaps or weaknesses. Provide ample space and enrichment
Digging or scratching at substrate or enclosure floor	Reptiles dig or scratch at the substrate or enclosure floor, attempting to burrow or escape	Evaluate substrate type and depth to meet the reptile's natural behaviour needs. Provide appropriate digging areas or hides. Ensure the enclosure is securely constructed and free from gaps or weaknesses
Pushing or testing enclosure doors or latches	Reptiles repeatedly push or test enclosure doors or latches in an attempt to open them	Ensure enclosure doors are securely locked and free from damage or wear. Use locking mechanisms or additional security measures. Monitor for signs of stress or agitation and address underlying causes
Attempting to squeeze through small gaps or vents	Reptiles attempt to squeeze through small gaps, vents, or openings in the enclosure	Inspect the enclosure for any gaps or openings that could allow escape. Seal gaps or weak points. Ensure ventilation systems prevent escape while maintaining airflow

(continued)

Table 9.1 (continued)

Behaviour	Description	Solution
Climbing or scaling enclosure walls or furniture	Reptiles climb or scale enclosure walls, furniture, or fixtures in an attempt to escape	Provide adequate climbing structures or branches. Ensure enclosure walls are smooth and free from ledges or footholds. Monitor for signs of stress or agitation and provide appropriate environmental enrichment
Testing or chewing on enclosure components	Reptiles test or chew on enclosure components, such as mesh or plastic	Use durable materials resistant to chewing or damage. Provide appropriate chewing toys or enrichment items. Monitor enclosure integrity regularly and repair any damage promptly
Restlessness or agitation near enclosure exits	Reptiles display restlessness or agitation near enclosure exits	Ensure enclosure doors are securely closed and latched. Minimise disturbances and provide a calm and stable environment. Monitor for signs of escape behaviour and address underlying causes promptly
Repeatedly testing enclosure boundaries	Reptiles repeatedly test the boundaries of the enclosure	Evaluate enclosure size and layout to ensure it meets the reptile's spatial needs. Provide adequate space and environmental enrichment. Monitor for signs of stress or agitation and address underlying causes
Inadequate space		
Increased aggression	Reptiles exhibit heightened aggression towards conspecifics or handlers due to territorial disputes or resource competition	Evaluate enclosure size and stocking density. Provide visual barriers and hiding spots. Consider separating incompatible individuals or providing larger enclosures
Reduced activity and exploration	Reptiles show decreased movement and exploration due to limited space for natural behaviours	Provide a larger enclosure with ample room for movement and natural activities. Add vertical components for better space use. Offer environmental enrichment to encourage exploration and physical activity
Stress-related behaviours	Reptiles exhibit stress-related behaviours such as hiding, pacing, or glass surfing in response to confinement	Assess environmental conditions and adjust enclosure size, layout, and furnishings to reduce stress. Provide hiding spots, visual barriers, and naturalistic features
Reduced feeding responses	Reptiles show decreased interest in food or exhibit reduced feeding responses due to stress or lack of space	Monitor feeding behaviour and adjust husbandry practices. Provide a larger enclosure with appropriate hiding spots and feeding stations. Offer a varied diet
Abnormal resting postures	Reptiles assume awkward or uncomfortable resting postures due to limited space	Ensure enclosure dimensions allow for proper thermoregulation and resting areas. Provide a variety of substrates and surfaces for comfortable resting

(continued)

Table 9.1 (continued)

Behaviour	Description	Solution
Excessive basking or soaking	Reptiles spend prolonged periods under heat lamps or in water, potential a coping mechanism for stress or discomfort	Monitor temperature gradients. Provide a larger enclosure with adequate basking and soaking areas. Offer environmental enrichment and hiding spots
Social withdrawal	Reptiles exhibit social withdrawal or isolation due to limited space for social interactions	Assess enclosure size and social dynamics. Provide hiding spots and visual barriers. Consider providing larger group enclosures or separate living spaces if necessary
Reduced reproductive behaviours	Reptiles show decreased reproductive behaviours or breeding success due to limited space for courtship and nesting	Provide a larger enclosure with appropriate nesting sites and hiding spots. Ensure compatible groupings and social dynamics. Monitor for signs of reproductive stress
Excessive vocalisation or agitation	Reptiles vocalise excessively or display signs of agitation due to confinement stress	Evaluate enclosure size and layout. Provide hiding spots and visual barriers. Minimise disturbances and ensure a quiet and calm environment
Reduced growth or development	Reptiles exhibit signs of stunted growth or delayed development due to inadequate space and limited opportunities for natural behaviours	Ensure enclosure dimensions allow for proper growth and development. Provide a larger enclosure with appropriate environmental conditions and enrichment
Poor environment		
Pacing or circling	Reptiles repeatedly move back and forth along the enclosure walls	Assess enclosure size and layout. Provide a larger habitat with ample space. Include environmental enrichment for natural exploration
Glass-surfing	Reptiles repeatedly slide against the glass walls of the enclosure	Ensure appropriate environmental conditions. Add visual barriers and hiding spots. Adjust lighting and temperature
Excessive hiding	Reptiles spend most of their time hiding in shelters or substrates	Review enclosure layout. Provide sufficient hiding spots. Adjust lighting and temperature. Consider species-specific needs for privacy and security
Aggression towards conspecifics	Reptiles display aggressive interactions with other individuals in the enclosure	Assess the compatibility of tank mates. Provide separate hiding spots. Increase enclosure size. Monitor interactions closely
Inadequate basking or hiding sites	Reptiles struggle to find suitable basking spots or hiding places	Ensure diverse temperature gradients and hiding spots. Monitor behaviour and adjust furnishings as needed
Excessive basking or soaking	Reptiles spend prolonged periods under heat lamps or in water	Check temperature gradients. Provide a variety of basking spots. Ensure access to clean water. Monitor for dehydration

(continued)

Table 9.1 (continued)

Behaviour	Description	Solution
Abnormal thermoregulation	Reptiles are unable to regulate their body temperature effectively	Assess placement and functionality of heat sources. Provide thermometers and temperature gradients. Monitor regularly
Lack of environmental enrichment	Reptiles display boredom or repetitive behaviours due to lack of stimulation	Introduce environmental enrichment items. Rotate regularly. Offer novel food items or foraging opportunities
Excessive vocalisation or stress signals	Reptiles vocalise excessively or display stress signals	Evaluate potential stressors. Provide a quiet environment. Offer hiding spots and visual barriers. Monitor behaviour closely
Abnormal reproductive behaviours	Reptiles display abnormal reproductive behaviours, such as nesting without a suitable substrate	Ensure the enclosure is equipped with appropriate nesting sites. Provide privacy during mating or nesting behaviours. Monitor and adjust enclosure furnishings
Humidity imbalance		
Excessive soaking or submersion	Reptiles spend prolonged periods soaking in water or submerging themselves	Monitor humidity levels within the enclosure using a hygrometer. Ensure humidity levels are appropriate for the species. Provide a humid hide or moist substrate to facilitate hydration and shedding
Avoidance of dry areas	Reptiles avoid dry areas of the enclosure, seeking out higher humidity zones	Evaluate enclosure design and substrate to ensure a range of microclimates. Provide hiding spots and shelters with varying humidity levels. Adjust misting or spraying frequency to maintain proper humidity gradients
Excessive drinking	Reptiles exhibit increased drinking behaviour	Monitor humidity levels and ensure they are within the appropriate range for the species. Provide access to clean, fresh water at all times. Offer a variety of water sources. Adjust humidity levels to promote proper hydration
Difficulty shedding	Reptiles experience difficulty shedding or display incomplete sheds	Maintain proper humidity levels within the enclosure, especially during shedding periods. Provide a humid hide or moisture chamber with a damp substrate. Monitor for signs of retained shed and offer assistance if necessary
Respiratory symptoms	Reptiles display signs of respiratory distress	Monitor humidity levels and ensure they are appropriate for the species. Provide proper ventilation within the enclosure. Seek veterinary attention if respiratory symptoms persist
Increased restlessness or agitation	Reptiles exhibit restless behaviour, pacing, or agitation	Evaluate humidity levels within the enclosure and adjust as needed. Ensure proper ventilation and airflow. Offer hiding spots and visual barriers to reduce stress and promote natural behaviours

(continued)

Table 9.1 (continued)

Behaviour	Description	Solution
Lethargy or reduced activity	Reptiles become lethargic or exhibit decreased activity levels	Monitor humidity levels and ensure they are within the appropriate range for the species. Provide access to clean water for drinking and soaking. Check for signs of dehydration and seek veterinary attention if necessary
Skin abnormalities or lesions	Reptiles display signs of skin irritation, dryness, or lesions	Maintain proper humidity levels within the enclosure. Provide a humid hide or moisture chamber with a damp substrate. Monitor for signs of infection or dermatitis and seek veterinary assistance if needed
Excessive shedding or shed retention	Reptiles experience frequent shedding or display signs of shed retention	Monitor humidity levels, especially during shedding periods. Provide a humid hide or moisture chamber. Offer regular misting or soaking to maintain skin hydration and promote healthy shedding
Abnormal drinking behaviour	Reptiles exhibit abnormal drinking behaviour, such as excessive or reduced drinking	Monitor humidity levels and ensure they are within the appropriate range for the species. Provide access to clean, fresh water at all times. Offer a variety of water sources. Monitor for signs of dehydration and seek veterinary attention if necessary
Light imbalance		
Nocturnal species exhibiting daytime activity	Nocturnal reptiles are active during the day	Ensure that the enclosure is adequately dark during the night and provide hiding spots to mimic natural darkness. Use dim red or blue lights for night-time observation to avoid disturbing the reptile's circadian rhythm
Diurnal species displaying nocturnal activity	Diurnal reptiles are active during the night	Ensure that the enclosure is adequately lit during the day with full-spectrum lighting that mimics natural sunlight. Provide hiding spots or shelters where the reptile can retreat from darkness if needed. Avoid exposing the reptile to artificial light sources during the night
Aggression or agitation due to constant lighting	Reptiles display aggression, restlessness, or agitation due to constant exposure to bright lighting	Implement a proper light cycle with a day-night cycle that mimics natural lighting patterns. Use timers to regulate lighting duration and intensity. Provide hiding spots or shaded areas within the enclosure to allow the reptile to retreat from bright light when necessary

(continued)

Table 9.1 (continued)

Behaviour	Description	Solution
Excessive basking or hiding due to inadequate lighting gradient	Reptiles spend excessive time basking or hiding due to inadequate lighting gradients	Ensure that the enclosure has a proper lighting gradient with areas for basking, hiding, and resting. Use basking bulbs or heat lamps to provide localised heat and UVB lighting. Monitor temperatures and adjust lighting fixtures accordingly to maintain optimal conditions
Lethargy or reduced activity due to insufficient lighting	Reptiles display lethargy, reduced activity, or poor appetite due to insufficient lighting	Provide full-spectrum lighting with appropriate UVB wavelengths to support natural behaviours and physiological processes. Ensure that UVB bulbs are replaced regularly according to manufacturer recommendations. Monitor for signs of vitamin D deficiency or metabolic bone disease and consult a veterinarian if needed
Abnormal shedding patterns due to inadequate UVB exposure	Reptiles experience abnormal shedding patterns or retained shed due to inadequate UVB exposure	Ensure that the enclosure has sufficient UVB lighting to support proper shedding and vitamin D synthesis. Provide opportunities for the reptile to bask under UVB bulbs for adequate exposure. Monitor humidity levels to prevent excessive dryness or dehydration during shedding
Stress-related behaviours due to erratic lighting change	Reptiles display stress-related behaviours, such as hiding, pacing, or glass surfing, in response to erratic lighting changes	Establish a consistent light cycle with gradual transitions between day and night phases. Avoid sudden changes in lighting intensity or duration that could cause stress. Monitor reptile behaviour closely and adjust lighting settings as needed to promote well-being
Reproductive dysfunction due to inadequate lighting	Reptiles experience reproductive dysfunction, such as decreased breeding activity or egg production, due to inadequate lighting	Provide proper lighting conditions, including appropriate photoperiod and UVB exposure, to support reproductive behaviours and hormonal regulation. Mimic seasonal changes in lighting to stimulate natural breeding cycles. Monitor for signs of reproductive stress or abnormalities and consult a veterinarian if needed
Abnormal colouration or pigmentation due to UVB deficiency	Reptiles exhibit abnormal colouration or pigmentation patterns due to UVB deficiency	Ensure that the enclosure has adequate UVB lighting to support proper pigmentation and colouration. Provide opportunities for the reptile to bask under UVB bulbs for adequate exposure. Monitor for signs of UVB deficiency, such as faded or dull colouration, and consult a veterinarian if needed

(continued)

Table 9.1 (continued)

Behaviour	Description	Solution
Repetitive behaviours or stereotypies due to environmental stress	Reptiles display repetitive behaviours or stereotypies, such as pacing or self-injurious behaviours, due to environmental stress related to lighting conditions	Evaluate and address any underlying stressors in the reptile's environment, including lighting conditions, enclosure size, enrichment, and social interactions. Ensure that lighting parameters are appropriate for the species and mimic natural lighting patterns. Provide environmental enrichment and behavioural stimulation to promote mental and physical well-being
Thermal imbalance		
Seeking excessive heat	Reptiles spend prolonged periods under heat sources or in basking areas, potentially leading to overheating	Check temperature gradients within the enclosure to ensure adequate thermal gradients. Adjust heat sources or provide additional heat gradients if necessary. Ensure proper ventilation to prevent overheating. Monitor reptile behaviour closely and offer cooler areas for thermoregulation
Avoiding heat sources	Reptiles avoid basking areas or heat sources, potentially leading to inadequate thermoregulation	Check temperature gradients and adjust heat sources to provide appropriate basking temperatures. Ensure that basking areas are not too hot or uncomfortable for the reptile. Offer alternative basking spots or adjust lighting to encourage thermoregulation behaviour
Excessive soaking	Reptiles spend excessive time soaking in water, potentially due to insufficient heat or humidity in the enclosure	Monitor humidity levels within the enclosure and ensure they are within the appropriate range for the species. Check temperature gradients and adjust heating elements to maintain proper ambient temperature and humidity. Provide a variety of basking spots and hiding areas to facilitate thermoregulation
Lethargy or reduced activity	Reptiles exhibit decreased activity levels or become lethargic, potentially due to thermal stress	Check temperature gradients to ensure they meet the reptile's species-specific requirements. Provide a range of temperature gradients within the enclosure to allow for thermoregulation. Monitor for signs of overheating or hypothermia and adjust heating accordingly. Offer environmental enrichment to encourage activity

(continued)

Table 9.1 (continued)

Behaviour	Description	Solution
Pacing or restlessness	Reptiles exhibit restless behaviour or pace along enclosure walls, potentially due to discomfort from thermal imbalance	Evaluate temperature gradients and ensure they provide adequate options for thermoregulation. Check for potential hot spots or areas of thermal stress within the enclosure. Provide hiding spots and visual barriers to reduce stress and encourage natural behaviour
Increased vocalisation or aggression	Reptiles display increased vocalisation or aggression, potentially as a response to thermal discomfort	Monitor temperature gradients and ensure they allow for proper thermoregulation. Evaluate potential stressors in the environment, such as inadequate heat or overcrowding. Provide hiding spots and visual barriers to reduce aggression and stress. Seek veterinary advice if aggression persists
Abnormal resting postures	Reptiles assume unusual or awkward resting postures, potentially due to thermal discomfort	Check temperature gradients and adjust heating elements to ensure proper thermoregulation. Provide a variety of basking spots and hiding areas to allow the reptile to choose its preferred temperature. Monitor for signs of discomfort or stress and adjust enclosure conditions accordingly
Reduced appetite or digestive issues	Reptiles exhibit reduced appetite or experience digestive problems, potentially related to thermal stress	Ensure proper temperature gradients and humidity levels to support digestion and metabolic processes. Offer a varied diet appropriate for the reptile's species. Monitor feeding behaviour and adjust environmental conditions as needed to promote healthy digestion
Increased respiratory rate	Reptiles display rapid or laboured breathing, potentially as a response to thermal stress	Monitor temperature gradients and ensure they allow for comfortable thermoregulation. Provide proper ventilation within the enclosure to maintain air quality. Check for signs of respiratory infections or other health issues and seek veterinary assistance if respiratory symptoms persist
Excessive faecal output or dehydration	Reptiles exhibit increased faecal output or signs of dehydration	Check temperature gradients and humidity levels to ensure proper hydration and metabolic function. Provide access to clean water for drinking and soaking. Monitor for signs of dehydration or metabolic imbalances and seek veterinary attention if necessary
Overcrowding		

(continued)

Table 9.1 (continued)

Behaviour	Description	Solution
Aggression towards conspecifics	Reptiles display aggressive behaviour with other animals as biting, chasing, or territorial disputes	Evaluate the size and compatibility of tank mates. Provide sufficient space and hiding spots to minimise competition and territorial conflicts. Consider separating incompatible individuals or reducing stocking density to alleviate aggression
Increased stress-related behaviours	Reptiles exhibit stress-related behaviours, such as hiding, pacing, glass surfing, or decreased appetite	Monitor for signs of stress and overcrowding, such as changes in behaviour or appetite. Provide hiding spots and visual barriers to reduce stress. Ensure adequate space and environmental enrichment to promote natural behaviours
Reduced basking or resting space	Reptiles struggle to find suitable basking spots or resting areas due to overcrowding	Evaluate enclosure size and furnishings to provide sufficient basking and resting spots for all individuals. Ensure adequate temperature gradients and hiding spots. Consider expanding the enclosure or reducing stocking density
Competition for food	Reptiles compete aggressively for food or display reduced feeding behaviours	Monitor feeding behaviour and see that food is available and accessible to all individuals without competition. Provide multiple feeding stations or offer food in separate containers. Consider feeding reptiles separately if competition persists
Increased restlessness or agitation	Reptiles display increased restlessness, pacing, or erratic movements due to overcrowding	Evaluate enclosure size and layout to provide adequate space for natural behaviours and movement. Ensure environmental enrichment and hiding spots. Consider reducing stocking density or providing larger enclosures if space allows
Excessive vocalisation	Reptiles vocalise excessively, displaying signs of distress or agitation	Identify and address potential stressors, such as overcrowding or competition for resources. Provide a quiet and calm environment. Consider separating incompatible individuals or reducing stocking density to alleviate vocalisation
Increased aggression towards handlers	Reptiles display aggression towards handlers or exhibit defensive postures during handling	Minimise handling and disturbances to reduce stress and aggression. Provide appropriate training and handling techniques. Consider separating aggressive individuals or reducing stocking density to minimise aggression

(continued)

Table 9.1 (continued)

Behaviour	Description	Solution
Reduced growth or development	Reptiles show signs of stunted growth or delayed development due to overcrowding and limited resources	Monitor growth rates and overall health of reptiles. Ensure adequate nutrition, space, and environmental conditions for proper growth. Consider reducing stocking density or providing larger enclosures to promote healthy growth
Increased risk of injury or disease	Reptiles exhibit signs of injury, illness, or compromised immune function due to overcrowding and stress	Monitor for signs of injury or disease, such as wounds, abrasions, or respiratory symptoms. Provide appropriate veterinary care and treatment. Consider reducing stocking density and improving husbandry practices to minimise risks
Reduced reproductive success	Reptiles exhibit decreased reproductive behaviours or reduced breeding success due to overcrowding and stress	Monitor reproductive behaviours and breeding success rates. Provide appropriate nesting sites, hiding spots, and environmental conditions to support reproduction. Consider reducing stocking density or providing separate breeding enclosures

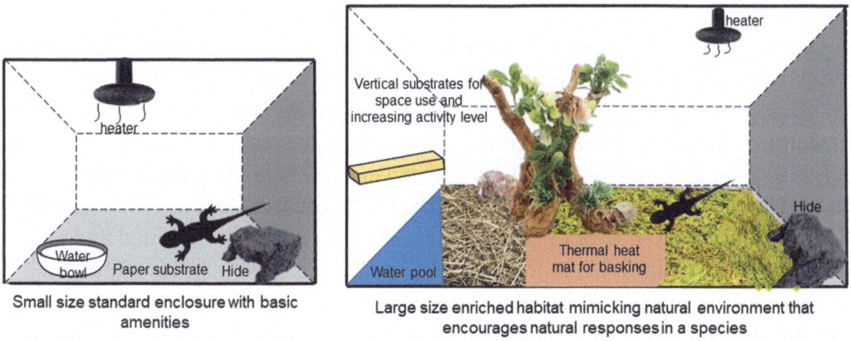

Fig. 9.1 Illustration of a standard vs. an enriched environment used for a lizard that can be practised by pet owners, zoo authorities, and research personnel

animal welfare. This has been tested in several species, including corn snakes, from enriched environments that display more interaction with novel objects (Hoehfurtner et al. 2021). Corn snakes in enriched environments are more efficient in using sensory faculties as they discriminate between familiar and unfamiliar human odours (Nagabaskaran et al. 2021). Such responses are hardly explored but have immense implications for human–reptile bonding. Thus, enrichment in the form of provisioning physical objects keeps the reptiles motivated and active, as in many studies, snakes have been seen to spend more time off the ground than on the floor of the enclosure (Rose et al. 2014). Overall, environmental enrichment can improve the cognitive capabilities of reptiles and keep them healthy. Environmental enrichment has been successfully used to reduce stereotypic behaviours in zoo mammals (Shyne 2006), improve welfare and decrease stress, increase activity and exploration, and improve body condition to reduce aggression, as seen in amphibians (Hurme et al. 2003; Michaels et al. 2014) and fishes (Näslund and Johnsson 2016). Enrichment leads to neurological benefits, improved cognitive flexibility, and improved immune function (Hannan 2014; Gelfo 2019).

A long-standing 'apocryphal' view about reptiles, which is also the reason for poor quality in reptile husbandry, is that reptiles are sluggish, sedentary, and reticent, driven by instinct and hence require minimalistic space. As a result, most of the husbandry practices unwittingly avoid considering the adequate space requirement of the reptile inhabitants. This misconception is a reflection of less knowledge about the natural history of animals. Such poor information leads to poor decision-making as seen in the Great Plains rat snake (*Elaphe guttata*), where the recommended enclosure size is only 0.000024 ha, while the actual area covered by this species in the wild is 10.17 ha (Griswold 2001; Sperry and Taylor 2008). Other examples of reptiles showing significant movement include arboreal monitors documented to have daily movement ranges greater than 186 m. Home ranges for some skink are even larger such as 1 ha, for box turtles it is 40 ha, for Indigo snakes it is 158 ha, and sea turtles that travel thousands of kilometres (Warwick et al. 2019). Therefore, spatial needs of reptiles is a crucial consideration while strategising husbandry practices and designing enclosures.

One of the essential senses of reptiles is olfaction—the least explored option for enrichment. Many reptiles are territorial, and males actively defend territory and obtain information about neighbouring and/or intruding rivals. Brown wall lizards (*Podarcis liolepis*) showed a significant reduction in escape behaviour, reduction in locomotion and faster habituation to enclosure, such as spending more time on the perches (Londoño et al. 2018) when the enclosure was enriched with olfaction from conspecifics. Olfactory enrichment needs to be properly exploited and can be a major focus to improve reptile husbandry and keep captive reptiles motivated.

However, there are studies which have found little impact of enrichment on reptile wellbeing. A recent study showed that leopard geckos (*Eublepharis macularius*) maintained in standard conditions were more willing and motivated to interact with novel environmental enrichments than individuals reared in enriched biotypes (Zieliński 2023, Fig. 9.2). Similarly, studies by Borgmans et al. (2018) on green anole (*Anolis carolinensis*) and Rosier and Langkilde (2012) on eastern fence lizards (*Sceloporus undulatus*) have shown no effect of enrichment. This hints at the genetic predisposition of the need for environmental and behavioural diversity in species as well. A challenge of environmental enrichment is that it cannot be definite, especially owing to the vast range of sensory faculties and physiological demands. Such modifications need to keep in mind the species-specific behavioural repertoire such that animals have a hold on their environment and are free to manifest the full range of behaviour. The enriched environments such as biotypes (a bioactive miniature environment with living elements that work together as an ecosystem) are still 'controlled deprivation' as no captive condition can replicate natural habitat. Burghardt (1999) states that 'all captive environments deprive animals of

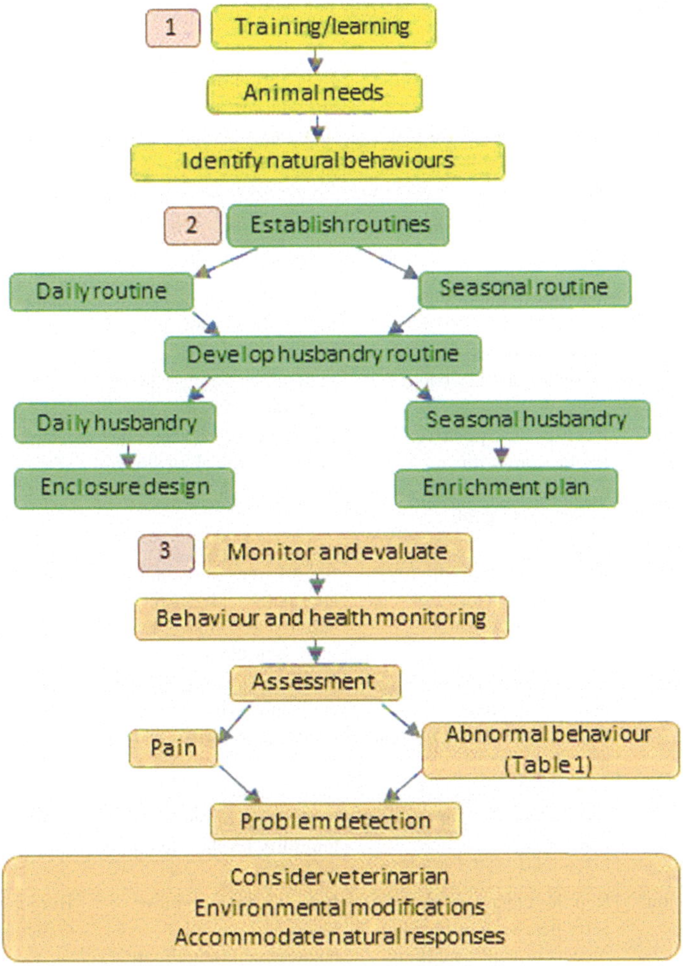

Fig. 9.2 Illustration of the standard practice for animal husbandry through three basic steps showing the importance of knowing the natural behaviour of the animal to provide optimum habitat. We have to realise that an *animal has its own world* in which it is optimally adapted and functioning. The behavioural senses are suited for their niche and need to be given priority while maintaining them in captivity

natural stimuli and these restrictions have varying, and often unpredictable, consequences on the welfare of captive animals'. Therefore, it is necessary to improve animals' living conditions and mental health (Table 9.2; Fig. 9.2).

Table 9.2 List of behaviours that can be used as indicators for good welfare

Behaviour	Description	Significance	References
Active exploration	Reptiles actively explore their enclosure, investigating objects and their surroundings	Indicates curiosity, mental stimulation, and comfort in their environment	Bartolomé et al. (2023)
Healthy appetite	Reptiles display regular and enthusiastic feeding behaviour	Shows signs of good health, proper digestion, and a stress-free environment	Bartolomé et al. (2023)
Regular basking	Reptiles frequently bask under heat sources and display appropriate thermoregulatory behaviour	Indicates proper thermoregulation, good health, and a well-maintained temperature gradient	Benn et al. (2019)
Social interactions (species-appropriate)	Reptiles have positive interactions with conspecifics, such as social grooming or gentle nuzzling	Suggests comfort with tank mates and a lack of stress or aggression	Clegg et al. (2015)
Engaging with enrichment items	Reptiles interact with toys, climbing structures, or other environmental enrichments	Shows signs of mental stimulation, physical activity, and engagement with their environment	Bartolomé et al. (2023)
Natural hiding and resting	Reptiles regularly use hides and resting areas, alternating with periods of activity	Indicates a well-designed enclosure that meets their needs for security and rest	Spitzer et al. (2022)
Normal shedding	Reptiles shed regularly and completely without retained skin	Shows signs of good health, appropriate humidity levels, and proper hydration	Benn et al. (2019)
Play behaviour (in some species)	Reptiles engage in play-like activities, such as pushing objects or gentle nudging	Indicates high comfort levels and mental engagement though rare	Burghardt (2015)
Calm and relaxed posture	Reptiles often display relaxed body language, such as stretched-out limbs or relaxed tails	Shows signs of a stress-free environment and overall well-being	Benn et al. (2019)
Clear eyes and smooth skin	Eyes are bright and alert; skin is smooth and free of lesions	Indicates good health and proper environmental conditions	Benn et al. (2019)
Appropriate defecation	Reptiles have regular and healthy bowel movements	Shows signs of proper digestion, hydration, and overall health	Benn et al. (2019)

(continued)

Table 9.2 (continued)

Behaviour	Description	Significance	References
Engaging in species-specific behaviours	Reptiles display natural behaviours specific to their species, such as burrowing, climbing, or soaking	Indicates a well-designed habitat that allows for natural behaviour expression	Miller et al. (2020)
Absence of stereotypic behaviours	Reptiles lack repetitive, purposeless behaviours such as pacing or head bobbing	Suggests a well-enriched environment and mental well-being	Meredith et al. (2016)
Response to human interaction	Reptiles respond calmly and curiously to caretakers, such as approaching or watching them	Indicates trust and comfort with human presence	Padalino and Menchetti (2021) and Benn et al. (2019)

9.4 Conclusion and the Way Forward

Proper care and enrichment of animals in captivity are necessary for understanding unique ethological needs and recognising that their perceptual and cognitive worlds significantly differ from those of humans. Neglecting this principle leads to ill-informed husbandry practices that incur physical injury and induce mental distress among captive animals. This is particularly relevant to reptiles, as mounting evidence suggests their capacity for sensory experiences and emotional states. Enriching their environment must transcend superficial remedies and fundamentally address the origins of abnormal behaviours arising from captive conditions (Table 9.1).

Reptiles possess intricate sensory adaptations that mediate interactions with the environment. Key adaptations include chemosensory mechanisms, infrared radiation detection, and tactile sensitivity to various textures. Each of these adaptations plays a pivotal role in shaping the reptile's perception of the world around them, which is fundamentally distinct from human sensory experiences. Among these sensory modalities, olfaction is often pivotal in reptile exploratory behaviours and to date remains an underexplored avenue in enrichment strategies. Reptiles exhibit reliance on olfactory cues for environmental assessment. By understanding the chemical cues that trigger natural behaviours and responses in reptiles, tailored olfactory enrichment can be designed to promote species-specific exploration, foraging, and overall well-being. This approach aligns with the call for ethologically informed husbandry practices.

The use of auditory enrichment is almost absent in practice. However, it can be a valuable addition to the captive environment of reptiles. While reptiles primarily rely on visual and olfactory stimuli, auditory enrichment can help create a more dynamic and stimulating environment for reptiles. Playing recorded natural sounds from the reptile's natural habitat, such as the sounds of running water, rain, or the calls of local wildlife, helps reduce stress and makes the reptile feel more at home. Another important consideration, depending on the reptiles' microhabitat, could be harnessing the sounds of the substrate, such as the natural rustling of dried leaves, paper, or sand, when the reptile moves through them. This mimics the senses in the wild and can encourage exploration. Other such options could be feeding sounds and social calls to provide a more personalised form of enrichment.

The existing body of knowledge regarding reptile husbandry and welfare is considerably inadequate, and needs further exploration. Species-specific requirements, encompassing physical and psychological dimensions, remain largely understudied. Further research must look into multifaceted enrichment strategies, including but not limited to social, occupational, and nutritional enrichment. Moreover, a more comprehensive investigation of sensory enrichment, encompassing taste, smell, thermal gradients, and visual stimuli is essential to construct a holistic framework for reptile well-being.

Zoos across the world historically are anthropocentric and need to undergo a paradigm shift in approach to captive animal care. While entertainment and education remain valuable, a more naturalistic and cognitive well-being outlook for animals is crucial. This requires a departure from traditional practices towards fostering environments that align with the animals' natural habitats and sensory experiences. Enhancing the welfare of captive reptiles necessitates a comprehensive re-evaluation of enrichment strategies. Acknowledging their distinctive sensory perceptions and cognitive experiences, particularly their reliance on olfaction, is paramount. By bridging the gap in our understanding of reptile ethology and implementing a holistic approach to enrichment that caters to various sensory modalities and innate behaviours, we can genuinely improve the lives of reptiles in captivity. Ultimately, our moral obligation as custodians demands a shift from anthropocentrism to a profound commitment to the mental well-being of these sentient beings.

References

Azevedo C, Cipreste C, Young R (2007) Environmental enrichment: a GAP analysis. Appl Anim Behav Sci 102:329–343. https://doi.org/10.1016/j.applanim.2006.05.034

Bacon H (2018) Behaviour-based husbandry—a holistic approach to the management of abnormal repetitive behaviors. Animals 8:103. https://doi.org/10.3390/ani8070103

Bartolomé A, Carazo P, Font E (2023) Environmental enrichment for reptiles in European zoos: current status and perspectives. Anim Welf 32:e48. https://doi.org/10.1017/awf.2023.43

Benn AL, McLelland DJ, Whittaker AL (2019) A review of welfare assessment methods in reptiles, and preliminary application of the welfare quality® protocol to the pygmy blue-tongue skink, Tiliqua adelaidensis, using animal-based measures. Animals 9(1):27. https://doi.org/10.3390/ani9010027

Bloomsmith M, Brent L, Schapiro S (1991) Guidelines for developing and managing an environmental enrichment program for nonhuman primates. Lab Anim Sci 41:372–377

Bonier F, Martin PR, Moore IT, Wingfield JC (2009) Do baseline glucocorticoids predict fitness? Trends Ecol Evol 24(11):634–642. https://doi.org/10.1016/j.tree.2009.04.013

Borgmans G, Palme R, Sannen A, Vervaecke H, Damme RV (2018) The effect of environmental provisioning on stress levels in captive green anole (*Anolis carolinensis*). Anim Welf 27(1):35–46. https://doi.org/10.7120/09627286.27.1.035

Broom D (2014) Sentience and animal welfare. Anim Sentience 2016:005

Burghardt G (2015) Play in fishes, frogs and reptiles. Curr Biol 25:R9. https://doi.org/10.1016/j.cub.2014.10.027

Burghardt GM (1999) Deprivation and Enrichment in Laboratory Animal Environments. J Appl Anim Welf Sci 2:263–266. https://doi.org/10.1207/s15327604jaws0204_1

Case BC, Lewbart GA, Doerr PD (2005) The physiological and behavioural impacts of and preference for an enriched environment in the eastern box turtle (*Terrapene carolina carolina*). Appl Anim Behav Sci 92(4):353–365. https://doi.org/10.1016/j.applanim.2004.11.011

Clegg I, Borger-Turner J, Eskelinen H (2015) C-well: the development of a welfare assessment index for captive bottlenose dolphins (*Tursiops truncatus*). Anim Welf 24(3):267–282. https://doi.org/10.7120/09627286.24.3.267

Conde DA, Colchero F, Gusset M, Pearce-Kelly P, Byers O, Flesness N, Browne RK, Jones OR (2013) Zoos through the lens of the IUCN red list: a global metapopulation approach to support conservation breeding programs. PLoS One 8(12):e80311. https://doi.org/10.1371/journal.pone.0080311

Gelfo F (2019) Does experience enhance cognitive flexibility? An overview of the evidence provided by the environmental enrichment studies. Front Behav Neurosci 13:150

Griswold WG (2001) Captive care and breeding of the corn Snake, *Elaphe guttata*. J Herpetol Med Surg 11(4):35–40. https://doi.org/10.5818/1529-9651.11.4.35

Hannan AJ (2014) Environmental enrichment and brain repair: harnessing the therapeutic effects of cognitive stimulation and physical activity to enhance experience-dependent plasticity. Neuropathol Appl Neurobiol 40(1):13–25. https://doi.org/10.1111/nan.12102

Hewson CJ (2006) Environmental enrichment for captive animals. Can Vet J 47(10):992

Hill SP, Broom DM (2009) Measuring zoo animal welfare: theory and practice. Zoo Biol 28(6):531–544. https://doi.org/10.1002/zoo.20276

Hoehfurtner T, Wilkinson A, Nagabaskaran G, Burman OHP (2021) Does the provision of environmental enrichment affect the behaviour and welfare of captive snakes? Appl Anim Behav Sci 239:105324. https://doi.org/10.1016/j.applanim.2021.105324

Hurme K, Gonzalez K, Halvorsen M, Foster B, Moore D, Chepko-Sade B (2003) Environmental Enrichment for Dendrobatid Frogs. J Appl Anim Welf Sci 6:285–299. https://doi.org/10.1207/s15327604jaws0604_3

Lambert H, Carder G, D'Cruze N (2019) Given the cold shoulder: a review of the scientific literature for evidence of reptile sentience. Animals 9(10):821. https://doi.org/10.3390/ani9100821

Lance VA, Grumbles JS, Rostal DC (2001) Sex differences in plasma corticosterone in desert tortoises, *Gopherus agassizii*, during the reproductive cycle. J Exp Zool 289(5):285–289

Londoño C, Bartolomé A, Carazo P, Font E (2018) Chemosensory enrichment as a simple and effective way to improve the welfare of captive lizards. Ethology 124:674–683. https://doi.org/10.1111/eth.12800

Martínez Silvestre A (2014) How to assess stress in reptiles. J Exotic Pet Med 23(3):240–243. https://doi.org/10.1053/j.jepm.2014.06.004

Melfi VA (2009) There are big gaps in our knowledge, and thus approach, to zoo animal welfare: a case for evidence-based zoo animal management. Zoo Biol 28(6):574–588. https://doi.org/10.1002/zoo.20288

Mellen J, Sevenich MacPhee M (2001) Philosophy of environmental enrichment: past, present, and future. Zoo Biol 20(3):211–226. https://doi.org/10.1002/zoo.1021

Meredith JB, Mallory DG, Devan MS, Abigail SK (2016) Does enrichment improve reptile welfare? Leopard geckos (*Eublepharis macularius*) respond to five types of environmental enrichment. Appl Anim Behav Sci 184:150–160. https://doi.org/10.1016/j.applanim.2016.08.003

Michaels CJ, Downie JR, Campbell-Palmer R (2014) The importance of enrichment for advancing amphibian welfare and conservation goals: a review of a neglected topic. Amphib Reptile Conserv 8(1):7–23

Miller LJ, Vicino GA, Sheftel J, Lauderdale LK (2020) Behavioral diversity as a potential indicator of positive animal welfare. Animals 10(7):1211. https://doi.org/10.3390/ani10071211

Moore MC, Thompson CW, Marler CA (1991) Reciprocal changes in corticosterone and testosterone levels following acute and chronic handling stress in the tree lizard, *Urosaurus ornatus*. Gen Comp Endocrinol 81(2):217–226. https://doi.org/10.1016/0016-6480(91)90006-R

Nagabaskaran G, Burman OHP, Hoehfurtner T, Wilkinson A (2021) Environmental enrichment impacts discrimination between familiar and unfamiliar human odours in snakes (*Pantherophis guttata*). Appl Anim Behav Sci 237:105278. https://doi.org/10.1016/j.applanim.2021.105278

Näslund J, Johnsson JI (2016) Environmental enrichment for fish in captive environments: effects of physical structures and substrates. Fish Fish 17(1):1–30. https://doi.org/10.1111/faf.12088

Padalino B, Menchetti L (2021) The first protocol for assessing welfare of camels. Front Vet Sci 7:631876. https://doi.org/10.3389/fvets.2020.631876

Rose P, Evans C, Coffin R, Miller R, Nash S (2014) Using student-centred research to evidence-base exhibition of reptiles and amphibians: three species-specific case studies. J Zoo Aquar Res 2(1):25–32. https://doi.org/10.19227/jzar.v2i1.23

Rosier R, Langkilde T (2012) Behavior under risk: how animals avoid becoming dinner. Nat Educ Knowl 2:8

Schuett G, Taylor E, Kirk E, Murdoch W (2004) Handling stress and plasma corticosterone levels in captive male Western diamond-backed rattlesnakes (*Crotalus atrox*). Herpetol Rev 35:229–233

Shyne A (2006) Meta-analytic review of the effects of enrichment on stereotypic behavior in zoo mammals. Zoo Biol 25(4):317–337. https://doi.org/10.1002/zoo.20091

Spain M, Fuller G, Allard S (2020) Effects of habitat modifications on behavioral indicators of welfare for Madagascar Giant hognose snakes (Leioheterodon madagascariensis). Anim Behav Cogn 7(1):70–81. https://doi.org/10.26451/abc.07.01.06.2020

Sperry JH, Taylor CA (2008) Habitat use and seasonal activity patterns of the Great Plains Ratsnake (*Elaphe guttata emoryi*) in Central Texas. Southwest Nat 53(4):444–449. https://doi.org/10.1894/PAS-05.1

Spitzer HB, Meagher RK, Proudfoot KL (2022) The impact of providing hiding spaces to farmed animals: a scoping review. PLoS One 17(11):e0277665. https://doi.org/10.1371/journal.pone.0277665

Sueur C, Pelé M (2019) Importance of living environment for the welfare of captive animals: behaviours and enrichment. In: La Fondation Droit Animal, Ethique et Sciences. https://www.fondation-droit-animal.org/proceedings-aw/importance-of-living-environment-for-the-welfare-of-captive-animals/. Accessed 25 Aug 2023

Tylan C, Camacho K, French S, Graham SP, Herr MW, Jones J, McCormick GL, O'Brien MA, Tennessen JB, Thawley CJ, Webb A, Langkilde T (2020) Obtaining plasma to measure baseline corticosterone concentrations in reptiles: how quick is quick enough? Gen Comp Endocrinol 287:113324. https://doi.org/10.1016/j.ygcen.2019.113324

Warwick C, Frye FL, Murphy JB (eds) (1995) Health and welfare of captive reptiles. Springer, Dordrecht

Warwick C, Arena P, Lindley S, Jessop M, Steedman C (2013) Assessing reptile welfare using behavioural criteria. In Pract 35(3):123–131. https://doi.org/10.1136/inp.f1197

Warwick C, Arena P, Steedman C (2019) Spatial considerations for captive snakes. J Vet Behav 30:37–48. https://doi.org/10.1016/j.jveb.2018.12.006

Zieliński D (2023) The effect of enrichment on leopard geckos (Eublepharis macularius) housed in two different maintenance systems (rack system vs. Terrarium). Animals (Basel) 13(6):1111. https://doi.org/10.3390/ani13061111

Population Studies of Pit Vipers (Reptilia: Serpentes: Viperidae: Crotalinae) in the Southern Western Ghats of Tamil Nadu, India

10

Albert Rajendran

Abstract

Population aspects of pit vipers, *Craspedocaphalus travancoricus*, *Craspedocaphalus peltopelor* and *Hypnale hypnale*, were studied in the southernmost Western Ghats of Tamil Nadu, India, comprising Kalakad Mundanthurai Tiger Reserve and the contiguous Courtallam range, covering an area of 1000 sq km. The information gained from marked pit vipers provides the core of this study and is the main basis for conclusions regarding population size, structure, activity, home range and movements. The high population size and small home range of these ambush hunters, with abundant prey base in this area, reveal a suitable habitat in a moderately undisturbed tropical forest.

Keywords

Pit vipers · Population size · Structure · Home range · Western Ghats

10.1 Introduction

Snakes have been animals of obscurity in India from time immemorial. The immense ecological role of snakes in controlling agricultural pests, especially rodents, is conveniently ignored while killing them at the very sight. A proper understanding and scientific approach have become important to reveal their beneficial role in the ecosystem.

Studies of most Indian snakes, in the late nineteenth and twentieth centuries, dealt with taxonomical and morphological descriptions for neo-nomenclature and identification, with distributional and brief ecological notes (Günther 1864;

A. Rajendran (✉)
Research Department of Zoology, St. John's College, Tirunelveli, Tamil Nadu, India

S. Sridhara et al. (eds.), *Animal Behavior in the Tropics*,
https://doi.org/10.1007/978-981-97-7324-4_10

Boulenger 1890; Wall 1899, 1905; Abercromby 1914; Rao 1917; Henry 1925; D'Abreu 1933; Smith 1943; Mahendra 1984). Smith's (1943) volume on 'Serpentes' is a remarkable basic reference material to Indian snakes in general, even now. Field guides on Indian snakes worth mentioning are Rajendran (1967), Whitaker (1978), Daniel (1983, 2002) and Whitaker and Captain (2004). The doctoral thesis on the family of Uropeltid snakes (Rajendran 1973, 1985) is one of the foremost detailed studies of any particular group of Indian snakes.

The pioneering works of snake ecology were done in the USA by Fitch (1949, 1960, 1963a, b, 1965, 1975) in different species of snakes, including the pit viper copperhead. Other noteworthy studies in the USA include Barbour et al. (1969), Bennett and Dawson (1976), Godley (1980), Semlitsch et al. (1981), Mushinsky et al. (1982), Reinert et al. (1984) and Mushinsky (1987). An important landmark in the ecological studies of snakes is the book *Snakes: Ecology and Evolutionary Biology* (Seigel et al. 1987) that stresses the importance of methodology and techniques in snake ecology and deals with topics from systematics to status and conservation of snakes, authored by different eminent workers. One finds a few studies on the population ecology of snakes from Australia, Britain, Canada, China, Costa Rica, Japan and Sweden (Prestt 1971; Tanaka et al. 1971; Gregory and Stewart 1975; Freedman and Catling 1978; Nickerson et al. 1978; Andren and Nilson (1983); Madsen 1984; Shine and Lambeck 1985; Shine 1986; Shine et al. 2002).

Much information on population of Indian snake species is anecdotal and frequently based on specimens collected from widely scattered localities over many years, and conveniently listed in the IUCN status as common, rare, endangered, vulnerable or data deficient. Long-term field studies are limited to easily traceable large-sized snakes such as Indian rock python (*Python molurus*) (Bhupathy and Vijayan 1989; Bhupathy 1990; Bhatt and Choudury 1993) and radio-telemetric study on king cobra (*Ophiophagus Hannah*) (R Whitaker, pers. discussion). There is a dearth of studies on Indian snake populations found in tropical dense forests, especially pit vipers. Four studies on the populations of Indian pit vipers are noteworthy (Sawant et al. 2010a, b; Sawant and Jadhav 2013; Virushali 2019); still, they reveal only the relative abundance and not the density per unit area of the habitat concerned.

As predators of small vertebrates, including both primary consumers and those of higher trophic levels, pit vipers are recognized as key species in the local ecosystem. These snakes were rarely seen performing normal activities, and even when they were found, observing them proved remarkably unrewarding. A pit viper found by chance usually lay motionless for long periods, merely resting in a coiled position waiting for the prey to come near, the sluggish behavior characteristic of the pit vipers, and termed as 'sit and wait' predators (Schoener 1971). As regards snakes, the term 'ambush hunter' is often used to replace 'sit-and-wait' (Mushinsky 1987). When the snake finally began to move, it might soon be irretrievably lost because of the perfection with which it blended with its background and the dense concealing vegetation and other cover in the situations it frequented.

Compared with the literature on the snakes' closest relatives, the lizards, where long-term field studies have advanced to experimental stages, studies on the dynamics of snake populations struggle with basic descriptions (Parker and Plummer 1987). The reason for this retardation is largely due to the less tractable nature of snakes. Such characteristics make it difficult to dedicate the extended time and energy needed for an accurate assessment of population processes (Tinkle 1979). Only a few recent studies on Indian pit vipers are on the ecology with an emphasis on abundance, foraging and habitat selection (Sawant et al. 2010a, b; Sawant and Jadhav 2013; Virushali 2019).

The present study is based primarily on records obtained in the southernmost regions of the Western Ghats of Tamil Nadu, India. The study area comprises Kalakad Mundanthurai Tiger Reserve (KMTR) and the contiguous Courtallam range of Tirunelveli Wildlife Sanctuary, about 1000 sq km. The information gained from marked pit vipers, recaptured after substantial intervals, provides the core of this study and is the main basis for conclusions regarding population size, structure, activity, home range and movements.

Two earlier studies in India, identified individual snakes through their patterns on the body (Bhupathy 1990: blotches of *Python molurus*; and Virushali 2019: dorsal head pattern of *Trimeresurus malabaricus*). However, my study of marked individuals is the first and foremost in Indian pit vipers and will be of interest chiefly to specialists as well as enthusiastic naturalists. It is hoped that it will also have practical value, especially for the people living in the hilly tracts of tea, cardamom and coffee plantations, where occasionally a pit viper may injure humans and domestic animals by venomous bite. The information gathered concerning movements, population size, seasonal habits and prey items provides a background essential for the effective management of the whole ecosystem. Hence, the present study on the population of pit vipers in the Western Ghats would offer valuable basic information necessary for further ecological studies on tropical snakes.

10.2 Study Area

The study area lies within a broad ecological unit of the southernmost Western Ghats (Fig. 10.1). It forms part of the Agasthiamalai Biosphere Reserve, encompassing rich tropical forest ecosystems of high biodiversity, that fall within the Kanyakumari, Tirunelveli and Tenkasi districts of Tamil Nadu and the Thiruvananthapuram and Kollam districts of Kerala in the southernmost Western Ghats and constitutes an important biogeographical 'hot spot' of high endemism of flora and fauna (Myers et al. 2000). The eastern part of this hill ranges is known as the Tinnevelly hills, south of Shencottah gap, extending up to the Thiruvannamalai peak and the foothills, the southern border of Kalakad Mundanthurai Tiger Reserve (KMTR). This ecological unit has been named 'Ashambu Hills' in ancient literatures.

It lies between latitudes of 8°25′ and 8°60′ N and longitudes of 77°15′ and 77°35′ E, with undulating mountainous topography and steep terrain, covering 1000 sq km, out of which KMTR of 895 sq km is in the southern part and the rest is the Courtallam range in the north. Previously, Kalakad and Mundanthurai were two different

Fig. 10.1 Study area. (Southern Western Ghats of Tamil Nadu, south of Shencottah gap)

sanctuaries with a 15 km connecting corridor of evergreen forest at Kakachi and a vast stretch of tea and a few cardamom plantations in between. In April 1988, they were merged to form Kalakad Mundanthurai Tiger Reserve (KMTR) to mark the southernmost home of the Indian Tiger, *Panthera tigris* (Plates 10.1 and 10.2).

KMTR in general has scrub jungle from the foothills (about 40 m MSL) up to about 100 m MSL. The lower slopes have tropical dry deciduous forests up to about 350 m MSL. On the upper reaches are found wet evergreen forests, *Ochlandra* reed-brakes and peaks interspersed with extensive grasslands meandering with shola forests. Courtallam hills have scrub jungle at the foothills (100 m MSL) and immediately merging with semi-evergreen and evergreen vegetation deciduous forests in between. The altitude of the study area ranges from about 100 m to 1835 m MSL. The major peaks are Kailas Parvatham (1177 m), Agasthiar Peak (1835 m), Kannunni (1660 m), Thiruvannamalai Peak (1600 m) and Varaiattu Mottai (1570 m).

There are three major rivers flowing eastwards, namely, Chittar from Courtallam hills, Tamiraparani from Mundanthurai and Manimuthar from Kalakad regions. Tamiraparani is the major perennial river originating from the drainage streams connected at the base of Agasthiamalai and merging with smaller rivers such as Pambar, Mylar, Karaiyar, Servalar, Manimuthar, Peyar, Ullar, Kouthalaiyar and so on. It flows into the plains of Tirunelveli, Tenkasi and Thoothukudi districts. Pachaiyar, Kodumudiyar, Nambiar and Naraikadu are smaller rivers of Kalakad region that are perennial sources of water in the upper reaches but have a definite period of dry months as they flow in the plains. The area enjoys both southwest and northeast

Plate 10.1 A view of the evergreen patch in KMTR

Plate 10.2 A view of the evergreen vegetation adjoining tea plantation in Nalumukku

monsoon rains with an annual average rainfall of 4300 mm. As many as 17 smaller rivers flow within KMTR (Fig. 10.2), and so the reserve is aptly called as a 'River Sanctuary' (Johnsingh 2001). The area forms part of the Western Ghats of India,

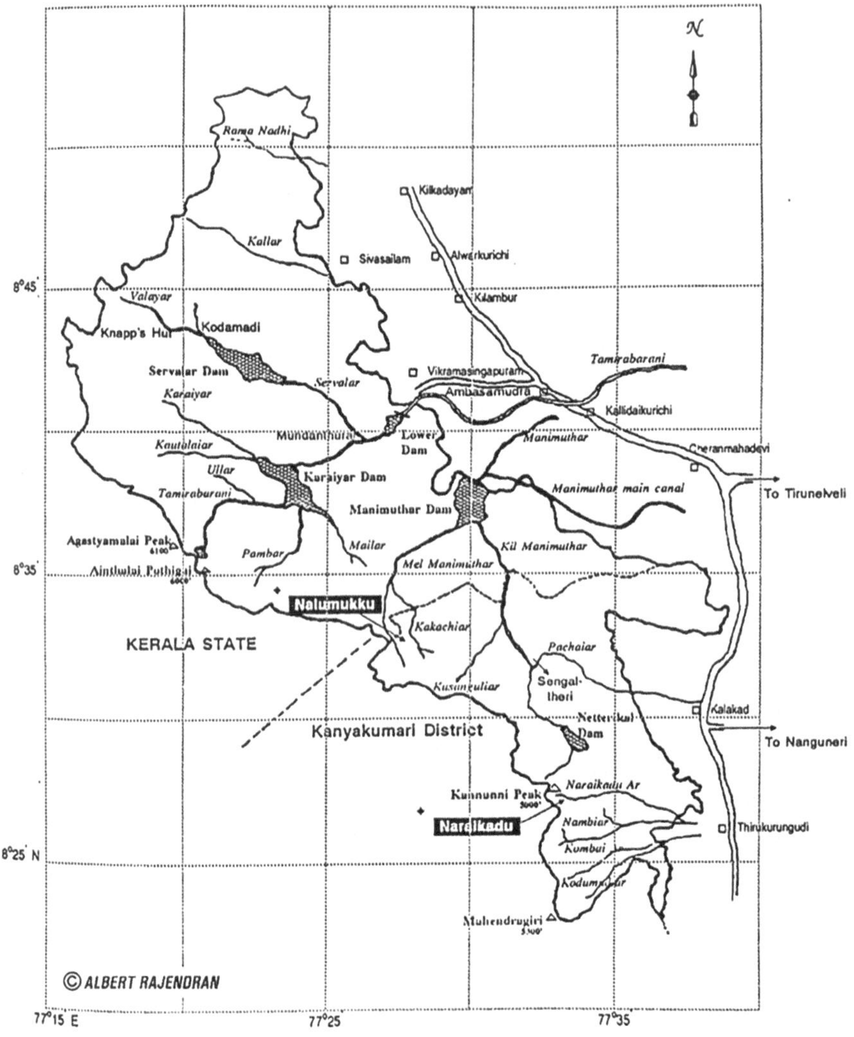

• Naraikadu and Nalumukku are the intensive study areas in KMTR

Fig. 10.2 Map of Kalakad Mundanthurai Tiger Reserve

consisting of black gneissic rocks in the underlying strata accounting for is old origin and geologic stability.

The diverse edaphic and climatic features of the area have led to rich biodiversity of fauna and flora. Six different major types of Indian forests occur in this area (Champion and Seth 1968; Legris and Blasco 1969; Parthasarathy and Mahadevan 1987). Out of 4390 species of flora (175 families and 1256 genera) recorded in the old Madras Presidency, this area accounts for 2105 species (157 families and 872 genera) of which 32 are endangered and about 60 are endemic (Henry and

Subramanyam 1981; Nair and Henry 1983). Among the fauna, most species are endangered, including major vertebrates. This area of southern Western Ghats with favourable climatic, topographical and ecological conditions with floral and faunal assemblage forms a conductive habitat for the pit vipers under study. Three intensive study areas were selected for further field studies.

10.2.1 Intensive Study Area 1 (Naraikadu: 800–1200 m)

This area covers the deciduous mountain ranges and the evergreen forests along the Naraikadu River up to the '*Grey Forest*', an undisturbed private estate of about 40 acres belonging to the Dohnavur Fellowship, within the core area of KMTR (WebPeploe 1942). The evergreen vegetation begins from 800 m upwards to the foot of the peaks of Kottangathatti and Kannunni. The peaks and immediate environs have extensive grasslands meandering with shola forests. There is dense undergrowth of shrubs and herbaceous plants. Naraikadu river flows eastwards from the base of these peaks through the estate downwards.

10.2.2 Intensive Study Area 2 (Nalumukku: 1300 m)

This study area comprises a monoculture of tea plantation adjoining evergreen forest vegetation at 1300 m, along with coffee and cardamom plantations. It lies between Kalakad and Mundanthurai sanctuaries of KMTR where a 15 km corridor of evergreen belt exists connecting the two sanctuaries at Kakachi which is about 5 km east of Nalumukku. The lease period of these plantations is getting over by 2028, and it is already declared as part of KMTR. The estates are managed by Bombay Burmah Trading Corporation. The plantations are more than a century-old, generated by clearing the original deciduous and evergreen forests of about 2000 hectares ranging from 500 m to 1300 m altitude, above Manimuthar Dam. Bordering areas around the plantations have original evergreen forest patches.

10.2.3 Intensive Study Area 3 (Therkumalai, Courtallam Range: 500–900 m)

This area (Fig. 10.3) covers the adjoining surroundings of Main Falls (200 m MSL) proceeding upwards along the Chittar river through Shembagadevi (300 m), Honey falls (450 m) and into the Therkumalai Estate (500–900 m), above into Mylodai. The overall vegetation in this riverine forest of Courtallam Hills presents features of interest from the point of view of ecology as well as of floristics (Mudaliar and Sundararaj 1954; Nair and Nayar 1986). There is no marked distinction of scrub, deciduous or evergreen vegetation along the course of the river. This situation is attributed, perhaps, to the topographic position of the hill range and the operation of the consequent climatic factors (Swamy and Govindarajulu 1956).

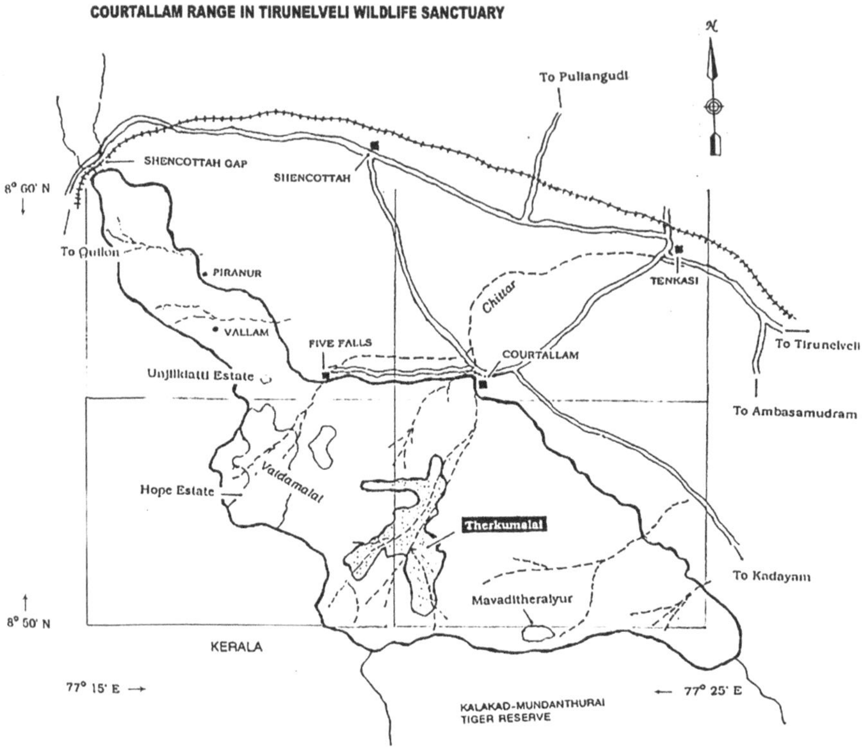

Fig. 10.3 Map of Courtallam range

There is evidence of evergreen species of flora even at the mouth of the Main Falls (200 m) merging with deciduous species. On either sides of the river on the slopes, the scrub is mixed with deciduous species up to Shembagadevi (300 m).

Systematic Position of Pit Vipers: The classification follows McDowell (1987)

Class: Reptilia, Order: Squamata, Sub-order: Serpentes, Infra-order: Alethiniophidia, Super-family: Colubroidea, Series: Opisthoglypha, Family: Viperidae, Sub-family: Crotalinae.

10.3 Description

Indian pit vipers have undergone many changes in the nomenclature from original description (Günther 1864; Boulenger 1890; Smith 1943; McDiarmid et al. 1999; Malhotra and Thorpe 2004; Mallik et al. 2021; Uetz and Hošek 2002). India is home to more than 28 species of pit vipers, restricted to the high rainfall areas of the Western Ghats, the Himalayas, the northeast and the Andaman Islands; of these, 9 species are found in the Western Ghats, of which 7 are endemic to Western Ghats, 1 found in Western and Eastern Ghats and 1 in Sri Lanka (Uetz and Hošek 2002).

The Travancore rock pit viper (*Craspedocephalus travancoricus*) and the large-scaled pit viper (*Craspedocephalus peltopelor*) are found usually between 500 m to 2000 m, and the hump-nosed pit viper (*Hypnale hypnale*) prefers lower elevations (150–1000 m). *C. travancoricus* and *C. peltopelor* are strictly endemic to the southernmost Western Ghats, south of the Shencottah gap of Tamil Nadu and Kerala (Mallik et al. 2021). *Hypnale hypnale* has its range in most parts of Indian Western Ghats up to Goa and Sri Lanka (Smith 1943; Maduwage et al. 2009). These are Crotaline Viperidae with short maxilla not extending anterior to prefrontal articulation, dentition reduced to a pair of front fangs, a pair of thick-walled venom glands and a deep loreal (heat-sensory) pit lodged in a fossa on prefrontal and maxilla on the side of the face (Bullock and Fox 1957). The head is broad with a distinct neck. The body is cylindrical with smooth or feebly keeled scales, and the tail is moderately short and prehensile.

10.3.1 Travancore Rock Pit Viper

Craspedocephalus travancoricus (Mallik et al. 2021)—Plate 10.3 Eye with vertical pupil as in other crotalines; upper surface of the head covered with smooth small scales as that of the neck scales; supraocualrs narrow; 9 supralabials; 11 infra labials. Dorsally, the neck scales are smaller than those of the body. There are 21 dorsal scale rows on the body for most of its length and 23 in the posterior mid-body, and from a point approximately one-fifth of the distance to the vent, there are only 15 rows. The dorsal one or two median scales are feebly keeled. The ventrals range between 144 and 149. The double row of sub-caudals ranges between 54 and 57 in

Plate 10.3 Travancore rock pit viper (*Craspedocephalus travancoricus*)

males and 44 and 48 in females. The total length ranges from 105 mm to 522 mm in length, with definite proportional variation in snout-vent length to tail length ratio among males and females.

Four colour morphs of *C. travancoricus* are recorded: green, blue, brown and orange (Virushali 2019). The reason for this colour variation is not known, as they are seen as a mixed population in the same habitat. In my study area, green and blueish morphs are common, one orange was seen, and brown was not encountered. In a green morph, the dorsal colour of the head is covered with dark brown to light brown, with scales bordered with pale yellowish green. The pattern of the small blotches on the head is found to be individually different. The body is faded brown or green with yellow markings. The tail has yellowish-green bands on a dark brown to black background. The snake camouflages extremely well, with the overall colour and pattern, to its usual microhabitat, namely, dark green wet rocks with mosses and lichens along the border of the river course or among lichen-clad low branches.

10.3.2 Large-Scaled Pit Viper

Craspedocephalus peltopelor (Mallik et al. 2021)—Plate 10.4. Eye with a vertical pupil as in other crotalines; nostril small in the nasal shield; broad snout; upper head shields very large and smooth; supraoculars very large; 7 or 8 supralabials; 2 rows of large temporal scales, feebly keeled and almost twice as large as the neck scales. The neck scales are slightly smaller than those of the body. The scales on the neck and the body are strongly keeled. There are 17 dorsal scale rows in the anterior body, 12–15 rows in the middle part and 9–10 rows on the posterior portion towards

Plate 10.4 Large-scaled pit viper (*Craspedocephalus peltopelor*)

the vent; the median 10 rows of the body scales are the largest. The ventrals range between 148 and 152. The double sub-caudals are 55–63 in males and 44–56 in females. The total length ranges from 123 mm to 640 mm with male–female proportional variation in snout-vent to tail length.

The colour is uniformly bright or leaf green on the dorsal body and paler below. A whitish or pale yellowish line runs along the side of the body, from the first scale row up to the vent. The posterior half of the tail is dark brown to blackish. As the snake is usually arboreal climbing onto the lower branches of bushes, it exhibits a perfect camouflage in the green foliage.

10.3.3 Hump-Nosed Pit Viper (*Hypnale hypnale*) (Merrem 1820; Plate 10.5)

The snout tip is horizontally blunt and slightly turned up at the tip, giving it the name; neck scales are as small as the body scales; body scales are feebly keeled, especially at the posterior body. There are 17 dorsal scale rows on the body for its full length to the vent and in a few specimens and 15 rows on the posterior part of the body. Males and females do not vary in ventral counts; ventrals range from 138 to 158, and the sub-caudals range from 38 to 46 in males and 32 to 38 in females. The subcaudals are usually paired, but in a few specimens, the first 4–8 of them are single. The snakes range from 88 mm to 512 mm in length. Though the females have more proportionate snout-vent length, the tail length does not have significant variation among males and females.

Plate 10.5 Hump-nosed pit viper (*Hypnale hypnale*)

The body ground colour varies from dark greyish, rufous, light brown to dark brown. A series of dark angular 16–23 spots are seen on either side of the median dorsal line, sometimes alternating with those on the other side forming miniature butterfly shapes. The anterior angular line of each spot has a dark brown stripe. The spots are darker (blackish brown) in juvenile and young individuals. The ventral side is paler or dull yellowish, sparsely dotted. The top of the head is usually dark brown with a lighter post-ocular stripe. The tip of the tail is often pale yellow or orange. As the snake is mostly terrestrial, the colour pattern blends well with the microhabitat it is found, namely, the dry fallen leaf litter of the forest floor.

10.4 Population Trends

A viable population is one that maintains its genetic vigour and potential for evolutionary adaptation (Soulé 1987). A crucial parameter in determining the amount of genetic variability maintained by a population is its size (Lande and Barrowclough 1987). Estimates or measures of such numerical attributes of a population are useful in making comparisons of related species, of populations of the same species in different habitats or of the same population at different times.

All animals are characterized by the space patterns they adopt within their habitat. Resources such as food, shelter and mates are distributed in space, and so space itself becomes an important resource. The area in which animal moves during normal daily and seasonal activities is known as the home range. Knowledge of home range is useful when examining population dynamics, particularly population size, habitat utilization, foraging frequency and food habits. Most field studies of snake populations have employed the mark–recapture technique as in Fitch (1987) for population estimation since direct enumeration and observation are feasible only in limited circumstances such as aggregation in a hibernaculum.

Observations and data presented in this study were primarily based on records of marked individuals of snakes and subsequent recapture in the intensive study areas. Activity field study was done by continuous visual observations of actual sightings of one *Craspedocephalus travancoricus*, one *C. peltopelor* and one *Hypnale hypnale*. Diet composition was studied by analyzing the stomach contents of sample specimens, palpation and visual records.

10.5 Population Size

The data presented were gathered during a series of visits to the intensive study areas from 2010 to 2013. The study area (about 1000 sq km.) was thoroughly surveyed during the pre-census period for about 10 months to fix the intensive study areas, where reasonably viable populations of the pit vipers existed for continuous study and with logistic support for stay in the forests. Two quadrats of each 100 m × 100 m were laid in each intensive study area, covering all possible microhabitats in the undulating sloppy terrain, including streams, undergrowth, shrubs and trees. In Nalumukku, the quadrats covered the tea plantation as well as the adjoining evergreen vegetation.

On each census date, the quadrats were thoroughly searched through a visual encounter survey to locate the snakes, and observations were recorded for probable prey species and other ecological parameters. Stumpy wooden pegs were fixed at the point of each snake's prolonged stay, and bright nylon threads with white flags were tied around the pegs for easy subsequent recognition. Volunteers and field assistants were employed whenever needed, especially during night surveys. Individual snakes were sexed by a blunt aluminium probe, basic morphometric measurements such as snout-vent length, tail length, number of ventrals and sub-caudals were made, and each snake was marked uniquely by ventral scale clipping (Brown and Parker 1976). Each mark involved the excision of a V-cut on the left or right of the ventral scale. The snake was permanently marked by clipping of ventral scales to individually recognize the snake. Many combinations were possible by this unique marking. The snake was then released at the same point of capture. This technique of marking is quite simple, and the scar of the cut could be recognizable even after 2 years. Many combinations were possible by this unique marking. Fitch (1987) states that among a multitude of marking methods in snakes, clipping ventrals or sub-caudals is the most effective for long-term continuous study. This method was successful in this study as pit vipers were usually sedentary and sit-and-wait ambush predators that remained in the same foraging and resting area for many days or weeks (Mushinsky 1987) and were easily traceable in microhabitat on subsequent days within the vicinity of the previous site.

Population size estimates were made using the Peterson Index of mark-recapture method as in Robson and Regier (1964), and Pollock (1982). Statistically, this method requires that (1) the population is constant with no mortality or recruitment during the experimental period, (2) no immigration or emigration occurs, and (3) certain statistical assumptions are met concerning the randomness of capture and mixing of marked and unmarked individuals. Although these requirements are rarely fully satisfied in studies of animal populations, the method has been effectively utilized in estimating populations of snakes in the United States (Carpenter 1952; Fitch 1960, 1963a, b, 1965, 1975) in Canada (Freedman and Catling 1978) and in Brazil (Guimarães et al. 2014). Therefore, Peterson's estimate was calculated by multiple censusing, keeping the first record of snakes actually handled and marked as 'initial population', and the captured unmarked snakes at each time are marked. This has fairly avoided discrepancies in successive census dates. The technique is described in detail by Smith (1974).

Peterson's estimate was calculated by the following formula:

$$\text{Population } Size\, N = \frac{\begin{array}{c}\text{Initial Marked population}(M)\\ \times \text{Total Number in the successive capture}(C)\end{array}}{\text{Marked population in the successive capture}(R)}.$$

For each of the census dates, a Peterson estimate of the population size was calculated for the three species of snakes. In addition, a separate population estimate was made using the Schnabel index (Schnabel 1938) following the formula:

$$N = \frac{?MC}{?R},$$

where M is the number of marked individuals in the initial capture. An approximate standard error of the estimate was calculated with the number of sampling days, including the first marking date (pre-census date) for both Peterson and Schnabel estimates. Population estimate by the two methods was made for the three species of snakes for the 2 years put together. For each population estimate, 95% confidence intervals were also calculated.

For *Craspedocephalus travancoricus* (at Naraikadu), I did 10 census records in the first year (May 2010–May 2011) and 9 records in the second year (June 2011–May 2012); For *Craspedocephalus peltopelor* (at Nalumukku), I did 11 records in the first year (June 2011–June 2012) and 10 records in the second year (July 2012–June 2013); For *Hypnale hypnale* (at Therkumalai), I did 10 records in the first year (February 2011–February 2012) and 9 records in the second year (March 2012–February 2013).

Peterson's estimates of the population sizes of the three species of pit vipers are summarized in Table 10.1. The census dates are spread over two calendar years. The small standard errors are due to fairly good recapture percentages. The snakes

Table 10.1 Summarized population estimates for the three species of pit vipers

	Craspedocephalus travancoricus		*Craspedocephalus peltopelor*		*Hypnale hypnale*	
Method of population size estimate	Population size ± SE (95% limits)	95% Confidence interval	Population size ± SE (95% limits)	95% Confidence interval	Population size ± SE (95% limits)	95% Confidence interval
Peterson index						
1st year	65 ± 16	49–81	38 ± 7	31–45	41 ± 10	31–51
Second year	71 ± 18	53–89	57 ± 5	52–62	68 ± 11	57–79
For the 2 years	68 ± 8	60–76	47 ± 6	41–53	54 ± 8	46–62
Schnabel index						
1st year	64 ± 7	57–71	40 ± 5	35–45	43 ± 5	38–48
Second year	73 ± 7	66–80	57 ± 5	52–62	70 ± 6	64–76
For the 2 years	69 ± 8	61–77	49 ± 4	45–53	56 ± 4	52–60
Mean of the two methods						
1st year	65		39		42	
Second year	72		57		69	
For the 2 years	69		48		55	

especially adults do not have much dispersals from their respective home ranges in the intensive study areas. Table 10.1 shows that the approximate population size of *Craspedocephalus travancoricus* is 69 for an area of 1.6 ha (43 snakes/ha), that of *C. peltopelor* is 48 for an area of 2 ha (24 snakes/ha), and that of *Hypnale hypnale* is 55 for 1.6 ha (34 snakes/ha). Population sizes of Indian pit vipers by mark–recapture method were not studied earlier as per unit area. The studies of Sawant et al. (2010a, b) and Virushali (2019) on pit vipers in the Western Ghats of Goa and southernmost Maharashtra are limited to relative abundance.

The trend in the estimates and the summarized data (Table 10.1) revealed that the population sizes of the pit vipers studied do not have much fluctuation in this part of the Western Ghats. The recapture success is moderately good in my study, especially in the later part of the census, involving a period of 2 years of at least 19 census dates. The results obtained in my study too need further verification and testing throughout their geographic range. When several sets of figures have become available for the same area at different times, they have often shown drastic discrepancies (Turner 1977). Figures are influenced by the fact that within a species the amount and type of activity changes rapidly and progressively throughout the season (Gibbons and Semlitsch 1987). Among males, females (breeding and non-breeding), immatures and juveniles, behaviour and seasonal responses may be different from each other (Fitch 1987).

The pit vipers in the study revealed a high population size (Table 10.2). High variation in population estimates of snake populations is found to be common in many other snake census studies (Fitch 1960, 1975; Gregory 1974; Guimarães et al. 2014). Fitch (1975) considered high variability to be a feature of mark–recapture censuses involving snakes because of low recapture success and the secretive behaviour of most snakes. A long-term ecological study (almost 30 years) of the Shedao pit viper (*Gloydius shedaoensis*), an insular population endemic to the island of Shedao in northeastern China by Shine et al. (2002) has exposed the works of

Table 10.2 Comparison of population size for various Viperids

Species	Study area	Population size (no/ha)	Reference
Vipera berus	South Britain	14	Prestt (1971)
Vipera berus	An island, west of Sweden	4	Andren and Nilson (1983)
Vipera ursinii	Uzbekistan, Central Asia	7	Bogdanov (1965)
Crotalus cerastes	Mojave Desert, California	<1	Brown (1970)
Crotalus viridis	Central California	3	Fitch (1949)
Crotalus viridis	British Columbia	1	Preston (1964)
Crotalus viridis	British Columbia	2–3	Macartney (1985)
Crotalus horridus	Northeastern Kansas	<1	Fitch (1982)
Trimeresurus flavoviridis	Japan	3–9	Tanaka et al. (1971)

Chinese researchers, hitherto unknown to other parts of the world. A robust mark–recapture model on the golden lancehead pit viper in Brazil by Guimarães et al. (2014) is worth mentioning to follow for further validation, to have a good trend in the population dynamics for a longer period, especially in disturbed environments.

The relative abundance with high population size (number/ha) of the three species of pit vipers in the southern Western Ghats is due to a combination of (1) dense cover provided by the semi-evergreen and evergreen forest patches, (2) the apparent abundance of suitable food resources within the small home ranges of the snakes (i.e. abundance of frog species in the perennial streams, invertebrates such as soil arthropods in the thick undergrowth, skinks and agamid lizards on the forest floor and bushes, frogs among the leaf litter of the forest floor), (3) suitable areas for winter torpor (e.g. crevices among rocks and bottom trunks of large trees), and (4) probably a moderately stable and least-disturbed habitat.

10.6 Population Structure

An analysis of an animal population requires knowledge of the underlying sex ratio and age structure. The recorded data of body proportions (SVL and tail length) and the total length of all the individual snakes of the three species are not presented here. Table 10.3 summarizes data on sex ratios and the total body length of the snakes. The ratio of males to females for *Craspedocephhalus travancoricus* approximates 1:1, while for *C. peltopelor* and *Hypnale hypnale* the male–female ratio approximates 1:1.5.

Variations in the sex ratios among snakes are due to many factors. Sex ratios of snakes at birth (primary sex ratios) generally are not statistically different from 1:1 and observed sex ratios in the later age groups (secondary sex ratios) often deviate from equality (Parker and Plummer 1987). Shine and Bull (1977) related this to genetic mechanisms. Dispersal of young and adult males to long distances from the site of activity, sexual differences in habitat selection or activity patterns may also lead to variations in the sex ratios. The basic question is whether the deviations are real, reflecting the actual population structure. Studies on similar lines of deviations from an equal sex ratio are of interest and need comparisons of data from different localities.

Table 10.3 Mean body length of the three species and sex ratios

Species	Number examined	Mean body length (mm ± SD)		Proportion of males ± SE	Approximate M/F sex ratio
		Male	Female		
Craspedocaphalus travancoricus	66 *M* = 34 *F* = 32	383 ± 109	386 ± 120	0.515 ± 0.063	1:1
Craspedocaphalus peltopelor	54 M = 22 F = 32	370 ± 158	428 ± 182	0.407 ± 0.055	1:1.5
Hypnale hypnale	57 M = 23 F = 34	356 ± 118	375 ± 129	0.404 ± 0.054	1:1.5

The mean total length of females was slightly longer than males for the three species. Figure 10.4 shows that the length–frequency distributions for the three species are skewed toward adult sizes. The proportion of young and juveniles was quite small to that of adults. The low proportion of young individuals could have resulted from a combination of high infant mortality, the tendency of young to disperse too

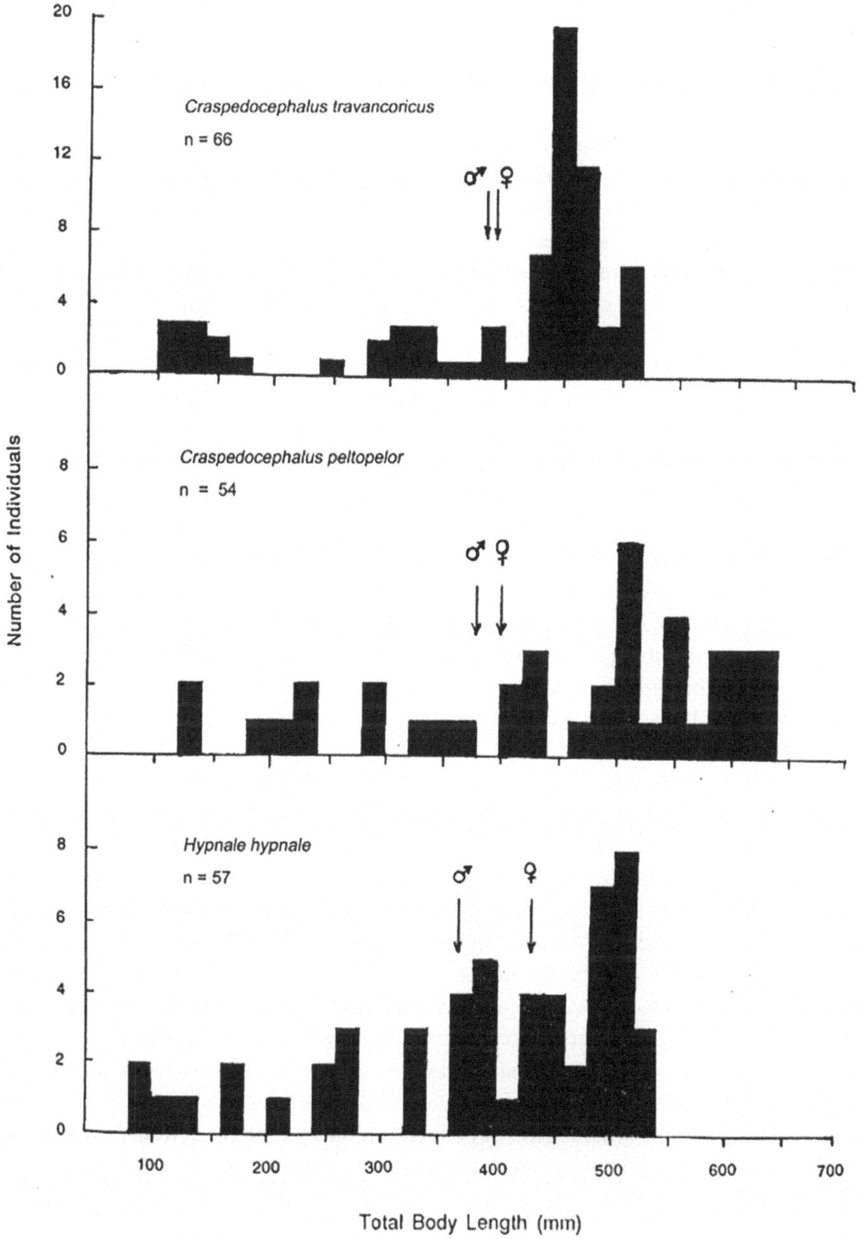

Fig. 10.4 Length–frequency distributions for the three species of pit vipers

far, as they feed on insects and others and because of behavioural differences between young and adult snakes.

The low proportion of young individuals in the population and differences in habits were also suggested by Fitch (1960) for copperheads in Kansas and Freedman and Catling (1978) for garter snakes in Canada. It is suggested that long-term capture and recapture studies year after year would give a concrete understanding of population growth, age, reproduction and recruitment.

10.7 Activity Field (Home Range and Movements)

Nickerson et al. (1978) used the term 'activity field' to describe activity-linked movement of snakes. Home range size was calculated following the minimum convex polygon method described by Jennrich and Turner (1969). Actual locations of sightings of the snakes were used to plot the activity areas. As the Indian Crotalines are ambush hunters with prolonged stays in one place and very slow moving, it was tried to continuously follow the snake day–night throughout the activity range. One male *Craspedocephalus travancoricus* in Naraikadu, one female *C. peltopelor* in Nalumukku and one female *Hypnale hypnale* at Therkumalai could be successfully followed to measure the home range size with sufficient sites in the convex polygon. The activity areas utilized by *C. travancoricus* and *H. hypnale* were measured in square metre, as they were almost terrestrial. The three-dimensional activity range in volume was estimated for *C. peltopelor*, as it was arboreal and moderately terrestrial, following the water displacement method (Nickerson et al. 1978). From the point of release, movements and nature of activity were recorded by actual visual sightings. Locations of prolonged stay in one place were marked by inserting flagged wooden pegs in the soil and the points of stay on rocks or fallen logs were painted. A compass was used to know the direction of the movements, a string was used to measure the linear distance between consecutive points, and peripheral points of stay were connected to fix the activity field.

The activity field was plotted to scale as convex polygons on graph paper, and the number of enclosed squares was counted, for measuring the home range size for *Craspedocephalus travancoricus* and *Hypnale hypnale*. For *C. peltopelor* of Nalumukku, in addition to the area of activity, a three-dimensional scale model of the snake's activity field was constructed, with a 3 mm cardboard. Thin needles were inserted into the cardboard at the appropriate sites at all vertical points, representing the snake's arboreal movements. Pieces of cardboard were cut into shapes so that the model became an enclosed multi-sided volume in the shape of the convex form used by the snake. It was then covered with a layer of polythene sheet and submerged into a graduated glass jar of water. The volume of displaced water was determined, and simple conversions gave the volume of the snake's activity field in cubic metres (m^3). For location data of single animals, estimates made using peripheral points give the maximum extent of the animal's range (Worton 1987).

10.8 Activity Field of *Craspedocephalus travancoricus*

The movements of a male *C. travancoricus* (SVL: 390 mm and tail length: 86 mm) at Naraikadu are summarized below.

The snake was initially found camouflaged on a wet rock covered with lichens, moss and herbaceous plants, under thick foliage tall trees on 2 May 2011 at 1100 h. It was captured, marked by ventral scale clipping and released at the same point of capture, after recording basic measurements. Immediately after release, it moved from the point of release, onto a dry fallen tree log across the river, 3.5 m from the previous location, and stayed coiled in its typical ambush posture and tail clinging to a small piece of peeled bark. On 8 May at 0425 h, it fed on a small frog which jumped onto the log from the water below. The next feed was on 7 June, and the prey was an *Indirana beddomii* frog which jumped onto to the rock near the snake. On 25 June, it moulted. On 4 July, it again fed on a large *Hylarana temporalis.* On 5 August, it moulted. On 7 September 2011, the snake returned to the same fallen tree log across the river, after 129 days. The activity was within the polygonal area, even after subsequent tracking for 280 days.

The total horizontal movement recorded: 76.7 m.
The total activity field utilized by the snake: 48.2 sq.m.

Being an ambush hunter, the snake had fewer movements from one point to another and had prolonged stays in the marked locations (Fig. 10.5). The movements of the snake correlated with the time of the day corresponding to temperature fluctuation, seasonal thermoregulatory purposes (basking), rainfall, waiting in ambush, and for feeding purposes. It was generally found merging with the substratum, either a rock full of moss, or a dark green tree stump with lichens or in root clumps, near moist places or along water edge.

The snake had once stayed along the periphery of the range on 28 October and had entered into winter torpor on 2 November 2011 in a deep crevice with leaf litter at the base of a tree, *Ervatamia coronaria,* when the mean air temperature was 17 °C and the crevice temperature 22 °C. On 27 January 2012, the snake came out of the crevice after a winter torpor for 85 days, presumably to avoid the cold and wet ambient conditions. Subsequently, after basking on a rock, when the air temperature was 20 °C, it ate a *Hylarana temporalis* on the same day.

On tracking the snake for 280 days, a certain activity pattern is revealed (see Table 10.4). The snake spends much time only resting motionless in ambush (66.39%) and the last time feeding (0.05%).

The movements of the snake from one point to another had been very slow and for short distances, the longest distance was 9.6 m on 2 June 2011, and the shortest distance was 1 m when it moved for basking after heavy rain on 28 October 2011 and to hide beneath a rock. The longest time spent at a single location was 85 days 18 h, from 2 November 2011 to 27 January 2012. The time recorded for the movements between different sites and for other activities reveals *Craspedocephalus travancoricus* a nocturnally active and occasionally diurnal, mostly terrestrial as well

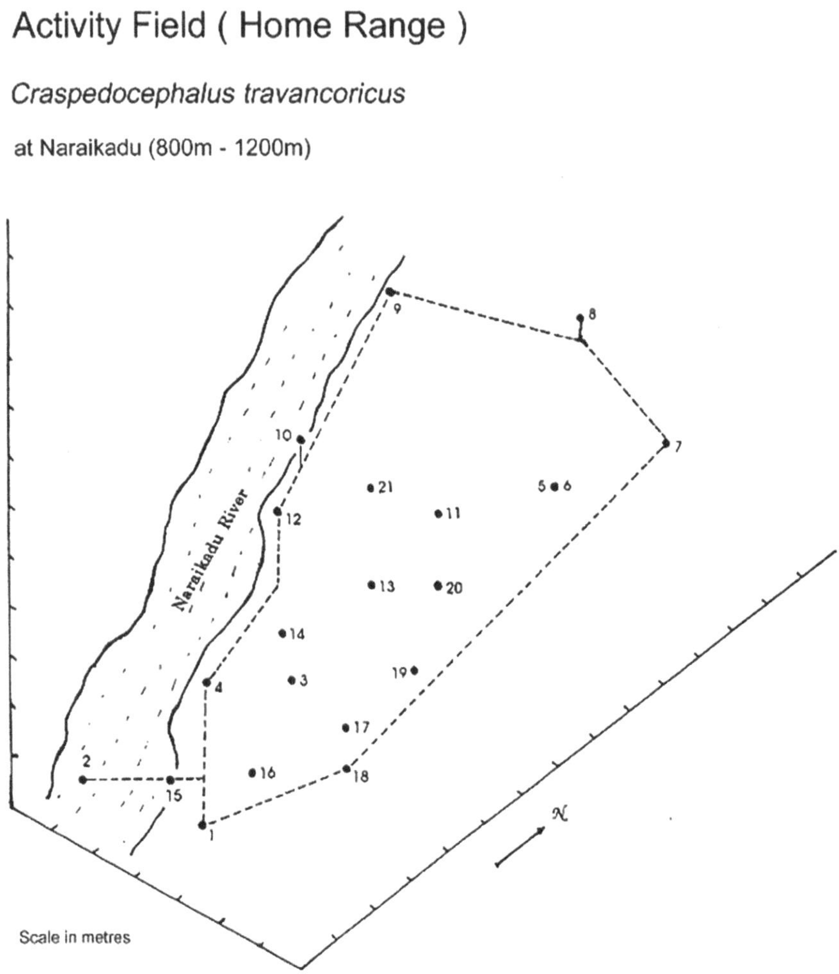

Fig. 10.5 Activity field of *C. travancoricus* at Naraikadu. Numbers 1–21 are the chronological sequence of the location sites where the snake had prolonged stays

Table 10.4 Activity pattern of a male *Craspedocephalus travancoricus* at Naraikadu, based on observations made during the measurement of activity field (from 2 May 2011 to 7 February 2012)

	Activity time (in min)	Percentage of activity (%)
Feeding	193	0.05
Moulting	125	0.03
Moving	280	0.07
Basking	11,457	2.84
Winter torpor	123,480	30.63
Resting in ambush	267,665	66.39

Total period of observation—403,200 (280 days)

as arboreal in low branches, ambush hunter. The activity field was overlapping with two other individuals along the river edge but not recorded for home range measurements.

10.9 Activity Field of *Craspedocephalus peltopelor*

The movements of a female *C. peltopelor* (SVL: 493 mm and tail length: 80 mm) at Nalumukku are summarized below. The snake was found on the canopy of a tea bush, basking in the sun on 28 November 2011 at 1150 h. At 1430 h on the same day, it moved to the canopy of the next tea bush, got down to the ground and to the edge of the tea garden and stayed coiled at the base of the stem. On 29 November 2011 at 1920 h, it crossed a 2 m wide open path from the tea edge to the adjacent evergreen vegetation, climbed a *Hopea utilis* tree along the projections of the trunk nodes, to about 2 m from the ground into a hollow crevice in the tree, in which fallen dry leaves had collected. It entered into winter torpor in the tree hollow where the temperature was 22 °C and the air temperature was 16 °C. The snake came out of the winter torpor on 2 February 2012 at 1530 h after 65 days when the air temperature was 20 °C.

From the tree hollow, it got down, crossed the open path into the tea garden and climbed on a tea bush. It moved on the tea canopy for some distance and lay flat, basking with its tail clinging to a twig. On 3 February 2012, it went into a mouse burrow and fed on a field mouse. Again on 17 March 2012 at 1425 h, it fed on a green lizard, *Calotes calotes*. On 26 March, it moulted the skin at 2235 h. The snake was monitored till 7 April 2012 when it reached about 1.5 m from its original location of capture and release after 130 days.

Total horizontal movement recorded: 50.0 m.
Total vertical movement recorded: 14.8 m.
The total area utilized by the snake: 54.1 sq.m.
The three-dimensional activity field: 99.5 m^3.

The snake had utilized the tea garden as well as the evergreen vegetation nearby for its activity. It had fewer movements from one point to another and had prolonged stays in the marked locations (Fig. 10.6). The movements had correlations with time of the day, temperature fluctuations (winter torpor), thermoregulatory purposes (basking), camouflage (merging along the twigs and leaves of the tea bush as well as in the evergreen vegetation) and for feeding purposes.

The movements of the snake were also very slow and for short distances, the longest distance covered from one point to another was 7 m, and the shortest distance was 1.5 m, for basking along the tea bushes in February 2012. The longest time spent at a single location was 65 days 5 h, in a tree hollow crevice at 2 m above ground level. The shortest time spent at a single location was 1 h 50 min, just after feeding a green lizard (*Calotes calotes*), before going under a rock ledge. After tracking the snake continuously for 130 days, the activity pattern showed that it

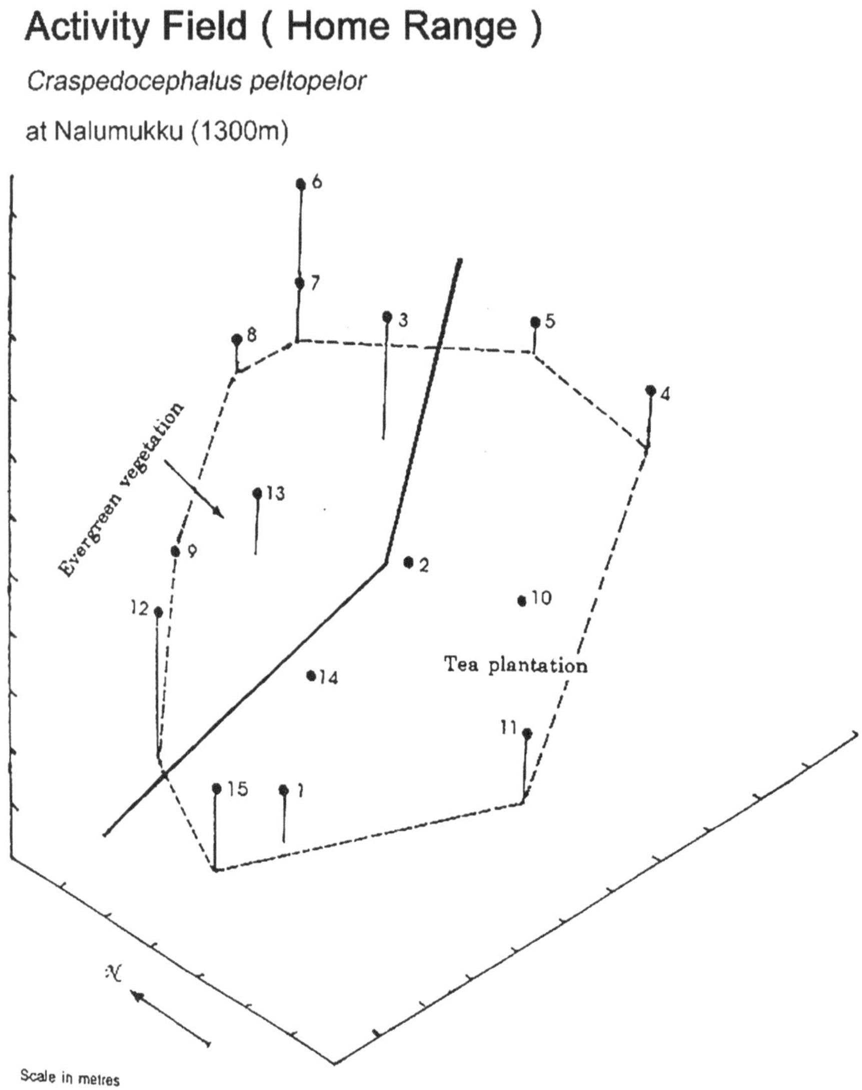

Fig. 10.6 Activity field of *C. peltopelor* at Nalumukku. Numbers 1–15 are the chronological sequence of the location sites where the snake had motionless prolonged stays. Numbers 1, 2, 4, 5, 10, 11, 14 and 15 indicate the sites utilized by the snake in tea plantation. Numbers 3, 6, 7, 8, 9, 12 and 13 indicate the sites utilized by the snake in the adjacent evergreen forest patch

spent much of its time in motionless ambush posture (45.30%) and winter torpor (50.00%) and very less time for feeding (0.07%), moving (0.07%) and moulting (0.04%). The time spent and pattern of the snake's different activities are summarized below in Table 10.5.

Table 10.5 Activity pattern of an adult female *Craspedouphalus peltopelor* at Nalumukku, based on observations made during home range measurement (from 28 November 2011 to 7 April 2012)

	Activity time (in min)	Percentage of activity (%)
Feeding	130	0.07
Moulting	65	0.04
Moving	135	0.07
Basking	6490	4.53
Winter torpor	93,600	50.00
Resting in ambush	84,870	45.30

Total period of observation—187,290 min (130 days)

The time recorded for the movements revealed that *C. peltopelor* is a predominantly arboreal, nocturnally active but with certain exceptions when it is diurnal too.

10.10 Activity Field of *Hypnale hypnale*

The movements of a female *H. hypnale* (SVL: 396 mm and tail length: 65 mm) at Therkumalai are summarized below. The snake was located among the leaf litter, about 1 m from the base of a huge *Terminalia* tree on 16th February 2012, at 0915 h. On release after marking, it stayed coiled at the base of the stem, with its tail wriggling for a few seconds. There were many skinks (*Ristella* sp.), small frogs (*Microhyla* sp.) and plenty of soil arthropods among the leaf litter on the forest floor. On 19 February at 1920 h, it moved 3 m on the forest floor, to another site. Further sites of prolonged stays were all within the leaf litter, around the tree trunk. The snake had fewer movements from one point to another and lay motionless in the marked locations (Fig. 10.7). On 5 April at 0815 h, at site 7, it caught and ate a skink. It moulted the skin on 28 April and drank a little water collected on a dry leaf. After 68 days of its previous feeding, on 13 June at 1815 h, at site 12, it ate two small frogs. On 8 September 2012, after 155 days, it came to site 16, about 3 m from site 6. The snake had very few movements within the leaf litter, and for short distances, especially for selecting ambush sites for feeding. The shortest distance moved between consecutive sites was 1.5 m, and the longest distance was 3 m. It was found extremely camouflaged with the substratum and displayed tail wriggling whenever any skink or a frog came within its reach; still, it had eaten only twice during the 155 days of observation, though its small home range had rich prey availability. Two other individuals of *H. hypnale* were also recorded in the vicinity of the above snake, overlapping the home range. A few individuals were seen lying in ambush posture along streams, waiting for frogs.

The total horizontal movement recorded: 34.5 m.
The total activity field utilized by the snake: 23.4 sq.m.

After tracking the snake continuously for 155 days, the activity pattern shows that it spends much of its time in motionless ambush posture (93.41%), basking

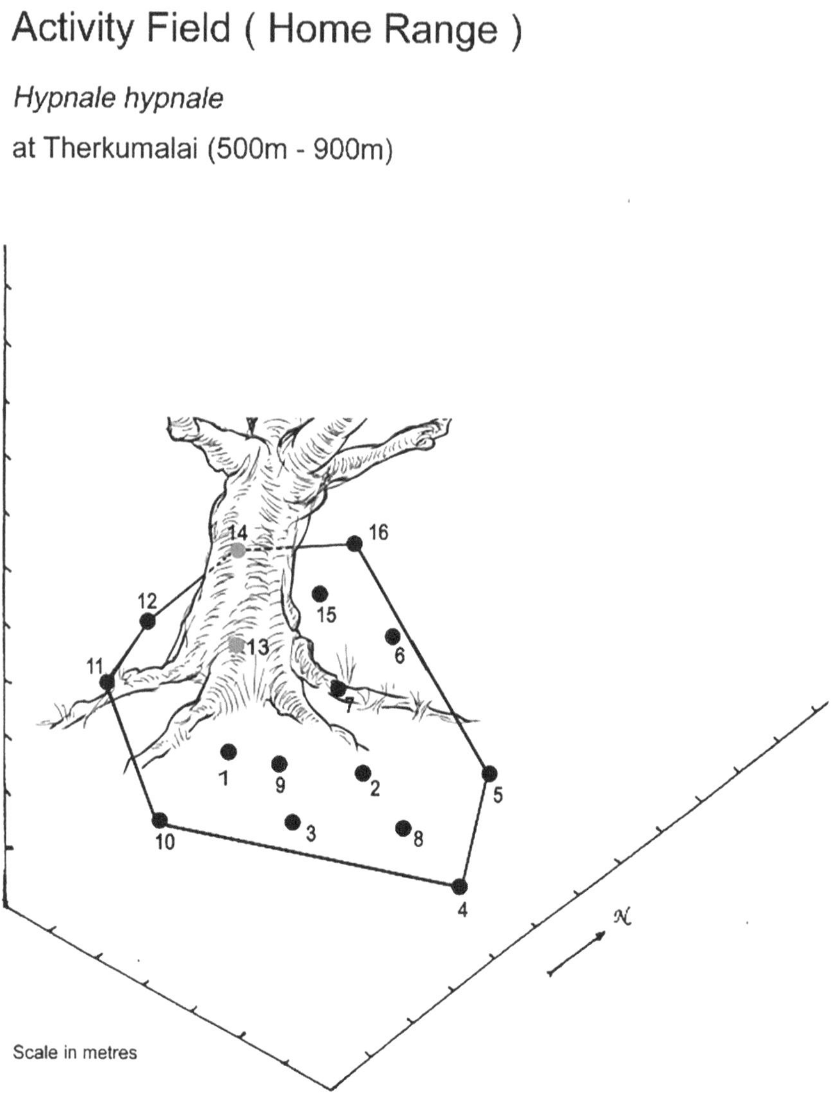

Fig. 10.7 Activity field of *Hypnale hypnale* at Therkumalai. Numbers 1–16 are the chronological sequence of the location sites where the snake had motionless prolonged stays

(6.45%) and very less time for feeding (0.07%), moulting (0.03%) and moving (0.06%). The time spent and pattern of the snake's different activities are summarized in Table 10.6.

The time recorded for the movements revealed that *H. hypnale* is predominantly terrestrial, nocturnally active and occasionally diurnal too. *H. hypnale* did not undergo winter torpor, as observed in *Craspedocephalus travancoricus* and *C. peltopelor*, as it is generally found in low elevations with higher temperatures, the highest temperature recorded in its habitat being 20 °C.

Table 10.6 Activity pattern of an adult female *Hypnale hypnale* at Therkumalai, based on observations made during home range measurement (from 16 February 2012 to 8 September 2012)

	Activity time (in min)	Percentage of activity (%)
Feeding	110	0.05
Moulting	70	0.03
Moving	130	0.06
Basking	14,400	6.45
Resting in ambush	208,490	93.41

Total period of observation—223,200 min (155 days)

The term 'Activity Field' has been referred to as the area of movements of an animal for its foraging, seasonal movement to escape extreme environment, daily activity movements and escape from enemies (Nickerson et al. 1978). Though many workers (Fitch and Shirer 1971; Henderson et al. 1976; Nickerson et al. 1978; Emerson et al. 2022) demonstrated that radiotelemetry was the most feasible means of studying the movements of ambush hunters of tropical forest snakes, the present study was possible because of relatively short movements of the three species, the lethargic nature of the snake, and the habit of lying motionless for long periods of time. However, radio-telemetric studies of pit vipers revealed comparisons among juvenile, young, males and females in the same as well as different localities.

Huey and Pianka (1981) demonstrated in the lacertid lizards of the Kalahari Desert, that ambush hunters move relatively short distances per move and per unit of time and move a small proportion of the time. Tables 10.4, 10.5 and 10.6 reveal similar observations of activity patterns of *C. travancoricus* in Naraikadu, *C. peltopelor* in Nalumukku and *H. hypnale* in Therkumalai. There is a definite non-linear home range area, covering its habitat utilization. Virushali (2019) tried to follow the movements of *Trimeresurus malabaricus* through linear points of GPS records, though for a limited time, in the Western Ghats regions of Maharashtra and Goa, India. Sawant et al. (2010a, 2010b) observed a higher number and population shift of *Hypnale hypnale* in the adjacent cashew plantations during post-monsoon, winter, and late summer seasons, in a highly disturbed zone in Goa. Individual snakes preferred different microhabitats within the broader habitat, mainly pertaining to prey availability and abundance. A valid analysis of the activity range and habitat utilization of most tropical snake species requires prolonged fieldwork (Nickerson et al. 1978).

The percentage of feeding time of the pit vipers is found limited (Tables 10.4, 10.5 and 10.6). *C. travancoricus* has spent only 0.05%, *C. peltopelor* (0.07%), and *H. hypnale* (0.07%) of the total activity recorded. This is supported by the fact that the foraging frequency is low in snakes because of low energetic requirements (Bennett and Dawson 1976) and low rates of energy flow (Gans and Pough 1982). Shine (1986) summarized data from 66 taxa to address the frequency of snake feeding and confirmed the statements of Bennett and Dawson (1976), and Gans and Pough (1982). The three species of pit vipers under study, resting in ambush for long periods and do not forage actively, and in this context, can be considered low-energy specialists, as Shine (1986) termed the Aurafura file snake.

Acknowledgements I owe my heartfelt gratitude to my father Late Dr. M.V. Rajendran for his profound knowledge and information on reptiles in general and KMTR in particular. He was the moving spirit behind this study and provided valuable suggestions. With Late Prof. Carl Gans of the University of Michigan, I had useful discussions relevant to my study. I thank my wife and colleague H.M. Mahilini for tolerating my long absence from home on fieldwork. As honorary wildlife warden (KMTR & Tirunelveli WL division), I have had useful discussions with the Tamil Nadu Forest Department personnel, especially the Field Directors, Pramode Kant, V.K. Melkhani, R.K. Ojha, H. Malleshappa, I. Anwardeen, and all the range forest officers, and other field staff. I am grateful to the PCCF & CWLW, Tamil Nadu Department of Forests & Climate Change, for the permission to this study in KMTR and Courtallam range. My special thanks are due to Dr. V. Chelladurai, Former Research Officer, Central Council for Research in Ayurveda and Siddha (AYUSH), Government of India, for helping me in the vegetation analysis, and for frequent accompanying during my field trips. Personnel assistance of volunteers and field assistants were sought for continuous monitoring of the snakes, and the following need special mention: Esakkial, Jeramiah, Devarual and Packianathan in Naraikadu; Ebenezar Babu (BBTC Estates), Chelladurai, my sons Nithin and Naren in Nalumukku; Ganesan and Palani in Therkumalai. I thank them all.

References

Abercromby AF (1914) Poisonous snakes of India and Ceylon. Spolia Zeylan 9:268–270

Andren C, Nilson G (1983) Reproductive tactics in an Island population of adders, *Vipera berus* (L.), with a fluctuating food resource. Amphib Reptil 4:63–79

Barbour RW, Harvey MJ, Hardin JW (1969) Home range, movements, and activity of the eastern worm snake, *Carphophis amoenus amoenus*. Ecology 50:470–476

Bennett AF, Dawson WR (1976) Metabolism. In: Gans C, Dawson WR (eds) Biology of the Reptilia, vol 5. Academic Press, London, pp 127–223

Bhatt K, Choudury BC (1993) The diel activity pattern of Indian python (*Python molurus molurus* Linn) at Keoladeo national park, Bharatpur, Rajasthan. J Bombay Nat Hist Soc 90:395–403

Bhupathy S (1990) Blotch structure in individual identification of the Indian python *Python molurus molurus* Linn and its possible usage in population estimation. J Bombay Nat Hist Soc 87:399–404

Bhupathy S, Vijayan VS (1989) Status, distribution and general ecology of the Indian python *Python molurus molurus* Linn in Keoladeo National Park, Bharatpur, Rajasthan. J Bombay Nat Hist Soc 86:381–387

Bogdanov OP (1965) Ecology of reptiles of Central Asia. Nauka Publ, Uzbek Soviet Socialist Republic, Tashkent. [In Russian]

Boulenger GA (1890) The fauna of British India, including Ceylon and Burma, Reptilia and Batrachia. Taylor & Francis, London, p 541

Brown TW (1970) Autecology of the sidewinder (*Crotalus cerastes*) at Kelso Dunes, Mojave Desert, California, PhD dissertation, University of California, LA

Brown WS, Parker WS (1976) A ventral scale clipping system for permanently marking snakes (Reptilia, Serpentes). J Herpetol 10:247–249

Bullock TH, Fox W (1957) The anatomy of the infra-red sense organ in the facial pit of pit vipers. J Cell Sci 3:219–234

Carpenter CC (1952) Comparative ecology of the common garter snake (*Thamnophis sirtalis*), the ribbon snake (*Thamnophis sauritus*), and Butler's garter snake (*Thamnophis butleri*) in mixed populations. Ecological monographs 22:235–258

Champion HG, Seth SK (1968) A revised survey of the forest types of India. Govt of India, New Delhi, p 404

D'Abreu EA (1933) The occurrence of the green pit viper (*Trimeresurus gramineus* Shaw) at Nagpur. J Bombay Nat Hist Soc 36:512

Daniel JC (1983) The book of Indian reptiles. Bombay Natural History Society, Mumbai, p 141

Daniel JC (2002) The book of Indian Reptiles & Amphibians. Oxford University Press, New Delhi, p 238
Emerson JD, DeSantis DL, Mata-Silva V, Wagler AE, Johnson JD (2022) Movement, home range size, and habitat use of eastern black-tailed rattlesnake (*Crotalus ornatus*) in the northern Chihuahuan desert. Herpetologica 78(2):110–118
Fitch HS (1949) Study of Snake populations in Central California. Am Midl Nat 41:513–579
Fitch HS (1960) Autecology of the copperhead. Univ Kansas Publ Mus Nat Hist 13:85–288
Fitch HS (1963a) Natural history of the racer (*Coluber constrictor*). Univ Kansas Publ Mus Nat Hist 15:351–468
Fitch HS (1963b) Natural history of the black rat snake (*Elaphe o. obsoleta*) in Kansas. Copeia 1963:649–658
Fitch HS (1965) An ecological study of the garter snake (*Thamnophis sirtalis*). Univ Kansas Publ Mus Nat Hist 15:493–564
Fitch HS (1975) A demographic study of the ringneck snake (*Diadophis punctatus*) in Kansas. Univ Kansas Publ Mus Nat Hist Misc Publ 62:1–53
Fitch HS (1982) Resources of a snake community in prairie-woodland habitat of northeastern Kansas. In: Scott NJ Jr (ed) Herpetological communities, vol 13. U.S. Fish and Wildlife Service, pp 83–97
Fitch HS (1987) Collecting and life-history techniques. In: Siegel RA, Collins JT, Novak SS (eds) Snakes: ecology and evolutionary biology. Mac Publishers, New York, pp 143–164
Fitch HS, Shirer HW (1971) A radio-telemetric study of spatial relationships in some common snakes. Copeia 1971:118–128
Freedman W, Catling PM (1978) Population size and structure of four sympatric species of snakes at Amherstburg, Ontario. Can Field Nat 92:167–173
Gans C, Pough FH (1982) Physiological ecology: its debt to reptilian studies, its value to students of reptiles. In: Gans C, Pough FH (eds) Biology of the Reptilia, vol 12. Academic Press, New York, pp 1–16
Gibbons JW, Semlitsch RD (1987) Activity patterns. In: Siegel RA, Collins JT, Novak SS (eds) Snakes: ecology and evolutionary biology. Mac Publ Co, New York, pp 396–421
Godley JS (1980) Foraging ecology of the striped swamp snake *Regina alleni* in southern Florida. Ecological monographs 50:411–436
Gregory PT (1974) Patterns of spring emergence of the red-sided garter snake (*Thamnophis sirtalis parietalis*) in the Interlake region of Manitoba. Can J Zool 52:1063–1069
Gregory PT, Stewart KW (1975) Long-distance dispersal and feeding strategy of the red-sided garter snake (*Thamnophis sirtalis parietalis*) in the Interlake of Manitoba. Can J Zool 53:238–245
Guimarães M, Munguia-Steyer R, Doherty PF Jr, Martins M, Sawaya RJ (2014) Population dynamics of the critically endangered golden lancehead pitviper, *Bothrops insularis*: stability or decline? PLoS One 9(4):e95203
Günther CLG (1864) The reptiles of British India. Ray Society, London. (reprint), Oxford & IBH Publ, New Delhi, p 452
Henderson RW, Nickerson MA, Ketcham S (1976) Short term movements of the snakes *Chironius carinatus*, *Helicops angulatus* and *Bothrops atrox* in Amazonian Peru. Herpetologica 32:304–310
Henry GM (1925) Notes on *Ancistrodon hypnale*, the hump-nosed viper. Spolia Zeylan 13:257–258
Henry AN, Subramanyam K (1981) Studies on the flora of Agastyamalai and surrounding regions in Tirunelveli district, Tamil Nadu. Bull Bot Survey India 23:42–45
Huey RB, Pianka ER (1981) Ecological consequences of foraging mode. Ecology 62:991–999
Jennrich RI, Turner FB (1969) Measurement of non-circular home range. J Theor Biol 22:227–237
Johnsingh AJT (2001) The Kalakad-Mundanthurai Tiger Reserve: a global heritage of biological diversity. Curr Sci 80(3):378–388
Lande R, Barrowclough GF (1987) Effective population size, genetic variation, and their use in population management. In: Soule ME (ed) Viable populations for conservation. Cambridge University Press, Cambridge, pp 87–124

Legris P, Blasco F (1969) Variabilité des facteurs du climat: Travaux de la section scientifique et technique. Institut Français de Pondichérry, Pondicherry

Macartney JM (1985) The ecology of the northern Pacific rattlesnake, *Crotalus viridis oreganus*, in British Columbia. MS thesis, University of Victoria, British Columbia

Madsen T (1984) Movements, home range size and habitat use of radio-tracked grass snake (*Natrix natrix*) in southern Sweden. Copeia 707–713

Maduwage K, Silva A, Manamendra-Arachchi K, Pethiyagoda R (2009) A taxonomic revision of the south Asian hump-nosed pit viper (Squamata: Viperidae: *Hypnale*). Zootaxa 2232:1–28

Mahendra BC (1984) Handbook of the Snakes of India, Ceylon, Burma, Bangladesh, and Pakistan. In: Annals of Zoology, Agra, p 412

Malhotra A, Thorpe RS (2004) A phylogeny of four mitochondrial gene regions suggests a revised taxonomy for Asian pitvipers (*Trimeresurus* and *Ovophis*). Mol Phylogenet Evol 32:83–100

Mallik AK, Srikanthan AN, Ganesh SR, Vijayakumar SP, Campbell PD, Malhotra A, Shanker K (2021) Resolving pitfalls in pit viper systematics—a multi-criteria approach to species delimitation in pit vipers (Reptilia, Viperidae, *Craspedocephalus*) of peninsular India reveals cryptic diversity. Vertebr Zool 71:577–619

McDiarmid RW, Campbell JA, Touré T (1999) Snake species of the world: a taxonomic and geographic reference, vol 1. Herpetologists' League, Washington, DC, p 511

McDowell SB (1987) Snake systematics. In: Siegel RA, Collins JT, Novak SS (eds) Snakes: ecology and evolutionary biology. Mac Publ Co, New York, pp 3–50

Merrem B (1820) Versuch eines Systems der Amphibian 1. (Tentamen Systematis Amphibiorum). J. C. Kreiger, Marburg, p 191

Mudaliar CD, Sundararaj D (1954) A short account of the flora of Tirunelveli district. Centenary Souvenier Madras Herbarium, Chennai, pp 57–63

Mushinsky HR (1987) Foraging ecology. In: Siegel RA, Collins JT, Novak SS (eds) Snakes: ecology and evolutionary biology. Mac Publ Co, New York, pp 302–334

Mushinsky HR, Hebrard JJ, Vodopich DS (1982) Ontogeny of water snake foraging ecology. Ecology 63:1624–1629

Myers N, Mittermeier RA, Mittermeier CG, da Fonseca GAB, Kent J (2000) Biodiversity hotspots for conservation priorities. Nature 403:853–858

Nair NC, Henry AN (1983) Flora of Tamil Nadu, India. Series I, II, III: Analysis. Bot Surv India, Coimbatore

Nair KKN, Nayar MP (1986) Flora of Courtallum, vol 1. Botanical Survey of India, Calcutta, p 2

Nickerson MA, Sajdak RA, Henderson RW, Ketcham S (1978) Notes on the movements of some neotropical snakes (Reptilia, Serpentes). J Herpetol 12:419–422

Parker WS, Plummer MV (1987) Population ecology. In: Siegel RA, Collins JT, Novak SS (eds) Snakes: ecology and evolutionary biology. Mac Publ Co, New York, pp 253–301

Parthasarathy N, Mahadevan A (1987) Floristic account of forest types in Kalakad reserve forest, Western Ghats, India. J Econ Taxon Bot 10:355–360

Pollock KH (1982) A capture-recapture design robust to unequal probability capture. J Wildl Manag 46:752–757

Preston WB (1964) The importance of the facial pit of the northern pacific rattlesnake (*Crotalus viridis oreganus*) under natural conditions in southern British Columbia. MS Thesis. University of Victoria, British Columbia

Prestt I (1971) An ecological study of the viper *Vipera berus* in southern Britain. J Zool 164:373–418

Rajendran MV (1967) Nam Nattu Pampugal (snakes of our land—in Tamil). Jeya Publ, Palayamkottai

Rajendran MV (1973) Studies in the Uropeltid snakes. PhD Thesis, Madurai Kamaraj University

Rajendran MV (1985) Studies in Uropeltid snakes. Publications Division, Madurai Kamaraj University, Madurai, p 132

Rao CRN (1917) Notes on Lachesis anamallensis and allied forms. Rec Ind Mus 13:11–15

Reinert HK, Cundall D, Bushar LM (1984) Foraging behavior of the timber rattlesnake, *Crotalus horridus*. Copeia 1984:976–981

Robson DS, Regier HA (1964) Sample size in Petersen mark-recapture experiments. Trans Am Fish Soc 93(3):215–226

Sawant NS, Jadhav TD (2013) Factors influencing habitat selection by arboreal pit vipers. Zoolog Sci 30(1):21–26

Sawant NS, Jadhav TD, Shyama SK (2010a) Distribution and abundance of pit vipers (Reptilia:Viperidae) along the Western Ghats of Goa, India. J Threat Taxa 2(10):1199–1204

Sawant NS, Jadhav TD, Shyama SK (2010b) Habitat suitability, threats and conservation strategies of hump-nosed pit viper *Hypnale hypnale* Merrem (Reptilia: Viperidae) found in Western Ghats, Goa, India. J Threat Taxa 2(11):1261–1267

Schnabel ZE (1938) The estimation of the total fish population of a lake. Am Math Mongr 45:348–352

Schoener TW (1971) Theory of feeding strategies. Annu Rev Ecol Syst 11:269–404

Seigel RA, Collins JT, Novak SS (1987) Snakes: ecology and evolutionary biology. Mac Publ Co, New York, p 529

Semlitsch RD, Brown KL, Caldwell JP (1981) Habitat utilization, seasonal activity, and population size structure of the southeastern crowned snake *Tantilla coronata*. Herpetologica 37:40–46

Shine R (1986) Ecology of a low-energy specialist: food habits and reproductive biology of the Arafura file snake (Acrochordidae). Copeia 1986:424–437

Shine R, Bull JJ (1977) Skewed sex ratios in snakes. Copeia 1977:228–234

Shine R, Lambeck R (1985) A radiotelemetric study of movements, thermoregulation and habitat utilization of Arafura file snakes (Serpentes: Acrochordidae). Herpetologica 41:351–361

Shine R, Sun L-X, Zhao E, Bonnet X (2002) A review of 30 years of ecological research on the Shedao Pitviper, *Gloydius shedaoensis*. Herpetol Nat Hist 9(1):1–14

Smith MA (1943) The Fauna of British India, Ceylon and Burma. In: Reptilia and Amphibia: Serpentes, Vol 3. Taylor and Francis, London, p 583

Smith RL (1974) Ecology and field biology. Harper and Row, New York, p p850

Soulé ME (1987) Introduction. In: Soule ME (ed) Viable populations for conservation. Cambridge University Press, Cambridge, pp 1–10

Swamy BGL, Govindarajulu E (1956) On a collection of plants from Courtallam. J Madras Univ Bull 26:427–449

Tanaka H, Wada Y, Oguma Y, Sasa M, Noboru Y, Ono T, Matsuschita N (1971) A method of population estimation for the habu, *Trimeresurus flavoviridis*. Snake 3:9–13

Tinkle DW (1979) Long-term field studies. BioScience 29:717

Turner FB (1977) The dynamics of populations of squamates, crocodilians and rhyncocephalians. In: Gans C, Tinkle DW (eds) Biology of the Reptilia, vol 7. Academic Press, New York, pp 157–264

Uetz P, Hošek J (2002) In: Uetz P, Freed P, Aguilar R, Hosek J (eds) THE REPTILE DATABASE. http://www.reptile-database.org/. Accessed 15 Nov 2023

Virushali SVN (2019) Evaluation of anti-tumor activity of venom of *Trimerusurus malabaricus* (Malabar pit viper) using an animal model: a bioanalytical approach. PhD Thesis, University of Mumbai

Wall F (1899) Notes on 26 specimens of the Pohur or Himalayan pit viper (*Agkistrodon himalayanus*). J Bombay Nat Hist Soc 12:411–414

Wall F (1905) A popular treatise on the common Indian snakes, part 1 (*Lachesis grramineus*, *Dryophis mycterizans*). J Bombay Nat Hist Soc 16:533–534

WebPeploe G (1942) Naraikadu, trees, shrubs, herbs, creepers, scandents, shrubs, mammals and reptiles, 6 vols. (Handwritten)

Whitaker R (1978) Common Indian Snakes: a field guide. Mac Co India, New Delhi

Whitaker R, Captain A (2004) Snakes of India, the Field Guide. Draco Books, Chennai, p 481

Worton BJ (1987) A review of models of home range for animal movement. Ecol Model 38:277–298

Chemical Signalling While Mating in Tropical Bats

11

Juliet Vanitharani

Abstract

Bats use a combination of visual, acoustic and olfactory signals to communicate. Breeding behavioural studies in a few bat species reveal that chemical signals from scent glands are involved in communication. Bats possess modified skin glands at distinct regions to secrete odoriferous compounds. The chemical compounds of the scent gland secretion, saliva and urine directly stimulate biochemical reactions. In bats, the integumentary glands empty their secretions outside the animal's body to expel the odour. These chemical signals aid the bats to communicate and mediate sexual behaviour during the breeding season. The scent-marking behavioural patterns of bats in their roosting sites are categorised as territory marking, settlement of pairs in the marked territory, grooming, fanning, body rubbing, sniffing and mating. The nocturnal and social lifestyles of bats suggest that besides acoustic signals, scent plays a key role in communication to achieve reproductive success. This chapter attempts to bring out the chemical signalling from a few common insectivores and frugivores bats in tropical ecosystems.

Keywords

Communication · Odoriferous compounds · Chemical signals · Scent glands

J. Vanitharani (✉)
Bat Research Laboratory, Zoology Department and Research Centre, Sarah Tucker College, Tirunelveli, Tamil Nadu, India

S. Sridhara et al. (eds.), *Animal Behavior in the Tropics*,
https://doi.org/10.1007/978-981-97-7324-4_11

11.1 Introduction

In mammals, scent production is strong and acts as a secondary sexual signal during mating season. The chemical odorants have efficacy over long and short distances and are effective in darkness, energetically efficient and long-lasting (Rich and Hurst 1998). This is one of the main causes of reproductive success (Kamran and Kerth 2003). Bat as a mammal does a variety of scent-marking behaviours to advertise their physical fitness (Bradbury 1977), accelerate the onset of puberty in females (Eisenberg and Kleiman 1972), identify the potential males (Brooke and Decker 1993), mark their pups (Leon 1983; Esser and Schmidt 1989) and advertise the social and reproductive status to the colony mates (Fenton 1985). To study the scent-marking activity, samples were collected from a few temple-dwelling common insectivorous bats such as *Hipposideros speoris* (Schneider's leaf-nosed bat 1800), *Hipposideros ater* (dusky leaf-nosed bat), and *Taphozous melanopogon* (black-bearded tomb bat) and fruit bats such as *Rousettus leschenaulti* Desmarest 1820 (fulvous fruit bat) and the tent-making *Cynopterus sphinx* (greater short-nosed fruit bat) from the tent roost of an ornamental palm tree of the garden. Three tagged individuals under each sex of the harem colony were selected. Samples of urine, saliva and anal and integument secretions of different glandular regions of males and females during breeding and non-breeding seasons were subjected to GC-MS for analysis. Males that defend and roost with groups of females (harem holders) have significantly different chemical profiles from males found roosting in all-male groups (bachelors). These results suggest that chemical signals have the potential to communicate both mating status and individual identity and thus could be used to mediate interactions among individuals within harem.

Several earlier reports from different bat species (Brooke and Decker 1993, 1996) in *Noctilio leporinus*; Voigt and Helversen (1999) in *Saccopteryx bilineata*; Bloss et al. (2002) in *Eptesicus fuscus*; Lily (2005) in *Hipposideros speoris, Taphazous melanopogan* and *Roussetus leschenaulti*; Sofia (2007) in *Hipposideros ater*; Vanitharani et al. (2007) in *Cynopterus sphinx*; Voigt and Helversen (1999), Voigt et al. (2008), Caspers and Voigt in *S. bilineata*; Adams et al. (2018) in Greater spear-nosed bat, *Phyllostomus hastatus* and Chaverri et al. (2018) have documented that during mating season, bats use integumentary glandular secretion for sexual attraction by way of territory-marking, self-anointment, body-rubbing and wing-flapping behaviours. A literature review on chemical communication in bats has instigated to look into the chemical signalling system in bats.

11.2 Biological-Odour-Influenced Sexual Behaviour

Biological odours produced from urine, faeces, saliva, vaginal mucus, sweat and scent glands are important in a variety of behavioural interactions in mammalian species (Guiraudie et al. 2003). Research suggests that secretions from specialised glands, often in combination with urine and saliva, are responsible for species recognition and in conveying information about sex and colony membership. Such chemical cues are seen in microorganisms, insects, rodents, farmyard and domestic animals and primates to play a significant role in sexual behaviour (Archunan 2003). In bats, of all the types of communication, the olfactory seems successful (Scully et al. 2000a). Earlier chemical cue studies on bats by Herreid (1960), Mainoya and Howell (1977), McCracken (1984), Gustin and Mac Cracken (1987), Kunz (1990), Brooke and Decker (1993, 1996), Voigt and Helversen (1999) and Scully et al. (2000b) confirmed that in social life of bats, chemical signals are important mode of gathering information, advertisement of social and reproductive status, territory marking and kin recognition. The females of mammalian species except humans are receptive to their males only during the oestrus period. This readiness is informed through chemical signals. The present study review also confirmed the chemical odorant emissions from the integumentary secretions, urine and saliva act as efficient informers in bat's communal life.

11.3 Scent-Marking and Mating Behaviour

Generally, during the breeding season, the potent males attract the responding female by exhibiting various courtship behaviours. They emit chemical signals from integumentary glands and disperse the specific odorous compounds all over their body regions through self-grooming. Self-grooming helps maintain the social bond between the pair and reduces mate selection tension (Voigt and Helversen 1999). The males rub the glandular secretions with urine over their own body, on their conspecifics and also on the mating territory of their day roost. Such rubbing behaviour has been noticed in velvet free-tailed bat (*Molossus molossus*) (Haussler 1987), grey-headed flying fox (*Pteropus poliocephalus*) (Ratcliffe 1932; Nelson 1965), Mexican free-tailed bat (*Tadarida brasiliensis Mexicana*) (Davis et al. 1962) and phyllostomid bats (*Phyllostomus hastatus*) (Valdivieso and Tamsitt 1964).

Self-anointment of glandular sticky secretions not only aids in releasing odour to the environment to attract their partner (Bloss et al. 2002) but also helps the animal look more attractive to their conspecifics (Bowyer and Kitchen 1987). Bats use a special way of dispersing body fragrance through wing-flapping and fanning behaviour. The potential males fan the anointed odour by fast fluttering and emanate the scent towards the awestruck hanging female in the marked territory. Such wing-flapping behaviour has been noticed in many bats species like *Saccopteryx*, *Cyttarops*, *Peropteryx* and *Cormura* (Robbins and Sarich 1988); *Cynopterus* sp. (Nowak 1994); *H. speoris*, *T. melanopogon* and *Rousettus leschenaultii* (Lily 2005), *H. ater* (Sofia 2007); and *C. sphinx* (Vanitharani et al. 2007). This mesmerisation through the chemical fragrance makes the mating partners settle in the marked territory for a prolonged period of more than 24–48 h.

11.3.1 Scent-Marking, Mate Selection and Mating Behaviour in *Hipposideros speoris*

During the breeding season, males and females of *H. speoris* show peculiar scent-marking behaviour in the day roost (Plate 11.1). The potential males make specific territory-marking in the day roost with the anal glandular secretions and urine (Plate 11.1a). The males always settle in that marked spot and try to attract the females through various gestures. The dominant male while resting and grooming scratch the frontal regions and disperse the secretions all over the body. Meanwhile, the responding female and the other potent male intruders try to move close to the dominant male's territory. The self-anointed dominant male hangs alone in the marked territory and flaps the left wing and the right wing repeatedly one after another to disperse the scent. The odour lured the responding females to settle around him in the marked territory (Plate 11.1b).

To distract the female's attention towards the other male intruders, the dominant male fans the fragrance of the self-anointed odour (Plate 11.1c) and also drags the responding female by holding her with his phalanges on wing with a thumb claw, which restricts the responding female to stay within his marked territory (Plate 11.1d, e). Other non-selected females are chased away by the dominant male. In the same way, the responding female conveys her sexual responses and readiness to mate through the odoriferous enormous secretions of her anal glands. The male bends its head towards the ventral side to anoint the dorsal region of the female with his frontal secretions (Plate 11.1f) and very often flaps the wings to fan the anointed male odour towards the responding female. The mating pair is confined to the marked territory and stays close together for several hours. Both partners are involved in body rubbing and sniffing behaviour. After this long-term (several hours) courtship behaviour, the male slowly creeps over the female, sits on the back, rubs the glandular secretions over the female and then mates.

11.3.2 Scent-Marking, Mate Selection and Mating Behaviour in *Hipposideros ater*

During the breeding season, both sexes of *H. ater* show series of courtship behaviour exhibited by a combination of activities like sniffing the genitalia, licking, grooming and huddling as regular pre- and post-copulatory behaviours (Plate 11.2). The dominant males of *H. ater* settle in their own marked territory of the roost with the harem members (Plate 11.2a), perform self-grooming and wing-flapping to disperse their odours and thereby attract the females to respond. Initially, inter-individual space was maintained by the harem members. Then the dominant male drags the selected female (Plate 11.2b) and embraces her with his wings for a prolonged time. Selected partners get settled in the marked territory (Plate 11.2c). Dominant male repeatedly rubs and by wing-flapping fans the chemical on their partner to mark the female with their integumentary secretions and saliva (Plate 11.2d, e). Such settlement behaviour continues for a few days. The engaged pairs are settled in the marked territory. After the partner's bondage is confirmed, the female allows the male to creep over (Plate 11.2f, g). Embraced mating occurs roughly for 80–95 s.

Plate 11.1 (**a**) Dominant male mark the territory with glandular secretions and urine. (**b**) Attracted females settled around the marked territory. (**c**) Male fan the fragrance of the self anointed odour and restrict the responding female to settle in the marked territory. (**d**) Male drag the responding female by holding her with his phalanges. (**e**) Male intruders trying to attract the responding female. (**f**) The self anointed male sniff and creep over the female and rub the secretion on the dorsal region and show mating behavior

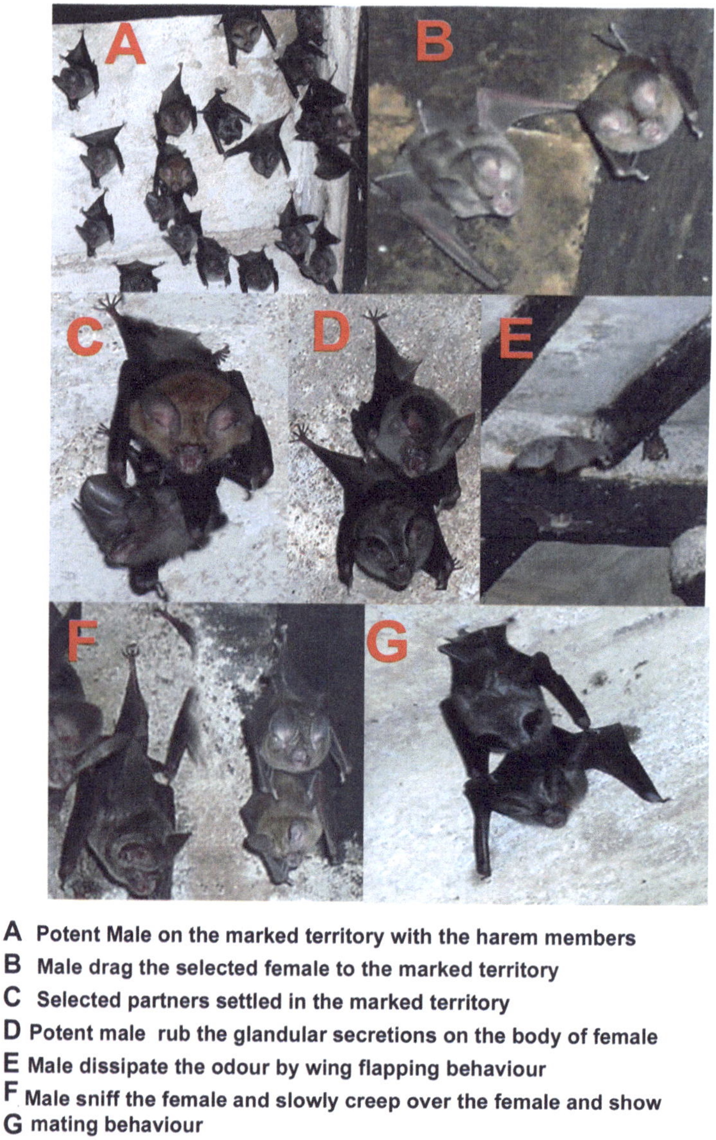

Plate 11.2 (**a**) Potent Male on the marked territory with the harem members. (**b**) Male drag the selected female to the marked territory. (**c**) Selected partners settled in the marked territory. (**d**) Potent male rub the glandular secretions on the body of female. (**e**) Male dissipate the odour by wing flapping behaviour. (**f**) Male sniff the female and slowly creep over the female and show G mating behaviour

11.3.3 Scent-Marking, Mate Selection and Mating Behaviour in *Taphozous melanopogon*

Mating season lasts for only a few weeks in the winter. Only during the breeding months (December and January) the adult males of *T. melanopogon* mingle with females and sub-adults and show a series of mating behaviours (Plate 11.3). Adult females show aggressive behaviour towards the mature males if they enter the roost area during non-breeding season. However, during the breeding season, females allow males to settle in their day roost. Dominant males try to select territories in their day roost. They use anal glandular secretions to mark their territory. The territory of the dominant males can be identified by the aggressive defence behaviour and the repeated settling of the male in the marked area. The odour dispersed from the dominant male and its marked territory attracts harem females to respond (Plate 11.3a).

The dominant male allows only the selected female to settle near the marked as well as mating site. Male dissipates the body fragrance through wing-flapping and impounds the attracted females in the marked territory (Plate 11.3b). When the responding female tries to move away from the mating site, the male shows a peculiar visual display of sideways movement, a distinctive crab-like movement towards the responding female, and stretches the wing slowly towards the attracted female and drags her with his legs and thumb claw (Plate 11.3c). Both partners settle for days together in the marked spot. The dominant male never allows the selected partner to move away. The male sniffs the responding female and after confirmation rubs the throat glandular secretions over the female's body (Plate 11.3d). During pairing, they begin to groom themselves, sniff each other and spread the secretions all over the body. This type of pair settling in the colony is obvious in the day roost during mating season. During the mating season, sub-adults aggregate and settle in the corner of the roost. The intruder male tries to attract the pair and amuse the responding female of the dominant male (Plate 11.3e). After both partners agree to the prolonged stay in the marked territory, the anointed dominant male slowly creeps over the mating partner, mounts on his selected female and begins to exhibit mating behaviour (Plate 11.3f).

11.3.4 Scent-Marking, Mate Selection and Mating Behaviour in *Rousettus leschenaulti*

In October and March alone, a remarkable pairing behaviour was observed in *R. leschenaulti* (Plate 11.4). The mating partners settle in various areas during the day and roost for several days. During this time, sub-adult bats roost on the periphery of the colony. Males exhibit territorial behaviour in the day roost only during the mating period. Mating partners become more opportunistic in feeding and never leave the roost till the mating process is over.

During the breeding season, the dominant males mark the territories by rubbing the frontal glandular secretions and licking the territory with saliva (Plate 11.4a).

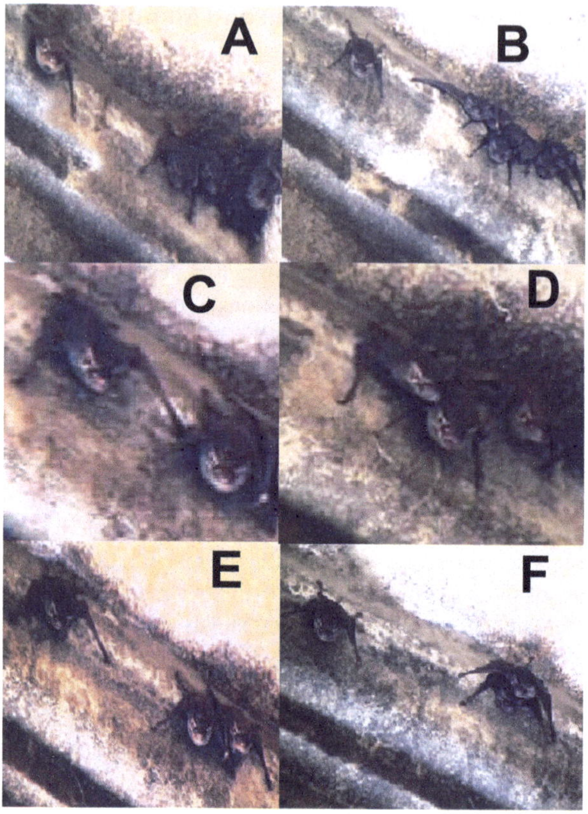

A Dominant male settle in his marked territory and attract the females through various gestures

B Male dissipate the body fragrance through wing flapping and impound the attracted females in the marked territory

C Dominant male drags the responding female by holding her with his thumb claw to the marked territory

D The male sniff the responding female and after confirmation rubs the throat glandular secretions

E Intruder male tries to attract the paired female

F Male slowly creeps over the mating partner and begins to exhibit mating behaviour

Plate 11.3 (**a**) Dominant male settle in his marked territory and attract the females through various gestures. (**b**) Male dissipate the body fragrance through wing flapping and impound the attracted females in the marked territory. (**c**) Dominant male drags the responding female by holding her with his thumb claw to the marked territory. (**d**) The male sniff the responding female and after confirmation rubs the throat glandular secretions. (**e**) Intruder male tries to attract the paired female. (**f**) Male slowly creeps over the mating partner and begins to exhibit mating behavior

The dominant male settles in the marked territory and dissipates the odoriferous self-anointed secretion through wing-flapping behaviour to attract the females (Plate 11.4b). Dominant male often licks the genital area and disperses the anal

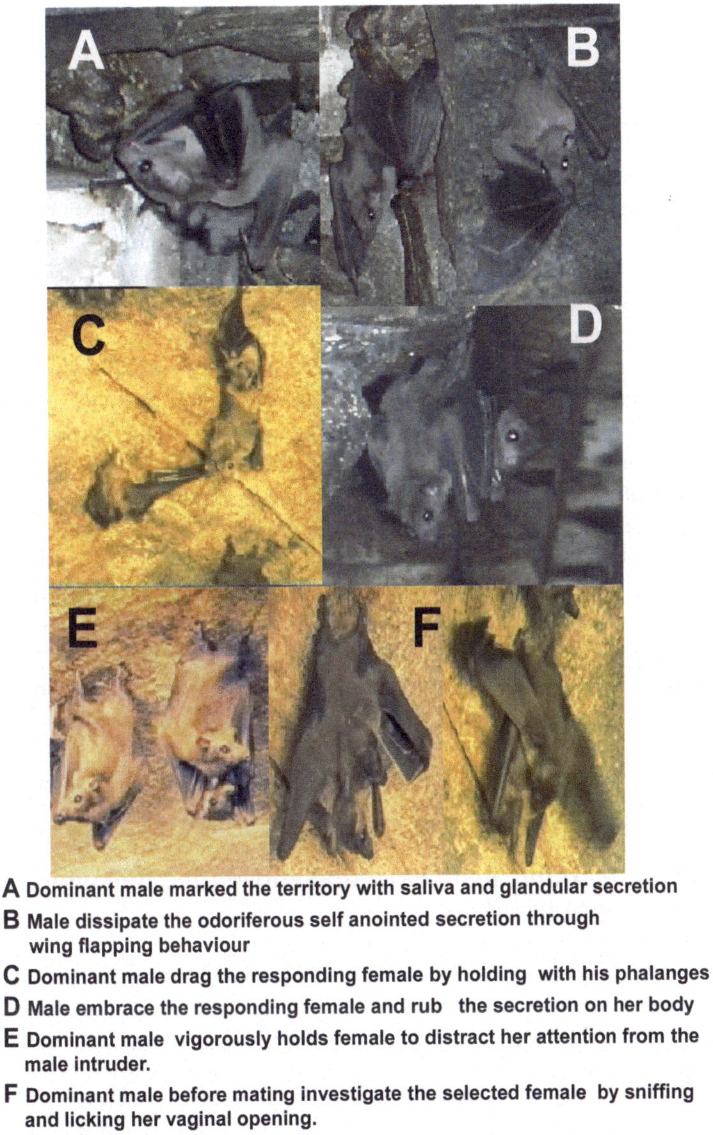

Plate 11.4 (**a**) Dominant male marked the territory with saliva and glandular secretion. (**b**) Male dissipate the odoriferous self anointed secretion through wing flapping behaviour. (**c**) Dominant male drag the responding female by holding with his phalanges. (**d**) Male embrace the responding female and rub the secretion on her body. (**e**) Dominant male vigorously holds female to distract her attention from the male intruder. (**f**) Dominant male before mating investigate the selected female by sniffing and licking her vaginal opening

glandular secretions over the body and make the environment fragment. The responding females initially settle together at the corner, opposite to the dominant male's marked territory. The dominant male attracts the females through various wing-flapping gestures. The selected responding female is pulled aggressively by the male accompanied by loud vocal calls before settling in the marked territory (Plate 11.4c). After prolonged display behaviour of the male, the responding female slowly moves to the marked territory and settles near the male. The male embraces the responding female and marks his partner by frequent facial rubbing with the frontal and neck glandular secretions on her body (Plate 11.4d). During this pair settlement, males very often lick the genital secretion and self-anoint till the responding female settles near the dominant male submissively.

Several other mature males settle around the mating pair and try to attract the female. To divert the female's attention from the activity of the male intruders, the dominant male drags the responding female aggressively to the marked site, hugs the female tightly and fans the anointed odour by wing-flapping behaviour (Plate 11.4e). The male sometimes even bites the female to confine her to his marked territory. The male often rubs the secretion on the female by body rubbing, licking and biting. During this courtship behaviour, intruder males interact vehemently that the dominant male reacts through his aggressive biting, pushing, chasing, fighting and screaming behaviour. Fighting intensified sometimes with gestures such as striking with wings and biting. At the same time, the dominant male vigorously holds the selected female to distract the attention of the female away from the intruder. During courtship, the male climbs over the female, investigates the selected female by sniffing, licks her vaginal opening (Plate 11.4f) and, after confirming her full response, grasps her folded wings and holds her neck with his jaws and starts copulation.

11.3.5 Scent-Marking, Mate Selection and Mating Behaviour in *Cynopterus sphinx*

During the breeding season in October, a harem colony roosted in the ornamental palm tree was selected as the study site (Plate 11.5). This tent-making fruit bat species are gregarious and typically roost in same-sex groups of 8–9 individuals (Plate 11.5a). The sexes remain separate until the mating season. During the mating season, three potent males select different leaves of the same palm tree as their selected mating territory and try to attract the females. The self-anointed dominant male marks the place with glandular secretion by way of rubbing and chewing the substratum of the palm leaf (Plate 11.5b). Intruder males also settle in their marked territory to distract the responding female (Plate 11.5c). Dominant male dissipates the odour by wing-flapping gestures (Plate 11.5d). This behaviour facilitated the liberation of the strong odours to fragrant that area. Male self-anointment induces the females to respond and settle around the marked territory. The dominant male succeeded in attracting and making the responding female to settle in the marked territory by way of showing various courtship behaviours like wing-flapping and statue-like gestures. The selected female is dragged by the dominant male

A Harem colony roosting in the tent made in an arnamental palm tree

B Self anointed dominant male marked the territory by chewing the leaf substratum with glandular secretions and trying to attract the female

C Intruder males settle in their marked territory to distract the responding female

D Dominant male dissipate the odour by wing flapping gestures

E After mate selection the mating partners settled in the marked territory and the male showing the rubbing behaviour to mark the female

F Male sniff the genitalia of the female prier mating

Plate 11.5 (**a**) Harem colony roosting in the tent made in an ornamental palm tree. (**b**) Self anointed dominant male marked the territory by chewing the leaf substratum with glandular secretions and trying to attract the female. (**c**) Intruder males settle in their marked territory to distract the responding female. (**d**) Dominant male dissipate the odour by wing flapping gestures. (**e**) After mate selection the mating partners settled in the marked territory and the male showing the rubbing behaviour to mark the female. (**f**) Male sniff the genitalia of the female prier mating

repeatedly and made to stay in the marked territory. After mate selection, the mating partners settle in the marked territory, and the dominant male starts the rubbing behaviour to make specific markings on the responding female's body (Plate 11.5e).

After marking, he starts the courtship behaviour by sniffing the genitalia of the female prior mating (Plate 11.5f).

11.4 Chemical Signals Through Odoriferous Secretions

Species-specific variations in 'bat odours' produced by the scent glandular secretions with unpleasant and pungent smell are due to the various lipid derivative compositions and the microbial action of the scent glands. This odour becomes intense during mating season and indicates their important role in reproduction. Variation in the chemical odours of the glandular secretions is due to the shape and the type of functional groups associated with the odour-producing organic compounds. If the functional groups are with single oxygen atoms (alcohol, aldehyde and ketone derivatives), they play a vital role in the production of strong odours for a prolonged period. In addition, other chemical compounds such as acid esters with oxygen atoms, alkynes, alkenes and specific compounds with sulphur and nitrogen composition provide strong odour-producing substances to the glandular secretions.

Gas- Chromatography and Mass- Spectrometry (GC-MS) analysis proved that odour-producing substances are present in studied bats are species-specific sexually dimorphic alcohols, aldehydes, acid with esters, ketone and steroid derivatives. These compounds also exhibit seasonal variations. In all the bat species, these substances provide important chemical signals during the mating season. In addition, fatty acids with high carbon numbers help as ingredients for chemical signalling during the mating season. These chemicals facilitate marking territories at the roosting site to attract potential partners during reproduction, individual recognition and the maintenance of social behaviour among the colony mates.

11.5 Odour Secretions as Mate Attractant

The fatty acid odorant compounds of the secretions are used as sex attractants in many animals during mate selection. Earlier studies on mate attractants showed that they are with mixed fatty acid compounds, and more number of carbon atoms were observed in the following animals: C16 fatty acid in silkworm moth (Wilson 1965); C9 fatty acid that acts as the queen substance, which is broadcasted by the queen bee (*Apis mellifera*) during spawning (Janzen et al. 1982); C15–C25 fatty acids in the occipital secretion of camel (*Camelus bactrianus*) (Ayorinde et al. 1982); C20 fatty acid in the civet cat (*Civettictis civetta*) (Xavier 1994); C16–C18 fatty acids in the preorbital secretions of oribi (*Ourebia orebi*) (Mo et al. 1995); C11 fatty acid in cockroach (*Periplanata americana*) (Nojima et al. 1999); C14–C18 fatty acid in the interdigital secretions of bontebok (*Damaliscus dorcas dorcas*) and C20 fatty acid in the blesbok (*Damalicusdor casphillipsi*) (Burger et al. 1999). Compounds like glycolipids, non-polar lipids and phospholipids are used to mark the partners during pairing in the fishing bat (*N. leporinus*) (Brooke and Decker 1993, 1996). A similar diagnostic role of lipid, fatty acid derivatives from the specialised integumentary glandular secretion of palm squirrel (*Funambulus palmayum*) has been proved by Balakrishnan and Alexander (1985) and Kairali and Alexander (1984) in house rat (*Rattus rattus*) and Guiraudie et al. (2003) in prepubertal pigs (PIGS *Sus scrofa*).

The GC-MS profile of the glandular secretions as mate attractants with mixed fatty acid compounds and more number of carbon atoms of the studied bat species are given in Table 11.1. *H. speoris* produces C14–C20 fatty acid secretions with octa decanoic acid in the frontal region of males and C8–C20 fatty acid with tridecane, tetradecane, octadecane and phenol compounds in the anal gland secretions of both sexes during the breeding season. *H. ater* as sex attractants use decanoic acid in the saliva and 1,3-epoxy-4-methyl pentane in the ear between breeding. In *T. melanopogon*, the facial region of the male and the female possess sex-attracting secretions during the breeding season with C30 decanoic acid and C28 octadecanoic acid in males and C20 octadecanoic acid in females. The neck glandular secretions of the male are more pronounced during breeding and enhanced with C12 and C30 octadecanoic acid, whereas the female has no such scent glands in the neck.

It is also confirmed that *R. leschenaulti* uses the following compounds as sex attractants. The males possess fatty acids with a higher number of carbon atoms in the frontal glandular secretions C16–C17 hexadecanoic acid and C14 decanoic acid and anal glandular secretion C20 decanoic acid, whereas females possess C12 octadecanoic in the neck region during the the breeding season.

The GC-MS profile of the glandular secretions of the three potent males of the *C. sphinx* has confirmed that they have a specific odoriferous fatty acid with ester compounds, octadecanoic acid, 1,2-benzenedicarboxylic acid, dioctyl ester and undecanoic acid in the integumentary glandular secretion of the frontal region, ear and neck region present only during mating season. Hexadecyl ester of octadecanoic acid, docosanoic acid, 1,2-benzenedicarboxylic acid and dinonyl ester are present in the urine. In addition, the study revealed the dominant male which was selected by the female possesses more odoriferous fatty acid derivatives than the other two competent males. Decanoic acid has been reported in both sexes during breeding and non-breeding season. The dominant male possessing 1,2-benzene dicarboxylic acid and bis(2-ethylhexyl) ester is not represented by the other two competing potent males. In general, low-quality sexual attractant-bearing subordinate males are rejected by the females. Jones and Nowell (1973) discussed that the female mice are not attracted by the low-quality sexual attractant odours of the male. Similarly, they avoid mating with the subordinates and reject males through the odour sense.

In addition, fatty acids with high carbon numbers present in males were also noticed in the glandular region of the females during the breeding seasons. This may be due to the rubbing of these secretory compounds on the selected females. Such remarkable similarity in possession of fatty acid compounds are noticed in the anal, glandular secretions of studied bat species has confirmed that males use decane and decanoic acid compounds as sex attractants and the markers of their selected females. The studies on these bats during non-breeding season confirm this type of fatty acid with high carbon number biochemical markers were absent.

Table 11.1 Fatty acid derivatives and ester compounds used by male bats during breeding season for mate attraction

S. no.	Species	Chemical compounds	Chemical signal emitters
Mate attractants of Hipposideros speoris male during breeding season			
1	Potent male	2, Phenyl-1, 3-dioxan-5-yl ester of Octadecanoic acid	Frontal glandular secretions
		Isopropyl myristate-aliphatic ester	
		4-(Phenylazo)-2-[2-(tetrahydro-2-oxo-3-furanyl) ethoxyl] ethyl ester of benzoic acid	
		Bis (2 methyl propyl) ester of 1, 2-benzene dicarboxylic acid	Anal glandular secretions
		Butyl octyl ester of 1, 2-benzene dicarboxylic acid	
		Bis (2-ethyl hexyl) ester of 1, 2 Benzene dicarboxylic acid	
2	Intruder male 1	Hexyl-diethyl ester of Propanedioic acid-aliphatic ester	Frontal glandular secretions
3	Intruder male 2	1, 2, 3-Propane ethyl ester Octadecanoic acid	Anal glandular secretions
Mate attractants of Hipposideros ater male during breeding season			
1	Potent male	Decanoic acid	Urine, saliva
		3-Methyl heptyl acetate -ester	Urine
		Methyl-2-hydroxy dodecanoate-ester	Saliva
		3-Butenyl propyl ether	Anal integumentary glandular secretion
2	Intruder male 1	Decanoic acid	Urine
3	Intruder male 2	Nil	Nil
Mate attractants of Taphozous melanopogon male during breeding season			
1	Potent male	Decanoic acid	Facial glandular secretions
		3-[(1-Oxohexa decyl) oxy] propyl ester of octa decanoic acid	
		Bis (2-ethyl hexyl ester of 1, 2 Benzene dicarboxylic acid	
		2-[(Oxo dodecyl) oxy]-1, 3-propanediyl ester of Octadecanoic acid	Neck glandular secretions
		3-(1-Oxohexa decyl) oxy-2-(1-oxotetradecyl) oxy propyl ester of Octadecanoic acid	
		Bromomethyl) methyl bis (hydroxyl imino) heptyl ester of Propanedioic acid	
		Decyl ester of acetic acid	Anal glandular secretions
		1-Hexadecyl ester of acetate	

(continued)

Table 11.1 (continued)

S. no.	Species	Chemical compounds	Chemical signal emitters
2	Intruder male 1	Octadecyl ester of Octadecanoic acid	Facial glandular secretions
		3-Phenyl-methyl ester of 4-Isothiazole carboxylic acid	Neck glandular secretions
3	Intruder male 2	Decyl ester of acetic acid	Facial glandular secretions
Mate attractants of Rousettus leschenaultii male during breeding season			
1	Potent male	Hexadecyl ester of octadecanoic acid	Frontal glandular secretions
		2-Hydroxy methyl ester of Hexadecanoic acid	
		2-(Octadecyl oxy) ethyl ester of 9-Octadecanoic acid	
		Dimethyl ester of Dodecaneioic acid	
		Tetradecyl ester of Octa decanoic acid	Neck glandular secretions
		1, 2, 3-Propane triyl ester of Octa decanoic acid	
		3-Phenyl-methyl ester of 4-Isothiazole carboxylic acid	
		Diisodecyl ester of 1, 2-Benzenedicarboxylic acid	
		Diisooctyl ester of 1, 2-benzenedicarboxylic acid	Anal glandular secretions
		5-Heptadecyl-methyl ester of –2-Thiophene carboxylic acid	
2	Intruder male 1	Methyl ester of 15-Tetracosenoic acid	Frontal glandular secretions
		1-[(Dodecyl) methylethane] ester of Octa decanoic acid	Neck glandular secretions
3	Intruder male 2	2methyl ester of 9-Octadecanoic acid	Frontal glandular secretions
		Dioctyl ester of 1, 2 benzene dicarboxylic acid	Neck glandular secretions
		2-Cyano-3-(3-bromo phenyl)-ethyl ester of –2-Propenoic acid	Anal glandular secretions
Mate attractants of Cynopterus sphinx male during breeding season			
1	Dominant male	Octa decanoic acid	Integumentary glandular secretion of frontal region
		1,2-Benzene dicarboxylic acid, bis (2-ethyl hexyl) ester	Integumentary glandular secretion of ear region
		1,2-Benzene dicarboxylic acid, dioctyl ester	Saliva
2	Intruder male 1	Undecanoic acid	Integumentary glandular secretion of neck region
		Hexadecyl ester of octadecanoic acid	Urine
3	Intruder male 2	Docosanoic acid	Urine
		1,2-Benzene dicarboxylic acid, dinonyl ester	Urine

The behavioural studies clearly established that the chemical cues clearly inform the females about the genetic compatibility of the potential males and in males to attract and mark the corresponding females (Wedekind et al. 1995). The present study report also confirmed fatty acid odorants were used as sex attractants during mate selection.

11.6 Odour Secretions as Territory Markers

Territory marking includes the marking of territorial boundaries. In mammals, males scent mark the boundaries of their harem territories with glandular secretions and urine to defend territory against intruders (Gosling 1982; Miquelle 1991) to reduce the agonistic interaction between the individuals (Rozenfeld and Denoel 1994) and convey the information of the male's physical condition and the quality of genome (Lingington 1983; Hurst and Barnard 1992; Hurst and Rich 1999). Similar observations in bat behaviour have been noted by Bradbury and Vehrencamp (1976), Voigt and Helversen (1999) and Voigt et al. (2001).

The chemical compounds with functional groups carrying oxygen, sulphur, nitrogen atoms and steroid derivatives play a vital role in the production of strong odour for a prolonged time and are used to mark territory. The possession of such type of markers is also reported in the preorbital secretion of musk ox (Flood et al. 1989), scent gland of rat (Kannan 1998), interdigital secretion of blesbok (Burger et al. 1999) and exocrine saccules of dwarf hamsters (Burger et al. 2001), and these animals use these compounds to mark home range territory.

Bats generally use such glandular secretions to mark their mates and territory. The absence of such compounds in females clearly indicated that these odoriferous compounds are the male markers. Such chemical markers of the studied bat species are given in Table 11.2. *H. speoris* possesses such chemical markers in their frontal gland as (Phenyl azo)-2-[2-(tetra hydro-2-oxo-3-furanyl) ethoxyethyl ester of Benzoic acid and 2, 3-Dimethoxy-1-nitro-4-pentachlorophenyl Naphthalene. Similarly male anal gland of *T. melanopogon* has 1-[3-(octadecyloxy)propoxyl]-9-octadecane as a marker.

Male frontal gland of *R. leschenaulti* possesses 2-(octadecyloxy)ethyl ester of 9-octadecanoic acid, and they use this as a specific marker to mark their partners. These specific markers are present only in the glandular secretion of males and not found in females during breeding and non-breeding seasons. The chemical markers, 2-methyl l-2-Isobutyl-1,3-benzoxathiole and 2-thiozolamine of the frontal secretions, and 1,4-benzodiazepine, pregnane tetrol, hydroxy androstane, and hydroxymethyl cholesterol of the anal secretions are used by potent males during mating season on the mating territory, which facilitated to liberate the strong odours and fragrance.

Table 11.2 Functional group with oxygen, sulphur and nitrogen atom and steroids used by male bats during breeding season to mark the territory

S. no.	Species	Chemical compounds	Chemical signal emitters
Territory markers of male Hipposideros speoris during breeding season			
1	Dominant male	2, 3-Dimethoxy-1- nitro-4- pentachlorophenyl naphthalene	Frontal
		3, 6-Dimethoxy- methyl- 2-phenyl, 4-Quinolinone	
		Androstan-17-one	
		1,2,4,5-Tetrachloro-3, 6, -bis (trichloro methyl) Benzene	
		2, 2, 3, 3, 4, 4, 6, 6- Octachloro 1, 1-Biphenyl	
		Cholestan-3-one	Anal
		Tetradecane	
		4-Chloro octane	
		1-Octadecanethiol	
		1-Chloro Octadecane	
		1-Iodo Tridecane	
		3, 5-Dimethyl Octatriacontane	
2	Intruder male 1	3-Methyl butyl Benzoate	Frontal
		1-Chloro-1-buten-3-yne	
		Tetradecane	Anal
		1-(2-Octa decyloxy ethoxy)-9-Octadecane	
3	Intruder male 2	2-Methyl Tridecane	Anal
		Heptadecane	
Territory markers of male Hipposideros ater during breeding season			
1	Dominant male	2,2,5-Trimethly hexane 3,4-dione	Urine
		3-Methyl-3-butanol	Neck
		5-methyl-2-heptanamine	Ear
		2-Propanamine	Frontal
2	Intruder male 1	1,6-Anhydro 3,4 dideoxy Beta-D-gluco hexopyranose	Neck
		Alpha D-Arabinopyranose	Face
3	Intruder male 2	3-Methyl heptyl acetate,	Urine
		Alpha D-Arabinopyranose	Frontal
Territory markers of male Taphozous melanopogon during breeding season			
1	Dominant male	Cholesta-7,9(11)-dien-3-ol	Facial
		2-Bromo-cyclic,1,2-ethane diyl, Cholestan-3-one	
		2, Butyl,1-octanol	Neck
		2-(1-Methyl ethylidene)-cyclic 1,2-ethanediyl acetal) Cholestan-3-one	
		Cholestan-3-one	Anal
		25 Ethyl-24-hydroxy-3-beta-methoxy-4, 4, 14, alpha-tri methyl * cholest-9(11)-ene	
		Estra-1,3, 5 (10)-trien-17-one	
		1-[3-(Octa decyloxy) propoxyl]-9 Octadecane*	

(continued)

Table 11.2 (continued)

S. no.	Species	Chemical compounds	Chemical signal emitters
2	Intruder male 1	Cholestan-3-one	Neck
		1, 3, 5, 2, 4, 6-Triazatriphos-phorine	Anal
3	Intruder male 2	1, 1-Di (2-Methoxy phenyl azo Benzene)	Facial
		Prostaglandin	Anal
Territory markers of male Rousettus leschenaultii during breeding season			
1	Dominant male	1, 3 Bis (1, 1-dimethyl ethyl)-5-methyl Bbenzene	Frontal
		5-Undecen-4-one	
		Cholestan-3-one	
		3-Chloro-17 beta,Hydroxy-acetate androsta-2, 4, 6-triene-2-carboxal dehyde	
		3,20-Diamine-Pregnane	
		6-MethylOctadeeane	Neck
		2-MethylPentadecane	
		2 (Dimethyl amino)-1-phenyl-5-nonanone	Anal
		1, 1-(1-Fluoro,1,2-ethene diyl) bisbenzene	
		Cholesta-2,8-dien-6-01	
		N-Octyl-1-Octananmine	
2	Intruder male 1	13-Decyl-13-heptyl Tritriacontane	Frontal
		1-Penta bromo 2-Benzene	
		Triacontane	Neck
		1-(2, 4, 6-Trimethyl Phenyl)-3methyl-1-butene	Anal
3	Intruder male 2	Cyano acetate	Frontal
		1, 1-Bi phenyl, 4, 4-dimethyl	
Territory markers of male Cynopterus sphinx during breeding season			
1	Dominant male	2-Methy l-2-Isobutyl-1,3-benzoxathiole	Frontal
		2-Thiozolamine	Frontal
		1,4-Benzodiazepine	Neck
		Pregnane tetrol	Anal
		Hydroxy androstanone	Anal
		Hydroxy methyl cholesterol	Anal
2	Intruder male 1	Tetraphenyl-1,3-Oxathiol	Anal
3	Intruder male 2	N-Butyl butanamide	Neck

11.7 Odour Secretions for Individual Recognition

The earlier study revealed that functional groups such as alcohol, aldehyde and ketone derivatives give information about male dominance and individual recognition (Martin and Lopenz 2000). Table 11.3 clearly explains alcohol, aldehyde and ketone derivatives uphold individual identity among the colony members. The tabulated chemicals illustrated that each member of the colony has its own specific

Table 11.3 Alcohol, aldehyde and ketone derivatives uphold individual identity among the colony members

S. no	Species	Chemical compounds	Chemical signal emitters
Alcohols, aldehydes and ketones used by Hipposideros speoris for individual identity			
1	Dominant male	2, 3-Dimethoxy-1-nitro-4-pentachlorophenyl Naphthalene	Frontal
		3-Methyl butyl Benzoate	
		2, 4-Bis (1-1-dimethylethyl) Phenol	Anal
		Bis (2-ethyl hexyl)	
2	Potent male 1	3, 6-Dimethoxy-methyl-2-phenyl, 4-Quinolinone	Frontal
3	Potent male 2	,2,4,5-Tetrachloro-3, 6, -bis (trichloro methyl) Benzene	Frontal
4	Female I	3-Methyl-3 butenol	Urine
5	Female II	Neoheptanol	Anal
6	Female III	2-Ethyl 1-butanol	Anal
Alcohols, aldehydes and ketones used by Hipposideros ater for individual identity			
1	Dominant male	2,2,5-Trimethyl hexane-3,4-dione	Urine
		2,3-Epoxy hexanol	Urine
		5-Hexen-2-one	Urine
		3-Methyl 3 butenol	Neck
		5-Methyl-5-hexen-2-one	Neck
		2,2,5-Trimethyl hexane-3,4-dione	Ear
		N-Hexanal	Neck
		Hexyn-3-ol	Anal
2	Potent male 1	3-Octen-2-one	Saliva
		2-Methyl-1-pentanol	Neck
3	Potent male 2	N-Hexanal	Urine
		5-Methyl-5-hexen-2-one	Anal
4	Female I	3-Methyl-3 butenol	Urine
		5-Metyl 1-5-hexen-2-one	Urine
		7-Methyl-4-decane	Saliva
		Neoheptanol	Anal
5	Female II	4-Methyl-1-Octene	Urine
		2,2,5-Trimethyl hexane-3,4-dione	Anal, ear
6	Female III	Iso butenyl carbinol	Ear
		3-Hexane-2,5-diol	Neck
		2-Ethyl 1-butanol	Anal
Alcohols, aldehydes and ketones used by Taphozous melanopogon for individual identity			
1	Dominant male	2-Bromo-cyclic 1,2-ethane	Facial
2	Potent male 1	1, 1-Di (2-Methoxy phenyl	Facial
3	Potent male 2	2, Butyl,1-octanol	Neck
4	Female I	25 Ethyl-24-hydroxy-3-beta-methoxy-4, 4, 14, alpha-Tri methyl	Anal
5	Female II	1-Hexadecyl	Facial

(continued)

Table 11.3 (continued)

S. no	Species	Chemical compounds	Chemical signal emitters
6	Female III	1, 1-Di (2-Me thoxy phenyl azo Benzene	Facial
Alcohols, aldehydes and ketones used by Rousettus leschenaultii for individual identity			
1	Dominant male	1, 3 Bis (1, 1-dimethyl ethyl)-5-methyl benzene	Facial
		1, 1-Bi phenyl, 4, 4-dimethyl	
2	Potent male 1	13-Decyl-13-heptyl	Neck
3	Potent male 2	1, 1-Bi phenyl, 4, 4-dimethyl	Neck
4	Female I	2 (Dimethyl amino)-1-phenyl	Anal
5	Female II	6-MethylOctadeeane	Anal
6	Female III	2-MethylPentadecane	Anal
Alcohols, aldehydes and ketones used by Cynopterus sphinx for individual identity			
1	Dominant male	Cholan-3-ol	Neck
		2-Oxopropanal	Frontal
		Benzopyranone	Anal
2	Potent male 1	3-Methyl-3-butenol	Frontal
		Cycopentanol	Anal
3	Potent male 2	3,4-Epoxy-2-Hexanone	Neck
4	Female I	5-Andrstan-12-one	Neck
		3-Methyl-2-pentanone	Ear
5	Female II	3-Methyl-4-heptanone	Ear
6	Female III	2-Butyl octanol	Anal
		3,4-Epoxy-2-octanone	Neck

chemical compounds not possessed by other individuals. Flood et al. (1989) noticed such biochemical compounds from the preorbital secretion of musk ox and blesbok *Damaliscusd orcasphillipsi* (Burger et al. 1999). These compounds play a vital role in individual recognition. An additional recent report by Burger et al. (2001) also revealed that the ventral glandular secretions and preorbital glandular secretions of hamsters, *Phodopussung orussungorus*, contain aldehydes and ketones. These secretions help in home range marking through which they elicit a high level of individual recognition.

11.8 Odour Secretion for the Maintenance of Social Behaviour

Chemical compounds like acids with ester, alkanes, alkenes and specific compounds with sulphur and nitrogen composition are proven odour-producing substances. Earlier reports suggest that these compounds, especially help in group recognition in a colony and also aid in sex identity in a social system (Scully et al. 2000a, b).

Compounds like acid ester with phenols and ethanol compounds are present in the male frontal and anal gland of *H. speoris*, and such chemicals are: male frontal secretions with 2-phenyl-1, 3-dioxan-5-yl ester and 4-phenyl azo-2-([2-(tetrahydro-2-oxo-3-furanyl) ethoxy] ethyl ester of benzoic acid and male anal: bis (2-methyl propyl) ester of 1, 2-benzene dicarboxylic acid, butyl octyl ester of 1, 2-benzene dicarboxylic acid and bis (2-ethyl hexyl) ester of 1, 2-benzene dicarboxylic acid. Similarly, *T. melanopogon* also possesses the same type of esters in the neck glandular secretions. They are (2-[(oxododecyl) oxy]-1, 3-propanediyl ester of octadecanoic acid; 3-(1-oxohexadecyl) oxy-2- (1-oxotetradecyl) oxy propyl ester of octadecanoic acid and (bromomethyl) methyl bis 5, 6 (hydroxyimino) heptyl ester of propanedioic acid.

In *R. leschenaulti*, the male frontal secretions have hexadecyl ester of octadecanoic acid and dimethyl ester of dodecanedioic acid, whereas the females possess methyl ester of 15-tetracosanoic acid. There are also specific compounds with the sulphur present in the male frontal, and anal secretions that perform similar functions like esters. Such compounds in the male frontal region are 2-iodothiophene 1, 3-bis (1, 1, demethyl ethyl) 5-methyl benzene, and that of the female anal is 5-heptadecyl-methyl ester of 2-thiophene carboxylic acid. In *C. sphinx*, 2,4-dimethyl decane, trimethyl pentane,2,9-dimethyldecane, tetramethyl and undecene are chemical compounds possessed by anal, frontal, neck and ear glandular secretions of all the adult male members of the colony both during breeding and non-breeding seasons. Similarly, 2,3-dimethylpentane, 2,3,4-trimethyl heptanes and dodecane,4,6,8-trimethyl nonene are chemicals possessed by frontal and ear glandular secretions of all the adult female members. These compounds help induce social behaviour to discriminate familiar individuals from unfamiliar, self-recognition and determination of complex relationships among the colony members. The absence of such similar compounds in the competent males clearly indicated that they are from another social group and are intruders during mating season to attract the females (Table 11.4).

Such chemical compounds eliciting social behaviour have been noticed in reindeer (*Rangifer tarandus*) (Brundin et al. 1978), guinea pigs (*Cavia porcellus*) (Wellington et al. 1979), antelope (*Oreotragus oreotragus*) (Burger et al. 1997) and liolaemus lizard (*Liolaemus chiliensis*) (Carlos et al. 2001). The reports of Barnard et al. (1991) have inferred that the females of the house mice (*Mus domesticus*) recognised the close relatives through this male attractant and avoid mating. Similarly, they can recognise the major histocompatibility of deleterious genes through the male odours (Lingington 1983; Kunz and Fenton 2003).

11.9 Specific Chemical Cues

The present study has confirmed the glandular secretion mixed with faeces and urine spread on the floor of the roost and together with bat body odour dispersed the 'pungent' smell to the surroundings of the roost. Similar factors have been reported in other animals like rabbits, hamsters and goats. In rabbits (*Oryctolagus cuniculus*), lipid extracts from the anal gland and of the inguinal gland have the 'rabbity' odour (Mykytowycz 1968). The specific odour of the hamster (*Phodopus sungorus*)

Table 11.4 Alkane and alkene organic compounds exhibited by the colony members maintain the colony structure and related social behaviour

S. no	Species	Chemical compounds	Chemical signal emitters
Chemical compounds possess by all the members of the Hipposideros speoris colony both during breeding and non-breeding season			
1	Male adult	1-Chloro-1-buten-3-yne	Frontal glandular secretions
		2,4-Bis (1-1-dimethylethyl) Phenol	Anal glandular secretions
2	Female adult	1-Chloro-1-buten-3-yne	Frontal glandular secretions
Chemical compounds possess by all the members of the Hipposideros ater colony both during breeding and non-breeding season			
1	Male adult	Eicosyl trichlorosilane	Saliva
		2-Methyl-nonane	
		2,3,3-Trimethyl pentane	
		2,2-Dimethyl butane	Urine
		1,3-Epoxy-4-Methyl pentane	Ear
		2,2,5-Trimethyl hexane-3,4-dione	
		Tetradecane	Face
		Pyrrolidine	Anal
		Isotetradecane	
2	Female adult	1-3 Epoxy-4-methyl pentane	Face
		2-Chloro octane	Neck
		3-Ethyl octane	
		Octacosane	
		Hexa decane	
		n-Tetradecane	Anal
		n-penta decane	
		1-Iodo-2-Methyl undecane	
Chemical compounds possess by all the members of the Taphozous melanopogon colony both during breeding and non-breeding season			
1	Male adult	2, Butyl,1-octanol	Neck
		Cholesta-7,9(11)-dien-3-ol	Facial
2	Female adult	1, 1-Di (2-Methoxy phenyl azo Benzene	Facial
		1, 3, 5, 2, 4, 6-Triazatriphos-phorine	Anal
Chemical compounds possess by all the members of the Rousettus leschenaultii colony both during breeding and non-breeding season			
1	Male adult	2-Iodo Thiophene	Frontal gland
		13-Decyl-13-heptyl Tritriacontane	
		1-Penta bromo 2-Benzene	
		Cholestan-3-one	
		6-MethylOctadeeane	Neck gland
		2-MethylPentadecane	
		1-(2, 4, 6-Trimethyl phcnyl)-3mcthyl-1-butene	Anal gland
		1, 1-(1-Fluoro,1,2-ethene diyl) bisbenzene	

(continued)

Table 11.4 (continued)

S. no	Species	Chemical compounds	Chemical signal emitters
2	Female adult	1, 1-Bi phenyl, 4, 4-dimethyl	Frontal gland
		Cholestan-3-one	
		Triacontane	Neck gland
		2 (Dimethyl amino)-1-phenyl-5-nonanone	Anal gland

is due to the peculiar fatty acid compositions of the buccal secretions (Burger et al. 2001). Wool house et al. (1993) also reported that the cloacal glands of goat (*Trichosorus vulpeculakerr*) possess fatty acid components responsible for the characteristic odour of male goat. Similarly, in bats, Gudger (1945) remarked musky sweet odour from the roosting site of *N. leporinus*. Voigt and Helversen (1999) have noticed a sweet smell from the roost of *S.bilineata*. Scully noticed a spicy odour from most of the phyllostomid bats. Lily and Vanitharani (2005a, b) have remarked about the specific pungent smell from the roost of *H. speoris*. Earlier chiropteralogists also suggested specific fatty acid compounds like octadecanoic acid, tetradecanoic acid, alkene and steroid derivatives of integumentary anal glandular secretions are species-specific (Sugiyama et al. 1981; Brooke and Decker 1993; Bloss et al. 2002). The colony of *H. ater* has a characteristic pungent odour. During the nonbreeding season, they maintain a maternity colony during which the odour is intense and spreads into the environment as a mixture of fishy and milky odour along with a pungent smell. This is due to pyrrolidine, a nitrogen-containing compound and fatty acids like palmitic acid and decanoic acids produced by anal and salivary secretion of the sub-adult males and lactating mothers.

11.10 Role of Microbes in Chemical Signalling

Microbes also add volatile metabolites to an animal's scent profile. Morphological structures which host scent-producing microbes are common in many mammals (Starck 1958; Dapson et al. 1977; Scully et al. 2000a, b). These microbes are involved in scent production through metabolic degradation in the scent glandular secretions to produce strong smell (Dapson et al. 1977; Studier and Lovoie 1984; Voigt et al. 2005a, b, 2008). Bacteria like *Corynebacterium xerosis* even modify odourless androstene steroids and testosterone into several odoriferous metabolites (Rennie et al. 1990, 1991). This scent helps bring mating success (Wyatt 2003). One of the prominent examples is the microbial involvement in the armpit region of humans to produce odourless compounds (Rennie et al. 1990, 1991; Gower et al. 1994). Bacterial composition influences the anal glands of lions, foxes and cows and causes the production of volatile fatty acids responsible for the species-specific odour (Albone 1977; Albone et al. 1977). German (1976) observed that the bacteria in scent glands convert the glandular secretions into odoriferous chemical compounds like carboxylic acids.

Similarly, odoriferous fatty acid fractions are produced by chemical hydrolysis and microbial degradation of the glandular secretion (Wool House et al. 1993).

11.11 Role of Microbes in Odour Production

There are studies to support microbial involvement in the odour production of glandular odour secretions in bat species: *Tadarida brasiliensis*, *Molossus bondae* and *Eptesicus fuscus* (Dapson et al. 1977). During grooming, the bats spread the microbes to colonise in skin, external glands and mucous membranes. Studier and Lovoie (1984) say this behaviour in bats facilitates kin, colony mate and conspecifics recognition. *Staphylococcus aureus* and *Proteus mirabilis* bacteria in the chin and the wing region of *Noctilio leporinus* presumably are influenced by the animal's major histocompatibility complex. Table 11.5 shows the isolated microorganisms in the glandular region of *Hipposideros speoris*, *Taphozous melanopogon* and *Rousettus leschenaulti*. The present review analysis showed microbial involvement in the odour production is present in the frontal region of *H. speoris*, facial region of *T. melanopogon* and neck region of *R.leschenaulti*. Voigt et al. (2005b) reported that bacteria *Pseudomonas aeruginosa* and *Staphylococcus aureus* in the composition of the wing pouches of *S. bilineata* produces a very strong odour and influences the male's scent to produce mating success. Even a small-sized male has fitness benefits over larger males because of this odour influence (Voigt et al. 2008). The same microbe *Pseudomonas aeruginosa* is present in the anal region of *T. melanopogon* and facial and anal region of female *R. leschenaulti*. The existence of phenol products in the glandular secretions provides a good habitat for these microbes (Burger et al. 2001). The microbial fermentation role of these bacteria in the glandular products plays a vital role in the production of a scent profile that is specific to the carrier. Because of the important role of the microbial flora in the glandular secretions, the animals are careful in the maintenance of microbial richness in the glandular

Table 11.5 Isolated microorganisms in the glandular region of *Hipposideros speoris*, *Taphozous melanopogon* and *Rousettus leschenaulti*

Glandular region		*Hipposideros speoris*	*Taphozous melanopogon*	*Rousettus leschenaultii*
Face	Male	*Staphylococcus aureus*, *Bacillus pumilus*	*Staphylococcus aureus*	*Pseudomonas aeruginosa*
	Female	*Proteus mirabilis*, *Brevibacillus brevis*	*Pseudomonas aeruginosa*	*Staphylococcus aureus*
Neck	Male	No gland	*Bacillus pumilus*	*Bacillus alvei*
Anal	Male	*Bacillus alvei*	*Pseudomonas aeruginosa*, *Bacillus pumilus*, *Staphylococcus albus*	*Bacillus alvei*, *Pseudomonas aeruginosa*
	Female	*Staphylococcus aureus*	*Pseudomonas aeruginosa*, *Bacillus licheniformis*	*Pseudomonas aeruginosa*, *Staphylococcus albus*

organs. This has been observed in *S. bilineata* (Voigt et al. 2005b). Maintenance of microflora in the glands of bats is done by the repeated grooming, scratching and licking behaviour of the bats.

11.12 Conclusion

This review strongly suggests that these identified biochemical compounds of male and female scent glands, urine and saliva are involved in mate attraction and individual recognition important factors to bring reproductive success in a race. Males recognise oestrous females, and the females recognise the potential males through chemical signals and express the responses in their behavioural activities. In addition, these breeding behaviours keep the individuals close to each other and maintain social bonding and vital aspect in the maintenance of colonial life. Chemical signals play an important role during breeding. They coordinate and integrate many of the activities, which in turn bring reproductive success.

Bats are one of the keystone species in the ecosystem. They need to be protected to ensure a sustainable ecosystem. Studies of chemical communication make important contributions to understanding the social behaviour and mating biology of bats. Identified chemical constituents of the odour may help manufacture artificial scents to lure the temple bats to alternative roosting sites. The anti-odorous compounds can be synthesised from these bat odours and thus may help evacuate bats from temple instead of mass killing them during temple renovation and cleaning.

References

Adams MD, Li Y, Gerald SW (2018) Male scent gland signals mating status in greater spear-nosed bats, *Phyllostomus hastatus*. J Chem Ecol 44:975

Albone E (1977) Ecology of mammals: a new focus for chemical research. Chem Br 13:92–99

Albone ES, Gosden PE, Ware GC (1977) Bacteria as a source of chemical signals in mammals. In: Muller-Schwarze D, Mozell MM (eds) Chemical signals in vertebrates. Plenum Press, New York, pp 35–43

Archunan G (2003) Pheromones, chemical signals for reproductive behaviour. In: Annamalai R, Narayanan M, Vanitharani J (eds) Proceedings 28th Conference of Ethol. Soc. India, pp 38–42

Ayorinde F, Wheeler JW, Wemmer C, Mustaugh J (1982) Volatile components of the occipital gland secretions of the bacterian camel, *Camelus bactrianus*. J Chem Ecol 8:177–183

Balakrishnan M, Alexander KM (1985) Sources of body odour and olfactory communication in some Indian mammals. Ind Rev Life Sci 5:277–313

Barnard CJ, Hurst JL, Aldhous P (1991) Of mice and kin the functional significance of kin bias in social behaviour. Biol Rev 66:379–430

Bloss J, Terry EA, Bloss M, Wendy RH, Kunz TH (2002) Potential use of chemical cues for colony mate recognition in the big brown bat, *Eptesicus fuscus*. J Chem Ecol 28:799–814

Bowyer RT, Kitchen DW (1987) Significance of scent-marking by Roosevelt elk. J Mammal 68:418–423

Bradbury JW (1977) Social organisation and communication. In: Wimsatt WA (ed) Biology of bats, vol 3. Academic Press, New York, p 651

Bradbury JW, Vehrencamp SL (1976) Social organisation and foraging in Emuballonurid bats. I Field studies. Behav Ecol Sociobiol 2:1–17

Brooke AP, Decker DM (1993) Scent—marking and the use of odorous secretions by the fishing bat, *Noctilio leporinus*. Bat Res News 34:103
Brooke AP, Decker DM (1996) Lipid compounds in secretions of fishing bat *Noctilio leporinus*. J Chem Ecol 22:1411–1428
Brundin A, Anderson G, Anderson K, Mossing T (1978) Short chain aliphatic acids in the interdigital gland secretion of reindeer, *Rangifer tarandus* and their discrimination by reindeer. J Chem Ecol 4:613–622
Burger BV, Yang TP, LeRoux M, Brand WF, Cox AJ, Hart PF (1997) Mammalian exocrine secretions. XI: constituents of preorbital secretions of antelope, *Oreotragus oreotragus*. J Chem Ecol 23:2383–2400
Burger BV, Nell AE, Spies HSC, Le Roux M, Bisalke RC, Band PAJ (1999) Mammalian exocrine secretions XII: constituents of interdigital secretions of Blesbok, *Damaliscus dorcas phillipsi*. J Chem Ecol 25:2057–2084
Burger BV, Smit D, Spies HSC, Schmidt C, Schmidt U, Telistina AY (2001) Mammalian exocrine secretions XVI: constituents of secretion of supplementary sacule of dwarf hamster, *Phodopus sungorus sungorus*. J Chem Ecol 27(6):1277–1288
Carlos AE, Antonieta L, Hermann MN (2001) Chemical composition of precloacal secretions of Liolaemus lizards. J Chem Ecol 27(8):1677–1690
Chaverri G, Ancillotto L, Russo D (2018) Social communication in bats. Biol Rev 93:1938
Dapson RW, Studier EH, Buckingham MJ, Studier AL (1977) Histochemistry of odoriferous secretions from integumentary glands in three species of bats. J Mammal 58:531–535
Davis RB, Hereid CF, Short HC (1962) Mexican free-tailed bat in Texas. Ecological monographs 82:311–346
Eisenberg JF, Kleiman DG (1972) Olfactory communication in mammals. Annu Rev Ecol Syst 3:1–32
Esser KH, Schmidt U (1989) Mother-infant communication in the lesser spear-nosed bat, *hylloslomus discolor* (Chiropera: Phyllostomidae) evidence for acoustic learning. Ethology 82:156–168
Fenton MB (1985) Communication in the Chiroptera. Indiana University Press, Bloomington, p 161
Flood PF, Abrams SR, Muir GD, Rowell JE (1989) Odour of the musk ox. J Chem Ecol 15:2207–2247
German ML (1976) A mechanism for individual recognition by odour in *Herpestes auropanetatus* (Carnivara : Vivessidae). Anim Behav 24:141–145
Gosling LM (1982) A reassessment of the function of scent-marking in territories. Z Tierpsychol 60:89–118
Gower DB, Holland KT, Mallet AI, Rennie PJ, Watkins WJ (1994) Comparision of 16-androstene steroid concentrations in apocrine sweat and axillary secretions: inter conversions of 16-androstenes by the axillary microflora—a mechanism for axillary odour production in man? J Steroid Biochem Mol Biol 48:409–418
Gudger EW (1945) Fisherman bats of the Caribbean region. J Mammal 26:1–15
Guiraudie G, Pageat P, Cain AH, Patricia NLM (2003) Functional characterization of olfactory binding proteins for appeasing compounds and molecular cloning in vermonasal organs of prepubertal pigs. Chem Senses 28:43–63
Gustin MK, Mac Cracken GF (1987) Scent recognition between females and pups in bat *Tadarida brasiliensis mexicana*. Anim Behav 35:13–19
Haussler U (1987) Male social behaviour in *Molossus molossus* (Molossidace). In: Hanak V, Horacek I, Gaisler J (eds) European bat research. Charles University Press, Prague, pp 125–130
Herreid CF (1960) Comments on the odour of bats. J Mammal 41:396
Hurst JL, Barnard CJ (1992) Kinship and social behaviour in wild house mice: effects of social group membership and relatedness on the responses of dominate males towards juveniles. Behav Ecol 3:196–206
Hurst JL, Rich TJ (1999) Scent marks as competitive signals of male quality. In: Johnston RE, Muller Schwarze D, Sorensen PW (eds) Advances in chemical communication in vertebrates. Plenum Press, New York, pp 209–226

Janzen DH, Devries PJ, Higgins ML, Kimsey LS (1982) Seasonal and site variation in male bees arriving at chemical baits in deciduous forest and rain forest. Ecology 63:66–74
Jones RB, Nowell NW (1973) Aversive effects of the urine of a male mouse upon the investigatory behaviour of its defeated opponent. Anim Behav 21:707–710
Kairali R, Alexander KM (1984) Specialised integumentary glands of rat *Rattus rattus*. Ind Acad Sci 93:535–542
Kamran S, Kerth G (2003) Sections of the integumentary gland contain information about individuality and colony membership in the Bechteins bat. Anim Behav 65:363–369
Kannan S (1998) Histomorphology, biochemistry and chemical identification of rat scent glands in relation to rodent pest management. PhD Thesis, Bharathidasan University. Tiruchirapalli, India
Kunz TH (1990) Ecological and behavioural methods for the study of bats. Smith Sonian Institution Press, London and Washington, DC
Kunz TH, Fenton MB (2003) Bat ecology. In: Sensory ecology and communication in the Chrioptera. University of Chicago Press, Chicago and London, pp 90–127
Leon M (1983) Chemical communication in mother-young interactions. In: Vandenberg JG (ed) Pheromones and reproductions in mammals. Academic Press, New York, pp 39–77
Lily D (2005) Scent gland histology biochemistry and chemical signaling during the mating behaviour of three species of temple bats. PhD Thesis submitted to Manonmaniam Sundaranar University Tirunelveli, Tamil Nadu, India
Lily D, Vanitharani J (2005a) Olfactory communication through scent-marking behaviour during breeding in *Hipposideros speoris* (Schneiders leaf nosed bat 1800). J Theor Exp Biol 1(3):121–125
Lily D, Vanitharani J (2005b) Chemical signaling in *Hipposideros speoris* (Schneider's leaf nosed bat 1800). J Theor Exp Biol 1(4):183–192
Lingington S (1983) Social preference for partners carrying 'good genes' in wild house mice. Anim Behav 31:325–333
Mainoya JR, Howell KM (1977) Histology of the frontal sac in three species of leaf nosed bats (Hipposideridae). Afr J Ecol 15:147–155
Martin J, Lopenz P (2000) Chemoreception, symmetry and mate choice in lizards. Proc R Soc Lond 267:1265–1269
McCracken GF (1984) Communal nursing in Mexican free-tailed maternity colonies. Science 223:1090–1091
Miquelle DG (1991) The function of scent urination in moose. Am Nat 138:460–477
Mo WP, Burger BV, LeRoux M, Spies HSC (1995) Mammalian exocrine secretions IX. Constituents of preorbital secretion of oribi, *Ourebia orebi*. J Chem Ecol 21:1191–1215
Mykytowycz R (1968) Territorial marking by rabbits. Sci Am 218:116–126
Nelson JE (1965) Behaviour of Australian Pteropodidae (Megachiroptera). Anim Behav 13:544–575
Nojima S, Nishida R, Sakuna M, Kuwahra Y (1999) A glandular gift in the genus cockroach, *Blattella germanica*. J Insect Behav 12:627–640
Nowak RM (1994) Wlakers bats of the world. Johns Hopkins University Press, Baltimore, pp 1–287
Ratcliffe FN (1932) Notes on the fruit bats (*Pteropus sps.)* of Austrila. J Anim Ecol 1:32–57
Rennie PJ, Gower DB, Holland KT, Mallet AI, Watkins WJ (1990) The skin microflora and the formation of human axillary odour. J Cosmet Sci 12:197–207
Rennie PJ, Gower DB, Holland KT (1991) *In-vitro* and *in-vivo* studies of human axillary odour and the cutaneous microflora. Br J Dermatol 124:596–602
Rich TJ, Hurst JL (1998) Scent marks as reliable signals of the competitive ability of mates. Anim Behav 56:727–735
Robbins LW, Sarich VM (1988) Evolutionary relationships in the family, Emballonuridae (Chiroptera). J Mammal 69:1–13
Rozenfeld FM, Denoel A (1994) Chemical signals involved inspacing behaviour of breeder bank voles, *Clethrionomys glareolus*. J Chem Ecol 20:803–813

Scully WMR, Fenton MB, Salenaddin J (2000a) A histological examination of holding sacs and scent glandular organs of some bats, Emballonuridae, Hipposideridae, Phyllostomidae, Vespertilionidae and Molossidae. J Can Zool 98:613–623
Scully WMR, Fenton MB, Salenuddin J (2000b) Histological examination of holding sacs and scent glandular organs of some bats. Can J Zool 78:613–623
Sofia E (2007) Studies on roost selection, population dynamics reproduction and feeding behavior of *Hipposideros ater*,(dusky leaf nosed bat). PhD Thesis submitted to Manonmaniam Sundaranar University Tirunelveli, Tamil Nadu, India
Starck D (1958) Beitrag zur kenntnis der aimtaschen und anderer Hautdrusenorgane von *Saccoptexyx bilineata,* Temminck 1838 (Chiroptera—Emballonuridae). J Morphol 99:3–25
Studier EH, Lovoie KH (1984) Microbial involvement in scent production in Noctilionid bats. J Mammal 65:711–714
Sugiyama T, Sasada TH, Masaki J, Yamastha K (1981) Unusual fatty acids with specific odour from nature goat. Agric Biol Chem 45:2625–2657
Valdivieso D, Tamsitt JR (1964) The histology of the chest gland of the pale spear-nosed bat. J Mammal 45:536–539
Vanitharani J, Vijaya M, Lily D (2007) Scent glands and chemical signaling in *Cynopterus sphinx* (short nosed fruit bat) during breeding season. J Theor Exp Biol 3:67–75
Voigt CC, Helversen VO (1999) Storage and display of odour by male *Saccopteryx bilineata* (Chiroptera:Emballonuridae). Behav Ecol Sociobiol 47:29–40
Voigt CC, Von Helverson O, Michnes R, Kunz TH (2001) The economics of maintanence in the sac-winged bat, *Saccopteryx bilineata*. Behav Ecol Sociobiol 50:31–36
Voigt CC, Barbara C, Stephanie S (2005a) Bats, bacteria and bat smell: sex-specific diversity of microbes in a sexually selected scent organ. J Mammal 86(4):745–749
Voigt CC, Heckel G, Mayer F (2005b) Sexual selection favours small and symmetric males in the polygynous greater sac-winged bat, *Saccopteryx bilineata* (Emballonuridae, Chiroptera). Behav Ecol Sociobiol 47:29–40
Voigt CC, Behr O, Caspers B, von Helversen M, Knornschild FM, Nagy M (2008) Songs, scents, and senses: sexual selection in the greater sac-winged bat, *Saccopteryx bilineata*. J Mammal 89(6):1401–1410
Wedekind C, Seebeck T, Bettens F, Paepke AJ (1995) MHC dependent mate preference in humans. Proc R Soc B Ser Lond 260:245–249
Wellington JL, Byrne KJ, Preti G, Beauchamp GK, Smith AB (1979) Perineal scent gland of wild and domestic Guinea pigs. A comparative chemical and behavioural study. J Chem Ecol 5:737–751
Wilson EO (1965) Chemical communication in the social insects. Science 149:1064–1071
Wool House AD, Weston RJ, Hamilton BH (1993) Analysis of secretions from scent producing glands of brush tail possum, *Trichosorus vulpecula kerr*. J Chem Ecol 20:239–253
Wyatt T (2003) Pheromones and animal behaviour. Cambridge University Press, Cambridge
Xavier F (1994) A study on the civet cat, as a sustainable wild life resource. M.Phil. thesis, University of Kerala, Trivandrum

Olfactory Learning in Greater Short-Nosed Fruit Bat (*Cynopterus sphinx*)

12

K. Emmanuvel Rajan, Murugan Mukilan,
and D. Mary Rajathei

Abstract

The greater short-nosed fruit bat, *Cynopterus sphinx*, feeds on a variety of fruits, flowers, and leaves. Bats discriminate ripe and unripe fruits of the same species by olfactory cues. Like other neotropical bats, *Cynopterus sphinx* modifies the leaves/stems of creeper plants and trees as "day roost-tent". Males construct these tents and prefer to stay with females and their young. An earlier radio-tracking study revealed that after sunset, *C. sphinx* reaches the previously visited foraging area using spatiotemporal memory and starts searching for food sources with the aid of olfactory cues. Experiments are designed to understand how *C. sphinx* learn and recognize their food. A series of experiments demonstrate that neuromodulation regulates the downstream signaling molecules in olfactory learning. Specifically, the fine-tuned level of serotonin (5-hydroxytryptamine [5-HT]) in the olfactory bulb (OB) activated the 5-HT1A receptor and serotonin transporter (SERT) to regulate the signaling molecules. Signaling molecules associated with the neuronal circuits in different brain regions, which selectively drive the expression of genes that sorting the preference and aversive olfactory

K. E. Rajan (✉)
Department of Animal Science, School of Life Sciences, Bharathidasan University, Tiruchirappalli, Tamil Nadu, India

M. Mukilan
Department of Biotechnology, Sri Ramakrishna College of Arts and Science, Coimbatore, Tamil Nadu, India

D. M. Rajathei
Department of Biotechnology and Bioinformatics, Bishop Heber College, Tiruchirappalli, Tamil Nadu, India

S. Sridhara et al. (eds.), *Animal Behavior in the Tropics*,
https://doi.org/10.1007/978-981-97-7324-4_12

memory. This review focuses on the molecular mechanism of learning and memory in the greater short-nosed fruit bat (*Cynopterus sphinx*).

Keywords
Cynopterus sphinx · Olfactory learning · Neurotransmitter · Serotonin · Immediate early genes · MicroRNAs

12.1 Introduction

The order Chiroptera is classified into two suborders: Megachiroptera and Microchiroptera with 202 genera under 18 families (Wilson and Reeder 2005). More than 900 species are recorded in Microchiroptera, whereas Megachiroptera includes one family (Pteropodidae) with 186 species recorded (Wilson and Reeder 2005). The Megachiropterans feed on flowers and fruits; nectar and pollen are confined to old-world tropics. The Microchiropterans, as the name implies, are much smaller in size, but they far exceed the Megachiropterans in number and distribution (Nowak 1994). The most striking difference between Megachiropterans and Microchiropterans is that the former orient primarily by vision and olfaction, while the latter orient by echolocation. Only a few megabat species belong to the genus *Rousettus* and *Eonycteris* that produce tongue-clicking for orientation (Holland et al. 2004; Raghuram et al. 2009). They depend on vision and olfaction to find food, avoid hurdles, and maneuver in their environment (Acharya et al. 1998; Luft et al. 2003; Sanchez et al. 2006; Elangovan et al. 2006). With vision, olfaction is more likely to play a significant role in successful foraging and to discriminate between ripe and unripe fruits (Acharya et al. 1998; Luft et al. 2003; Sanchez et al. 2006).

12.2 Learning and Memory

Behavioral testing of animals permits obtaining information on the functional role of neuroactive molecules/neurotransmitters on social behavior, perception, learning, and memory. Behavior refers to specific responses of animals to endogenous (e.g., the action of neuroactive molecules/neurotransmitters) or environmental signals (light/sound/odorant). Specific olfactory stimuli elicit specific behavioral responses, which are implicated in mother–offspring interaction, appetitive, aversive, sexual selection, and other social behavior (Ravel et al. 1994; DeRosa and Hasselmo 2000). Olfactory stimuli have powerful information over other stimuli, especially for survival (including food and predators) across species ranging from insects to mammals. Perceptual learning (including olfactory learning) is a form of implicit memory, which is associated with specific odorants and their discrimination based on previous experience (Wilson et al. 2004). Memory is classified into short term (1–3 h) and long term (>24 h), which is dependent on RNA and protein synthesis (Emptage and Carew 1993; Izquierdo et al. 1998; McGaugh 2000).

Memory formation and retrieval are processed in three different stages: (1) encoding (processing received information), (2) storage (establishing a permanent record of the encoded information), and (3) retrieval (recalling the stored information in response to associated cues/response).

12.3 Olfactory Learning and Memory

Olfactory memory is examined based on the response to the odorant, which is more stable for a long period of time (Roediger et al. 2008); however, it has few limitations to study than other memories, that is, auditory, visual, tactile, and gustatory (Herz and Engen 1996). In addition, olfactory memory completely differs from other forms of memory and highly resistant to forgetting. Many factors such as the strength of odor stimulus, the context of training, duration, and timing of training, and neurotransmitters/other molecules influence learning and memory through interaction with different brain regions. Neural transmission is most commonly regulated by the synthesis and release of small neuropeptides, neuroactive molecules, and neurotransmitters (OB und Halbach and Dermietzel 2006). Most often, altering the level of neurotransmitters in the central nervous system (CNS) is a choice of method to study the role of neurotransmitters in learning and memory formation (Miranda 2007; Morris et al. 1986; Hasselmo 1999).

12.4 Neurotransmission in Olfactory Learning

Signal transmission is established between the neurons through the specialized junctions called synapses. In chemical synapse, neurotransmitters released from presynaptic region binds at postsynaptic neuron receptors to transfer the information and activate signaling process (Vizi 2000; Vizi and Kiss 1998; Umbriaco et al. 1995). Activation of presynaptic neurons alters the membrane polarity and triggers the release of neurotransmitters at the synaptic cleft; they diffuse and bind with the postsynaptic neuron to transfer the information (Kalsner and Westfall 1990; Wu and Saggau 1997). Change in the action potential is dependent on the influx of Ca^{2+} channels (so-called L-type channels), which is linked to the release of neurotransmitters. The elevation of Ca^{2+} in the presynapsis binds to synaptic vesicles, changes their polarity, and migrates towards the docking protein for the process of exocytosis and neurotransmitter release. Following the release of neurotransmitters at the inter-synaptic cleft, the neurotransmitters bind to the neurotransmitter-specific postsynaptic receptor complex (Uchihashi et al. 1998; Nakai et al. 1999).

Neurotransmitters are two major groups: (1) biogenic amines (Acetylcholine (Ach), serotonin [5-hydroxytryptamine (5-HT)], Dopamine (DA), Nor-epinephrine (NE), epinephrine, histamine) and (2) small amino acids (γ-amino butyric acid, glutamate, glycine). Serotonin (5-hydroxytryptamine [5-HT]) structure was first identified and reported by Maurice M. Rapport in 1948. 5-HT is placed under biogenic amines (indolamines or monoamine transmitters) and is synthesized from tryptophan, which regulates the biosynthesis of serotonin by acting as a rate-limiting factor (Mössner et al. 2006). Tryptophan, an essential amino acid, crosses the blood-brain barrier through an active transport system. Because of its receptor diversity and divergence in transduction pathways, 5-HT regulates several biological functions, including food intake, rhythm, sexual behavior, emotional states, learning, and memory through its different receptors (Buhot et al. 2000; Bjorvatn et al. 2002; Farias et al. 2010). The availability of serotonin at the synaptic cleft critically regulates both short-term facilitation (STF) and long-term facilitation (LTF) through the activation of signaling molecules and associated pathways.

12.5 Activity-Dependent Olfactory Learning

Activity-dependent neural transmission directs the changes in synaptic structure and function and further activates the cellular signaling from extracellular to cellular and nucleus, such changes triggers the modification at RNA/protein level for every specific activity. Such changes and the associated level of synaptic function are considered a fundamental mechanism in the biology of learning and memory (Sweatt 2001; Kind and Neumann 2001). Based on this, experimental models have been developed to understand synaptic strength and its subsequent effect on response by providing repeated similar electrical stimulation. Which mimic the physiology, of the neuron, the observed changes increase synaptic strength referred to as long-term potentiation (LTP) and decline the strength referred to as long-term depression (LTD) (Malenka and Nicoll 1999; Thiels et al. 2002). Accepting the reported mechanism, further studies indicate that LTP is essential at the postsynaptic level to induce/suppress the synaptic response to activation/inhibition of the signaling mechanism. For the process of memory formation or consolidation, repeated LTP is required to activate more than one phase of transcription. Studies in mammalian models showed that long-term synaptic plasticity is required for long-term facilitation (LTF), as well as LTP and LTD, but short-term plasticity may not provide the same facilitation (Castellucci et al. 1986; Frey et al. 1996). Across the animal species, behavioral studies have demonstrated that de novo protein synthesis is required for the formation of long-term memory (LTM) (Bourtchouladze et al. 1998; Igaz et al. 2002; Rosenblum et al. 1993; Schafe and LeDoux 2000; Scharf et al. 2002).

12.6 Signaling Mechanism in Olfactory Learning

Exposure to a new stimulus results in the activation of RNA-dependent protein synthesis through the activation of neuronal signaling pathways. Various neuronal signaling molecules like protein kinase A (PKA), mitogen-activated protein kinase (MAPK)/extracellular regulated kinase (ERK), cyclic adenosine monophosphate (AMP) response element binding protein (CREB), immediate early genes (IEGs) (early growth response gene 1: *Egr-1*, *C-fos*, and *C-jun*), and postsynaptic density (PSDs) proteins were involved in the acquaintance of new information in the form of odor during the process of olfactory learning (Emptage and Carew 1993; Bailey et al. 1996; Impey et al. 1998; Roberson et al. 1999; Vanhoutte et al. 1999). The mitogen-activated protein kinase (MAPK)/extracellular regulated kinase (ERK) cascade has a potential role in initiating the transcription of specific genes (Sweatt 2001). Earlier studies demonstrated that activated MAPK/ERK translocates into the nucleus, interacts with cyclic AMP response element binding protein (CREB) and CCAAT-enhancer-binding protein (C/EBP), and initiates the transcription of immediate early genes (IEGs) (Treisman 1996). Other than learning and memory formation, CREB is also involved in the circadian rhythms, addiction, development, differentiation, and regeneration of the nervous system (Barco and Kandel 2006; Carlezon et al. 2005; Dragunow 2004; Mioduszewska et al. 2003; Persengiev and Green 2003; Silva et al. 1998).

Several studies have reported that CREBs is a universal modulator, which involves the regulation of synaptic plasticity and LTP in the process of memory formation (Barco et al. 2002; Bito et al. 1996, Impey et al. 1996; Josselyn and Nguyen 2005; Segal and Murphy 1998). Phosphorylation/activation of CREB on Ser-133 excites the expression of immediate early genes (Hardingham et al. 2001; Athos et al. 2002). Activated immediate early genes (IEGs) act as transcription factors and bind to regulatory elements to activate or repress the transcription of late-response genes (Alberini et al. 1994; Goelet et al. 1986; Taubenfeld et al. 2001a, b). The IEGs are two different types of effctor (e.g., *Arc/Arg3.1*) and regulators (*Zif-268*, *C-fos*, and *C-jun*). The former directly induces the cellular changes, which is considered marker for cellular response, whereas the latter regulates the expression of late-response genes (Christy and Nathans 1989; Swirnoff and Milbrandt 1995). IEGs directly or indirectly fine-tune the expression of synaptic proteins, synaptic plasticity, and memory formation.

12.7 Olfactory Learning in *Cynopterus sphinx*

The greater Indian short-nosed fruit bat (*C. sphinx*) feeds on fruits, flowers, and leaves (Bhat 1994; Marimuthu et al. 1998, 1999). Earlier captive and field studies reported that mothers carry their pups during foraging and later accompany their juveniles during foraging flights (Gopukumar et al. 2003; Elangovan et al. 2007). Mother–pup interactions during the early postnatal and weaning periods help pups develop foraging skills. Interestingly, the pups learn about the palatable and non-palatable fruits by their association with their mother during foraging. Later on, they develop an independent food preference. If oactive molecules, and neurotra the pups are separated during early foraging, they fail to develop food preferences or test the novel food (Ganesh et al. 2012). These studies demonstrate that the interaction between *C. sphinx* mothers and pups during early life is an adaptive plasticity period to develop food preference and foraging skills.

12.8 Role of Serotonin in *C. sphinx* Olfactory Learning

The olfactory bulb (OB) is the first region in the olfactory system of the brain, which receives dense serotonergic fibers and where the first synaptic interaction occurs for odorant stimulation. The study reported that depletion of serotonin in the OB using an analog of serotonin 5,7-dihydroxytryptophan (5,7-DHT) before exposure to the novel odor causes a deficit in recognition of the novel odor (not available fruit in the wild) and learning. Further, the level of deficit in olfactory learning is dependent on the level of depleted serotonin. Specifically, the process of learning is classified into an attempt towards the novel odor (initial recognition), landing on the food with the novel odor (feeding attempt), and a feeding bout (successful feeding). Notably, in all stages of learning, the 5,7-DHT-treated bats showed poor performance compared to the other bats. The subsequent question is the role of the 5-HT receptor in the olfactory circuit. There are seven major classes of 5-HT receptors; all of the receptors are G-coupled, except 5-HT3, which is the only ionotropic receptor. Screening the expression of all receptors showed that the 5-HT_{1A} receptor alone responded to novel odor exposure and was suppressed when 5-HT was depleted in OB. Earlier, in another animal model, a common signaling mechanism was demonstrated in olfactory learning/memory. Specific analyses of Ganesh et al. (2010) describe that novel odor stimulation failed to activate the signaling molecules including *Erk*1/2 and cyclic AMP response element binding protein 1 (*Creb*1) and their subsequent phosphorylation if the level of serotonin depleted more than 50% in OB. This observation concluded that in OB, the level of 5-HT is the key regulator to initiate a cAMP signaling cascade to in the neuronal circuit involved in olfactory learning/recognition and memory formation in *C. sphinx*. This raises an interesting concept of the 5-HT regulating olfactory input and downstream signaling in the OB.

In the signaling cascade, phosphorylated CREB in the OB activates early growth response gene-1 (*egr-1*) in the downstream pathway. Therefore, the expression of *egr-1* was silenced in OB using the antisense (AS) oligodeoxynucleotides (ODN) method. Bilateral infusion of AS-ODN of *egr-1* in OB suppressed the expression of *egr-1* even though the level of 5-HT in OB was normal and exposed to odor. Parallel behavioral analysis showed that their feeding attempts and bouts are not comparable to those of other bats. Further analysis revealed that reduced level of *egr-1* failed to coordinate with the olfactory stimulus for the phosphorylation of calcium/calmodulin kinase (CaMKII) and postsynaptic density protein-95 (PSD-95) in the OB. As reported earlier, along with CaMKII, PSD-95, and synaptotagmin-1 (SYT-1), other regulatory molecules, including microRNAs (miRs), are also involved in the olfactory neuronal circuit. When AS-ODN of *egr-1* was infused into OB, the expression pattern of miR-132 and miR-148a was altered. Interestingly, these two miRs show opposite patterns of expression in which either one may act as a negative regulator. Surprisingly, infusion of miR-132 AS-ODNs in the OB before exposure/training to the novel odor reduced the expression of CaMKII, CREB, IEGs (*Egr-1*, *C-fos*, and *C-jun*), and PSD-95. Notably, it negatively regulates the expression of miR-148a, which demonstrates the balancing act of miR-132 and 148a in the regulation of olfactory learning and memory in *C. sphinx* (Ganesh et al. 2012; Mukilan et al. 2015).

12.9 Preference and Aversive Odor Learning in *C. sphinx*

All the earlier studies described the molecular mechanism in olfactory learning and memory, which measured the neuronal activity in OB by using expression of immediate early response genes *Egr-1* and *C-fos*. However, the response of IEGs to the preference odor (PO)/aversive odor (AO) in different brain regions, that is, OB, hippocampus (HP), and amygdala (AMY). On the other hand, odor-mediated changes in methylation of the promoter and a negative regulator protein phosphatase-1 (PP-1) were also taken for analysis. Different time point analyses after an olfactory stimulus (30 and 60 min) PO showed a similar expression pattern of *Egr-1* and *C-fos* in the OB, HP, and AMY. Note that AO also induces the expression of *C-fos* immediately (within 30 min) in OB, HP, and AMY; however, it started to decrease within 60 min only in OB and HP but not in AMY. The methylation status of *C-fos* and the action of PP-1 may act as regulators to control expression. The extended two-dimensional gel electrophoresis analysis showed that miR-132 in OB regulates the protein response to a novel odor stimulus. Among them, hypoxanthine phosphoribosyltransferase-1 (HPRT-1) is known to be involved in purine metabolism and influences serotonin transmission. Further analysis indicates that inhibition of miR-132 suppresses the expression of HPRT, serotonin transporter (SERT), and *N-methyl*-d-*asparate* (NMDA) receptors (2A,B). These analyses suggest that miR-132 regulates the process of olfactory learning and memory formation through signaling of SERT and NMDA receptors that is possibly associated with *PRTFDC1*–HPRT interaction. Initially, *C. sphinx* responds to both AO and PO, but their subsequent response to training and ability to recall is only associated with PO food (Mukilan et al. 2018a, b, Rajan, 2021).

12.10 Conclusion

This review focused on behavioral neurobiology, especially the molecular mechanism of olfactory memory in a mammalian model, the fruit bat *Cynopterus sphinx*. An animal model (fruit bat) revealed that mother–pup interactions during the early foraging period may provide an opportunity for pups to learn about novel food sources. During olfactory learning, 5-HT in OB through its receptor activates *Erk*1, *Creb*1, and IEGs (*Egr-1* and *C-fos*). Further, suppression of *egr-1* expression with specific antisense (AS) oligodeoxynucleotides (ODN) down-regulates the expression of synaptic proteins such as synaptotagmin-1 (SYT-1), calcium/calmodulin kinase type II (CaMKII), postsynaptic density protein-95 (PSD-95), and phosphorylation of CaMKII. On the other hand, microRNA-138 and microRNA-148a act as positive or negative regulators of IEGs, and their differential expression in different brain regions. Silencing the microRNA-138 reduced the odor-induced expression of CaMKII, CREB, IEGs (*Egr-1*, *C-fos*, and *C-jun*), and PSD-95 and negatively regulates miR-148a. Odor-mediated regulation of protein phosphatase-1 (PP-1) acts on IEG expression in different brain regions, and its dynamic regulation orchestrates the gating of preference odor (PO) and aversive odor (AO) memory. This chapter

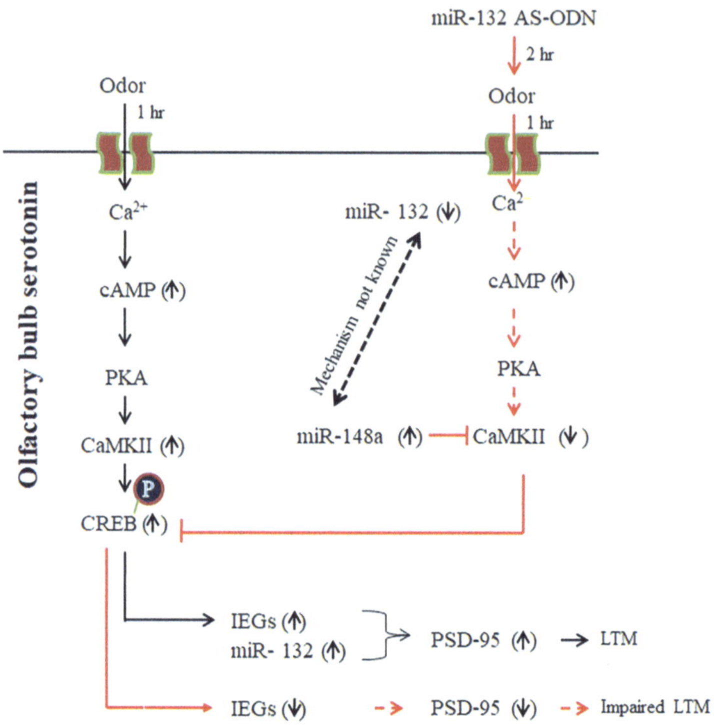

Fig. 12.1 Diagram showing the possible mechanism of serotonin-mediated signaling pathway involving *Cynopterus sphinx* olfactory learning and memory

explains that the level of 5-HT in the OB is the first line of activation in the neuronal circuit of olfactory learning and memory (Fig. 12.1).

References

Acharya KK, Roy A, Krishna A (1998) Relative role of olfactory cues and certain non-olfactory factors in foraging of fruit-eating bats. Behav Process 44:59–64

Alberini CM, Ghirardi M, Metz R, Kandel ER (1994) C/EBP is an immediate-early gene required for the consolidation of long-term facilitation in Aplysia. Cell 76:1099–1114

Athos J, Impey S, Pineda VV, Chen X, Storm DR (2002) Hippocampal CRE-mediated gene expression is required for contextual memory formation. Nat Neurosci 5:1119–1120

Bailey CH, Bartsch D, Kandel E (1996) Toward a molecular definition of long-term memory storage. Proc Natl Acad Sci U S A 93:13445–13452

Barco A, Kandel ER (2006) Role of CREB and CBP in brain function. Chapter in transcription factors in the nervous system: development, brain function and disease. Wiley-VCH, Hoboken, NJ

Barco A, Alarcon JM, Kandel ER (2002) Expression of constitutively active CREB protein facilitates the late phase of long-term potentiation by enhancing synaptic capture. Cell 108:689–703

Bhat HR (1994) Observations on the food and feeding behavior of *Cynopterus sphinx* Vahl. (Chiroptera, Pteropodidae) at Pune, India. Mammalia 58:363–370

Bito H, Deisseroth K, Tsien RW (1996) CREB phosphorylation and dephosphorylation: a Ca2+– and stimulus duration-dependent switch for hippocampal gene expression. Cell 87:1203–1214

Bjorvatn B, Gronli J, Hamre F, Sorensen E, Fiske E, Bjorkum AA, Portas CM, Ursin R (2002) Effects of sleep deprivation on extracellular serotonin in hippocampus and frontal cortex of the rat. Neuroscience 113:323–330

Bourtchouladze R, Abel T, Berman N, Gordon R, Lapidus K, Kandel ER (1998) Different training procedures recruit either one or two critical periods for contextual memory consolidation, each which requires protein synthesis and PKA. Learn Mem 5:365–374

Buhot MC, Martin S, Segu L (2000) Role of serotonin in memory impairment. Ann Med 32:210–221

Carlezon WA Jr, Duman RS, Nestler EJ (2005) The many faces of CREB. Trends Neurosci 28:436–445

Castellucci VF, Frost WN, Goelet P, Montarolo PG, Schacher S, Morgan JA, Blumenfeld H, Kandel ER (1986) Cell and molecular analysis of long-term sensitization in Aplysia. J Physiol 81:349–357

Christy B, Nathans D (1989) DNA binding site of the growth factor-inducible protein *Zif268*. Proc Natl Acad Sci U S A 86:8737–8741

DeRosa E, Hasselmo ME (2000) Muscarinic cholinergic neuromodulation reduces proactive interference between stored odor memories during associative learning in rats. Behav Neurosci 114:32–41

Dragunow M (2004) CREB and neurodegeneration. Front Biosci 9:100–103

Elangovan V, Yuvana Satya Priya E, Marimuthu G (2006) Olfactory discrimination ability of the short-nosed fruit bat *Cynopterus sphinx*. Acta Chiropterol 8:247–253

Elangovan V, Yuvana Satya Priya E, Raghuram H, Marimuthu G (2007) Wing morphology and flight development in the short-nosed fruit bat Cynopterus sphinx. Zoology (Jena) 110(3):189–196

Emptage NJ, Carew TJ (1993) Long-term synaptic facilitation in the absence of short-term facilitation in Aplysia neurons. Science 262:253–256

Farias FM, Passos CS, Arbo MD, Zuanazzi JA, Steffen VM, Henriques AT (2010) Monoamine levels in rat striatum after acute intraperitoneal injection of strictosidinic acid isolated from *Psychotria myriantha* Mull. Arg. (Rubiaceae). Phytomedicine 17:289–291

Frey U, Frey S, Schollmeier F, Krug M (1996) Influence of actinomycin D, a RNA-synthesis inhibitor, on long-term potentiation in rat hippocampal neurons in vivo and in vitro. J Physiol 490:703–711

Ganesh A, Bogdanowicz W, Haupt M, Marimuthu G, Rajan KE (2010) Role of olfactory bulb serotonin in olfactory learning in the greater short-nosed fruit bat, *Cynopterus sphinx* (Chiroptera: Pteropodidae). Brain Res 17:108–117

Ganesh A, Bogdanowicz W, Balamurugan K, Ragu Varman D, Rajan KE (2012) Egr-1 antisense oligodeoxynucleotide administration into the olfactory bulb impairs olfactory learning in the greater short-nosed fruit bat *Cynopterus sphinx*. Brain Res 1471:33–45

Goelet P, Castellucci VF, Schacher S, Kandel E (1986) The long and short of long-term memory: a molecular framework. Nature 322:4419–4422

Gopukumar N, Nathan PT, Doss PS, Prakash A, Emmanuel K, Balasingh J, Marimuthu G, Kunz TH (2003) Early ontogeny of foraging behaviour in the short-nosed fruit bat *Cynopterus sphinx* (Megachiroptera): preliminary results. Mammalia 67:139–145

Hardingham GE, Arnold FJL, Bading H (2001) Nuclear calcium signaling controls CREB-mediated gene expression triggered by synaptic activity. Nat Neurosci 4:261–267

Hasselmo ME (1999) Neuromodulation: acetylcholine and memory consolidation. Trends Cogn Sci 3:351–359

Herz RS, Engen T (1996) Odor memory: review and analysis. Psychon Bull Rev 3:300–313

Holland RA, Waters DA, Rayner JMV (2004) Echolocation signal structure in the megachiropteran bat *Rousettus aegyptiacus* Geoffroy 1810. J Exp Biol 207:4361–4369

Igaz LM, Vianna MR, Medina JH, Izquierdo I (2002) Two time periods of hippocampal mRNA synthesis are required for memory consolidation of fear-motivated learning. J Neurosci 22:6781–6789

Impey S, Mark M, Villacres EC, Poser S, Chavkin C, Storm DR (1996) Induction of CRE-mediated gene expression by stimuli that generate long lasting LTP in area CA1 of the hippocampus. Neuron 16:973–982

Impey S, Smith DM, Obrietan K, Donahue R, Wade C, Storm DR (1998) Stimulation of cAMP response element (CRE)-mediated transcription during contextual learning. Nat Neurosci 1:595–601

Izquierdo I, Barros DM, Mello e Souza T, de Souza MM, Izquierdo LA, Medina JH (1998) Mechanisms for memory types differ. Nature 393:635–636

Josselyn SA, Nguyen PV (2005) CREB, synapses and memory disorders: past progress and future challenges. Curr Drug Targets CNS Neurol Disord 4:481–497

Kalsner S, Westfall TC (1990) Presynaptic receptors and the question of autoregulation of neurotransmitter release. Ann N Y Acad Sci 604:1–652

Kind PC, Neumann PE (2001) Plasticity: downstream of glutamate. Trends Neurosci 24:553–555

Luft S, Curio E, Tacud B (2003) The use of olfaction in the foraging behaviour of the golden-mantled flying fox, *Pteropus pumilus*, and the greater musky fruit bat, *Ptenochirus jagori* (Megachiroptera : Pteropodidae). Naturwissenschaften 90:84–87

Malenka RC, Nicoll RA (1999) Long-term potentiation—a decade of progress. Science 285:1870–1874

Marimuthu G, Rajan KE, Johnkoil Raj A, Suthakar Isaac S and Balasingh J (1998). Observations on the foraging behaviour of a tent roosting Megachiropteran bat Cynopterus sphinx. Biotropica 30(2):321–324.

McGaugh JL (2000) Memory: a century of consolidation. Science 87:248–251

Mioduszewska B, Jaworski J, Kaczmarek L (2003) Inducible cAMP early repressor (ICER) in the nervous system—a transcriptional regulator of neuronal plasticity and programmed cell death. J Neurochem 87:1313–1320

Miranda MI (2007) Changes in neurotransmitter extracellular levels during memory formation. In: Bermúdez-Rattoni F (ed) Neural plasticity and memory: from genes to brain imaging. CRC, Boca Raton, FL

Morris RG, Anderson E, Lynch GS, Baudry M (1986) Selective impairment of learning and blockade of long-term potentiation by an N-methyl-D-aspartate receptor antagonist, AP5. Nature 319:774–776

Mössner R, Walitza S, Geller F, Scherag A, Gutknecht L, Jacob C, Bogusch L, Remschmidt H, Simons M, Herpertz-Dahlmann B, Fleischhaker C, Schulz E, Warnke A, Hinney A, Wewetzer C, Lesch KP (2006) Transmission disequilibrium of polymorphic variants in the tryptophan hydroxylase-2 gene in children and adolescents with obsessive-compulsive disorder. Int J Neuropsychopharmacol 9:437–442

Mukilan M, Ragu Varman D, Sudhakar S, Rajan KE (2015) Activity-dependent expression of miR-132 regulates immediate-early gene induction during olfactory learning in the greater short-nosed fruit bat, *Cynopterus sphinx*. Neurobiol Learn Mem 120:41–51

Mukilan M, Bogdanowicz W, Marimuthu G, Rajan KE (2018a) Odour discrimination learning in the Indian greater short-nosed fruit bat (*Cynopterus sphinx*): differential expression of *Egr-1*, *C-fos* and PP-1 in the olfactory bulb, amygdala and hippocampus. J Exp Biol 221(Pt 12):jeb175364

Mukilan M, Rajathei DM, Jeyaraj E, Kayalvizhi N, Rajan KE (2018b) MiR-132 regulated olfactory bulb proteins linked to olfactory learning in greater short-nosed fruit bat *Cynopterus sphinx*. Gene 671:10–20

Nakai T, Milusheva EA, Baranyi M, Uchihashi Y, Satoh T, Vizi ES (1999) Excessive release of [^{3}H] noradrenaline and glutamate in response to simulation of ischemic conditions in rat spinal cord slice preparation: effect of NMDA and AMPA receptor antagonists. Eur J Pharmacol 366:143–150

Nowak R (1994) Walker's bats of the world. The Johns Hopkins University Press, Baltimore, MD

OB und Halbach and Dermietzel R (2006) Neurotransmitters and neuromodulators: Handbook of receptors and biological effects. Wiley-VCH Verlag GmbH & Co. KGaA. ISBN:9783527313075 (Print); 9783527609963 (online)

Persengiev SP, Green MR (2003) The role of ATF/*CREB* family members in cell growth, survival and apoptosis. Apoptosis 8:225–228
Raghuram H, Thangadurai C, Gopukumar N, Nathar K, Sripathi K (2009) The role of olfaction and vision in the foraging behaviour of an echolocating megachiropteran fruit bat, *Rousettus leschenaulti* (Pteropodidae). Mammal Biol 74:9–14
Rajan KE, Marimuthu G (1999) Localization of prey by the Indian false vampire bat *Megaderma lyra*. Mammalia 63:149–158
Rajan KE, (2021) Olfactory learning and memory in the greater short-nosed fruit bat Cynopterus sphinx: the influence of conspecifics distress calls. J Comp Physiol A 207, 667–679
Ravel N, Elaagouby A, Gervais R (1994) Scopolamine injection into the olfactory bulb impairs short-term olfactory memory in rats. Behav Neurosci 108:317–324
Roberson ED, English JD, Adams JP, Selcher JC, Kondratick C, Sweatt JD (1999) The mitogen-activated protein kinase cascade couples PKA and PKC to cAMP response element binding protein phosphorylation in area CA1 of hippocampus. J Neurosci 19:4337–4348
Roediger HL, Zaromb FM, Goode MK (2008) A typology of memory terms. In: Byrne JH (ed) Concise learning and memory: the Editor's selection. Academic Press, New York
Rosenblum K, Meiri N, Dudai Y (1993) Taste memory: the role of protein synthesis in gustatory cortex. Behav Neural Biol 59:50–56
Sanchez F, Korine C, Steeghs M, Laarhoven LJ, Cristescu SM, Harren FJM, Dudley R, Pinshow B (2006) Ethanol and methanol as possible odor cues for Egyptian fruit bats (*Rousettus aegyptiacus*). J Chem Ecol 32:1289–1300
Schafe GE, LeDoux JE (2000) Memory consolidation of auditory Pavlovian fear conditioning requires protein synthesis and protein kinase a in the amygdala. J Neurosci 20:RC96
Scharf MT, Woo NH, Lattal KM, Young JZ, Nguyen PV, Abel T (2002) Protein synthesis is required for the enhancement of long-term potentiation and long-term memory by spaced training. J Neurophysiol 87:2770–2777
Segal M, Murphy DD (1998) CREB activation mediates plasticity in cultured hippocampal neurons. Neural Plast. 6(3):1–7
Silva AJ, Kogan JH, Frankland PW, Kida S (1998) CREB and memory. Annu Rev Neurosci 21:127–148
Sweatt JD (2001) The neuronal MAP kinase cascade: a biochemical signal integration system subserving synaptic plasticity and memory. J Neurochem 76:1–10
Swirnoff AH, Milbrandt J (1995) DNA-binding specificity of NGFI-A and related zinc finger transcription factors. Mol Cell Biol 15:2275–2287
Taubenfeld SM, Milekic MH, Monti B, Alberini CM (2001a) The consolidation of new but not reactivated memory requires hippocampal C/EBPbeta. Nat Neurosci 4:813–818
Taubenfeld SM, Wiig KA, Monti B, Dolan B, Pollonini G, Alberini CM (2001b) Fornix-dependent induction of hippocampal CCAAT enhancer-binding protein [beta] and [delta] co-localizes with phosphorylated cAMP response element-binding protein and accompanies long-term memory consolidation. J Neurosci 21:84–91
Thiels E, Kanterewicz BI, Norman ED, Trzaskos JM, Klann E (2002) Long-term depression in the adult hippocampus in vivo involves activation of extracellular signal-regulated kinase and phosphorylation of Elk-1. J Neurosci 22:2054–2062
Treisman R (1996) Regulation of transcription by MAP kinase cascades. Curr Opin Cell Biol 8:205–215
Uchihashi Y, Bencsics A, Umeda E, Nakai T, Sato T, Vizi ES (1998) Na^+ channel block prevents the ischemia-induced release of norepinephrine from spinal cord slices. Eur J Pharmacol 346:145–150
Umbriaco D, Garcia S, Beaulieu C, Descarries L (1995) Relational features of acetylcholine, noradrenaline, serotonin and GABA axon terminals in the stratum radiatum of adult rat hippocampus (CA1). Hippocampus 5:605–620
Vanhoutte P, Barnier JV, Guibert B, Pages C, Besson MJ, Hipskind RA, Caboche J (1999) Glutamate induces phosphorylation of Elk-1 and CREB, along with c-fos activation, via an extracellular signal-regulated kinase-dependent pathway in brain slices. Mol Cell Biol 19:36–146

Vizi ES (2000) Role of high-affinity receptors and membrane transporters in nonsynaptic communication and drug action in the central nervous system. Pharmacol Rev 52:63–89

Vizi ES, Kiss JP (1998) Neurochemistry and pharmacology of the major hippocampal transmitter systems: synaptic and nonsynaptic interactions. Hippocampus 8:566–607

Wilson DE, Reeder DM (eds) (2005) Mammal species of the world: a taxonomic and geographic reference, 3rd edn. The Johns Hopkins University Press, Baltimore

Wilson DA, Fletcher ML, Sullivan RM (2004) Acetylcholine and olfactory perceptual learning. Learn Mem 11:28–34

Wu L-G, Saggau P (1997) Presynaptic inhibition of elicited neurotransmitter release. Trends Neurol Sci 20:204–212

Roosting Behaviour of Indian Flying Fox (*Pteropus medius* Temminck)

13

Akshay Kumar Chakravarthy and Ashutosh Dey

Abstract

Bat roosts are on the decline in and around cities and towns in India. Bats are emigrating towards rural areas feeding mostly on cultivated lands and orchards for food. While feeding, bats may cause economic losses to cultivators. This has made farmers repel this group of animals. Bats play a vital role in the ecology of forests, seed dispersal, regeneration and pollination. Roosting behaviour in urban and rural areas poses different challenges for this group of animals. Bats, especially the frugivorous ones, are adapting to new situations. Conventionally, bat roosts are protected by public sentiments. Certain modern tools and techniques are provided for studying bats.

Keywords

Bats · Roosts · Behaviour · Conservation

13.1 Introduction

Bats are unique among mammals of their size because of their long lives, low fecundity, maternal care and slow development (Findley 1993). Bats are somewhat like birds ecologically. Bats are obviously K-strategists with relatively constant population size, greater competitive ability, iteroparity (producing offspring more than once in a lifetime) and greater energetic efficiency. There are 4200 living species of

A. Kumar Chakravarthy (✉)
Division of Entomology and Nematology, Indian Institute of Horticultural Research, Bengaluru, Karnataka, India

A. Dey
Environmental Management and Policy Research Institute, Bengaluru, Karnataka, India

S. Sridhara et al. (eds.), *Animal Behavior in the Tropics*,
https://doi.org/10.1007/978-981-97-7324-4_13

mammals, and almost 1000 of them are bats belonging to the order Chiroptera, sub-orders Megachiroptera (fruit bats) and Microchiroptera (insectivorous bats). Of these 1000 species, 25% are frugivorous, and 88% are exclusively tropical. The principal habitat of frugivorous bats include fruit canopy, open clearings and forest tracts. Droppings from frugivorous bats contain fruit pulp and seeds which help in seed dispersal. Fruit bats are physiologically and behaviourally adapted for feeding on different kinds of fruits. *Pteropus* species utilizes flowers/plants of 26 genera and fruits of 62. Fruits of *Ceiba* species attract 11 genera of Megachiroptera, and the fruits of *Ficus* sp. attract at least 13. They are also adapted to feed on guava, sapota, banana, litchi, jackfruit, mango, cashew and areca crops and in turn help in seed dispersal and pollination. Bat damaging fruits at times may result in crop damage. Bats were recorded in huge colonies ranging from 500 to 1000 in and around Bangalore, India. Changes in tropical ecosystems have often been limited to changes in plant community, composition and structure despite the numerous contributions made by the animals. Due to urbanization and modern agriculture, vast areas of forest land are disturbed by humans. This in turn had an effect on the bat roosts. The major threat to bat roosts is the destruction of habitat and tree roosts. Due to these disturbances, the huge bat roosts are being displaced and disturbed, resulting in the fragmentation of the original roosts into smaller roosts (100–500 bats).

In urban areas, generally roosting trees include *Ficus species*, *Delonix regia*, mango and so on, and in rural areas, roosting trees include *Ficus species*, *Terminalia*, *Casuarina*, and so on near cultivated lands. This situation forced the bat populations to depend on cultivated lands and plantation crops. During this process, bats may cause damage to the fruits by biting off the fruit, causing economic losses. This made the farmers develop a negative impact on this group of animals, and attempts were made to destroy the roosts near the cultivated tracts. Instead, efforts must be made to protect crops without adversely affecting the bat population. This is a serious problem. Because bats play a vital role in the basic ecology of forests and seed dispersal and regeneration of forests, there is a need for the conservation of bats. Bats are important pollinators and seed dispersers in tropical forests and have shared a long evolutionary history with angiosperms. Megachiropterans feed upon at least 145 genera of fruits in 30 families of plants (Mickleburgh et al. 1992). The most important families are *palmae* (16 genera), *Anacardiaceae* (10 genera) and *Sapotaceae* (8 genera). Generally, fruits are consumed when ripe, but this is not always so. In many places, fruit bats carry large seeded fruits and are key pollinators and seed dispersers. Over 443 plant products are derived from 163 plant species relying on bats for pollination or seed dispersal (Fujita and Tuttle 1991). These species provide substantial services via pollination, dispersal of plant parts and pest management (Mistry 2001). The bats (*Cynopterus sphinx*, *Pteropus medius* and *Rousettus leschenaultii*) pollinate more than 114 plant species and act as important pollen and seed vectors (Mickleburgh et al. 1992; Molur et al. 1998). The Indian flying fox (*Pteropus medius*, formerly *P. giganteus*), commonly known as the greater Indian fruit bat, is native to the Indian subcontinent. In Bihar, bats are known to keep destructive rodent species under control in cultivated crops, godowns and houses (Sinha 2002).

13.2 Role of Bats in Tropics

The bat populations across tropical areas are on the decline. Farmers protect fruit crops using pesticides and by physical/mechanical means of crop protection. Conservation of fruit bats is critically important (Mickleburgh et al. 1992). Because many species of fruit bats are dependent on primary forests and threatened by the large-scale destruction of forests in tropical areas, gaining knowledge of the processes of natural regeneration in disturbed habitats and the role animals play in these processes is of considerable conservation concern today.

The vast areas of disturbed land being created by human activity are disturbing and inimical to bats (Brown and Lugo 1990; Bawa and Seidler 1998; Holl and Kappelle 1999). Seed dispersal is a major way that animals contribute to successional changes in tropical ecosystems (Agoramoorthy 2000; Chakravarthy and Girish 2003; Chakravarthy et al. 2008b; Cox et al. 1991). Soil seed banks are often severely depleted in disturbed areas, making the deposition of seeds from undisturbed areas critical for succession to proceed (Chapman 1989; Franklin and Rey 2007). The role of bats in seed dispersal and regeneration of forests is an important function. German scientists have hit upon a novel yet cost-effective idea to revive reforestation in the tropics using bats as seed dispersers. They have designed bat roosts replicating large, hollow trunks to boost seed dispersal of a range of tropical plants. Indian bat fauna is rich and diverse with 112 species. India has more than 115 of the world's bats, including 13 Megachiroptera and 99 Microchireptera (Bates and Harrison 1997a, b).

India is home to several bat species. Bats roost on trees and huge caves in large numbers. Their numbers may range from 500 to 1000. When these traditional roosts are disturbed, the bats separate into small groups, and smaller-sized roosts are formed. However, the traditional roosts are more secure than the smaller-sized roosts probably formed out of the main bat roosts. The smaller roosts are amenable to disturbances. These smaller roosts are more susceptible to disturbances and are being displaced, which results in the fragmentation of the original populations. The major threat to bat roosts has been the destruction of habitat and trees. It is critically important to preserve the existing bat roosts.

Bats play a vital role in seed dispersal and regeneration of forests, and many trees depend on bats for pollination and seed dispersal. Bat roosts help preserve local, indigenous biodiversity elements. For the conservation of the Indian flying fox, a survey has to be made to identify the traditional bat roosts and protect them from disturbances. Bat groups account for 20% of mammal species worldwide, and about 50% of mammal species live in tropical forests. Across the world, ~15% of bat species are considered 'threatened' by the IUCN. A total of 18% are data deficient, and 57% have unknown populations, with 80% of global bat species at risk of extinction. Many of the lesser-known taxa such as bats have received little protection and face considerable threats and higher rates of endangerment. India's bat fauna is rich and diverse. India has more than 11% of the world's bats, including 13 Megachiroptera and 99 Microchiroptera (Bates and Harrison 1997a, b). Bats can incur economic loss while feeding on fruits (Srinivasulu and Srinivasulu 2001). But this is negligible compared to beneficial roles. In India, meat and other parts of the body of fruit bats are used to cure respiratory disorders and menstrual problems and are used as food. Fruit bats are often referred to as messengers of Lord Shiva. Bats conserve rainforests too. For instance, *Latidens salimalii* Thonglongya is the prime seed disperser in the rainforest, facilitating the restoration of tall fruiting trees of Agasthiyarmalai range of the southern Western Ghats (Vanitharani et al. 2005).

13.3 Observations

An effort was made to estimate the number of bats at the roost. Two observers armed with binoculars (80 × 30 and 70 × 50) counted the number of bats clinging to tree branches by going around the roosts, prior to an evening flight. The two observers counted the bats independently and separately. The numbers were averaged. At 6:20 p.m., few bats were observed making foraging flights. Four observers placed in four directions counted the number of bats issuing out of the roosting trees. The counts of each observer were added to make the size of the roost. The counts were recorded continuously for a week and averaged.

Feeding and foraging behaviour of bats were recorded on four dates from December 2007 to January 2008 using a four-wheel drive vehicle, binoculars 80 × 30 and telescopes when bats fed on fruits, twigs and flowers in and around the roost (5 km^2). Opportunistic observations were also recorded on numbers, reproductive status and threats to the roost.

These bats started breeding in December. The young ones are born from February to April. The adult bats weighed up to 1 kg ($n = 6$). Bats foraged on trees like Niligiri (*Eucalyptus*), 3–4 species of *Ficus*, banyan tree, hippie (*Madhuka*) and sapota (*Achras zapota*), and they feed on the young shoots and buds of these trees and fruits of *Ficus*. In general, the bats were observed leaving the roost for foraging by 6:00 p.m.. The bats were observed making recognizant flights around the trees and then dispersing in all directions for foraging. The young ones were left at the roost, and the mother bats foraged to nearer places to return to the roost early. The bats were observed drinking the water in nearby ponds by surface diving. Thus, the area around the roost provided shelter, food and breeding requirements for the fruit-eating bats. Surveys for the roosts of *P. medius* during the last 4 years indicated a sharp decline in Bangalore's urban and rural districts, Tumkooru, Chikkamagaluru, Shimoga and Hassan in south Karnataka. In this context, the roost size of 2600 individuals and their breeding assumes significance. Hunting of bats has been one of the great threats (Mickleburgh et al. 1992, 2001). People come from far-off places for hunting namely, Shimoga and Tumkur located 100–150 km away. The prospective hunters, who hail from outside the village, observe the flight paths, time, height and direction of fly ways. They erect large poles to hang the mist nets. The trapped bats are picked and sold at Rs. 200/bat. However, if the hunters are noticed by the villagers, they are chased away and punished. The hunters also use guns for killing. Apart from serving as food, the bats are used for medicinal purposes. Interaction with villagers revealed that on average, about 10% of bats are killed at this roost. Observations at other roosts revealed that about 30% of bats are eliminated by hunting alone in different parts of Karnataka. Bats play a crucial role in the ecosystem. A very simplistic example is fruit bats' task as flower pollinators and seed dispersers and insectivorous bats' task in controlling much of the insect pest populations. Although fruit bats damage a small percentage of agricultural crops, their role in forest regeneration more than compensates for this loss in the long term from the perspective of the greater good. Folivorus bats consume flowers and fruits affecting seed dispersal and pollination. Some of the adaptations of bat-pollinated syndrome include nocturnal anthesis, colour emission and presentation of rewards to the pollinators. Bats are an important pollinator for economical crops including durian, petai, and jackfruit. Nathan et al. (2005) studied bat-pollinated plants such as *Parkia biglandulosa*, *Bassia latifolia*, *Ceiba pentandra* and *Kigelia pinnata*. Bats helped in cross-pollination. Public awareness of the importance of flying foxes in pollination, seed dispersal and the benefits of their excrement (as natural organic manures) may create a better understanding to preserve their roosting habitats.

13.4 Heritage Roosts

The Indian flying fox is the largest bat found in south Karnataka and lives in close proximity to humans. Anecdotal accounts of *P. medius* suggest that some roosts have increasing populations, whereas others have decreasing ones. Isolated studies have provided insights into possible disturbances and their impacts. However, as a species, very little is known about *P. medius* population trends in the wild.

Development works around human habitations and expanding roads have in the recent past affected roosts by the roadside; increasing interest in bat meat and hunting for medicine have resulted in depletion in numbers in some roosts; disturbance to traditional roosts has resulted in bats finding alternate roosts nearby or farther away (Molur and Walker 1998). Bats have been known to shift roosts from old native trees such as the fig trees to introduced non-native trees such as *Eucalyptus* and *Casuarina*; some roosts are reported to have disappeared (Chakravarthy 2007). There is a small Muneshwara temple in the place where people worship the god, and by belief, the whole area is protected and conserved. Observations have revealed that bats are not of any nuisance value to the villagers as they do not damage the crops and feed on the sprigs of foliage of *Ficus*, *Madhuca latifolia*, *Eucalyptus* and other trees. People have a positive attitude towards these bats. Interactions with villages using the questionnaire revealed that all supported the idea of declaring the roost as a heritage site. The villagers had many reasons for declaring the site as a protected area. For conservation, the availability of high-quality roosting habitat is an important factor. Concept management practices will confer benefits to many species, although conservation efforts are directed towards one species.

The roosting site was in a village called Gidadapalya located 15 km north of Kunigal town (map), Tumkooru (13.3° N, 77.1° E), with about 75% under vegetation cover. The village has 150 houses, and farmers cultivate diversified crops like tomato, finger millet, areca nut, coconut, sapota and other vegetables. The village is surrounded by forest plantations of *Casuarina*, *Acacia* and indigenous tree species like *Ficus*, *Bauhinia*, *Samanea saman*, *Tamarindus indica* and other species. The roosting site extends to about one acre (10,000 sq m) and was covered with canopies of indigenous trees mentioned earlier. The canopy cover was almost 100%. The ground cover was diverse and surrounded by thick *Lantana* bushes. The bats (*Pteropus medius*) have been residing in this village for the past 85 years. The present roosting site is the third place that the bats have changed since in the first place the trees were cut as it was a private property. The bats shifted to the second place. Even at this place, the trees were cut. Later, the bats shifted to the present place where there were 30 *Ficus benghalensis* (banyan) trees and bamboo clumps. The land of the roost site belongs to Mr. Gangadhar, where bats have been roosting for the past 34 years. Mr. Gangadhar and the villagers were keen to declare the roost as a heritage site. It is believed that a rishi (saint) had advised Mr. Gangandhar's father to protect these bats. So this place is conserved/protected by villagers. The villagers are nature-loving that is evidenced by the maintenance of good vegetation cover (75%) in the village. A small temple is at the centre of the roost site. *Lantana* plants were manually planted to protect intervention of cattle and people inside the roost site so that bats are left undisturbed. *Lantana* bushes served as a buffer. There is a huge collection of leaf litter from banyan trees and bat guano (faecal droppings) at the site. It is collected in April–May after the rains, and around 10 tractor loads (each load costs Rs. 1000) are collected per year, and villagers use this manure for cultivating ragi, areca, coconut and other crops.

The reasons are as follows:

- The village has less population. Around 150 families have a conservation attitude and are well-vegetated, and villagers want to conserve the bats and trees.
- Bat roosts, especially of *P. medius*, are dwindling rapidly due to the cutting of trees and increasing interference from human populations in Karnataka. Traditional roosts have been drastically reduced.
- The village (Gidadapalya) has good vegetation all around and forest lands too, far and near.
- *Lantana* bushes act as a buffer zone and prevent interventions by cattle and humans to the roosting site.
- Many villagers are organic farmers. It will encourage farmers to go for organic farming and realize or derive multiple benefits.
- Other animals and birds inhabit the roost site. Mongoose (*Herpestes auropunctatus*), wild boars (*Sus scrofa*); snakes; birds such as peacocks, woodpeckers, chloropsis, pied cuckoo, owls and so on; hares; rodents and foxes are found here. These indigenous biodiversity elements are crucially important for the productivity of the village. The area has 30 *Ficus* trees. In addition to bats, other biodiversity components too are protected.
- The roost site will be conserved in the long run because the whole community is involved, and the participatory approach is sustained.

13.5 Roost Site

In ten districts of south Karnataka, *Pteropus medius* was recorded utilizing 40 species of trees for roosting, and a total of 72 roosts were recorded in the region. *Ficus*, *Samanea saman*, *Mangifera indica*, and *Eucalyptus* sp. were the dominant species utilized by bats for roosting (Chakravarthy 2009). The most frequently used-up trees by bats are shown in Table 13.1. The roosting trees were selected either in protected places like temples, government buildings and gardens or along the national highway roads in semi-urban areas and outskirts of Bangalore city. Of the 72 roosts recorded, 44 were found in the protected area (Table 13.2). Interestingly, in the protected areas, the canopies of roosting trees were well-spread and large, and generally, the trees were more than 50 years old. Trees selected along roadsides were having smaller canopies, and the trees were younger (10–20 years old). The protected roosts were traditional roosts, and the unprotected ones were newly formed roosts (Table 13.2). The unprotected roosts were categorized as disturbed roosts as the bats utilized the trees for shorter periods (few months to a year or two) than the undisturbed roosts that were found from the past few decades (Table 13.3). Mishra recorded nine species of trees utilized by *P. medius* in Delhi, and Kumar et al. recorded 18 species of trees in Uttar Pradesh. Pandian and Suresh (2021); Banack (1998) and Bates et al. (1994) reviewed the literature on bat roosts across India and recorded 13 species of trees in which bats *P. medius* were found roosting in 4 districts of north Tamil Nadu, South India. In their study, maximum numbers of bat roosts were found in *Pongamia piñnata* (Table 13.4). A negative correlation was detected between the numbers of *P. medius* and tree heights. However, the workers found a positive and significant relationship between the tree diameter at breast height and the canopy size of the roosting tree. The study of Pandian and Suresh (2021) further revealed that *P. medius* preferred trees with well-spread, larger dbh

Table 13.1 Nutrients and microbial composition of guano and bolus of flying fox (*Pteropus giganteus*) (range in parentheses)

Parameter	Guano	Bolus
Total nitrogen (%)	2.6 ± 0.5 (2–3.3)	3.3 ± 0.82 (2–4)
Total phosphorus (%)	4.2 ± 0.8 (3.1–5.2)	4. 3 ± 0.6 (3.5–5)
Potassium (%)	0.6 ± 0.04 (0.6–0.7)	0.7 ± 0.04 (0.6–0.7)
PH	7.3 ± 0.1 (7.1–7.4)	7.1 ± 0.3 (6.7–7.4)
Bacteria (cfu/g dry wt)	$29 \times 10^4 \pm 50$ ($25–32 \times 10^4$)	$48 \times 10^4 \pm 28$ ($46–50 \times 10^4$)
Actinomycetes (cfu/g dry wt)	$5.55 \times 10^4 \pm 7.8$ ($5–6 \times 10^4$)	$4.1 \times 10^4 \pm 7.8$ ($3.5–4.6 \times 10^4$)
Fungi (cfu/g dry wt)	$2.9 \times 10^4 \pm 3.5$ ($3.1–4.3 \times 10^4$)	$4.6 \times 10^4 \pm 3.5$ ($4.3–4.8 \times 10^4$)

Table 13.2 Classification of bat roosts based on the protection in south Karnataka

Roost type	Number of roosts
Protected roosts	44
Unprotected roosts	26

Table 13.3 Classification of bat roosts as traditional roosts and newly formed roots

Roost type	Number of roosts
Traditional roosts	50
Newly formed roost	20

Table 13.4 Classification of bat roosts based on the roosting tree

Sl. no.	Tree	Number of roosts
1	*Ficus* sp.	50
2	*Samanea saman*	15
3	*Mangifera indica*	9
4	*Eucalyptus*	8
5	Other tree species (silver oak, tamarind, casuarina, etc.)	32

and larger canopy size for roosting. The results of the above study corroborate with the findings of the study on *P. medius* in south Karnataka (Chakravarthy 2007, 2009; Chakravarthy et al. 2008a).

- Protected roost: The roosts located in safe places such as temples, mosques, churches, villages and so on are well-protected by the people because of religious beliefs. Roosts are located within the compound walls of government, private offices and even schools and hospitals that are well-guarded and protected.
- Unprotected roosts: The roosts located in places such as developing areas, away from the town, near the national highway and roadsides and any private land are constantly disturbed by pruning and cutting off most trees. So such roosts are not protected.
- There were contrasting features between the two categories of bat roosts (*t*-test at 5% level of significance).
- Traditional roost: The roosts have been present for over 50 years, and the bat population should be stable.
- Newly formed roost: The traditional roosts disturbed due to tree pruning and human disturbances will separate into smaller groups and may shift to nearby trees for roosting. Such roosts are called newly formed roosts.
- There were differences between the traditional and newly formed roosts, and this observation was based on a long-term study.
- Areas (6–8 sq km) around the bat roosts were surveyed to record the dominant tree species on which bats were found foraging and feeding.
- *Ficus* sp. was the dominant species on which the bats roosted. There were statistically significant differences between the number of bat roosts on *Ficus* sp. and other tree species in southern Karnataka.

13.6 Urban and Rural Roosts

It is important to understand how an animal adjusts to the new changes in the environment (Brooke et al. 2000). This is a major challenge in behavioural biology. Mobility is crucial for a wide range of behavioural processes and can lead to a measurable response to a combination of integral states and environmental changes such as an increase in competition, shifts in the distribution of resources and so on. Natural habitats are vanishing, and urban habitats are increasing. While most animals are negatively affected by this process, a few like Indian flying foxes may get adjusted to the changing habitat conditions. Bats can modify their behaviour to the city environment. Although this response is species-specific. There is little documentation on the movement of bats in urban and semi-urban areas.

Bats of the family Pteropodidae establish rapidly in urban areas as observed in Bangalore city, Karnataka, South India, with the Indian flying fox, *P. medius*. One hypothesis advanced for this is its preferable microclimatic conditions. City conditions mostly provide warmer temperatures preferred by these bats (Parris and Hazell 2005). Cities are also known to provide bats relatively free from predation (Parry-Jones and Augee 2001). Cities also provide easy navigation for bats because of landmarks, availability of food (fruits) years-round and clustered distribution of fruit trees in urban environments. One successful example of a bat exploiting resources in urban areas is the Egyptian fruit bat (*Rousettus aegyptiacus*) that feeds on the diversity of fruit plants. Korine et al. (1999) worked in Israel on *R. aegyptiacus* and found differences in foraging patterns between urban and rural areas. This is primarily because fruit trees exploited by bats showed differences in their distribution in urban and rural areas. The above workers used the global urbanization footprint criterion to differentiate between rural and urban foraging areas. In south Karnataka, frequent shifts in roosts in urban compared to rural areas were recorded. The urban roosts showed frequent splitting and coalesce in bat colonies than roosts in rural areas with respect to *P. medius*. Similarly, recorded that bats were more diverse while foraging in urban than rural environments in Israel. In their study area, bats in urban environments spent three times sites in rural areas. The above workers recorded bat foraging in urban areas were prone to switch foraging sites compared to bats in rural areas that returned to the same site, night after night. The workers presented two reasons for the changes in foraging in urban and rural sites: (1) increased competition for roost sites in urban landscapes and (2) switching sites in urban areas that may confer benefits to bats, for example, in diversification of diet. In south Karnataka, *P. medius* did not reveal any territorial behaviours in bats. In Israel too, social interactions at the foraging sites suggested that competition was not a driving force for bats to switch roost sites. If competition was a factor, then there would be a negative correlation between bat diversity and the time bat spent at a foraging site. However, this was not the case. Indeed, the workers found a positive correlation between the two variables. Using onboard sensors previously mentioned workers recorded continuous acceleration of nine tagged bats and found that most of the time (>80%), bats were resting on trees between feeding bouts. In Israel further hypothesized that bats have diversified and improved their diet in urban environments by switching roost sites and acquiring more properties and other nutrients. The fact that bats more frequently switched fruit species when moving between sites in urban areas supported the hypothesis of diet diversification. Another benefit bats derive from switching sites in urban environments is the exploration of new sites. This is important in a rapidly changing city. In the study area in Israel, workers found owls more abundant in rural than urban areas. Owls are predators of bats. Taking all the above factors into consideration, bats have to decide where to roost and where to forage.

13.7 Roost Size

Of 72 bat roosts monitored in south Karnataka, larger roosts (>1000 individuals) constituted about 14% were mostly relatively undistributed and well-protected. The distribution of bat roosts into five categories based on size is shown in Table 13.5. Bats in large roosts foraged in the surrounding cultivated and vegetated tracts. Smaller roosts (<1000 individuals) were generally prone to frequent disturbances, highly ephemeral and frequently shifted their place of roost. Often the split and coalescence occurred in these roosts. These roosts were noisy also (Chakravarthy 2009). Pandian and Suresh (2021) interacted with 80 respondents in Tamil Nadu and found that 87% of residents were concerned over the declining roost sizes. The residents considered roosting populations of bats important in their culture and life. The bat population size using the photographic estimation method at the 13 *E. spelaea* roosts (5 times per roosts) found inside Gunung Reng for 20 days. Bat population size estimate resulted in about 207–344 individuals of *E. spelaea* ($F(4.0) = 5.66$, p value <0.05). This estimate is a relatively modest population size of bats surrounded by human-dominated habitats. Several factors affected the population size; birth, death, immigration and emigration. Brautigam and Elmqvist (1990) worked on conservation of flying fox in Pacific islands. Meyers et al. (2000) and Mickelburgh et al. (2002) have advanced conservation measures for bats.

Roosting numbers of bats increase, temporal patterning (clustering) in their emergence behaviour, as measured by chi-square or G statistic. The statistic may increase due to an increase in clustering (Speakman et al. 1999).

Table 13.5 Distribution of bat roosts based on roost size

Roost size	
Number of bats	Number of roosts
Less than 500	36
500–1000	24
1000–1500	5
1500–2000	5
2000–2500	1
2500–3000	1

Table 13.6 Classification of bat roosts as disturbed and undisturbed

Roost type	Number of roosts
Disturbed roosts	26
Undisturbed roosts	46

- Roost size is based on the number of bats present on the roosting trees. The bat counts were recorded by four observers using binoculars.
- Statistical significant differences were found between less than 500 bat roost size and other categories by *t*-test at a 5% level of significance.
- Disturbed roost: The roosts that are frequently disturbed by human activities such as tree pruning, sound (noise) and light disturbances caused by traffic, hunting for food and medicinal use, and climatic disturbances, that is, high wind velocity, high-intensity light, heavy rainfall, and so on are considered as disturbed roosts.
- Undisturbed roosts: The roosts that are not disturbed by any one of the above reasons such as human disturbances and climatic disturbances are considered as undisturbed roosts. The protected roosts are generally undisturbed (Table 13.6).

13.8 Tools and Techniques

Non-availability of equipment for working with sounds (bats) beyond the range of human hearing (ultrasonic) hindered recording of bat echolocation. Most of the species of echolocating bats separate pulse and echo in time and cannot tolerate overlap between the outgoing pulse and returning echo (Schnitzler and Kalko 2001). These bats produce signals about 10% of the time; they are 'low-duty cycle'. The range of frequencies in their echolocation sounds varies from very high-pitched (sounds >200 kHz) to relatively low-pitched (sounds ~10 kHz), which are not ultrasonic. The high-frequency echolocation calls have the potential to give an echolocating bat more details about its target than lower-frequency calls (Griffin 1958; Simmons and Stein 1980) as shorter wavelengths. Atmospheric attenuation and spreading loss combine to limit the movement of sound waves in air reducing the effective range of echolocation.

For determining the bat population size, the photographic estimation method of the *E.speleae* roosts (5 times/roost) found inside Gunung Reng resulted in about 207 to 344 individuals in 20 days. This estimate was a relatively modest population size of bats in an isolated limestone karst surrounded by human-dominated habitat. A photographic estimation method was adopted, and pictures were taken at each roost five times every day during the study period. The camera was set up to flash feature for white balance mode. The dark cave conditions need lighting help from torch light and headlamp when taking the roosts picture. The standard of the DSLR camera was ISO-1800, with a shutter speed of f/14-f/22 in low aperture mode.

Roosting numbers of bat increase were measured by chi-square or G statistics while clustering in emergence behavior. The statistic may increase if clustering increases. For this, Speakman et al. (1999) developed a method to separate this effect using computer-modelled event streams. Using this method, it examined intra- and inter-roost variations in the temporal patterning of emergence of the pipistrelle bats. The 'corrected' expression of clustering decreased as roost size increased, which was opposite to the effect found with uncorrected expression, that predicted a priori from the bottleneck hypothesis.

The Echo Meter Touch permits recording calls heard on our phone or tablet. This gives a record of the places from where we recorded calls from the bat we have heard. If the touch has GPS functionality, then it also tags each call and displays them on Google Maps. Echo Meter 3 is the sound-held, ultrasonic detector to feature a sonogram display. This tool is technologically advanced and scientifically useful. It focuses on monitoring the movements of bats in minutes. Song Meter Mini Bat ultrasonic recorder is an innovative tool for detecting and recording bat vocalizations. The tool is useful in bat species inventory. Presence/absence surveys are used to detect endangered species and monitor habitat health. The bat recorder detects and captures quality ultrasonic vocalizations (Baranowski and Nita 2023). The Mini Song Meter has a weaker anti-alias filter reaching only 12 db of signal reduction at 60% of the sample rate. The Micro Song Meter has an anti-alias filter reaching only 6 db of signal reduction at 60% of the sample rate (Dowdy and Wilam 2019). Audiomoth 1.2 is a low-cost, full-spectrum acoustic logger, based on the Gecko processor range. This device has a GPS-synced clock module, which can be used to locate vocalizing bats (Hill et al. 2018).

Samania saman

(Rain Tree)

Fruit bat on bamboo tree

Fruit Bat on Bamboo Tree

Samanea saman (rain tree)

Fruit Bat on the roosting tree

Fruit bat on the roosting tree

Mangifera indica (mango tree with blossom)

Roosting tree

Indian flying fox (*Pteropus medius*)

Eucalyptus globulus

P.medius feeding on fruits of *Ceiba pentandra*

P. medius feeding on fruits of *Ceiba pentandra*

Ficus benghalensis

Ficus benghalensis

Albizzia lebbeek

Albizia lebbeck

Artocarpus hirsutus

Artocarpus hirsutus

References

Agoramoorthy G (2000) Population status of the Indian fruit bat, *Latidens salimalii* in Tamilnadu state, India. Final report. Fauna & Flora Preservation Society, Cambridge

Banack S (1998) Diet selection and resource use by flying foxes (*Genus pteropus*). Ecology 79(6):1949–1967

Baranowski K, Nita B (2023) Habitat loss for black flying foxes and implication for Hendra viruses. Landsc Ecol 39:1605–1618

Bates PJJ, Harrison DL (1997a) Bats of the Indian subcontinent. Harrison Zoological Museum Publication, Sevenoaks

Bates PJJ, Harrison DL (1997b) Bats of the Indian subcontinent. Harrison Zoological Museum, Kent, p 228

Bates PJJ, Harrison DL, Thomas NM, Muni M (1994) The Indian fruit bat *Latidens salimalii* thonglongya, (1972) (Chiroptera: Pteropodidae) rediscovered in southern India. Bonn Zool Beitr 45:89–98
Bawa KS, Seidler R (1998) Natural forest management and conservation of biodiversity in tropical forests. Conserv Biol 12(1):46–55
Brautigam A, Elmqvist T (1990) Conservating Pacific Island flying foxes. Oryx 24:81–89
Brooke AP, Solek C, Tualaulelei A (2000) Roosting behavior of colonial and sclitary flying foxes in American Samoa (Chiroptera: Pteropodidae). Biotropica 32(2):338–350
Brown S, Lugo AE (1990) Tropical secondary forests. J Trop Ecol 6(1):1–32
Chakravarthy AK (2007) Observations on roosts of Indian Flying fox, *Pteropus gigangetius* in Karnataka. Bat Net Newslett 8(1–2):10
Chakravarthy AK (2009) A roosting site for bats. People in conservation 2 (1) pp. 6
Chakravarthy AK, Girish AC (2003) Crop protection and conservation of frugivorous bats in orchards of Hill and coastal regions of Karnataka. Zoos Print J 18(8):1169–1171
Chakravarthy AK, Yeshwanth HM, Prasanna Kumar NR (2008a) Gaint fruit bat roosts at Kunigal Tumkur—case study for preservation as a heritage site. Tiger Paper
Chakravarthy AK, Yeshwanth HM, Vijay Kumar L, Kumar NRP (2008b) Giant Indian fruit bat (*Pteropus giganteus* brunnich) roost in Karnataka, South India: a case for preservation as a heritage site. In: 3rd International Appropriate Technology Conference—2008, November 12–15, 2008, Kigali, Rwanda
Chapman CA (1989) Primate seed dispersal: the fate of dispersed seeds. Anim Behav 58(4):787–795
Cox PA, Elmqvist T, Pierson ED, Rainey WE (1991) Flying foxes: strong interactors in South Pacific Island ecosystems: a conservation hypothesis. Conserv Biol 5(4):448–454
Dowdy NJ, Wilam EC (2019) Characteristics of tiger moth (Erebidae: Arctiinae) anti-bat sounds can be predicted from timbale morphology. Front Zool 16:24
Fenton MB, Brigham RM, Mills AM, Rautenbach IL (1985) The roosting and foraging areas of Epomophorus wahlbergi (Pteropodidae) and Scotophilus viridis (Vespertilionidae) in Kruger National Park, South Africa. J Mammal 66:461–46
Findley JS (1993) Bats: a community perspective. Cambridge University Press, p 161
Franklin J, Rey SJ (2007) J Trop Ecol 23:1–12
Fujita MS, Tuttle MD (1991) Flying foxes (Chiroptera: Pteropodidae): threatened animals of key ecological and economic importance. Conserv Biol 5(4):455–463
Griffin DR (1958) Listening in the dark: The acoustic orientation of bats and men. Yale Univer Press
Hill AP, Prince P et al (2018) Audio-moth: evaluation of a smart open acoustic device for monitopring biodiversity and the environment. Methods Ecol Evol 9(5):1199–1211
Holl KD, Kappelle M (1999) Tropical forest recovery and restoration. Trend Ecol Evol 14(10):378–379
Korine C, Izhaki I, Arad Z (1999) Is the Egyptian fruit-bat *Rousettus aegyptiacus* a pest in Israel? An analysis of the bat's diet and implications for its conservation. Biol Conserv 88(3):301–306. https://doi.org/10.1016/S0006-3207(98)00126-8
Meyers N, Mittermeier RA, Mittermeier CG, Da Fonseca GAB, Kent J (2000) Biodiversity hotspots for conservation priorities. Nature 403:853–858
Mickelburgh SP, Hutson AM, Racey PA (2002) A review of global conservation status of bats. Oryx 36:18–34
Mickleburgh SP, Hutson AM, Racey PA (1992) Old world fruit bats. An action plan for their conservation. IUCN/SSC Chiroptera Specialist Group, IUCN, Gland, pp 1–16
Mickleburgh SP, Hutson AM, Racey PA (2001) Microchiropteran bats, global status survey and conservation action plan. IUCN/SSC Chiroptera Specialist Group, IUCN, Gland, pp Ix–x
Mistry S (2001) Tropical ecosystems: structure, diversity and human welfare. In: Proceedings of the International Conference on tropical ecosystems. Oxford-1BH publications, New Delhi, p 791
Molur S, Walker S (1998) Conservation assessment and management plan (C.A.M.P) Workshop. In: Mammals of India report summary. Zoo Outreach Organization, India, Coimbatore

Molur S, Nameer PO, Walker S (1998) Report of the workshop "Conservation assessment and management plan for mammals of India". (BCPP-endangered species project). Zoo Outreach Organization/Conservation Breeding Specialist Group India, Coimbatore, p 176

Nathan T, Rulan PI, Marimuthu G (2005) Department of Animal Behaviour and Physiology School of Biological Sciences, Madurai Kamaraj University. Curr Sci 88:10–25

Pandian M, Suresh S (2021) Roosting habits and habitats of the Indian Flying Fox Pteropus medius Temminck, 1825 in the northern districts of Tamil Nadu. India Journal of Threatened Taxa 13(12):19675–19688. https://doi.org/10.11609/jott.7468.13.12.19675-19688

Parris, Hazell (2005) Biotic effects of climate change in urban environments: the case of the grey-headed flying-fox (*Pteropus poliocephalus*) in Melbourne, Australia. Biol Conserv 124(2):267–276

Parry-Jones KA, Augee ML (2001) Factors affecting the occupation of a colony site in Sydney, New South Wales by the Grey-headed Flying-fox, *Pteropus poliocephalus* (Pteropodidae). Austral Ecol 26:47–55

Schnitzler H-U, Kalko EKV (2001) Echolocation by Insect-Eating Bats. Bioscience 51(7):557–569. https://doi.org/10.1641/0006-3568(2001)051[0557,Ebieb]2.0.Co;2

Simmons JA, Stein RA (1980) Acoustic imaging in bat sonar: Echolocation signals and the evolution of echolocation. J Comp Physiol 135:61–84. https://doi.org/10.1007/BF00660182

Sinha YP (2002) Status and distribution of chiropteran diversity of Gangetic plains of Bihar. Records of the Zoological Survey of India, Calcutta

Speakman JN, Irwin N, Tallach R (1999) Stone effect of roost size on the emergence behaviour of pipistrelle bats. Anim Behav 58(4):787–795

Srinivasulu B, Srinivasulu C (2001) Magnitude of depredation on grapes by short nosed fruit bats *Cynopterus sphinx Vahl,* 1797 in Secunderabad, India. Curr Sci 80:14–15

Vanitharani J, Pearch MJ, Jeyapraba L, Annamalai R (2005) A review of the distribution and status of *Latidens salimalii* (Chiroptera: Pteropodidae) with new records from the Western Ghats, India. Vegetation 107(108):149–162

Synurbization in the Red-Vented Bulbul (*Pycnonotus cafer*) in India

14

Anil Kumar

Abstract

Urbanization affects the bird life in a number of ways. Ambient noise, lack of suitable nesting sites and foraging material in urban landscape often impose a selection pressure on birds. Analysis of the data revealed that in urbanized habitats, the red-vented bulbul (RVB; *Pycnonotus cafer*) often altered its breeding behaviour in a number of ways like selection of unusual sites for nesting (such as nesting in buildings, wall holes and electric fixings), exotic plants for nesting and polythene pieces, paper pieces, threads, rags, and the like as nesting material. Breeding success was higher in urban areas than in forests. Plasticity in song characteristics, namely, increases in minimum and dominant frequencies and amplitude shifts were also observed. Basically, the red-vented bulbul is a wild species, and in forest, it forages mostly upon a number of wild fruits, while in urban areas, this species is seen foraging upon a variety of food, including kitchen waste. Such shifts in behaviour are examples of rapid evolution and provide excellent opportunities to address scientific questions pertaining to behaviour and evolution.

Keywords

Urbanization · Nesting · Breeding behaviour · Frequency shift

---

A. Kumar (✉)
Gangetic Plains Regional Centre, Zoological Survey of India, Patna, Bihar, India

S. Sridhara et al. (eds.), *Animal Behavior in the Tropics*,
https://doi.org/10.1007/978-981-97-7324-4_14

14.1 Introduction

Synurbization denotes adaptations/adjustments of mainly birds and mammals to urban environments. It deals with ecological and behavioural plasticity and micro-evolutionary changes in animal populations under anthropogenic pressure (Luniak 2004). In most parts of the world, urbanization is spreading at a very fast rate (Meyer and Turner 1992). Most urban habitats are characterized by sparseness, patchiness, lack of usual vegetation structure, high levels of low-frequency (mostly below 2.00 kHz) anthropogenic noise and presence of non-native species (Goldstein-Golding 1991; Slabbekoorn and Ripmeester 2008; Dowling et al. 2012; Isaksson 2018). It imposes a selection pressure on flora and fauna at various levels. Species that can tolerate such conditions usually exhibit some adaptations/alterations at foraging as well as breeding levels (Møller et al. 2015). Birds are one of the most affected faunal groups. Studies showed that in urban habitats, birds became less diverse and usually dominated by omnivorous and frugivorous than insectivorous and carnivorous species and adapted to nesting and/or roosting in artificial structures such as buildings (Clergeau et al. 1998; Wheater 1999; Lim and Sodhi 2004).

Anthropogenic noise in urban habitats has an adverse impact on the transmission efficacy of the acoustic communication of animals (Katti and Warren 2004). The faunal elements, which use low-frequency vocal signals (ranging below 3.00 kHz), are especially vulnerable to the masking effect of anthropogenic noise. For successful survival, in urban habitats, birds adapt to their communication behaviour, as vocalizations play a key role in mate acquisition and defence of valuable resources such as territory for foraging and nesting (Searcy and Andersson 1986). So the acoustic interference of anthropogenic noise on the vocal signals of birds possibly may have a considerable impact on both inter- and intra-sexual selection. To cope with this, birds adopt a number of strategies for successful communication (Warren et al. 2006; Phillips et al. 2020). Some species increase signal redundancy (Brumm and Slater 2006), vocalize loudly (Brumm 2004), use more pure tones (Dubois and Martens 1984) and sing nocturnally in areas that are noisy during the day (Fuller et al. 2007). A positive correlation between anthropogenic noise level and song frequency has been reported in some avian species such as great tit (*Parus major*) (Slabbekoorn and Peet 2003), house finch (*Carpodacus mexicanus*) (Fernández-Juricic et al. 2005) and song sparrow (*Melospiza melodia*) (Wood and Yezerinac 2006). In the Indian context, such studies are almost negligible. It prompted me to reanalyze the data collected on breeding biology and acoustic communication of an avian species the red-vented bulbul (RVB; *Pycnonotus cafer*). The aim of this study was to understand the impact of urbanization on the breeding, foraging and acoustic communication in this species.

14.2 Material and Methods

RVB is a common, resident, widely distributed, perky tobacco-brown (with black head, black–brown throat and scale-like markings on back and breast; Fig. 14.1) species, commonly available in urban/suburban areas and scrub forests of the Indian subcontinent. It is a strict pairing, non-territorial, sexually alike species that breeds overall from March to September in the study area. Generally, it forms a cup-shaped nest in shrubs, small trees and thorny bushes, about 1–2 m from the ground level (Vijayan 1978; Ali 1996; Grimmett et al. 1998), while in some cases, it is reported to nest in an unusual manner, namely, nesting in tree hole (Sivasubramanian and Sundaramoorthy 1992), mud hole (Lamba 1976) and disused tube light (Mundkur 1984).

From April 1995 to June 2012, I collected data on select aspects of behavioural ecology and socio-biology of RVB at various locations mostly in north India. Some parts of this study such as foraging ecology (Bhatt and Kumar 2001), acoustic communication (Kumar 2004a), some socio-biological aspects of breeding (Kumar 2004b) and communication value of displays and postures (Kumar 2010) have been published. To understand the impact of urbanization on the behaviour of this species, the data has been reanalyzed and presented.

The breeding success of this species was studied in two breeding seasons, namely, April 1995 to September 1996 in and around Gurukul Kangri University campus, Haridwar (29°55'N, 78°8′E), India. The habitat of this area was urbanized, comprising human settlements, gardens, ornamental trees and shrubs. During the study period, over 4 km^2 area was surveyed almost twice/thrice in a week (with few exceptions) in the search of the nests and for recording breeding activities. The

Fig. 14.1 Photograph of red-vented bulbul showing the morphological characteristics

behavioural observations were made from 3 to 10 m distance from the nest with the help of a field binocular (7 × 50). Abandoned nests were also counted to calculate the nesting preference. Results were expressed as mean ± SE. Some observations and recordings of acoustic signals in seven individuals were made in and around Jodhpur (26°21′N, 73°06′E) from March to April 2001. The habitat of this area was open with sparse vegetation and noisy due to highway transportation and the location of an air force station. Production of strong ambient noise due to the take-off of jet fighter planes was common in this area. Observations were also carried out from March 2002 to March 2005 in Dehradun, Uttarakhand, mainly in the Wildlife Institute of India (WII) campus and surrounding area (30°26'N, 78°06′E). Observations were also made in the Zoological Survey of India (ZSI) campus, Indian Institute of Remote Sensing (IIRS) campus and the lower hills of Mussoorie. All sites were within a 30 km radius. During the study period, the study sites were surveyed about once/twice a week (with few exceptions), in search of the breeding activities such as nest site selection, nest formation, egg laying, incubation behaviour and recording of acoustic signals. From April 2005 to June 2012, the observations and recordings were limited, and observations were recorded in few locations in Arunachal Pradesh and Solan in Himachal Pradesh. From January 2016 to October 2022, detailed observations were made in the ZSI campus, Dehradun, and the outskirts of Dehradun town.

Songs of the bulbul were recorded in natural habitats using Sony CFS 1030S or Marantz PMD222/PMD670 digital sound recorders and JVC MZ-500 or Sennheiser ME-66/Audio-Technica AT815b microphones and analyzed using Avisoft-SASLab Pro. All birds recorded were unbanded (except a few). To avoid recording the same individual twice, each bird was recorded only when it was within view, and recordings were only made once on each occasion. For bio-acoustic analysis, spectrograms were displayed on a computer monitor, and measurements of variables were made using frequency and time coursers. All spectrograms were calculated using the following setting of SASLab: 512 FFT-length, 100% frame, FlatTop window and 75% time window overlap. In some cases, the temporal resolution was decreased or increased from 50% to 93.75%, as required. Software SPSS was used for statistical analysis. To understand the impact of ambient noise on the song characteristics, songs were analyzed and grouped into two categories: songs in forest habitats (based on the recordings of 12 individuals in Dehradun) and songs in urban/suburban habitats (based on the recordings of seven individuals in Jodhpur). Songs recorded in urban areas were further divided into the quiet phase (without background noise) and the noisy phase (with background noise produced by road and air transport). Analysis was carried out at the element level of songs. Songs of this species comprised discrete phrases further made up of elements.

14.3 Results and Discussion

14.3.1 Alterations in Breeding Behaviour and Foraging

In north India, the RVB breeds mainly from April to August and rear 2–3 broods. It forms a cup-shaped nest and prefers shrubs, small trees and thorny bushes for nesting, mostly 1–2 m from the ground level (Ali 1996; Grimmett et al. 1998; Kumar 1999).

Data collected on the breeding biology of this species at Haridwar (from April 1995 to September 1998) and Dehradun (from March 2002 to September 2004) showed that in urban/suburban areas, this species selected largely exotic and/or ornamental garden plants for nesting (Fig. 14.2). In Haridwar town, 14 species of plants (all exotic) were selected for nesting. The hedges of *Pithecellobium dulce* were the most preferred nesting sites (Kumar 1999). Later on, in Dehradun, it selected 28 species of plants for nesting (Kumar 2004b). The hedges of *Duranta repens* were the most preferred (17.86%). The percentage of nesting in exotic plants was higher (83.33%) than in endemic plants (16.67%). A total of 38.1% of nests were in thorny plants, and ornamental plants were also preferred (77.38%). The nest was composed of grass twigs, fine stems and leaf races, and the inner surface of the nest was lined with fine grasses, cattle hairs and so on. Pieces of papers, plastic threads, polythene, wool, cloths and rags were also used as nesting material in nests (Kumar 2004b).

14.3.2 Unusual Nesting

In urban areas, it was observed that RVB often nest in an unusual manner, inside/under the buildings on electric fixings/cables (Kumar 2004b). On 29 June 2000, I noticed a nest of this species in an unusual manner. The nest was placed in the front shade (made up of jute cloth) of a general store (*Kirana shop*) in Osian town, about 40 km west of Jodhpur. The shade of the shop was made up of a wooden frame, supported by wooden pillars, and it was covered with a thin jute cloth. On the joint of the wooden frame, the twigs of the nesting material were firmly joined together with the help of spider webs. The adult female was incubating the eggs without fear (Fig. 14.3) while the place was full of human activities. The nest was located about 2.1 m from the ground. People (with an average height of about 1.52 m) were moving just below the nest. In the absence of the bird, I observed three eggs in the nest.

Besides this, few other unusual nesting locations were also observed. In July 1997, one pair formed a nest in the broken cover of a tube light on a street pole at Haridwar. However, the nesting was disturbed by a rufous treepie (*Dendrocitta vagabunda*), within a few days after completion. The next year (in April 1998), the same pair (colour-banded) was also observed nesting at the same location. The result was unsuccessful nesting, due to predation. In the last week of April 1999, an unusual nest was observed. This time, the nest was located beneath the ventilator shade of a house in Haridwar town. The nest was firmly attached on one side with

Fig. 14.2 Nesting of red-vented bulbul in different types of exotic and/or ornamental garden plants. (**a**) Nesting in *Polyalthia longifolia*, (**b**) nesting in *Hibiscus rosa* sp., (**c**) nesting in the hedges of *Campsis grandiflora*, (**d**) nesting in the hedges of *Lawsonia inermis* and (**e** and **f**) nesting in the fruit bunch of *Musa paradisiaca*

Fig. 14.3 Nest of red-vented bulbul beneath the shade of a shop in Osian town, Jodhpur

Fig. 14.4 Nest of red-vented bulbul on the top of *Araucaria* plant (**a** and **b**) situated inside the lobby of a house, at WII campus, Dehradun

an electric rubber-coated wire, while the opposite side of the nest cup was attached with another parallel metallic wire (supporting wire for electric cable). Observations revealed that two nestlings were seen fed by parents, and after completion of the nestling period, both were fledged; nesting was successful.

In April 2002, one nest was seen on the top of the *Araucaria* plant, about 0.65 m from the ground, inside the lobby of a house (Fig. 14.4a, b). The owner of the house informed me that the same species (most probably the same pair) formed its nest at the same place last summer (i.e. in 2001), and the nesting was successful. In the same manner, in May 2003, one nest was seen on *Thuja orientalis* plant in front of a shop. This was a crowded place. However, nesting was successful. In May 2004, one nest was located on an ornamental creeper plant (*Philodendron* sp.) in the ZSI office building, Dehradun (Fig. 14.5). The nesting was successful. In April 2003, a nest was seen inside the door shade of a building (Genetics Laboratory adjacent to

Fig. 14.5 Nest of red-vented bulbul, on an ornamental creeper plant (*Philodendron* sp.) located inside the lobby of the ZSI office, Dehradun

the old guest house) at WII, Dehradun. This nest was placed on an angle-iron frame in a corner of the shade. The nesting was successful. In May 2004, a nest was seen at the same place, and the female was incubating at the time of the last observation (Kumar 2004b).

On the basis of earlier studies and present observations, it is inferred that the selection of unusual nesting sites seems not to be by chance; this diversion may be a part of the adaptive breeding strategy for successful reproduction as, in most unusual cases, the nesting was successful (Table 14.1). It was an interesting point that in most cases of nesting under buildings, adjacent natural patches of vegetation/bushes were available for nesting even though individuals constructed the nest under artificial structures. It seems that this initiative was taken by the bird to avail the selective advantage of nesting at such places, that is, less predation pressure. It was observed that at most of the urban sites, predators such as crows, monkeys, treepies and raptors hesitated to enter such places.

It was reported that the RVB selects mostly exotic and ornamental plants for nesting in urban/suburban habitats (Kumar 1999, 2004b). Often, nests in these garden plants were exposed to periodical cutting/dressing. It seems that heavy predation and disturbance by the dressing of the hedges and plants created a selection pressure to select comparatively safer places for nesting. It has been observed that in urbanized habitats, the RVB has altered breeding behaviour in a number of ways like selection of unusual sites for nesting, such as mud holes, tree holes, the broken plastic cover of a street tube light, exotic plants, along with the use of threads, polythene pieces (Fig. 14.6), paper pieces, rags and so on as nesting material (Kumar 1999, 2004b). It feeds upon a number of unusual foraging items. Basically, the RVB

Table. 14.1 Unusual nesting sites, their location, height from the ground level and nesting success in Red-vented Bulbul

Unusual nesting sites	Location of the study	Height from the ground level (m)	Nesting success	Reference
Mud -hole	Deharadun	[a]	[a]	Lamba (1976)
Disused tube light		[a]	[a]	Mundkur (1984)
Floating water hyacinth	Keoladeo National Park, Bharatpur	NA	Unsuccessful	Nanjappa (1989)
Tree- hole	Keoladeo National Park, Bharatpur	3	[a]	Sivasubramanian and Sundaramoorthy (1992)
Inside the deserted nest of pied myna	Keoladeo National Park, Bharatpur	8	[a]	Sivasubramanian and Sundaramoorthy (1992)
Vehicle	Ahmedabad	NA	Successful	Urfi and Jethua (1998)
Inside the lobby of a house on the top of *Araucaria* plant	WII campus, Dehradun	0.65	Successful	Kumar (2004b)
Road side on *Thuja* plant	Main road, Dehradun	About 1.4	Successful	Kumar (2004b)
On an ornamental creeper plant (*Plilodnderon* sp.)	ZSI office, Dehradun	About 1.5	Successful	Kumar (2004b)
Inside the door-shade of a building	WII campus, Dehradun	About 2.0	Successful	Kumar (2004b)
Inside the broken cover of a street tube light	Haridwar	4.5 m	Unsuccessful	Personal observations
Under ventilator shade	Haridwar	3.6	Successful	Personal observations
Beneath the front shade of a shop	Osian town, Jodhpur	2.1	[a]	Personal observations

NA not applicable
[a]Information not available

is a wild species, and in the forest, it forages upon a number of wild fruits (Vijayan 1978; Ali 1996), while in a suburban area, it was found feeding upon a variety of food, including kitchen waste, cooked vegetables and rice (Bhatt and Kumar 2001).

The use of leaves of *Medicago sativa* for feeding was also an interesting observation because, generally, birds do not use the leaves for feeding purposes (Fig. 14.7). Like red-vented bulbul, some other avian species have been reported to alter their foraging behaviour or diet in urban habitats. For example, house sparrows were used as pray to a greater extent (88%) in urban areas than in rural areas (only 3%) in the diet of tawny owl (*Strix aluco*) (Goszczyński et al. 1993). In another study, the common gull (*Larus canus*) showed a flexible response to changed environmental

Fig. 14.6 Use of polythene piece as nesting material

Fig. 14.7 Foraging of red-vented bulbuls on the leaves of *Medicago sativa*

conditions by successfully exploiting the benefits of their wide food range (Kubetzki and Garthe 2007).

14.3.3 Breeding Success

Of the 26 nests studied, 19.23% were rejected by the birds before egg laying, mainly due to disturbances created by monkeys at the nesting site or nests being exposed due to cutting/dressing of the garden plants. Out of 64 (in different nests), 34.27% of eggs were destroyed by predators. Uncontrolled cattle grazing also affected the nesting success because cattle often defoliated the nesting plants and exposed the nests to predators causing 4.68% egg loss. A total of 34.37% of eggs were destroyed during the incubation phase due to various factors (4.68%, 9.37% and 3.12% of eggs were eaten by snakes, treepies and crows, respectively, while 4.68% of eggs were destroyed by monkeys and 7.8% were destroyed by unknown factors. After hatching, a total of 25% loss was caused to nestlings due to heavy rains (1.56%), unknown infections (4.68%), crows (4.68%), mongoose (4.68%) and unknown factors (9.37%). Thus, at the end breeding period, 40.62% of nestlings fledged. It is an interesting point revealed by this data that the rate of breeding success (40%) in the present study in an urban/suburban habitat is much higher than in forest habitats (8.3%, reported by Vijayan 1978).

Similar trends have been reported in spotted munia (*Lonchura punctulata*) (Sharma et al. 2004). A significant difference in breeding success was reported in urban (48.7%) versus forest (31.6%) habitats as a result of relatively less predation pressure in urban habitats. RVB also selected introduced plant species such as *Thuja orientalis* and *Polyalthia longifolia* for nesting in urban habitats. This may be possibly due to the dense foliage canopy, which is useful in preventing predators. It contributes to successful nesting/high breeding success. European wood pigeon (*Columba palumbus*) (Cramp 1972; Tomialojc 1979), herring gull (*Larus argentatus*) (Monoghan 1979), blackbird (*Turdus merula*) (Snow 1958; Evans et al. 2010), mallard (*Anas platyrhynchos*) (Bentz 1985), magpie (*Pica pica*) (Kavanagh et al. 1989), great tit (*Parus major*) and the blue tit (*Cyanistes caeruleus*) (Bańbura and Bańbura 2012) have been reported to achieve a higher breeding success in the urban than forest habitats.

It is suggested that forest habitats have greater predation pressure compared to urban habitats. In urban areas, the predators have opportunities to avail food from other resources such as garbage and wastes of slaughterhouses/meat shops. Due to the availability of grains in urban habitats from crop fields, open railway yards, warehouses and kitchen waste, a number of granivorous birds, namely, sparrows, munias, doves and pigeons, breed abundantly in urban habitats. The availability of hatchlings/juveniles of these species (source of food for predators) probably decreases the dependability of predators on the eggs/nestlings of RVB.

Behavioural diversions have also been reported in a number of other species such as in Europe, where urbanization has existed for thousands of years and many birds have become human commensals such as rock doves, starlings and house sparrows

(Diamond 1986). The latter species is now so closely associated with man that its original niche is unknown. In some cases, the pre-adaptations are clear. Shining starlings (*Aplonis metallica*) transferred their nests from holes in trees to vents of air conditioners. Swiftlets (*Collocalia* sp.) that would otherwise nest in caves now use tunnels excavated by humans. Goldie's lorikeet (*Trichoglossus goldiei*), a rare species occurring in primary forest, arrived in the 1970s to feed on casuarina seeds in highland towns, where it is now the most abundant bird (Diamond 1986). Roof-nesting in birds, such as red-wattled lapwing (*Vanellus indicus*) (Kumar and Sharma 2011) and common gull (*Larus canus*) (Kubetzki and Garthe 2007) seems to be a behavioural adaptation in changing environment.

14.3.4 Alterations in Song Characteristics

This species has been reported to sing throughout the year (Kumar 2004a). It is known to produce discrete songs, composed of strophes (phrases) preceded and followed by temporal gaps. The duration of song bouts may range few minutes to 28 min. However, sometimes the inter-phrase gaps were quite long (more than 1 min). So the beginning and ending of a song bout were ambiguous. In a song bout, usually the same types of strophes are repeated several times in a stereotyped manner with minor structural variations of elements. Duration of strophes ranged from 0.2 to 0.83 s, and the gap between strophes was about 0.88–14.00 s. The range of frequency varied from 0.96 to 8.00 kHz (in most phrases, 1.5–4.5 kHz). It is reported that RVB used two categories of songs. Type A songs were discrete, simple, loud, stereotyped and mostly spontaneous and commonly produced throughout the year, probably to maintain pair bonds and contacts, while type B songs were rare, complex with irregular inter-phrase gaps and low amplitude, mate-oriented and most likely used for mating purposes with mate acquisition/courtship displays (Kumar 2010).

The acoustical features of songs in forests and quiet and noisy conditions in urban habitats were compared (Table 14.2). It can be inferred from the data that individuals used relatively higher minimum frequency during noisy conditions of

Table 14.2 Frequency-based song characteristics of red-vented bulbul in forest and urban habitats

Song characteristics	Forest habitat	Urban habitat: Quiet phase	Urban habitat: Noisy phase
	(N = 12, n = 46)	(N = 07, n = 27)	(N = 07, n = 37)
Minimum frequency (kHz)	1.34 ± 0.032 (0.86–1.89)	1.42 ± 0.044 (1.12–1.87)	1.64 ± 0.032 (1.31–2.06)
Maximum frequency (kHz)	2.96 ± 0.063 (2.21–4.31)	3.03 ± 0.082 (2.43–4.5)	2.93 ± 0.064 (2.37–4.06)
Dominant frequency (kHz)	2.15 ± 0.039 (1.59–2.75)	2.07 ± 0.064 (1.56–2.56)	2.24 ± 0.038 (1.87–2.75)

Note: N = number of individuals and n = number of song elements (2–5 elements per individual used for analysis). Results are expressed in mean ± SE, and ranges are given in parenthesis

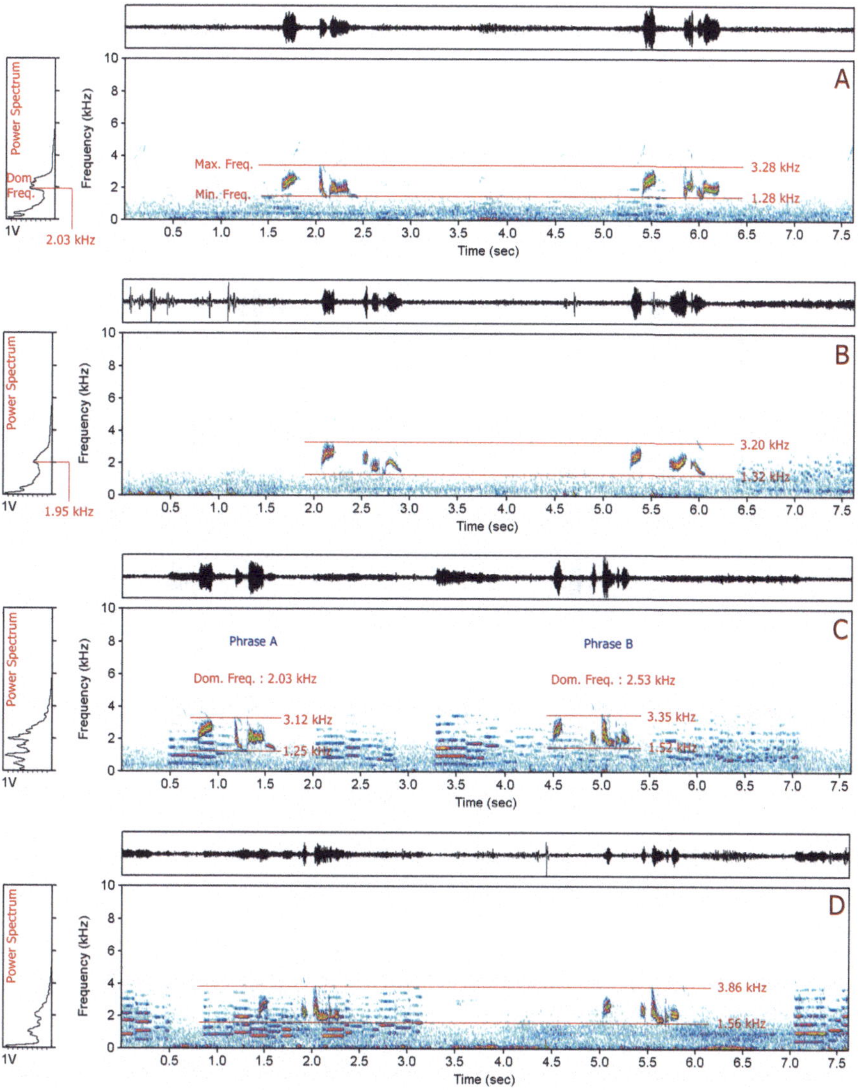

Fig. 14.8 Spectrograms showing the physical characteristics, structure and variations of songs recorded in quiet (**a** and **b**) and noisy (**c** and **d**) phases in an urban habitat of Jodhpur

the urban areas (Fig. 14.8) compared to forest (Fig. 14.9) and quiet conditions of urban habitats (Table 14.2). In urban areas of Jodhpur, the minimum frequency of songs in noisy conditions was significantly higher to quiet conditions (Mann–Whitney U test: $U = 225.5$, $Z = -3.74$, $P < 0.001$). However, maximum frequency did not show a significant difference ($P = 0.169$).

The dominant frequency in the noisy phase was also significantly higher than the quiet phase ($P = 0.024$). The minimum, maximum and dominant frequencies of the

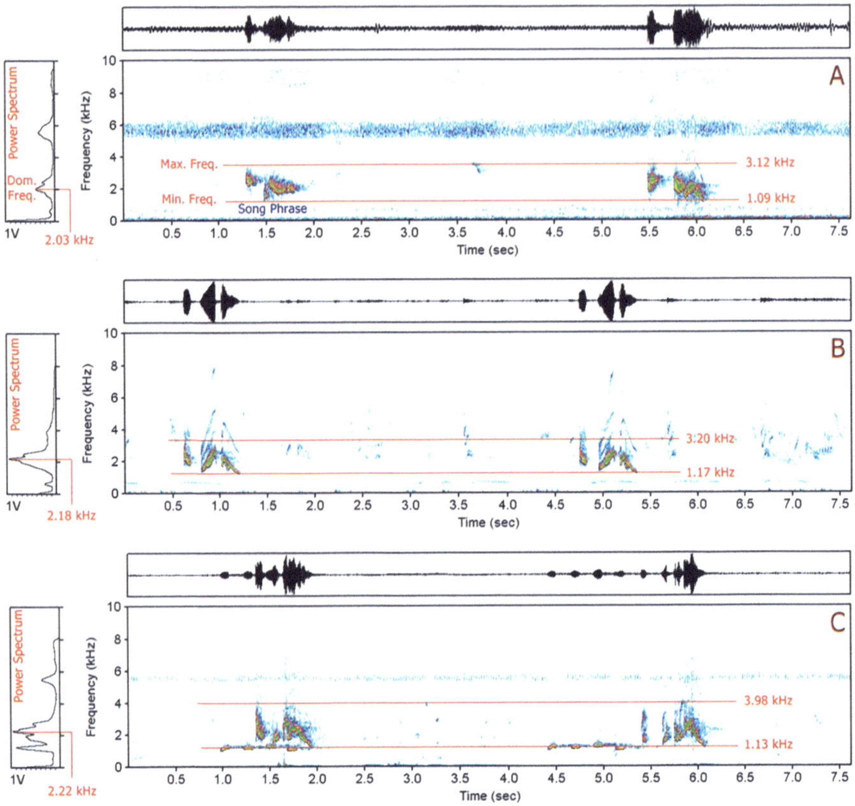

Fig. 14.9 Spectrograms showing the physical characteristics and structure of songs recorded in three different individuals (**a**, **b** and **c**) in forest habitats of Doon Valley

songs recorded in forest habitats did not show any significant difference from the songs recorded in a quiet phase of urban habitats ($P = 0.25$, 0.564 and 0.385, respectively). When compared the songs produced in the forest habitats with the songs produced in noisy phase of urban habitats, only the minimum frequency differed significantly ($P = <0.001$), while maximum and dominant frequencies were not significantly separable ($P = 0.65$ and 0.18, respectively). In few cases ($n = 4$), the amplitude of the signals was increased with increasing background noise, and when background noise increased enormously (i.e. during the take-off of jet fighters and prolonged buzzing of high-pressure horns of vehicles), the singing was stopped for that duration of time (Fig. 14.10).

The results of this analysis showed the singing of higher-pitched songs in noisy areas. Such trends have been reported in other birds also (Slabbekoorn and Ripmeester 2008; Bermúdez-Cuamatzin et al. 2009; Hu and Cardoso 2009). It may possibly help the birds manage the signal transmission effectively because urban noise is progressively louder at lower frequencies (Wood and Yezerinac 2006). It

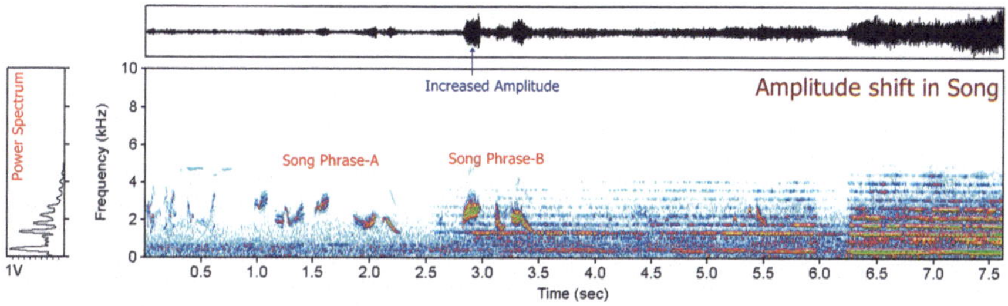

Fig. 14.10 Spectrogram is showing the amplitude shift of song during increasing background noise in urban areas

has been reported that most anthropogenic noise in urban areas occupies a frequency range below 2.0 kHz (Slabbekoorn and Peet 2003). In RVB, the minimum frequency of song in noisy conditions was raised by about 0.26 kHz compared to forest/quiet areas, most probably to increase the efficacy of the signals. It is reported that a slight increase in the minimum frequency of the vocal signals of birds in urban habitats supports coping with background noise (Hu and Cardoso 2009). Analysis of the data showed that compared to minimum frequency, the dominant frequency of songs in forest habitats was not much higher than the noisy phase in urban habitats (Table 14.2). The same findings have been reported in some avian species also (Hu and Cardoso 2009). It is suggested that the dominant frequency is less masked than the minimum frequency, so the selection to raise it should be poor. Moreover, such shifts in frequency impose a handicap in terms of energetics (Suthers et al. 1999).

To manage the masking effect of ambient noise in urban habitats, select avian species have been reported to adopt a number of strategies for successful communication (Dubois and Martens 1984; Brumm 2004; Brumm and Slater 2006; Fuller et al. 2007). These strategies/adaptations may be grouped into 'long-term genetic/ontogenic changes' such as differences in song structure between populations of the same species (Slabbekoorn and Smith 2002) and habitat-dependent signal differences as a result of lifetime plasticity (Hansen 1979; Slabbekoorn and Peet 2003) and 'short-term adjustments', that is, habitat-dependent signal differences in real time (Brumm and Slabbekoorn 2005). For the production of higher-pitched songs in noisy conditions, birds use either 'syllable pitch plasticity mode' (use of the same syllable in quiet and noisy areas but with a raised pitch in the latter) or 'differential syllable use mode' (use of low-frequency syllable types in quiet condition and higher-pitched type syllables in noisy conditions) (Bermúdez-Cuamatzin et al. 2009). This analysis revealed that RVB used different types of syllables/phrases (higher-pitched) in noisy conditions than quiet conditions (Fig. 14.8). Like RVB, some birds such as nightingales increase the amplitude of signals in noisy environments (Brumm 2004; Warren et al. 2006). It is known as 'Lombard effect' (Lombard 1911). It is observed in some birds and mammals (Brumm and Slabbekoorn 2005), in which the signaller increases the intensity of the signal in response to background noise.

14.4 Conclusion

The rapid developmental activities such as urbanization and consequent deforestation caused alterations in the natural habitats of the birds. These activities forced them to either disappear or adopt new niches. In such a changed landscape, the survival of a species depends on how fast it alters its foraging and breeding behaviour. The present study showed preliminary findings on behavioural plasticity in red-vented bulbul owing to urbanization. It may possibly help us understand how birds change their behaviour in changing environments to cope with the alterations and avail benefits in changed conditions. Such shifts in behaviour are examples of rapid evolution and provide opportunities to explore some basic problems such as sudden diversions in behaviour depending on genetic change or on culturally transmitted learning. Are the changes convergent or divergent in sympatric/sibling species? Over the years, the rate at which species evolve to occupy new niches can also be an interesting issue in evolutionary studies. Such studies on foraging and breeding behaviour in the changed landscape in economically important or threatened species can be useful in recommending their proper management and conservation measures to be adopted.

Acknowledgement I am grateful to Dr. Dhriti Banerjee, Director, ZSI, for kind support, encouragement and institutional facilities. I am grateful to Prof. Dinesh Bhatt (retired), Gurukul Kangri (deemed to be University), Haridwar; Dr. Q.H. Baqri, former officer-in-charge, ZSI, Jodhpur; and Director, WII, Dehradun, for encouraging and extending institutional facilities. I am thankful to Dr. Ajeet Singh (Haridwar), Dr. Neelesh Vashistha (Haridwar), Dr. Himmat Singh (Jodhpur) and many others for their cooperation at various levels during the study.

References

Ali S (1996) The book of Indian birds. Oxford University Press, Mumbai

Bańbura J, Bańbura M (2012) Blue tits *Cyanistes caeruleus* and great tits *Parus major* as urban habitat breeder. Intern Stud Sparrows 36:66–72

Bentz PG (1985) Studies on some urban mallard (*Anas platyrhynchos*) populations in Scandinavia. 1. Cause of death, mortality and longevity among Malma mallards as shown by ringing recoveries. Fauna Narv Ser C Cinclus 8:44–56

Bermúdez-Cuamatzin E, Ríos-Chélen AA, Gil D, Garcia CM (2009) Strategies of song adaptation to urban noise in the house finch: syllable pitch plasticity or differential syllable use? Behaviour 146:1269–1286

Bhatt D, Kumar A (2001) Foraging ecology of red-vented bulbul *Pycnonotus cafer* in Haridwar, India. Forktail 17:109–110

Brumm H (2004) The impact of environmental noise on song amplitude in a territorial bird. J Anim Ecol 73:434

Brumm H, Slabbekoorn H (2005) Acoustic communication in noise. Adv Study Behav 35:151–209

Brumm H, Slater PJB (2006) Ambient noise, motor fatigue, and serial redundancy in chaffinch song. Behav Ecol Sociobiol 60:475–481

Clergeau P, Savard J-PL, Menechez G, Falardeau G (1998) Bird abundance and diversity along an urban-rural gradient: a comparative study between two cities on different continents. Condor 100:413–425

Cramp S (1972) The breeding of urban woodpigeons. Ibis 114:163–171

Diamond JM (1986) Rapid evolution of urban birds. Nature 324:107–108
Dowling JL, Luther DA, Marra PP (2012) Comparative effects of urban development and anthropogenic noise on bird songs. Behav Ecol 23:201–209
Dubois A, Martens J (1984) A case of possible vocal convergence between frogs and a bird in Himalayan torrents. J Ornithol 125:455–463
Evans KL, Hatchwell BJ, Parnell M, Gaston KJ (2010) A conceptual framework for the colonisation of urban areas: the blackbirds *Turdus merula* as a case study. Biol Rev 85:643–667
Fernández-Juricic E, Poston R, De Collibus K, Morgan T, Bastain B, Martin C, Jones K, Treminio T (2005) Microhabitat selection and singing behaviour patterns of male house finches (*Carpodacus mexicanus*) in urban parks in a heavily urbanized landscape in the western U.S. Urban Habitat 3:49–69
Fuller RA, Warren PH, Gaston KJ (2007) Daytime noise predicts nocturnal singing in urban robins. Biol Lett 3:368–370
Goldstein-Golding EL (1991) The ecology and structure of urban greenspaces. In: Bell SS, McCoy ED, Mushinsky HR (eds) Habitat structure. Chapman and Hall, London
Goszczyński J, Jabloński P, Lesiński G, Romańowski J (1993) Variation in the diet of tawny owl *Strix aluco* L. along an urbanization gradient. Acta Ornithol 27(2):113–123
Grimmett R, Inskipp C, Inskipp T (1998) Birds of the Indian subcontinent. Oxford University Press, Delhi
Hansen P (1979) Vocal learning: its role in adapting sound structures to long-distance propagation, and a hypothesis on its evolution. Anim Behav 27:1270–1271
Hu Y, Cardoso GC (2009) Are bird species that vocalize at higher frequencies preadapted to inhabit noisy urban areas? Behav Ecol 20:1268–1273
Isaksson C (2018) Impact of urbanization on birds. In: Tietze D (ed) Bird species. Fascinating life sciences. Springer, Cham. https://doi.org/10.1007/978-3-319-91689-7_13
Katti M, Warren PS (2004) Tits, noise, and urban bioacoustics. Trends Ecol Evol 19:109–110
Kavanagh BP, Jerzak L, Gorsla W (1989) Factors affecting the breeding performance of the magpie (*Pica Pica*) in three European cities. In: Proceedings of the International Symposium of the Working Group on Grainivorous birds, pp 70–81
Kubetzki U, Garthe S (2007) Nests with a view: distribution, nest habitats and diets of roof-breeding common gulls (*Larus canus*) in northern Germany. Warerbirds 30(4):602–608
Kumar A (1999) Characteristics and significance of calls, songs and visual displays in two avian species *viz. Copsychus saularis* & *Pycnonotus cafer*. Ph.D. thesis, Gurukul Kangri University, Haridwar. p 105
Kumar A (2004a) Acoustic communication in the red-vented bulbul *Pycnonotus cafer*. An Acad Bras Cienc 76(2):350–358
Kumar A (2004b) Behavioural ecology and sociobiology of the red-vented bulbul *Pycnonotus cafer* in Dehradun, India. Ann For 12(1):125–135
Kumar A (2010) Communication value of displays and postures in Red-vented Bulbul *Pycnonotus cafer* (Aves: Pycnonotidae). Journal of Threatened Taxa 2(6):919–929
Kumar A, Sharma RK (2011) Observations on breeding behaviour and vocalizations in red-wattled lapwing, *Vanellus indicus* from northern India. J Exp Zool India 14(1):333–338
Lamba BS (1976) Red-vented bulbul *Pycnonotus cafer* nesting in a hole in a mud bank. J Bombay Nat Hist Soc 73:395
Lim HC, Sodhi NS (2004) Responses of avian guilds to urbanization in a tropical city. Landsc Urban Plan 66:199–215
Lombard E (1911) Le signe de l'elevation de la voix. Annales de Maladies de L'oreille et du Larynx 37:101–119
Luniak M (2004) Synurbanization adaptation of animal wildlife to urban development. In: Shaw WW, Harris LK, Vandruff L (eds) Proceedings of the 4th International Symposium on Urban Wildlife Conservation, Tucson, AZ, pp 50–55
Meyer WB, Turner BLII (1992) Human population growth and global land use/cover change. Annu Rev Ecol Syst 23:39–61

Møller AP, Díaz M, Grim T, Dvorská A, Flensted-Jensen E, Ibáñez-Álamo J, Jokimäki J, Mänd R, Markó G, Szymański P, Tryjanowskiothers P (2015) Effects of urbanization on bird phenology: a continental study of paired urban and rural populations. Climate Res 66:185–199. https://doi.org/10.3354/cr01344

Monoghan P (1979) Aspects of breeding biology of herring Gull *Larus argentatus* in urban colonies. Ibis 121:475–481

Mundkur T (1984) Unusual nesting sites of the red-vented bulbul. Hornbill 2:27–28

Nanjappa C (1989) An hitherto unrecorded nesting site of a red-vented bulbul *Pycnonotus cafer*. J Bombay Nat Hist Soc 86:102

Phillips JN, Rochefort C, Lipshutz S, Derryberry GE, Luther D, Derryberry EP (2020) Increased attenuation and reverberation are associated with lower maximum frequencies and narrow bandwidth of bird songs in cities. J Ornithol 161:593–608

Searcy AW, Andersson M (1986) Sexual selection and the evolution of song. Annu Rev Ecol Syst 17:507–533

Sharma RC, Bhatt D, Sharma RK (2004) Breeding success of the tropical spotted Munia *Lonchura punctulata* in urbanized and forest habitats. Ornithol Sci 3:113–117

Sivasubramanian C, Sundaramoorthy T (1992) Additional nesting sites of red-vented bulbul *Pycnonotus cafer* (Linn.). J Bombay Nat Hist Soc 89:257

Slabbekoorn H, Peet M (2003) Birds sing at a higher amplitude in urban noise. Nature 424:267

Slabbekoorn H, Ripmeester EAP (2008) Birdsong and anthropogenic noise: implications and applications for conservation. Mol Ecol 17:72–83

Slabbekoorn H, Smith TB (2002) Habitat-dependent song divergence in the little Greenbul: an analysis of environmental selection pressures on acoustic signals. Evolution 56:1849–1858

Snow DW (1958) The breeding of blackbird *Turdus merula* at Oxford. Ibis 100:1–30

Suthers RA, Goller F, Pytte C (1999) The neuromuscular control of birdsong. Philos Trans R Soc B 354:927–939

Tomialojc L (1979) The impact of predation on urban and rural woodpigeon (*Columba palumbus* L.) populations. Pol Ecol Stud 5(4):141–220

Urfi AJ, Jethua K (1998) Unusual nest location of red-vented bulbul *Pycnonotus cafer* (Linn.). J Bombay Nat Hist Soc 95:116

Vijayan VS (1978) Breeding biology of bulbuls, *Pycnonotus cafer* and *Pycnonotus luteolus* (class: Aves, family: *Pycnonotidae*) with special reference to their ecological isolation. J Bombay Nat Hist Soc 75:1090–1117

Warren PS, Katti M, Ermann M, Brazel A (2006) Urban bioacoustics: it's not just noise. Anim Behav 71:491–502

Wheater CP (1999) Urban habitats. Routledge, London, pp i–xii+175

Wood WE, Yezerinac SM (2006) Song sparrow (*Melospiza melodia*) song varies with urban noise. Auk 123:650–659

Foraging Ecology of Wintering Anatids: A Review

15

Arkajyoti Mukherjee

Abstract

Wetlands worldwide are rapidly degrading due to anthropogenic activities, impacting the biota, including waterbirds, that depend on these ecosystems. As waterfowls are an excellent indicator of wetland health, study on their abundance and ecology is gaining attention. This review work focuses on the foraging ecology of the wintering migratory and resident Anatids in India. Foraging habitat, predominant diet, foraging techniques and time of foraging of 37 Anatids (30 ducks, 4 geese and 3 swans) are mentioned in this chapter. As overall foraging strategies differ between breeding and wintering grounds largely, this work will assist conservation biologists in framing future strategies for the management of waterfowls.

Keywords

Waterfowl · Anatidae · Wintering ground · Foraging ecology · Diet

15.1 Introduction

Wetlands play several ecological functions like water purification, flood prevention, aquifer recharge, maintaining biodiversity and many others (Green and Elmberg 2013). However, wetlands have fast degraded in the past century due to the aggressive anthropogenic interferences and already over half of the world's wetland wealth has been lost (Ma et al. 2010; Quesnelle et al. 2013). Waterbirds, especially

A. Mukherjee (✉)
Department of Chemical Engineering, Jadavpur University, Kolkata, West Bengal, India

Ecotechnology Project Lab, Government College of Engineering and Leather Technology, Kolkata, West Bengal, India

S. Sridhara et al. (eds.), *Animal Behavior in the Tropics*,
https://doi.org/10.1007/978-981-97-7324-4_15

waterfowls, are excellent bioindicators of general ecological status of wetland ecosystems as they are easy to count (Amat and Green 2010). Therefore, recent studies not only focus on the studies of abundance of waterbirds but also focus on the detailed foraging studies of the waterbirds as it can aid in the management of both waterbirds and their habitat (Urfi 2017; Mukherjee et al. 2021b, 2022).

The Anatidae family contains majority of species in order Anseriformes comprising approximately 150 species of ducks, geese and swans and referred to as "waterfowl" (Backues 2015). A total of 42 species of ducks, geese and swans were recorded from India (Islam and Rahmani 2009). However, Urfi (2017) reported 36 Anatids from India, 8 out of which were threatened and classified their broad diet as mixed diet and planktivorous. Many migratory waterfowls migrate to India from the north and beyond the temperate latitudes following two major flyways: Central Asian Flyway (CAF) and East Asian-Australasian Flyway (EAAF) (Li et al. 2009; Dhanjal-Adams et al. 2017; CAF National Action Plan-India 2018). Kirby (2010) reported that on a flyway scale, the highest proportion (19%) of threatened migratory waterbirds use the EAAF while CAF is used by 14% of threatened and near-threatened waterbird species. Migratory waterfowl are typically winter visitors (between October and March) in the Indian subcontinent and residents share their foraging grounds with them during winter months. Both residents and migrants use myriads of foraging strategies to exploit foraging resources in the same wintering grounds. In this chapter, adaptive strategies of both resident and migratory waterfowls for meeting nutrient requirements in foraging grounds during winter months are described.

15.2 Foraging Ecology of Ducks

Family Anatidae is represented by a variety of ducks exhibiting a wide array of foraging strategies for resource utilization. Some of these are listed in the International Union for Conservation of Nature (IUCN) threatened category requiring proper conservation measures. Foraging habits, techniques, time and habitat of 30 species of ducks are mentioned in this section. A summary of foraging ecology of 37 species of Anatids is mentioned in Table 15.1.

15.2.1 Lesser Whistling Duck (*Dendrocygna javanica*)

This species is a resident, moving about locally with drought and flood and partial local migrant (Ali and Ripley 1978). These ducks are one of the most abundant Anatids found throughout India. This species mainly consumes grasses, wild and cultivated rice, seeds and shoots of water weeds (Carboneras and Kirwan 2020l). However, their diet is also complemented with small molluscs (mainly freshwater snails), small fish, frogs and worms (Kear 2005; Onwuka et al. 2020). Moreover, besides aquatic plants, terrestrial plants and vertebrates also are found to be part of their diet (Onwuka et al. 2020). Mukherjee et al. (2021a) did a painstaking study on foraging behaviour of this bird and reported that this species preferred shorelines with hydrophytic vegetation in wetlands for roosting and foraging. They also

Table 15.1 A summary of foraging ecology of Anatids

Sl. no	Name	IUCN status	Migration status	Diet	Foraging time	Foraging technique
Duck						
1	Lesser whistling duck	LC	R	H	N, C	Generalist
2	Fulvous whistling duck	LC	R	H	D, N	Filtering
3	Ruddy shelduck	LC	WM	O	D, N	Diving
4	Common shelduck	LC	WM	C	D, N	Filtering
5	Northern pintail	LC	WM	H	D, N	Upending
6	Green-winged teal	LC	WM	H	D, N	Filtering
7	Baikal teal	LC	WM	H	N	Filtering
8	Indian spot-billed duck	LC	R	H	D	Filtering
9	Eastern spot-billed duck	LC	WM	H	D, N	Filtering
10	Mallard	LC	WM	O	D, N	Filtering
11	Gadwall	LC	WM	H	D	Head-dip
12	Falcated duck	NT	WM	H	D	Filtering
13	Eurasian wigeon	LC	WM	H	D, N	Filtering
14	Garganey	LC	WM	H	N	Filtering
15	Northern shoveler	LC	WM	C	D	Filtering
16	Marbled teal	NT	WM	H	N	Filtering
17	Red-crested pochard	LC	WM	H	D, N	Diving
18	Common pochard	VU	WM	H	N	Diving
19	Ferruginous pochard	NT	WM	O	N	Diving
20	Baer's pochard	CR	WM	O	D	Diving
21	Tufted duck	LC	WM	O	D, N	Diving
22	Greater scaup	LC	WM	O	D, N	Diving
23	Cotton pygmy-goose	LC	R	H	D, N	Filtering
24	Knob-billed duck	LC	R	H	D	Grazing
25	White-winged duck	EN	R	O	D	Filtering
26	Common goldeneye	LC	WM	C	D	Diving
27	Smew	LC	WM	C	D	Diving
28	Goosander	LC	WM	C	D	Diving
29	Red-breasted merganser	LC	WM	C	D	Diving
30	White-headed duck	EN	WM	H	D, N	Diving
Goose						
31	Bar-headed goose	LC	WM	H	D	Grazing
32	Greylag goose	LC	WM	H	D	Grazing
33	Greater white-fronted goose	LC	WM	H	D	Grazing

(continued)

Table 15.1 (continued)

Sl. no	Name	IUCN status	Migration status	Diet	Foraging time	Foraging technique
34	Lesser white-fronted goose	LC	WM	H	D	Grazing
Swan						
35	Tundra swan	LC	WM	H	D	Neck-dip
36	Mute swan	LC	WM	H	D	Neck-dip
37	Whooper swan	LC	WM	H	D	Neck-dip

IUCN status: *LC* Least Concern; *NT* Near Threatened; *VU* Vulnerable; *EN* Endangered; *CR* Critically Endangered
Migration status: *R* Resident, *WM* Winter Migrant
Diet: *H* Herbivorous, *O* Omnivorous, *C* Carnivorous
Foraging time: *N* Nocturnal; *D* Diurnal; *C* Crepuscular

recorded eight foraging techniques used by this species namely, diving, upending, beak-dip, head-dip, neck-dip, filtering, picking and grazing. However, head-dip was the most preferred foraging technique followed by diving. Diurnal foraging is rare in this species and mainly foraging is observed at night, early morning and evening (Carboneras and Kirwan 2020l).

15.2.2 Fulvous Whistling Duck (*Dendrocygna bicolor*)

This species is also resident, nomadic and possibly also local migrant in part. However, these birds are scattered widely but sporadically in many areas; everywhere rather scarce and much less common than *D. javanica* (Ali and Ripley 1978). This species is predominantly herbivorous feeding on aquatic seeds, fruits, bulbs, leaf shoots, buds, grasses, rushes and occasionally small insects (BirdLife International 2022d). They also feed on cultivated rice fields (Hohman and Lee 2020). This species is principally a dabbling duck. However, they occasionally exhibit shallow water diving and upending (Hohman and Lee 2020). Both diurnal and nocturnal feeding was recorded in this species (Kear 2005). Intensive feeding in the early morning and late afternoon is prominent and nocturnal feeding in rice fields is also well documented (Dallmeier 1991).

15.2.3 Ruddy Shelduck (*Tadorna ferruginea*)

This species is a winter migrant to the Indian subcontinent, which arrives in early November and returns at the end of March. This species is omnivorous; besides consuming grain, tender shoots and tubers, it also consumes crustaceans, molluscs, aquatic insects, reptiles, small fish, frogs etc. (Ali and Ripley 1978; Salvador and Amat 2022). It is also known for consuming terrestrial insects, especially locusts (BirdLife International 2022e). However, wintering populations show dominant herbivory (www.allaboutbirds.org). Wintering populations often show a preference for streams and slow-flowing rivers beside large wetlands (Mukherjee et al. 2021b).

This species prefers to forage on the edges of waterbodies, especially in muddy zones (Nouidjem et al. 2021). Diving in shallow waters is the most preferred foraging technique, followed by occasional straining of food from surface of water (www.allaboutbirds.org). High percentage of both diurnal and nocturnal foraging is recorded in this species (Nouidjem et al. 2021).

15.2.4 Common Shelduck (*Tadorna tadorna*)

Like *T. ferruginea* these ducks are also exclusive winter migrants. Their diet mostly consists of invertebrates like molluscs, insects and crustaceans (Carboneras and Kirwan 2020d). Occasional foraging on small fishes and plant materials (algae, seed and grain) is also evident (Viain et al. 2011). Kear (2005) reported that in wintering grounds small crustaceans and insect larvae are especially favoured. These ducks move their bills in a scything action through mud to obtain food. They sieve primarily small organisms <8.0 mm long from the upper 2 cm of muddy surfaces of wetlands (Viain et al. 2011). Foraging by digging and dabbling is also recorded in this species. They also use head-dip and upending in shallow water to obtain food (Kear 2005). Wintering ducks spend half their time during the day in feeding (Mouloud et al. 2006) and nocturnal foraging in line with tidal rhythms is also prominent (Hotker 1995). Like other shelducks, these birds also prefer riverine habitats besides marshes and lakes (Ali and Ripley 1978; Grimmett et al. 2011).

15.2.5 Northern Pintail (*Anas acuta*)

This species is a winter visitor and is widespread throughout India during the winter months. Their diet consists of chiefly vegetable matter—grass, corns, shoots and seeds of aquatic plants, and wild and cultivated rice. However, wintering populations consume molluscs, worms, water insects and larvae to a lesser extent (<10% of total food consumed) (Ali and Ripley 1978; Bethke 1991; Clark et al. 2020). These ducks prefer shallow water zones of waterbodies for foraging (Mukherjee et al. 2021a) and also feed in dry fields by walking (Clark et al. 2020). Upending in shallow water is the most preferred foraging technique in this species followed by beak-dip, head-dip, neck-dip and filtering (Mukherjee et al. 2021a). Daily migration between roosting and foraging grounds is evident in this species. Nocturnal foraging is also recorded in this species (Casazza et al. 2012). Usually, rice fields adjacent to roosting grounds are preferred for nocturnal foraging (Parejo et al. 2019).

15.2.6 Green-Winged Teal (*Anas crecca*)

These ducks are also winter migrants to the Indian subcontinent and migrate in large flocks throughout India. The broad diet of this species consists of plant materials (shoots, tubers and seeds of aquatic plants, and grains of wild and cultivated rice), and aquatic insects, molluscs and crustaceans (Ali and Ripley 1978; Johnson et al.

2020). Euliss and Harris (1987) also reported that the diet of wintering ducks mostly consists of plant materials (62%). This species is an opportunistic forager and can shift their food preferences according to the availability (Nummi 1993) and wintering populations also forage in agricultural areas. Dabbling is the most preferred foraging technique (around 70%) in this species followed by head-dip and upending (Johnson and Rohwer 2000). This species has lamellae on inner edge of bill, which aids in capturing smaller food particles (Nummi 1993). It typically forages in shallow waters near shorelines and also on mudflats (Pöysa 1983). Nocturnal foraging is also prevalent in this species (Arzel et al. 2007).

15.2.7 Baikal Teal (*Sibirionetta formosa*)

This species is a very rare and sporadic winter visitor. Kear (2005) reported that wintering populations of this species are almost exclusively herbivorous. They forage on seeds, leaves and other vegetative parts of aquatic plants. Moreover, crops such as rice are also preferred by this species. Allport et al. (1991) recorded that this species is solely feeding on rice from paddy fields. Dabbling is the most preferred foraging technique on waterbodies followed by head-dip and upending. This species is highly nocturnal on foraging grounds and uses the waterbodies as diurnal roosting grounds (Carboneras and Kirwan 2020b). Foraging on waterbodies is evident in the afternoon time (Collar et al. 2001). Daily migration from roosting to foraging grounds (often 7–10 km away) during winter is also recorded in this species (Allport et al. 1991; Collar et al. 2001). These ducks mainly spend their winter season in freshwater or brackish wetlands, floodplains, meadows and rice fields (Fujioka et al. 2010).

15.2.8 Indian Spot-Billed Duck (*Anas poecilorhyncha*)

Two subspecies are present, namely, *Anas poecilorhyncha poecilorhyncha* and *Anas poecilorhyncha haringtoni*. *A. p. poecilorhyncha* is a widespread resident throughout Indian subcontinent, nomadic and occasionally migratory. However, *A. p. haringtoni* (earlier known as Burmese spot-billed duck) is distributed in Assam. Both subspecies show similar foraging habits. This species is chiefly herbivorous and they predominantly consume shoots, corms etc. of aquatic plants, seeds of sedges, grains of wild and cultivated rice etc. (Ali and Ripley 1978). However, they occasionally consume water insects, their larvae, worms and molluscs. Water-snail *Vioipara* was identified from the stomach content of this species (Ali and Ripley 1978). Dabbling (filtering), head-dip and upending are the most preferable foraging techniques in shallow water (Carboneras and Kirwan 2020j). They also use digging or grubbing for foraging in muddy shorelines and grazing on rice fields. Nearly 20% of diurnal time is devoted to foraging (Sharma et al. 2019) and they sometimes exhibit nocturnal foraging (Jiang et al. 2007). They prefer freshwater marshes, lakes, irrigation tanks and pools with emergent vegetation (Grimmett et al. 2011).

15.2.9 Eastern Spot-Billed Duck/Chinese Spot-Billed Duck (*Anas zonorycha*)

This species was formerly treated as conspecific with Indian spot-billed duck. However, unlike the Indian spot-billed duck, this species is a winter visitor and mostly winters in North-East India (Ali and Ripley 1978). Foraging ecology of this species is quite similar to *A. poecilorhyncha*. This species predominantly feeds on seeds and leaves of various plants, including rice. Occasional foraging on animal matters is also documented in this species (Del Hoyo et al. 2020). In shallow water, these ducks use filtering and upending (Shimada et al. 2006). Nocturnal foraging is also well documented in this species (Shimada et al. 2006; Jiang et al. 2007). Wintering birds prefer shallow freshwaters with abundant emergent vegetation (Del Hoyo et al. 2020).

15.2.10 Mallard (*Anas platyrhynchos*)

This species is a winter visitor to India. It breeds in small numbers in Kashmir (Ali and Ripley 1978). This species is an omnivorous and opportunistic feeder. Wintering populations are chiefly vegetarian and they prefer corms, seeds and shoots of Gramineae and other aquatic plants, wild and cultivated rice etc. Jorde et al. (1984) reported that almost 97% of the diet of wintering mallards is plant materials. However, they also consume molluscs, tadpoles, fish spawns, worms etc. to a lesser extent (Ali and Ripley 1978; Drilling et al. 2020). It is noteworthy that during breeding season this species becomes almost entirely carnivorous (Drilling et al. 2020). Commonly use four foraging techniques in shallow water, namely filtering, head-dip, neck-dip and upending (Guillemain et al. 2000a). Besides these, grazing on land and grubbing at the plants are also documented in this species (Johnson and Rohwer 2000). Wintering populations exhibit nocturnal foraging also (Jiang et al. 2007). They prefer freshwater marshes and red-edged lakes as roosting grounds during the daytime and often fly to flooded paddy fields and forage there at night time (Grimmett et al. 2011).

15.2.11 Gadwall (*Mareca strepera*)

This is one of our commonest wintering ducks and is found in large numbers throughout India. This species is predominantly herbivorous and consumes seeds, shoots, tubers of marsh plants and aquatic weeds, and grains of wild and cultivated rice. Very rarely it consumes insects, worms, molluscs etc. (Ali and Ripley 1978; Leschack et al. 2020). Plant makes up 95%–97% of the total diet in wintering grounds (Leberg 2016). This species is a typical surface feeder and feeds mainly by head-dip (Grimmett et al. 2011). However, it also uses upending in shallow waters for foraging and dives rarely for foraging. As compared to other dabbling ducks, gadwall tends to use deep water zones for foraging (Bolduc 2002) and the optimum water depth for foraging in this species is about 18–36 cm (Bolduc and Afton 2004). This species prefers freshwater and coastal marshes and lakes with extensive aquatic and emergent vegetation. Foraging ecology of this species depends on environmental parameters strongly (Mukherjee et al. 2023). Nocturnal foraging is not regular in this species (Lameris et al. 2021).

15.2.12 Falcated Duck (*Mareca falcata*)

This species is a rare but not infrequent winter visitor. This species is also a predominant herbivore and consumes seeds of various types (rice and grains) of grasses and green parts of aquatic vegetation (Carboneras and Kirwan 2020f). They supplement their diet with occasional intake of aquatic invertebrates like insects and soft-shelled molluscs (Johnsgard 1978). This species forages mainly by dabbling and upending in shallow waters (Grimmett et al. 2011) and also grazes on dry, arable lands. This bird forages more actively in the morning session (Abhinav and Dhadwal 2017). These ducks winter on the coast and also in large shallow waterbodies, rice fields and flooded meadows. Throughout the winter months, ducks use paddy fields for foraging during the night (Meng et al. 2019).

15.2.13 Eurasian Wigeon (*Mareca penelope*)

This species is also a widespread winter visitor to India. This species is also chiefly vegetarian and prefers corms, seeds, stems, rhizomes and shoots of grasses, sedges and marsh and aquatic plants and wild and cultivated rice (Ali and Ripley 1978; Carboneras et al. 2020a). They rarely consume water insects and their larvae, molluscs etc. During winter, this species also feeds on droppings of other birds like gulls (Elkins 1996). Ali et al. (2016) recorded that feeding is the major diurnal activity in this winter migrant. A similar observation was also recorded by Saker et al. (2016) as they reported nearly 40% time of their diurnal hours were allocated to feeding by wintering ducks. In winter, they feed both during day and night (Kear 2005). Like other non-diving ducks, this species feeds by walking about and grazing or grubbing on grassy edges of shallow lakes and in wet paddy fields, or by filtering and upending in shallow water (Ali and Ripley 1978; Carboneras et al. 2020a). Open lakes, reservoirs, rivers, pools, marshes, tidal creeks and saltmarshes are favourable roosting and foraging grounds for this species (Grimmett et al. 2011).

15.2.14 Garganey (*Spatula querquedula*)

This species is one of our commonest and most widespread migratory ducks. This species also exhibits predominant herbivory and mainly depends on seeds and tender shoots of marsh plants, grasses and sedges and grains of wild and cultivated rice. However, they also consume some animal matter like water insects, larvae, worms and molluscs. (Carboneras and Kirwan 2020h). Wintering populations have preferences for rice and grass (Kear 2005). Ali and Ripley (1978) identified shells of *Planorbis* sp. and *Melania tuberculata* among stomach contents in India. Dabbling, head-dip, picking and upending are preferable foraging techniques in this species and this species prefers shallow water zones of wetlands for foraging (Carboneras and Kirwan 2020h). These birds prefer wintering in freshwater wetlands and coastal lagoons with emergent vegetation (Grimmett et al. 2011). This species is a dominant nocturnal forager and if not disturbed this species exhibits diurnal foraging to some extent (Kear 2005).

15.2.15 Northern Shoveler (*Spatula clypeata*)

This species is one of the common winter visitors to India. This is a predominant carnivorous species. They consume a wide array of small nektonic animals like crustaceans, molluscs, water insects and larvae, fish spawns and worms. (Dubowy et al. 2020). Their diet also includes some quantity of vegetable matter, shoots and corms of aquatic weeds, seeds etc. (Ali and Ripley 1978). During winter, they also forage in rice fields (Fujioka et al. 2010). This bill morphology is well adapted for straining or filtering food items from surface water (Guillemain et al. 2000b; Kear 2005) and often feeds by sweeping the bill from side to side while swimming (Grimmett et al. 2011). It sometimes uses head-dip to rake up the bottom sediment and also upends in shallow waters. This species is mostly a diurnal feeder, but foraging on full moon nights is also well documented (Dubowy et al. 2020). Khemis et al. (2017) reported that about 27% time of diurnal hours were allocated to foraging in wintering ducks. This species prefers all types of shallow freshwater bodies as roosting and foraging grounds.

15.2.16 Marbled Teal (*Marmaronetta angustirostris*)

This species is chiefly a rare winter visitor to northern parts of the Indian subcontinent. Wintering populations of this species are almost entirely vegetarian and prefer shoots, tubers and seeds of aquatic plants (Ali and Ripley 1978). They rarely consume worms, molluscs, aquatic insects and larvae (Carboneras and Kirwan 2020n). Kear (2005) documented foraging behaviour of this species. It feeds by dabbling at surface waters, upending and diving (rarely) in shallow waters and mud filtering from shore. During winter they forage mainly during night and spend very less time foraging during diurnal hours. This species prefers shallow freshwater lakes and ponds with extensive emergent vegetation as foraging habitats (Grimmett et al. 2011).

15.2.17 Red-Crested Pochard (*Netta rufina*)

This species is a common winter visitor. This species is mostly vegetarian and prefers shoots, buds, rhizomes and seeds of submerged aquatic plants and grasses (like Sagittarius) (Ali and Ripley 1978). However, a small portion of their diet consists of water insects, molluscs, tadpoles and fish (Salvador et al. 2022b). These birds seldom steal food from other waterfowls (like gadwall) and even from fish (Salvador et al. 2022b). Wintering populations allocate about 10.9% to 16.6% of diurnal hours in feeding (Mukherjee et al. 2020). Nocturnal foraging is also recorded in this species (Salvador et al. 2022b). Diving is the most preferred foraging technique in this species followed by beak-dip, head-dip and upending (Mukherjee et al. 2020). Mukherjee et al. (2020) also recorded intersex variation in foraging techniques and also recorded variation in foraging techniques in altered habitat conditions (shallow water). Male ducks often bring food for female ones (Salvador et al. 2022b). This species is considered a link between pochards (Aythyinae) and dabbling ducks (Anatinae); hence, an interesting species from a phylogenetic point of view (Mukherjee et al., 2021). In winter, these birds use an extensive range of open water bodies dominated by submerged macrophytes and also in water bodies with fringing vegetation. However, very rarely they winter in rivers (Grimmett et al. 2011).

15.2.18 Common Pochard (*Aythya ferina*)

This species is also a winter visitor to the Indian subcontinent. This predominant herbivorous species chiefly consumes rhizomes, buds, shoots and seeds of aquatic plants, grasses and sedges (Kear 2005). They often supplement their diet with crustacea, molluscs, water insects and larvae, worms and occasionally tadpoles and small fish (Ali and Ripley 1978; Carboneras et al. 2020b). This species is a predominant diver and time allocated for diving is >80% in this species (Mukherjee et al. 2021a). This species also uses other foraging techniques like upending, neck-dip, head-dip and picking (Mukherjee et al. 2021a). Males are deeper divers than females and often displace females from food-rich zones (Marsden and Sullivan 2000). In winter, this species is crepuscular and often forages throughout the night (Kear 2005). They fly to their feeding grounds at dusk and return to the daytime refuge at dawn. This species prefers lakes and reservoirs with large open water for wintering (Grimmett et al. 2011).

15.2.19 Ferruginous Pochard (*Aythya nyroca*)

This species breeds in Kashmir and Ladakh and winter visitors in the rest of the subcontinent (Ali and Ripley 1978). This species is omnivorous but plant materials predominate the diet (BirdLife International 2022c). Preferred food materials are corms, shoots, leaves and seeds of aquatic plants; crustaceans, molluscs, water insects and larvae, worms, frogs and small fish (Carboneras and Kirwan 2020g).

Like other pochards, diving is the major foraging technique in this species (Draidi et al. 2019). However, it also employs other techniques like head-dip, neck-dip, upending and filtering in shallow waters (Carboneras and Kirwan 2020g). Wintering populations spend less time foraging (nearly 17% of diurnal hours) than resting in the daytime (Muzaffar 2004; Datta 2014). However, Houhamdi and Samraoui (2008) recorded that throughout winter 67% time of the nocturnal time budget is devoted to foraging. Similarly, Datta (2014) recorded that nearly 65% of nighttime budget was expended on feeding. Thus, this species is a predominant nocturnal feeder. These ducks prefer wintering in large lakes, freshwater pools, irrigation tanks, lagoons and coastal marshes with extensive submerged vegetation (Grimmett et al. 2011).

15.2.20 Baer's Pochard (*Aythya baeri*)

This species is an uncommon and erratic winter visitor to India. Very little information on the foraging ecology of this species is available. This is an omnivorous species and prefers both animal and plant materials (Carboneras and Kirwan 2020a). However, during winter aquatic plants and seeds are more preferred (www.planetofbirds.com). Diving is the major foraging technique and one can dive up to 40 s at a time (www.planetofbirds.com). Foraging is higher in the morning time in wintering populations (Chowdhury et al. 2012), and afterwards foraging drops to nearly 7% in the daytime. This species prefers wintering in large rivers, coastal lagoons, estuaries and lakes in small groups (Grimmett et al. 2011).

15.2.21 Tufted Duck (*Aythya fuligula*)

This species is also a common winter visitor to the Indian subcontinent. This species is an omnivore and prefers eating molluscs (www.animaldiversity.org). Besides molluscs they also consume water insects, amphibians and fishes. Among plant materials, they prefer mostly leaves, stems and roots of submerged aquatic plants and occasionally consume seeds (Ali and Ripley 1978; Carboneras and Kirwan 2020p). Watkins (1999) recorded these ducks consuming faeces of other waterfowl. Diving is mostly used foraging technique in this species and they often dive for long times (20 s to 1 min). (Mukherjee et al. 2021a; www.animaldiversity.org). Diving is the major foraging technique employed by this bird and more than 80% of time is allocated to this technique while foraging (Mukherjee et al 2023; Mukherjee et al. 2021a). Other minor techniques like neck-dip, head-dip, beak-dip, upending and picking are also less frequently used by this species for foraging (Mukherjee et al. 2021a). This species forages both day and night and wintering populations often exhibit high diurnal foraging at peak winters (Carboneras and Kirwan 2020p). Deep waters with clear surfaces and submerged vegetation are preferred foraging grounds for this duck (Mukherjee et al., 2021).

15.2.22 Greater Scaup (*Aythya marila*)

This species is a rare vagrant in winter (Ali and Ripley 1978). This omnivorous species prefers bivalves, snails, insects, crustaceans, larvae of insects and fishes along with vegetative parts of plants and seeds (Kessel et al. 2020). Wintering populations prefer eating fish (Clements et al. 2020). Diving is the major foraging technique in this species and this species dives usually in water <2 m deep (Kessel et al. 2020). Besides diving, this species also employs picking from bottom sediments. This species is primarily a diurnal feeder (Kessel et al. 2020). However, wintering populations show high nocturnal foraging (Nilsson 1970). This species winters mainly in brackish lagoons, estuaries, sheltered bays and large freshwater lakes (Snow and Perrins 1998; Grimmett et al. 2011).

15.2.23 Cotton Pygmy-Goose (*Nettapus coromandelianus*)

This species is resident to the Indian subcontinent and can be found throughout the country. This species shows predominant herbivory and consumes shoots, corms, seeds of aquatic plants (like Ipomoea, Hydrilla and Ruppia), grasses and grains of cultivated and wild rice (Mukherjee 1974). Besides these plant materials, they also consume water insects and their larvae, crustaceans, worms, small fishes etc. (Ali and Ripley 1978). They forage mainly by grazing in marshes and wet grasslands (Grimmett et al. 2011). However, in shallow water dabbling, wading and rarely diving are the choicest foraging techniques (Kear 2005; Carboneras and Kirwan 2020e). They forage mainly in the early morning and late at night (Carboneras and Kirwan 2020e). Preferable habitats of this species are well-vegetated freshwater pools, lakes, shallow lagoons, marshes, irrigation tanks, flooded paddy fields, rivers and streams (Wells 1999; Grimmett et al. 2011; Carboneras and Kirwan 2020e).

15.2.24 Knob-Billed Duck/African Comb Duck (*Sarkodiornis melanotos*)

This is a resident species and it migrates locally with water conditions. This species is largely vegetarian and prefers corms, shoots and seeds of aquatic and marsh plants, and grains of wild and cultivated rice (Ali and Ripley 1978). They also consume water insects and their larvae, molluscs, and occasionally frogs and fish (Brown et al. 1982; Safford and Hawkins 2013; Carboneras and Kirwan 2020k). They mainly obtain food by grazing in marshes and wet grasslands. However, in shallow water conditions, they employ dabbling, upending and wading (Safford and Hawkins 2013). Diurnal foraging is more prominent in this species and they use waterbodies for roosting during the night (Saranga et al. 2022). They prefer freshwater swamps, rivers, lakes, rice fields and inundated flood plains as foraging and roosting grounds (Kear 2005; Grimmett et al. 2011).

15.2.25 White-Winged Duck (*Asarcornis scutulata*)

This species resident species found in North-East India. This species is omnivorous and they consume seeds and green parts of aquatic plants (like Hydrilla), grain, rice, molluscs, insects and small vertebrates (fishes and frogs) (Carboneras and Kirwan 2020q). They employ dabbling and head-dip most times to obtain food from shallow waters and this species is an occasional diver (Carboneras and Kirwan 2020q). They are crepuscular in nature and forage actively at dusk and dawn (www.wwfindia.org). Feeding is more intense in the morning than evening (Green 1993). However, they also occasionally feed at night if moonlight is present (Green 1993). This species prefers small stagnant and slow-flowing freshwater wetlands in tropical forests for roosting and foraging (Grimmett et al. 2011). Saikia and Saikia (2011) reported that streams flowing through dense semi-evergreen forests with adequate vegetation cover near forest edges are the most preferred habitat of this species.

15.2.26 Common Goldeneye (*Bucephala clangula*)

This species is an occasional winter visitor and migrates to north India. This species is mainly carnivorous and prefers mainly aquatic invertebrates (insects, molluscs, crustaceans), fish and fish eggs (Ali and Ripley 1978). However, nearly a quarter of their diet consists of plant materials like seeds and tubers of aquatic plants (Bellrose 1980). Diving is the major foraging technique employed by this species (Eadie et al. 2020). Average diving duration is 25 s (Palmer 1976). This species is a predominant diurnal feeder, but some reports of nocturnal feeding are present (Palmer 1976). Shallow water along shorelines with clear surfaces is the preferred microhabitat for foraging for this species (Eadie et al. 2020). This species is very aggressive and dominates all other waterfowl when competing for foraging (www.allaboutbirds.org). This species winters in open waters in freshwater lakes and large rivers (Grimmett et al. 2011).

15.2.27 Smew (*Mergellus albellus*)

This species is an uncommon and sparse but regular winter visitor to northern India. This carnivorous species mainly consumes fish (Ali and Ripley 1978). Besides that, they consume crustaceans, molluscs, aquatic insects and their larvae, worms etc. (Carboneras and Kirwan 2020o) and occasionally consume some vegetable matter (seeds, leaves and roots). This species feeds mostly by diving in shallow water (1–4 m) and duration of diving is usually around 30 s (Kear 2005). This ducks dive in quick succession. This species is a diurnal feeder and usually scans food by head-dip (Grimmett et al. 2011). This species winters in freshwater lakes, rivers and Himalayan streams (Grimmett et al. 2011).

15.2.28 Goosander (*Mergus merganser*)

This species breeds in Ladakh and is chiefly a winter visitor to the Indian subcontinent and this species in the eastern Himalayan foothills. This species consumes almost exclusively fish (up to 125 or 150 mm long), for seizing which the serrated bill is specially adapted (Ali and Ripley 1978). It also consumes aquatic invertebrates (insects, molluscs and crustaceans), frogs, small mammals, birds and plants along with a number of pebbles (Del Hoyo et al. 1992; Pearce et al. 2020). This species consumes food by diving and prefers to feed in water zones with depths <4 m (BirdLife International 2022a). Cooperative hunting is often observed in this species (Ali and Ripley 1978; Zonfrillo 2009). This species is a bimodal diurnal feeder with morning and evening peaks (Sjöberg 1985) and sometimes forages after sunset (Pearce et al. 2020). This species winters in streams, rivers, lakes and very rarely coastal waters (Grimmett et al. 2011).

15.2.29 Red-Breasted Merganser (*Mergus serrator*)

This species is a vagrant winter visitor to winter visitor. Foraging ecology and habitat requirements of this species are quite similar to goosander (Ali and Ripley 1978). Like goosander, this species also feeds mainly on small fishes (Craik et al. 2020). Other foraging resources are crustaceans, insects, amphibians and some floating plant materials (Ali and Ripley 1978; Vandendriessche et al. 2007). Major foraging techniques are cooperative herding, individual search by head-dip, shallow diving and deep diving (Bowles 1980). This species is also a diurnal feeder and foraging is highest in the early morning and evening sessions (Sjöberg 1985). This species winters in large lakes, rivers and coastal waters (Grimmett et al. 2011).

15.2.30 White-Headed Duck (*Oxyura leucocephala*)

This species is a sparse winter visitor to the Indian subcontinent. Wintering populations of this species are predominantly herbivorous (Ali and Ripley 1978). Besides plant materials (corms, seeds and leaves of aquatic plants) this species also consumes insects and their larvae, crustaceans, molluscs (Johnsgard 2010). This species forage mainly by diving (Salvador et al. 2022a) and prefers open water for foraging. This species also employs surface feeding and picking occasionally (Amat and Sanchez 1982). Both diurnal and nocturnal foraging are recorded in this waterfowl (Green and Hughes 1996; Green et al. 1999). However, wintering population forages more actively at night (nearly 90% of nocturnal hours) than daytime (nearly 22% of diurnal hours). They prefer wintering in large fresh waters, lakes and brackish lagoons with extensive submerged aquatic vegetation (Grimmett et al. 2011; Halassi et al. 2016).

15.3 Foraging Ecology of Goose

Four geese namely the bar-headed goose, greylag goose, greater white-fronted goose and lesser white-fronted goose are regular or sparse winter visitors to India. All these species show predominant herbivory and grazing is their preferred foraging technique.

15.3.1 Bar-Headed Goose (*Anser indicus*)

This species is a predominant winter visitor to India. The diet of this species largely consists of grasses. Besides this, they also feed on tubers, tender shoots of wheat, gram, grains, nuts and roots (Ali and Ripley 1978). Vaidyanathan (2022) reported that these birds sporadically hunt for insects, small fishes and crustaceans. Grimmett et al. (2011) reported that these species also consume agricultural crops such as rice, wheat, corn and barley. Major forging technique is grazing both in land and water (Carboneras and Kirwan 2020c). This species is a predominant diurnal feeder (www.animalia.bio). This species prefers sites with high vegetation cover and low vegetation height as their foraging sites. These sites must be close to the water and at lower elevations (Yanfeng et al. 2012).

15.3.2 Greylag Goose (*Anser anser*)

This species is also a winter migrant to the Indian subcontinent. This species is almost exclusively vegetarian. During the winter months, they mainly consume grass and shoots of winter crops such as wheat and gram (to which they may do considerable local damage) and paddy left over in the stubbles after harvesting; also, aquatic weeds and tubers etc. (Ali and Ripley 1978). Sibly (1981) reported that these geese have inefficient digestive systems so they almost exclude animal diet. Like bar-headed goose, they use grazing as a preferable foraging technique. However, while swimming they sometimes use upending and head-dip and probing techniques for foraging (Amat et al. 1991; Carboneras and Kirwan 2020i). Mainly this species is a diurnal feeder and prefers morning and afternoon sessions (Carboneras and Kirwan 2020i). Greylag geese prefer arable lands more frequently as foraging grounds and also winters in swamps, lakes and coastal lagoons (Grimmett et al. 2011; Polakowski and Kasprzykowski 2016).

15.3.3 Greater White-Fronted Goose (*Anser albifrons*)

This species is a sparse and rare winter visitor to the Indian subcontinent. This is herbivorous. Birds feed on grasses and grains (Ali and Ripley 1978). They consume a wide range of plant parts like leaves, roots, tubers, wood, bark, stems, seeds, grains, nuts and fruits by grazing (Ely et al. 2020; www.animaldiversity.org). These ducks are markedly diurnal feeders and very rare nocturnal feeders (Zhong et al. 2022). This species wintering population prefer meadows, paddy fields and mudflats as their foraging grounds (Fan et al. 2020).

15.3.4 Lesser White-Fronted Goose (*Anser erythropus*)

This species is also a rare and sparse winter visitor like greater white-fronted goose (Ali and Ripley 1978). This species is also herbivorous. They feed mainly on grasses, roots, stems, leaves and fruits of aquatic and terrestrial plants along lake shores, rivers, wetlands and marshes (Kear 2005; BirdLife International 2022b), and grazing on land is the most preferred foraging technique (Carboneras and Kirwan 2020m). However, grasses comprise more than half of their diet (Karmiris et al. 2009) and forage diurnally. Kear (2005) also reported that wintering populations supplement their diet by consuming agricultural grains.

15.4 Foraging Ecology of Swans

Three species of swans recorded from the Indian subcontinent and all of them are rare stragglers in winter (Ali and Ripley 1978). Those species are Tundra swan (*Cygnus columbianus*), Mute swan (*Cygnus olor*) and Whooper swan (*Cygnus cygnus*). These three species have similar foraging ecology and all of them are predominantly herbivores (prefer submerged vascular plants and grasses) (Ali and Ripley 1978; Limpert et al. 2020; Ciaranca et al. 2020; Carboneras and Kirwan 2020r). Wintering populations consume agricultural crops. Some animal matter (like molluscs) are also present in their diet (Limpert et al. 2020). Foraging techniques like head-dip, neck-dip, upending, diving and grazing on land are employed for foraging (Ali and Ripley 1978). These birds forage throughout the diurnal hours and sometimes at night during the full moon (Limpert et al. 2020).

15.5 Conclusion

Many migratory and resident birds share wetlands in India during winter months. They use a wide array of foraging strategies for survival. Moreover, migratory waterfowl behave very differently in these wintering grounds from their breeding grounds (Hobson et al. 2014). This chapter enumerated the foraging requirements and foraging strategies of Anatids, which will aid in framing future conservation strategies for protection of wintering grounds.

References

Abhinav C, Dhadwal DS (2017) Falcated duck *Mareca falcata*: an addition to the avifauna of Himachal Pradesh. Indian BIRDS 13(6):154–156

Ali S, Ripley SD (1978) Handbook of the birds of India and Pakistan together with those of Nepal, Sikkim, Bhutan, Bangladesh and SriLanka. In: Divers to hawks, vol 1, 2nd edn. Oxford University Press, Delhi, pp 1–382

Ali E, Ismahan H, Houhamdi M (2016) Time budget patterns and complementary use of a Mediterranean wetland (Tonga, north-East Algeria) by migrant and resident waterbirds. Riv Ital Ornitol Res Ornithol 86:19–28. https://doi.org/10.4081/rio.2016.268

Allport G, Poole C, Park E, Jo S, Eldridge M (1991) The feeding ecology, requirements and distribution of Baikal teal *Anas formosa* in the Republic of Korea. Wild 42:98–107

Amat J, Green A (2010) Waterbirds as bioindicators of environmental conditions. In: Biological monitoring in freshwater habitats. https://doi.org/10.1007/978-1-4020-9278-7_5

Amat AJ, Sanchez A (1982) Biología y ecología de la Malvasia *Oxyura leucocephala* en Andalucía. Doñana Acta Vert 9:251–320

Amat JA, Garcia-Criado B, Garcia-Ciudad A (1991) Food, feeding behaviour and nutritional ecology of wintering Greylag geese Anser anser. Ardea 79(2):271–282

Arzel C, Elmberg J, Guillemain M (2007) A flyway perspective of foraging activity in Eurasian Green-winged teal, Anas crecca crecca. Can J Zool 85:81–91. https://doi.org/10.1139/z06-201

Backues KA (2015) Chapter 16. Anseriformes. In: Miller RE, Fowler ME (eds) Fowler's zoo and wild animal medicine, vol 8. W.B. Saunders, Philadelphia, pp 116–126. https://doi.org/10.1016/B978-1-4557-7397-8.00016-5

Bellrose FC (1980) Ducks, geese, and swans of North America. Revised edition. Stackpole Books, Harrisburg, PA

Bethke RW (1991) Seasonality and interspecific competition in waterfowl guilds-a comment. Ecology 72(3):1155–1158

BirdLife International (2022a) Species factsheet: *Mergus merganser*. Downloaded from http://www.birdlife.org on 13/12/2022. Recommended citation for factsheets for more than one species: BirdLife International (2022) IUCN Red List for birds. Downloaded from http://www.birdlife.org on 13/12/2022

BirdLife International (2022b) Species factsheet: *Anser erythropus*. Downloaded from http://www.birdlife.org on 13/12/2022. Recommended citation for factsheets for more than one species: BirdLife International (2022) IUCN Red List for birds. Downloaded from http://www.birdlife.org on 13/12/2022

BirdLife International (2022c) Species factsheet: *Aythya nyroca*. Downloaded from http://www.birdlife.org on 13/12/2022. Recommended citation for factsheets for more than one species: BirdLife International (2022) IUCN Red List for birds. Downloaded from http://www.birdlife.org on 13/12/2022

BirdLife International (2022d) Species factsheet: *Dendrocygna bicolor*. Downloaded from http://www.birdlife.org on 13/12/2022. Recommended citation for factsheets for more than one species: BirdLife International (2022) IUCN Red List for birds. Downloaded from http://www.birdlife.org on 13/12/2022

BirdLife International (2022e) Species factsheet: *Tadorna ferruginea*. Downloaded from http://www.birdlife.org on 13/12/2022. Recommended citation for factsheets for more than one species: BirdLife International (2022) IUCN Red List for birds. Downloaded from http://www.birdlife.org on 13/12/2022

Bolduc F (2002) Effects of structural marsh management and salinity on sediments, hydrology, invertebrates, and waterbirds in marsh ponds during winter on the gulf coast Chenier plain. PhD Dissertation, vol 23. Louisiana State University, Baton Rouge, p 897

Bolduc F, Afton AD (2004) Relationships between wintering waterbirds and invertebrates, sediments and hydrology of coastal marsh ponds. Waterbirds 27:333–341

Bowles WF Jr (1980) Winter ecology of red-breasted mergansers of the Laguna Madre of Texas. Master's Thesis, Corpus Christi State University, Corpus Christi, TX

Brown LH, Urban EK, Newman K (1982) The birds of Africa, vol 1. Academic Press, London and New York

CAF National Action Plan-India (2018) India's National Action Plan for conservation of migratory birds and their habitats along central Asian flyway (2018–2023). Ministry of Environment, Forest and Climate Change, Government of India, pp 1–8

Carboneras C, Kirwan GM (2020a) Baer's Pochard (*Aythya baeri*), version 1.0. In: del Hoyo J, Elliott A, Sargatal J, Christie DA, de Juana E (eds) Birds of the world. Cornell Lab of Ornithology, Ithaca, NY. https://doi.org/10.2173/bow.baepoc1.01

Carboneras C, Kirwan GM (2020b) Baikal Teal (*Sibirionetta formosa*), version 1.0. In: del Hoyo J, Elliott A, Sargatal J, Christie DA, de Juana E (eds) Birds of the world. Cornell Lab of Ornithology, Ithaca, NY. https://doi.org/10.2173/bow.baitea.01

Carboneras C, Kirwan GM (2020c) Bar-headed goose (*Anser indicus*), version 1.0. In: del Hoyo J, Elliott A, Sargatal J, Christie DA, de Juana E (eds) Birds of the world. Cornell Lab of Ornithology, Ithaca, NY. https://doi.org/10.2173/bow.bahgoo.01

Carboneras C, Kirwan GM (2020d) Common Shelduck (*Tadorna tadorna*), version 1.0. In: del Hoyo J, Elliott A, Sargatal J, Christie DA, de Juana E (eds) Birds of the world. Cornell Lab of Ornithology, Ithaca, NY. https://doi.org/10.2173/bow.comshe.01

Carboneras C, Kirwan GM (2020e) Cotton pygmy-goose (*Nettapus coromandelianus*), version 1.0. In: del Hoyo J, Elliott A, Sargatal J, Christie DA, de Juana E (eds) Birds of the world. Cornell Lab of Ornithology, Ithaca, NY. https://doi.org/10.2173/bow.copgoo1.01

Carboneras C, Kirwan GM (2020f) Falcated Duck (*Mareca falcata*), version 1.0. In: del Hoyo J, Elliott A, Sargatal J, Christie DA, de Juana E (eds) Birds of the world. Cornell Lab of Ornithology, Ithaca, NY. https://doi.org/10.2173/bow.falduc.01

Carboneras C, Kirwan GM (2020g) Ferruginous duck (*Aythya nyroca*), version 1.0. In: del Hoyo J, Elliott A, Sargatal J, Christie DA, de Juana E (eds) Birds of the world. Cornell Lab of Ornithology, Ithaca, NY. https://doi.org/10.2173/bow.ferduc.01

Carboneras C, Kirwan GM (2020h) Garganey (*Spatula querquedula*), version 1.0. In: del Hoyo J, Elliott A, Sargatal J, Christie DA, de Juana E (eds) Birds of the world. Cornell Lab of Ornithology, Ithaca, NY. https://doi.org/10.2173/bow.gargan.01

Carboneras C, Kirwan GM (2020i) Graylag Goose (*Anser anser*), version 1.0. In: del Hoyo J, Elliott A, Sargatal J, Christie DA, de Juana E (eds) Birds of the world. Cornell Lab of Ornithology, Ithaca, NY. https://doi.org/10.2173/bow.gragoo.01

Carboneras C, Kirwan GM (2020j) Indian spot-billed duck (*Anas poecilorhyncha*), version 1.0. In: del Hoyo J, Elliott A, Sargatal J, Christie DA, de Juana E (eds) Birds of the world. Cornell Lab of Ornithology, Ithaca, NY. https://doi.org/10.2173/bow.isbduc1.01

Carboneras C, Kirwan GM (2020k) Knob-billed Duck (*Sarkodiornis melanotos*), version 1.0. In: del Hoyo J, Elliott A, Sargatal J, Christie DA, de Juana E (eds) Birds of the world. Cornell Lab of Ornithology, Ithaca, NY. https://doi.org/10.2173/bow.knbduc1.01

Carboneras C, Kirwan GM (2020l) Lesser whistling-duck (*Dendrocygna javanica*), version 1.0. In: del Hoyo J, Elliott A, Sargatal J, Christie DA, de Juana E (eds) Birds of the world. Cornell Lab of Ornithology, Ithaca, NY. https://doi.org/10.2173/bow.lewduc1.01.4

Carboneras C, Kirwan GM (2020m) Lesser white-fronted goose (*Anser erythropus*), version 1.0. In: del Hoyo J, Elliott A, Sargatal J, Christie DA, de Juana E (eds) Birds of the world. Cornell Lab of Ornithology, Ithaca, NY. https://doi.org/10.2173/bow.lwfgoo.01

Carboneras C, Kirwan GM (2020n) Marbled teal (*Marmaronetta angustirostris*), version 1.0. In: del Hoyo J, Elliott A, Sargatal J, Christie DA, de Juana E (eds) Birds of the world. Cornell Lab of Ornithology, Ithaca, NY. https://doi.org/10.2173/bow.martea1.01

Carboneras C, Kirwan GM (2020o) Smew (*Mergellus albellus*), version 1.0. In: del Hoyo J, Elliott A, Sargatal J, Christie DA, de Juana E (eds) Birds of the world. Cornell Lab of Ornithology, Ithaca, NY. https://doi.org/10.2173/bow.smew.01

Carboneras C, Kirwan GM (2020p) Tufted duck (*Aythya fuligula*), version 1.0. In: del Hoyo J, Elliott A, Sargatal J, Christie DA, de Juana E (eds) Birds of the world. Cornell Lab of Ornithology, Ithaca, NY. https://doi.org/10.2173/bow.tufduc1.01
Carboneras C, Kirwan GM (2020q) White-winged duck (*Asarcornis scutulata*), version 1.0. In: del Hoyo J, Elliott A, Sargatal J, Christie DA, de Juana E (eds) Birds of the world. Cornell Lab of Ornithology, Ithaca, NY. https://doi.org/10.2173/bow.whwduc1.01
Carboneras C, Kirwan GM (2020r) Whooper swan (*Cygnus cygnus*), version 1.0. In: del Hoyo J, Elliott A, Sargatal J, Christie DA, de Juana E (eds) Birds of the world. Cornell Lab of Ornithology, Ithaca, NY. https://doi.org/10.2173/bow.whoswa.01
Carboneras C, Christie DA, Kirwan GM (2020a) Eurasian Wigeon (*Mareca penelope*), version 1.0. In: del Hoyo J, Elliott A, Sargatal J, Christie DA, de Juana E (eds) Birds of the world. Cornell Lab of Ornithology, Ithaca, NY. https://doi.org/10.2173/bow.eurwig.01
Carboneras C, Kirwan GM, Sharpe CJ (2020b) Common Pochard (*Aythya ferina*), version 1.0. In: del Hoyo J, Elliott A, Sargatal J, Christie DA, de Juana E (eds) Birds of the world. Cornell Lab of Ornithology, Ithaca, NY. https://doi.org/10.2173/bow.compoc.01
Casazza ML, Coates PS, Miller MR, Overton CT, Yparraguirre DR (2012) Hunting influences diel patterns in habitat selection by northern pintails *Anas acuta*. Wildl Biol 18:1–13
Chowdhury S, Lees A, Thompson P (2012) Status and distribution of the endangered Baer's Pochard *Aythya baeri* in Bangladesh. Forktail 28:57–61
Ciaranca MA, Allin CC, Jones GS (2020) Mute swan (*Cygnus olor*), version 1.0. In: Billerman SM (ed) Birds of the world. Cornell Lab of Ornithology, Ithaca, NY. https://doi.org/10.2173/bow.mutswa.01
Clark RG, Fleskes JP, Guyn KL, Haukos DA, Austin JE, Miller MR (2020) Northern pintail (*Anas acuta*), version 1.0. In: Billerman SM (ed) Birds of the world. Cornell Lab of Ornithology, Ithaca, NY. https://doi.org/10.2173/bow.norpin.01
Clements S, Dorr B, Davis B, Roy L, Engle C, Hanson-Dorr K, Kelly A (2020) Diets of scaup occupying baitfish and sportfish farms in eastern Arkansas. Food Webs 23:e00141. https://doi.org/10.1016/j.fooweb.2020.e00141
Collar NJ, Andreev AV, Chan S, Crosby MJ, Subramanya S, Tobias JA (eds) (2001) Threatened birds of Asia: the BirdLife international red data book. BirdLife International, Cambridge
Craik S, Pearce J, Titman RD (2020) Red-breasted merganser (*Mergus serrator*), version 1.0. In: Billerman SM (ed) Birds of the world. Cornell Lab of Ornithology, Ithaca, NY. https://doi.org/10.2173/bow.rebmer.01
Dallmeier F (1991) Whistling-ducks as a manageable and sustainable resource in Venezuela: balancing economic costs and benefits. In: Robinson JG, Redford KH (eds) Neotropical wildlife use and conservation. University of Chicago Press, Chicago, IL, pp 266–287
Datta T (2014) Time-activity budgets of wintering ferruginous duck, *Aythya nyroca*, at Gajoldoba wetland, Jalpaiguri, India. Turk J Zool 38:538–543
Del Hoyo J, Elliott A, Sargatal J (1992) Handbook of the birds of the world. In: Ostrich to ducks, vol 1. Lynx Edicions, Barcelona, Spain
Del Hoyo J, Collar N, Kirwan GM (2020) Eastern spot-billed duck (*Anas zonorhyncha*), version 1.0. In: del Hoyo J, Elliott A, Sargatal J, Christie DA, de Juana E (eds) Birds of the world. Cornell Lab of Ornithology, Ithaca, NY. https://doi.org/10.2173/bow.spbduc.01
Dhanjal-Adams KL, Klaassen M, Nicol S, Possingham HP, Chadès I, Fuller RA (2017) Setting conservation priorities for migratory networks under uncertainty. Conserv Biol 31(3):646–656
Draidi K, Bakhouche B, Lahlah N, Djemadi I, Bensouilah M (2019) Diurnal feeding strategies of the ferruginous duck (*Aythya nyroca*) in Lake Tonga (northeastern Algeria). Ornis Hungarica 27:85–98. https://doi.org/10.2478/orhu-2019-0005
Drilling N, Titman RD, McKinney F (2020) Mallard (*Anas platyrhynchos*), version 1.0. In: Billerman SM (ed) Birds of the world. Cornell Lab of Ornithology, Ithaca, NY. https://doi.org/10.2173/bow.mallar3.01
Dubowy PJ, Carboneras C, Kirwan GM (2020) Northern Shoveler (*Spatula clypeata*), version 1.0. In: Billerman SM (ed) Birds of the world. Cornell Lab of Ornithology, Ithaca, NY. https://doi.org/10.2173/bow.norsho.01

Eadie JM, Mallory ML, Lumsden HG (2020) Common goldeneye (*Bucephala clangula*), version 1.0. In: Billerman SM (ed) Birds of the world. Cornell Lab of Ornithology, Ithaca, NY. https://doi.org/10.2173/bow.comgol.01

Elkins N (1996) Eurasian Wigeons feeding on bird droppings. Br Birds 89(10):451

Ely CR, Dzubin AX, Carboneras C, Kirwan GM, Garcia EFG (2020) Greater white-fronted goose (*Anser albifrons*), version 1.0. In: Billerman SM (ed) Birds of the world. Cornell Lab of Ornithology, Ithaca, NY. https://doi.org/10.2173/bow.gwfgoo.01

Euliss NH, Harris SW (1987) Feeding ecology of northern pintails and Green-winged teal wintering in California. J Wildl Manag 51(4):724–732. https://doi.org/10.2307/3801733

Fan Y, Zhou L, Cheng L et al (2020) Foraging behavior of the greater white-fronted goose (*Anser albifrons*) wintering at Shengjin Lake: diet shifts and habitat use. Avian Res 11:3. https://doi.org/10.1186/s40657-020-0189-y

Fujioka M, Lee SD, Kurechi M, Yoshida H (2010) Bird use of rice fields in Korea and Japan. Waterbirds 33 (Special Publication 1):8–29

Green AJ (1993) The biology of the white-winged duck *Cairina scutulata*. Forktail 8:65–82

Green A, Elmberg J (2013) Ecosystem services provided by Waterbirds. Biol Rev Camb Philos Soc 89:105–122. https://doi.org/10.1111/brv.12045

Green AJ, Hughes B (1996) Action plan for the white-headed duck (*Oxyura leucocephala*). In: Heredia B, Rose L, Painter M (eds) Globally threatened birds in Europe: action plans. Council of Europe, and BirdLife International, Strasbourg, pp 119–145

Green AJ, Fox AD, Hughes B, Hilton GM (1999) Time activity budgets and site selection of white-headed ducks *Oxyura leucocephala* at Burdur Lake, Turkey in late winter. Bird Study 46(1):62–73. https://doi.org/10.1080/00063659909461115

Grimmett R, Inskipp C, Inskipp T, Allen R (2011) Birds of the Indian subcontinent. Oxford University Press, London

Guillemain M, Fritz H, Blais S (2000a) Foraging methods can affect patch choice: an experimental study in mallard (*Anas platyrhynchos*). Behav Processes 50:123–129. https://doi.org/10.1016/S0376-6357(00)00095-4

Guillemain M, Fritz H, Guillon N, Guillon N (2000b) Foraging behavior and habitat choice of wintering northern Shoveler in a major wintering quarter in France. Waterbirds 23:353. https://doi.org/10.2307/152217

Halassi I, Elafri A, Belhamra M, Houhamdi M (2016) Repartition et abondance de l'erismature a tete blanche *Oxyura leucocephala* dans les zones humides du nord-est Algerien. Alauda 84(1):23–32

Hobson KA, Wilgenburg SLV, Faaborg J, Toms JD, Rengifo C, Sosa AL, Aubry Y, Aguilar RB (2014) Connecting breeding and wintering grounds of neotropical migrant songbirds using stable hydrogen isotopes: a call for an isotopic atlas of migratory connectivity. J Field Ornithol 85(3):237–257

Hohman WL, Lee SA (2020) Fulvous whistling-duck (*Dendrocygna bicolor*), version 1.0. In: Billerman SM (ed) Birds of the world. Cornell Lab of Ornithology, Ithaca, NY. https://doi.org/10.2173/bow.fuwduc.01

Hotker H (1995) Activity rhythms of Shelducks (*Tadorna tadorna*) and waders (Charadrii) on the North Sea coast. J Ornithol 136(2):105–126

Houhamdi M, Samraoui B (2008) Diurnal and nocturnal behaviour of ferruginous duck *Aythya nyroca* at lac des Oiseaux, Northeast Algeria. Ardeola 55(1):59–69

Islam Z, Rahmani A (2009) Ducks, geese and swans of India: their status and distribution. Oxford University Press, BirdLife International, RSPB, BNHS, Oxford

Jiang S, Ge Z-M, Pei E-L, Xu X-J, Sang L-L, Wang T-H (2007) Waterfowl nocturnal behavior at the artificial wetlands behind the Chongming Dongtan seawall in winter. Chin J Zool 42(6):21–27

Johnsgard P (1978) Ducks, geese and swans of the world. University of Nebraska Press, Lincoln and London

Johnsgard PA (2010) Ducks, geese, and swans of the world. DigitalCommons @University of Nebraska, Lincoln

Johnson W, Rohwer F (2000) Foraging behavior of green-winged teal and mallards on tidal mudflats in Louisiana. Wetlands 20:184–188. https://doi.org/10.1672/0277-5212(2000)020[0184:FBOGWT]2.0.CO;2

Johnson K, Carboneras C, Christie DA, Kirwan GM (2020) Green-winged teal (*Anas crecca*), version 1.0. In: Billerman SM (ed) Birds of the world. Cornell Lab of Ornithology, Ithaca, NY. https://doi.org/10.2173/bow.gnwtea.01

Jorde DG, Krapu GL, Crawford RD, Hay MA (1984) Effect of weather on habitat selection and behavior of mallards wintering in Nebraska. Condor 86:258–265

Karmiris I, Kazantzidis S, Panagiotopoulou M (2009) Note on the diet of Lesser White-fronted Goose. In: Conservation of the Lesser White-fronted Goose on the European migration route-Final Report of the EU LIFE Nature project 2005–2009, pp 68–70. Edition: WWF Finland Report 27 & NOF Raportserrie Report No 1–2009

Kear J (ed) (2005) Ducks, geese and swans. General chapters, and species accounts (*Anhima to Salvadorina*), vol 1. Oxford University Press, Oxford

Kessel B, Rocque DA, Barclay JS (2020) Greater Scaup (*Aythya marila*), version 1.0. In: Billerman SM (ed) Birds of the world. Cornell Lab of Ornithology, Ithaca, NY. https://doi.org/10.2173/bow.gresca.01

Khemis MDEH, Boumaza O, Bensaci E, Amari H, Kheireddine B, Ali E, Hanane S, Bouslama Z, Houhamdi M (2017) Diurnal behavior and pairing chronology of the northern Shoveler wintering in unprotected remnant wetlands of North-Eastern Algeria. Zool Ecol 27:11. https://doi.org/10.1080/21658005.2016.1278321

Kirby J (2010) Review of current knowledge of bird flyways, principal knowledge gaps and conservation priorities. In: Review 2: CMS scientific council: flyway working group reviews. Just Ecology, Berkeley, Gloucestershire, pp 1–137

Lameris TK, Dokter AM, van der Jeugd HP, Bouten W, Koster J, Sand SHH, Westerduin C, Nolet BA (2021) Nocturnal foraging lifts time constraints in winter for migratory geese but hardly speeds up fueling. Behav Ecol 32(3):539–552

Leberg P (2016) 2017 coastal master plan: attachment C3-6—gadwall, *Anas strepera*, habitat suitability index model. Version II. Coastal Protection and Restoration Authority, Baton Rouge, Louisiana, pp 1–18

Leschack CR, McKnight SK, Hepp GR (2020) Gadwall (*Mareca strepera*), version 1.0. In: Billerman SM (ed) Birds of the world. Cornell Lab of Ornithology, Ithaca, NY. https://doi.org/10.2173/bow.gadwal.01

Li ZWD, Bloem A, Delany S, Martakis G, Quintero JO (2009) Status of Waterbirds in Asia—results of the Asian Waterbird census: 1987–2007. Wetlands International, Kuala Lumpur, Malaysia

Limpert RJ, Earnst SL, Carboneras C, Kirwan GM (2020) Tundra swan (*Cygnus columbianus*), version 1.0. In: Billerman SM (ed) Birds of the world. Cornell Lab of Ornithology, Ithaca, NY. https://doi.org/10.2173/bow.tunswa.01

Ma Z, Cai Y, Li B, Chen J (2010) Managing wetland habitats for waterbirds: an international perspective. Wetlands 30(1):15–27

Marsden SJ, Sullivan MS (2000) Intersexual differences in feeding ecology in a male-dominated wintering Pochard *Aythya ferina* population. Ardea 88(1):1–7

Meng F, Li H, Wang X et al (2019) Size matters: wintering ducks stay longer and use fewer habitats on largest Chinese lakes. Avian Res 10:27. https://doi.org/10.1186/s40657-019-0167-4

Mouloud B, Houhamdi M, Samraoui B (2006) Status and diurnal behaviour of the Shelduck *Tadorna tadorna* in the Hauts Plateaux, Northeast Algeria. Wild 56:65–78

Mukherjee AK (1974) Nests and eggs of Indian pygmy-geese. J Bombay Nat Hist Soc 71(2):192–194

Mukherjee A, Pal S, Mukhopadhyay SK (2020) Diurnal time activity budget and foraging techniques of red crested pochard (*Netta rufina*), wintering at wetlands of West Bengal, India. Turk J Zool 44:424. https://doi.org/10.3906/zoo-2003-23

Mukherjee A, Bandopadhyay A, Pal S, Mukhopadhyay SK (2021a) Foraging habitats and foraging techniques of five wintering waterfowl in light of genetic distances. Russ J Ecol 52(6):567–577. https://doi.org/10.1134/S1067413622010088

Mukherjee A, Pal S, Das P, Mukhopadhyay SK (2021b) Mid-winter diversity of waterbirds in West Bengal, India. J Bombay Nat Hist Soc 118:156. https://doi.org/10.17087/jbnhs/2021/v118/147142

Mukherjee A, Pal S, Adhikari S, Mukhopadhyay SK (2022) Physical habitat attributes influence diversity and turnover of waterbirds wintering at wetlands on central Asian and east Asian-Australasian flyways in eastern India. Wetlands 42:50. https://doi.org/10.1007/s13157-022-01559-1

Mukherjee A, Pal S, Das P (2023) Time activity budget and foraging behavior: important determinants of resource sharing and guild structure in wintering waterbirds. Eur J Wildl Res 69:30. https://doi.org/10.1007/s10344-023-01648-4

Muzaffar SB (2004) Diurnal time-activity budgets in wintering ferruginous Pochard *Aythya nyroca* in Tanguar Haor, Bangladesh. Forktail 20:25–27

Nilsson L (1970) Food-seeking activity of south Swedish diving ducks in the non-breeding season. Oikos 21:145–154

Nouidjem Y, Zoubiri A, Merouani S, Fateh M (2021) Ecology and behavioral response of Ruddy Shelduck (*Tadorna ferruginea*) in the wetland complex of Oued Righ valley (Algerian Sahara). Alger J Biosci 2(1):1–5

Nummi P (1993) Food-niche relationships of sympatric mallards and Green-winged teals. Can J Zool 71:49–55

Onwuka M, Rajpar M, Zakaria M (2020) Food variety of lesser whistling duck in Malaysian lakes. Pertanika J Sci Technol 28:1329–1343. https://doi.org/10.47836/pjst.28.4.11

Palmer RS (1976) Handbook of north American birds. Waterfowl. Part 1, vol 2. Yale University Press, New Haven, CT

Parejo M, Gutiérrez J, Navedo J, Soriano-Redondo A, Abad-Gómez J, Villegas A, Corbacho Amado C, Sanchez-Guzmán J, Masero J (2019) Day and night use of habitats by northern pintails during winter in a primary rice-growing region of Iberia. PLoS One 14:e0220400. https://doi.org/10.1371/journal.pone.0220400

Pearce J, Mallory ML, Metz K (2020) Common Merganser (*Mergus merganser*), version 1.0. In: Billerman SM (ed) Birds of the world. Cornell Lab of Ornithology, Ithaca, NY. https://doi.org/10.2173/bow.commer.01

Polakowski M, Kasprzykowski Z (2016) Differences in the use of foraging grounds by Greylag goose Anser Anser and white-fronted goose Anser Albifrons at a spring stopover site. Avian Biol Res 9:265–272. https://doi.org/10.3184/175815516X14739467542441

Pöysa H (1983) Morphology-mediated niche organization in a guild of dabbling ducks. Ornis Scand 14:317–326

Quesnelle PE, Fahrig L, Lindsay KE (2013) Effects of habitat loss, habitat configuration and matrix composition on declining wetland species. Biol Conserv 160:200–208

Safford RJ, Hawkins AFA (eds) (2013) The birds of Africa, vol 8. The Malagasy Region. Christopher Helm, London

Saikia PK, Saikia MK (2011) Present distribution, status, and ecology of white-winged wood duck and hornbills in Nameri National Park, considering the tropical forest disturbances of Assam. Zoos Print 26:1–11

Saker H, Rouaiguia M, Talai-Harbi S, Mouslim B, Bouslama Z, Houhamdi M (2016) Diurnal time budget of the Eurasian Wigeon (*Anas penelope*) at lac des Oiseaux (northeast of Algeria). J Entomol Zool Stud 4(4)

Salvador A, Amat JA (2022) Ruddy Shelduck (*Tadorna ferruginea*), version 3.0. In: Billerman SM, Bridwell MA, Keeney BK (eds) Birds of the world. Cornell Lab of Ornithology, Ithaca, NY. https://doi.org/10.2173/bow.rudshe.03

Salvador A, Amat JA, Green AJ (2022a) White-headed duck (*Oxyura leucocephala*), version 3.0. In: Billerman SM, Keeney BK (eds) Birds of the world. Cornell Lab of Ornithology, Ithaca, NY. https://doi.org/10.2173/bow.whhduc1.03

Salvador A, Amat JA, Özgencil IK (2022b) Red-crested Pochard (*Netta rufina*), version 2.0. In: Billerman SM (ed) Birds of the world. Cornell Lab of Ornithology, Ithaca, NY. https://doi.org/10.2173/bow.recpoc.02

Saranga BGPT, Wijeweera WPSN, Kannan R (2022) An update on the status and distribution of the knob-billed duck Sarkidiornis melanotos in Sri Lanka, with notes on its behaviour. Indian Birds Monogr 5:48–52

Sharma C, Bhardwaj R, Sharma V, Mathur P (2019) Time budget pattern of three Antidae family members (Ruddy Shelduck, Bar-headed goose and Indian spot billed duck) from Foy Sagar wetland Ajmer, Rajasthan. Res J Chem Environ Sci 7(3):59–66

Shimada T, Azuma A, Kumagai T (2006) Eastern spot-billed duck. Bird Res News 3(10):3–5

Sibly RM (1981) Strategies of digestion and defecation. In: Townsendand CR, Calow P (eds) Physiological ecology: an evolutionary approach to resource use. Blackwell Scientific, Oxford, pp 109–139

Sjöberg K (1985) Foraging activity patterns in the goosander (*Mergus merganser*) and the red-breasted merganser (*M. Serrator*) in relation to patterns of activity in their major prey species. Oecologia 67(1):35–39. https://doi.org/10.1007/BF00378448

Snow DW, Perrins CM (eds) (1998) The birds of the Western Palearctic. Concise Edition. Non-passerines, vol 1. Oxford University Press, Oxford

Urfi A (2017) Foraging ecology of aquatic birds: implications for conservation intervention and suggestions for future research. In: Wetland Science, pp 129–141. https://doi.org/10.1007/978-81-322-3715-0_7

Vaidyanathan S (2022) Bar-headed goose: the high-flying migrator. www.natureinfocus.in

Vandendriessche S, Stienen EWM, Vincx M, Degraer S (2007) Seabirds foraging at floating seaweeds in the Northeast Atlantic. Ardea 95(2):289–298

Viain A, Corre F, Delaporte P, Joyeux E, Bocher P (2011) Numbers, diet and feeding methods of common Shelduck *Tadorna tadorna* wintering in the estuarine bays of Aiguillon and Marennes-Oléron, western France. Wild 61:121–141

Watkins G (1999) Moorhen drinking liquid goose faeces. Br Birds 92(4):204

Wells DR (1999) The birds of the Thai-Malay peninsula. Non-passerines, vol 1. Academic Press, London

Yanfeng ZG, Jun W, Zhenji. (2012) Diurnal activity patterns and environmental factors on behaviors of Bar-headed geese Anser indicus wintering at Caohai Lake of Guizhou, China. Acta Ecol Sin 32:7280–7288. https://doi.org/10.5846/stxb201205180746

Zhong Y, Cheng L, Fan Y, Zhou L, Song Y (2022) The foraging window for greater white-fronted geese (*Anser albifrons*) is consistent with the growth stage of *Carex*. Diversity 14(11):943. https://doi.org/10.3390/d14110943

Zonfrillo B (2009) Goosanders taking bread. Br Birds 102(9):509–510

Migration in Tropical Birds

16

S. Sushanth

Abstract

Bird migration is a spectacular behavioural phenomena. Migration represents complex spatio-temporal patterns, and it balances between survival and fitness. Over the past decades, research on migration has made great studies, yet gaps remain. The studies and approaches to understanding migration have undergone transformations and state-of-the-art methods have emerged. World Migratory Day is an annual awareness-raising campaign commemorated every year on 8th October. It is to highlight the importance of conserving migratory birds and their habitats. It is a global effort and an effective tool to create awareness of the threats to migratory birds, their ecological importance and the need for international cooperation. Artificial light is increasing globally and affecting several bird species. Bird Cast introduced 3-day migration forecasts and released a major expansion called Migration Dashboard that permits one to see actual migration patterns.

This data is compiled with real-time radar data to create continental maps and waves of migration; migration is a multidiscipline activity and integrating them is necessary to understand the process and develop effective strategies to conserve migratory birds.

S. Sushanth (✉)
Biopsychology Laboratory, Vijnana Bhavan, University of Mysore, Mysuru, India

S. Sridhara et al. (eds.), *Animal Behavior in the Tropics*,
https://doi.org/10.1007/978-981-97-7324-4_16

Keywords

Birds · Migration · Technologies and conservations

16.1 Introduction

Animal behaviour is an interesting, deep and fascinating topic not just for researchers and scientists but also for the common people. Whatever an animal does is considered animal behaviour like communication, learning, sexuality, foraging, predation, emotions, etc. Understating it is important for both animals and humans. Much research and experiments have been and are being carried out to understand why an organism is doing, what it is doing; from the tiniest insect's pheromones to a gigantic Whales' communication, we have discovered and learnt many things.

Migration is certainly one of the most spectacular and fantastic natural phenomena occurring on planet Earth. It is a characteristic, seasonal and behavioural movement that not only requires a specific route and timing but also remarkable navigational skills and abilities. In fact, Migration is not only limited to one particular clade of organism but is seen across all major groups of animal kingdom that is right from the smallest butterflies to the huge Orcas, and indeed is a very difficult and meticulous process which relies on a lot of different aspects and cues like sun, stars, landmarks, streams, light and many more. Ornithologists across the world use the term bird migration to signify a regular movement of birds each year between two separate breeding and wintering areas. During this process, birds cover hundreds or thousands of kilometres, and some can even cross inhospitable areas such as seas, deserts, high mountain ranges or even a continent.

One of the most commonly used definitions for the term migration was proposed by Kennedy (1966). Migratory behaviour is persistent and straightened-out movement affected by the animal's own locomotory exertions or by its active embarkation upon vehicle. It completely depends on some of the temporary inhibition responses but promotes their eventual disinhibition and recurrence (Adkins 2017).

Historically, migration has been in human interest and has been recorded as much as 3000 years ago by ancient Greek authors including Aristotle and Homer. Information on bird migration is in the Book of Job of the species such as swallows, storks, and turtle doves (Adkins 2017). Nowadays scientific studies have used techniques like bird ringing and satellite tracking using GPS tagging (Adkins 2017).

Migration can either be 'Obligatory' (Individuals must and should migrate and it occurs repeatedly like clockwork) or 'Facultative' (Individuals will be able to choose whether to migrate or not). Within migratory species or within a population, not each and every individual migrates (Newton 2010). When all of the individuals migrate, it is called 'Complete migration', when some of the individuals decide to stay then it is called 'Partial migration' and the difference between migratory and non-migratory individuals is based on age or sex is termed as 'differential migration' (Adkins 2017).

Fig. 16.1 Bar-headed Geese *Anser indicus* in Magdi Lake, Karnataka, India migrated all the way from Mongolia. (Photo: Harsha KM)

Migration is all about distance and based on this, animals are classified as short-distance migrants and long-distance migrants. Why migration takes place? The major triggers for the migration to happen may be food availability, climate, mating, etc. Not everything that moves from one place to another can be considered as migration. Migration should be a process which should occur annually or seasonally such as the bar-headed goose *Anser indicus*, which comes to India all the way from Mongolia from November to April flying above the Himalayas (Graham et al. 2015; Ebird.org) (Fig. 16.1).

16.2 Bird Migration

Bird Migration is one of the most dramatic forms of seasonal behaviour which happens through one of the most unique modes of transportation: 'Flying'. Most of the birds are capable of flying, because, they have developed a very distinguished feature through the course of evolution, that is 'Feathers' (Devaraju et al. 2016). With the help of these feathers and many other adaptations like flight muscles, fat reservoirs, pneumatic bones, etc., birds can fly very long distances to reach their destinations in order to fulfil all their needs. Because of their ability to fly from one place to another, birds are pre-adapted for the process of migration (Newton 2010) as they have some major advantages like mileage and distance coverage. The major advantage of flight is their speed, this form of migratory locomotion is much faster than all other kinds such as walking, running, or swimming. As the result of bird migration, the pattern and distribution of birds are regularly changed both on local and global scales. Migration is not only limited to the birds that can only fly; most species of penguin (*Spheniscidae*) migrate by swimming. These routes taken by the birds can cover more than 1000 km (620 mi). Dusky grouse *Dendragapus obscurus* undertake altitudinal migration almost completely by walking only. It has been recorded that Emus *Dromaius novaehollandiae* in Australia perform long-distance migration by foot during drought season (Adkins 2017).

Approximately 1,800 bird species over 12,800 world's bird species migrate long distances each and every year in response to the seasons and climate (Adkins 2017). Roughly one-fourth of Indian bird species are outsiders, which is about 320 different bird species. About 70 of them are vagrants and many are local migrants, Ones who move and migrate from one place to another within Indian limits (Kothari 2007). Many of these migrations are in the direction north-south, with a greater number of species feeding and breeding in high northern latitudes during summer and later on moving about several 100 km down south for the winter season. Some species even extend this strategy to migrate annually between the Northern and Southern Hemispheres. For example, The Arctic tern *Sterna paradisaea* is famous for migration, flying from its Arctic breeding grounds in North to the Antarctic and back again each year, in which the total distance would be over 19,000 km (12,000 mi), giving it two summers every year. The average distance travelled by an Arctic tern will be the equivalent of going around the planet earth 60 times in its lifetime (Egevang et al. 2010; Adkins 2017).

Using different locations and areas during different times of the year, these migratory birds have been able to successfully use those areas to colonize which offer them favourable conditions and plenty of food resources (Pulido 2007). Birds will avoid many different kinds of uncertainties like harsh temperatures, low food availability, short days, etc. One main factor that favours the evolution of migration is to escape from the intra- and inter-specific competition in their habitats and also to avoid predation and parasites (Alerstam et al. 2003). Even though females of most bird species are smaller than the males, they often tend to be more migratory than the males (Ketterson and Nolan 1983; Berthold 1996, 2001). The cause for this kind of difference in migratory tendencies just between males and females is attributed to the low competitiveness of juvenile birds and females, resulting in low survival on the breeding grounds during the season of non-breeding when the food is low and scarce (Pulido 2007).

Migratory birds are categorized as Winter Visitors such as bar-headed goose *Anser indicus; Summer visitors* such as blue-cheeked bee-eater *Merops persicus;* Passage migrants like Amur falcon *Falco amurensis*, common cuckoo *Cuculus canorus;* and Vagrants like Lesser white-fronted goose *Anser erythropus*, graylag goose *Anser anser*. These categories are based on time of arrival and the type of visit. Most bird populations migrate long distances along a flyway that is they maintain the same particular path every year and will not change the route or destination unless it is really necessary for survival and reproduction.

There are about eight bird migration flyways in the world and they are Black Sea/Mediterranean flyway, East Asian-Australasian flyway, Mississippi Americas flyway, West Asian-East African flyway, Atlantic Americas flyway, Central Asian flyway, East Atlantic flyway and Pacific Americas flyway (MacKinnon et al. 2012).

Fig. 16.2 Black-winged stilt *Himantopus himantopus* flying low above water during migration. (Photo: Harsha KM)

The primary trigger and motivation for bird migration is the food resources and their availability. For example, Siberian crane *Leucogeranus* comes from Siberia to the Tropical region as the food availability will be high in tropics and low in Siberian region during winter. Birds will migrate for plenty of other reasons such as to escape the extreme temperatures, to breed and nest, climate, etc. (Adkins 2017). This is a typical example of leap-frog migration (Adkins 2017). Many species of fully migratory bird species often show leap-frog migration (birds that nest at higher latitudes spend the winter at lower latitudes), and many of them also show alternative chain migration, where populations 'slide' more evenly north and south without reversing the order (Adkins 2017).

Many of the bird species migrate in flocks. For huge number of birds, flying and migrating together in flocks helps to reduce energy consumption and also their distance mileage will substantially increase (Adkins 2017). Birds like geese (*Anser*) fly long distances by making a V-shaped formation, this helps them to conserve their energy when compared to flying alone. It is found in radar studies that red knots *Calidris canutus* and dunlins *Calidris alpina* will fly 5 kmph (3.1 mph) faster when flown in flocks than compared to flown alone (Adkins 2017).

Bird migration by flying occurs at varying altitudes. Each species of bird flies and migrates at a different height and altitude The highest migrating birds as geese and the lowest migrating birds as Petrels. Some species of bird fly at different altitudes at different locations, that is, their flight height and altitude depend on what type of land they are flying above. For example, Seabirds fly very low when they are above water, but they increase their altitude when crossing over land, and the completely reversed pattern is seen when it comes to land birds such as Black -winged stilts (Fig. 16.2).

16.3 Orientation and Navigation Mechanisms

Bird migration is completely based on cues, their adaptations and modifications or bodily changes. This indeed helps them to orient and navigate themselves in the open sky, which is the key aspect required to safely reach desired destination both within and between the habitats and also within certain time. Birds respond to the complex environment and changing climate by positioning themselves correctly moving forward and reaching their destination.

16.4 Navigation

All migratory birds will accomplish their goal of migration by travelling from point A to B, but they do not do this in the same manner. Some birds might just fly non-stop from A to B, some might take one or few stopovers, some birds will migrate during day and some at night, etc. All the findings of the navigational strategies used by them are grouped into three levels of ability (Bingman and Cheng 2005; Ronacher 2008).

- Piloting.
- Compass orientation.
- True navigation.

16.4.1 Piloting

Piloting is the ability to find a goal by means of familiar landmarks (Goodenough et al. 2010), it is basically a navigation using local and known landmarks (Yasukawa and Martínez 2014).

16.4.2 Compass Orientation

The term refers to the ability of an individual to fly and head in the migratory direction without using and landmarks but instead they use Sun, Star, and even magnetic field of earth as compass. This is seen in many different species of birds. This can be explained by simply relocating the bird which uses compass orientation to a far different location. If the bird again flies in the same direction, it indicates compass orientation.

When the immature European starlings *Sturnus vulgaris* were displaced from their starting point for the experimental purpose, they flew in the same direction as their parents flew and even went for same amount of distance (Perdeck 1967) that is they migrated in a parallel path as they were displaced before and, they did not reach the normal destination but travelled in the same direction and same amount of distance as their parents did (Goodenough et al. 2010).

Most of the birds migrating for the first time will reach their destination just like the ones who have been migrating for many years even though they will have no knowledge about where their destination is actually located. They all are being guided by an inherited programme that will let them know to fly in which direction and for what duration. This innate programme is referred to as vector navigation (Berthold 2001; Bingman et al. 2006). Besides their usage as a compass orientation during migration, they can also be used to improve a different type of navigation which is called path integration or dead reckoning (where a bird somehow integrates the migratory flight path information like direction, sound, duration, distance travelled, smell, twists and turns taken or even Earth's magnetic field).

16.4.3 True Navigation

True navigation is the ability of a bird or an animal to develop and establish a reference to its migratory destination. In a normal bird migration, regardless of where they start their migratory journey, they will always end up at their destination and without the use of landmarks (Bingman and Cheng 2005; Ronacher 2008). This conveys that the bird will reach its destination no matter where it starts the journey from, even if the bird is displaced from its starting point, the bird will change its direction while migrating and end up in a normal destination (Goodenough et al. 2010).

Not all birds, but only a small number of them show true navigational ability. One of the best examples of this is the Homing Pigeon *Columba livia* and certain other birds like swallows. Oceanic seabirds too are known to have true navigational ability (Goodenough et al. 2010). An experiment was done by Klaus Schmidt-Koenig and Hans Schlichte in 1972, where they put frosted lenses on the homing pigeons, through which the pigeons were able to see only nearby objects and not the far ones at all. They reach their home as accurately as the normal ones did and even with the accurate orientation. This confirmed their ability to true navigation and they did not require landmarks for their migration. However, when these pigeons reached their general area of the home loft, they were often found unable to find the loft itself so there, maybe, landmarks that are important to pinpoint the exact home loft location but not to determine the direction to the home (Koenig and Schlichte 1972; Goodenough et al. 2010).

16.5 Bird Migratory Cues

There are many different types of cues like visual cue, magnetic cue, electrical cue, etc. Every migratory bird uses some or the other cues. Some might use just a single cue, and some might use multiple. Any given species usually has several navigational mechanisms available, when one type of mechanism is not being able to be used due to some reason, there will always a backup. In a deeper aspect, in order for a navigational system to work properly, more than one or several sensory systems might be involved. At any given time, any given species, while migrating and navigating, usually has several navigational mechanisms available.

16.5.1 Visual Cue

Visual cues are the things that help the bird to orient and navigate itself while migrating. It could range from landmarks like hills, mountains and streams to celestial bodies like the sun, stars and polarized light. Birds use objects as cues which are available during their migration time and period. Diurnal migrating birds use cues like sun and polarized light and nocturnal migrating birds use cues like stars and moon. For example, pelicans, cranes, swallows, etc. are diurnal migrating birds, and nocturnal migrating birds are owls, flycatchers, cuckoos, etc.

Most of the diurnal migrating birds use the Sun as primary cue for migration i.e. the Sun as a celestial compass. They can determine the direction of compass from the sun's position. Due to the rotation of the Earth, it seems as of like the Sun is moving through the open sky, that is at the rate of 15° per hour. The path or the course that the sun seems to be taking and moving through the open sky depends on the observer's latitude and the season, but it is completely predictable, and this is how they use the sun as a compass.

The usage of visual cue as sun for the migratory orientation is highly complex as the sun's position will not be still in one place but will be moving through the sky. The Sun appears to be moving at a rate of 15° per hour (Goodenough et al. 2010). That is why, a bird migrating by flying straight for 1 h towards its destination and also navigating by keeping and maintaining a constant angle between the sun and its flight path would be 15° off of its actual destination. Therefore, some species of birds will take only short trips of migration so that, these kinds of errors due to the sun's motion would become inconsequential (Goodenough et al. 2010). Because these birds will not adjust their flight path and course with the sun. But when the sun is being used as a cue for a longer period of migration, the bird must adjust and compensate for the sun's apparent motion. Bird must calculate and measure the passage of time and also correctly adjust its angle with the sun's position constantly throughout the complete migrating journey. For example, a bird at 9 a.m., wishing to migrate south, might keep the sun at an angle of about 45° towards the left of its body and path.

Even when the sun's location is not visible because of the cloud coverage, these birds have special ability to infer the sun's location from the pattern of scattered polarized light in the sky. This is kind of a pattern created by the interaction between the light emitted by the sun and the atmosphere of the earth (Brines and Gould 1982). Basically, light emitted by the sun is unpolarized (it will be vibrating in all possible planes) but when the sunlight passes through atmosphere it becomes polarized by the molecules and particles of air. The degree and the direction of the polarization depend on the sun's position (Goodenough et al. 2010).

So, a plane is properly defined by three points:

1. The sun (the source of the photon),
2. The zenith (where the photon was scattered) and.
3. The eye of the observer.

The light reaching which will be reaching the animal's eye from the point zenith has a net polarization orthogonal to this exact plane (which is oriented east-west in this example). Moreover, the polarizing degree of the light which will range from 0% to almost 100% in the UV light varies entirely with the light scattering angle. It is absolutely minimum at 0° and 180° and total maximum at 90°. In theory, the consequential bulls-eye pattern of dark and light which is given below would permit the organism to infer the position of the sun even when it is hidden behind thick and dark clouds.

The most unfortunate thing is 'reality is much less wonderful', that is mostly for one thing, a photon most probably scatters more than once while coming in the route towards the observers or bird's eye, particularly at lower elevations; only polarization that happens above the horizon is very sufficiently reliable for the orientation purpose. The atmospheric conditions like humidity, heat, evaporation, and pollution will all increase the probability and chances of the above problem and that is secondary scattering. Organic compounds that are evaporated even tend to rotate the polarization (again this is a problem that is most evident at lower elevations) (Brines and Gould 1982; Gould 1982a, b).

Birds are capable of orienting themselves to the light polarization patterns, and some nocturnal migrating birds (for the reasons of avoiding being predated and for the need to reduce overheating as much as possible, most birds migrate during night) will frequently use the twilight polarization as their initial and primary cue before setting off on the long evening and nocturnal migratory journey (Gould and Gould 2012).

The timing of bird migration seems to be controlled by the bird under the primary influence of the changes in the length of the day and the night (Adkins 2017). Also, the primary physiological cue that is required in order for a bird to migrate is information regarding day and night length. These changes in the duration of day and night are also closely related to changes in the hormones in the body of the migratory bird, which was found during the experiment of migratory restlessness carried out by Johann Friedrich in 1795 (Adkins 2017). Short-distance bird migrants such as Wallcreeper *Tichodroma muraria* may not need this kind of information but instead they use knowledge about local and surrounding environment's weather conditions. But many long-distance migrating birds like merlin *Falco columbarius* appear to be pre-programmed genetically in order to respond to the changing length of day and night (Adkins 2017).

16.5.2 Magnetic Cue

Many birds' migration is mainly dependent on the earth's magnetic field. It helps to find the direction in which they want to fly, in navigation and also in orientation (reviewed in Wiltschko and Wiltschko 2005). Wiltschko and Wiltschko 2007 suggest that non-migratory birds were the first ones to get magnetic compass during the course of evolution. The ability to use Earth's magnetic field as a compass has its advantages. Unlike celestial cues, it is constant year-round, day and night. The Earth's magnetic field strength varies from region to region, that is on a global scale it will be the weakest at the equator and strongest at the poles and on a regional scale it depends upon the sediment thickness, bedrock of that region, etc.

Birds are capable of sensing earth's magnetic field in multiple ways. One of the ways of detecting magnetic field involves particular type of cells present on the upper layers of bill having magnetite (Lovette and Fitzpatrick 2016). Another way of sensing magnetic field is that some migratory birds have special proteins in their eyes which will help and allow them to detect the earth's magnetic field. Proteins are called cryptochromes, and they are CRY1, CRY2, and CRY4. They are associated with their bodies' internal circadian rhythm. There happens a paramagnetic interaction between the light and the cryptochromes present in the eye (Gould and Gould 2012). Cryptochromes will absorb the blue and green light and it requires a minimum amount of light (somewhat about at least quarter of the moon visible in the clear night sky). But under red light, these birds will be disoriented (Yasukawa and Martínez 2014).

16.5.3 Other Cues

Birds use their sense of smell (Lovette and Fitzpatrick 2016) to navigate and migrate but there is a debate and some doubt about this. The effective use of odours of its colony by tube-nosed seabird to reach back to the colony and also to their home burrows. Other birds like vultures have a keen sense toward smell and odours and will use the olfactory information which will help them to find their way (Lovette and Fitzpatrick 2016).

Another debatable cue which is used by some birds is through the detection of sound. Birds like domestic pigeon *Columba livia* can detect sound waves of frequencies which are as low as 0.05 Hertz. This frequency is not at all in the hearing range of humans. These kinds of low-frequency sound waves travel extremely long distances, it carries critical information like flight direction, presence or absence of mountains, bad weather conditions, etc. which are required in order to migrate (Lovette and Fitzpatrick 2016). According to Hagstrum 2000, a study which was carried out on homing pigeons showed that pigeons using their ability to detect the sound cue can make them fatally disoriented by the large supersonic aircraft's powerful sonic booms.

16.5.4 Migratory Flight

The major advantage of this unique mode of locomotion is its speed, which is much faster than any other mode of locomotion. One more positive note is that flying allows them to cross the most inhospitable areas which would be nearly impossible to cross by other modes of locomotion like oceans of terrestrial and land birds. All the birds will fly either by flapping or gliding through the air covering longer distances in a short period of time. Nevertheless, the migratory journey which they take and go through will depend on body size, mass, wing shape, muscle power, and other aspects.

The speed of a bird's flight can be measured by a flying device travelling next to it or by radar. Radar gives us the most accurate measurements and values and can be directly applied to birds during migration. In addition to values that get measured, flight speed can be calculated theoretically from the aerodynamics principles on the basis of the birds' body mass, wingspan, and the total wing area (Pennycuick 1969, 2008).

The relation between body weight and speed of the flight arises because each unit of the birds' wings' area will be carrying more amount of weight in larger birds. Since the ratio of surface area to volume is less. That is why, the air will flow faster under the wing by providing the necessary lift for the bird, which will make the larger birds fly faster. For example, a 10 g bird will cruise in flapping flight at about 32 km per hour, 20 g bird at about 40 km per hour, 100 g bird at around 43 km per hour, and 1000 g bird at around 54 km per hour (Bruderer and Boldt 2001). Bird wings will vary between short and broad or long and thin, based on their needs for their day-to-day activity (Rayner 1990). In addition to this, a bird can vary speed of its flight and frequency of its wingbeat based on the situation and some can even switch their type of flight like flapping and gliding. The final result is that some birds' species can change and alter their speed of flight in mid-air by more than threefold (Bruderer and Boldt 2001) and wind will add some extra variations.

All the birds flying and migrating will have a particular speed at which the requirement for flapping its wings will be minimal, and flying even a bit slower or faster than that will be costly for birds (Pennycuick 1969, 1975). This might explain why birds that are on migration usually do not fly at their maximum speed which they are capable of, because flying at full speed needs more energy. Birds are usually more likely to fly at a 'minimal power speed'. This gives them some extra mileage in their flight distance or at a somewhat faster 'maximum range speed,' which gives them the longest mileage.

Birds that are migrating are known to fly by using two types of flight principles, they are flapping flight and soaring flight (Sapir et al. 2011). Flight progress is made through continuous self-propulsion using wings during the powered flapping flight (Pennycuick 1969). During the process of soaring/gliding, the birds' wings will be held in a still position which will be outstretched (Fig. 16.3) and the movement headway is attained by using the energy accessible in the environment (Pennycuick 1972). Soaring particularly has two different phases that are alternating, ascent soaring and descent soaring. Ascent soaring is attained by a bird circling in the convective updraughts in which they also gain some potential energy and descent soaring is achieved by bird using its potential energy to move in an onward direction by dropping in the air (Sapir et al. 2011). The above-mentioned flight modes are dependent on the size of the bird and the meteorological and environmental conditions experienced while flying (Sapir et al. 2011).

Fig. 16.3 Amur Falcon *Falco amurensis* migrating by Soaring flight. (© Vijayalakshmi Rao)

16.5.5 Power Requirement for Migration

Power is the basic thing which is needed for flapping their wings in order to fly and migrate. Power requirement for a bird to fly is directly and indirectly related to aerodynamics, wing area, body mass, etc. Using aerodynamics models, researchers have calculated the relation between power requirement and the speed of the flight for birds over a wide range of varied body weights. The power (P) required for a bird to flap its wings in flight at maximum range speed increases with body mass (M) roughly according to the proportion $P = M^{1.17}$ (Pennycuick 1975). However, as larger will have lower metabolic rates (in proportion to mass 0.75), flying at maximum range speed will be significantly more costly for larger birds than smaller ones. The power needed by a bird to flap its wings in order to fly is usually expressed as a function of its basal metabolic rate (BMR), which is the lowest rate of energy and power consumption that an inactive bird can normally attain.

16.5.6 Adaptations

Adaptiveness of a bird is the sense of probability heightening of its survival or reproduction for a particular habit of migration that is inferred from the costs and benefits of a certain feature and characteristics of wing and from the correlated migratory evolution with that of wings and feathers (Bowlin et al. 2010; Rappole 2013). It is one of the major factors which help birds to migrate and cover such long distances during their migratory journeys. A full range of body variations and adaptations can be seen and witnessed among different populations of same species or even in the same population of same species and also in different species.

There are some general features and adaptations that are common in all migratory birds like for example deposition of fat and energy reserves or moulting—and these features are seen in most of the migratory bird species. These features will help us to define and understand migration and the migratory individuals (Dingle 2006; Dingle and Drake 2007). Yet none of these features and adaptations are unique to migratory birds and also present in non-migratory birds (Piersma et al. 2005).

Starting internally, they would need to completely alter their metabolism which will help them to meet all the demands of migration. They also would be controlling and changing their sleeping patterns in nocturnal migrants, and for this, they require special physiological adaptation. The physiological control of a bird's migration is very less understood (Holberton and Dufty 2005) But one of the most remarkable and fascinating structural differences internally that is seen in migratory birds when compared to non-migratory ones is the development of their hippocampus, the part of their brain which is associated with the activity of spatial memory. This is evident in the migratory populations and non-migratory populations of the same species (Cristol et al. 2003). In a study where they compared the plasma levels of corticosterone in two groups of gray catbirds *Dumetella carolinensis,* ones which were captured during the moult and the ones that were getting prepared for the migratory flight - The results showed that the birds that were in preparation for migratory flight had higher baseline amounts of corticosterone than the ones which were captured during the moult (Holberton et al. 1996).

Majority of the birds start to prepare for their migration by becoming hyperphagic, this is a condition in which due to the altered and variations in hormone levels, they start consuming lots of food to store more amount of energy through the accumulation of fat. Because fat is an ideal fuel for migratory birds, by storing them conveniently under the skin and in many cavities and several organs efficiently they look bit bigger and puffier. This excess amount of body fat will be used as an energy source while they are migrating (Rappole 2013).

Birds which double their lean body weight to migrate, should theoretically increase their maximum flight speed at the beginning of migration by 1.4 times and to attain this, their power output by flight muscles should increase by 2.8 times (Pennycuick 1969). During migration, the birds' body mass decreases gradually through the consumption of fuel, and the birds' cruising speed and power output are expected to gradually decrease so as to attain maximum possible range of flight. This reduction of flight speed during migration has been confirmed in radio-tracked Brent geese (Green and Alerstam 2000).

They will start to fly in flocks. This will help them to reduce energy excretion while migrating. This will also help them from the risk of predation and also increase the chances of finding a food source. The classic example of this is ducks and geese. Several studies have shown that these birds will conserve about 12–20% of their energy while flying in flocks in V formation than if flying alone (Fig. 16.4).

Fig. 16.4 Northern pintails *Anas acuta* flying in a line forming a V-shape. (Photo: Shivaprakash Adavanne)

16.5.7 Extremities

Considering in global scale, newly fledged young of many seabird species just directly launch themselves alone onto a long migratory journey of thousands of kilometres across open water and end up in a relatively precise target—a small set of islands in the South Pacific, for instance—regardless of wind and weather (Gill et al. 2005; McCaffery 2008).

There are many birds in tropics region which do extreme things in terms of migration like garganey *Spatula querquedula* migrates between India and Europe which is about several thousand kilometres one-way. A garganey *Spatula querquedula* had even surpassed the migrating distance of 7,700 km which was between Nigeria and India. The average speed at which these ducks' cruises is between 65 and 80 kmph or 40 to 50 mph, it can go even up to 95 kmph or 55 to 60 mph. Once a white stork *Ciconia ciconia* was captured in Tamilnadu which was ringed in Germany and it had migrated a distance of approximately 7000 km (Kothari 2007), bar-headed goose *Anser indicus* flying above the Himalayas which is about 14,000 ft above sea level where the oxygen will be very thin, bar-tailed godwit, Eurasian woodcock *Scolopax rusticola* travel about 2400 km or 1500 miles in just a single flight from Himalayas to the hills of western ghats and South India (Graham et al. 2015). The record so far seems to be held by geese, who were observed through a telescope from Dehradun, flying at a mind-boggling height of 8,830 m (29,500 ft), even higher than Mount Everest (Kothari 2007).

16.5.8 Longest Distance Migration

No one ever forgets to mention this record-breaking bird when they are talking about long-distance migration. Arctic tern *Sterna paradisaea* is the one which performs by far the longest migration of all the birds in the world. They migrate between Arctic and Antarctic waters every year (Greenland of North to Weddell Sea of South), which covers an average round trip distance of more than 90,000 km or 55,923 miles (they are somewhat a passage migrant for the whole Tropical region). They do this kind of extreme migration to follow the summer sun and daylight. They Prefer to glide in the air for most duration of their migratory journey. They are so lightweight so much so that, they let the ocean breeze carry them for great distances which they migrate without having to use a lot of energy by flapping its wings. They can also sleep and eat while they are gliding and migrating. By doing so, they could fly 1,000 miles a day if they did not need to fuel up in between (Newton 2008; Adkins 2017).

Arctic terns *Sterna paradisaea* are the ones which experience the highest amount of daylight per year than any other living organism in this whole world, as they shuttle and travel between the poles by experiencing continuous exposure to daylight of both the Northern and Southern Hemispheres. These birds remarkably can live to be more than 30 years, which means that, if we add up the complete distance travelled by them in their lifetime, their total journey will be equivalent to going to the moon and back, more than three times.

16.5.9 Longest Non-Stop Flight Migration

The most impressive of all non-stop migratory birds is undertaken by the Bar-tailed godwits *Limosa lapponica* (Fig. 16.5). They are also the best land bird to migrate over water. The one which broke all the previous records by shattering the limits of long-distance non-stop migratory flight is known by its tag number that is "234684", which flew for the distance of 13,560 km which is 8,435 miles over the duration of 11 days and 1 h (started on October 13th, 2022). It was from Alaska to the Tasmania, Australia. This was a complete non-stop migration, which means flying for 11+ days without stopping for food or rest. This broke the world record for the longest non-stop migration by a bird.

Birds generally, can change and alter their body and metabolism drastically by shrinking their internal organs size, adding lots of extra weight and burning through their fat reserves, barely sleeping and many more. But, to understand how bar-tailed godwits particularly manage their magnificent improbable journey, we have to first understand their energy consumption, reservoir, and flight costs.

Fig. 16.5 Bar-tailed godwit *Limosa lapponica*. (© Vijayalakshmi Rao)

16.5.10 High-Altitude Migration

This is kind of a migration that requires whole new different levels of adaptations. Bar-headed geese (Fig. 16.6) fly at extremely high altitudes where it is completely challenging to lowland humans and other animals. Naturalists' first sighted bar-headed geese migrating over the Himalayas, (which is the highest mountain range in the entire world) was a fascinating thing for everyone. They are usually found anywhere from Mongolia to the Tibetan plateau during the summer season where they breed and raise their offspring and migrate to Indian subcontinent during autumn.

The main problems of flying and migrating at high altitudes are like cost of climbing to that high level altitudes (which is greatest in large birds), thin air, extreme cold, low oxygen levels on high mountain ranges, and strong winds funnelling through the valleys. These are the reasons why many birds take a migratory route which would be around the mountain areas rather than a more direct route from above the mountain (Bruderer 1999) like for example, rosy starling *Pastor roseus*, which comes from the same location as bar-headed geese but will not fly above the Himalayas but instead fly around. Migrants that have been captured and studied in high altitudes often have more amount of fat and longer wingspan than the birds of the same species that are captured and studied from nearby lowland areas. Usually, most migratory birds reach an altitude of 5,000 to 6,000 m during their migration, but the bar-headed goose flies above the Himalayan mountains at an altitude of up to 9,000 m (Scott and Milsom 2007). There, at that height, the pO_2 level is only about one-third of the pO_2 of sea level. The levels of hypoxia at these high elevations, (even the lowest at which Bar-headed Geese cross the mountains)

Fig. 16.6 Bar-headed goose *Anser indicus*. (Photo: Sushanth S)

are very much sufficient to reduce the maximal O_2 uptake rates in humans substantially. Therefore, bar-headed goose should sustain high rates of O_2 consumption which is needed for flapping its wings. As the temperature there will be very low, they require extra metabolic energy for thermogenesis. Water balance in their body is also very challenging as the water tends to freeze at lower temperatures and also the present at those altitudes will be extremely dry. Crossing the Himalayas also requires them to ascend for many hours continuously, sustaining the longest sustained rates of climbing flight ever recorded to date and possibly even into headwinds (Graham et al. 2015).

16.5.11 Vagrancy

Vagrancy is any species (not only birds) that turns up at a place rarely. The individual is termed as vagrant/vagrants. Some of the vagrants are the ones who will be escaped cage birds and some of them are free-living birds that would have arrived at a place because for many different reasons such as flying in the wrong direction during their migration or hitching a ride on a ship along the sea. In most of the cases vagrants will not belong to a species that are normally kept as pets in captivity. Vagrancy is a common phenomenon that happens which are encountered and recorded all over the world in all kinds of living biological species. Almost all the vagrants that will be recorded are from a migratory population and most of the time in their first year after birth.

There can be apparently lots of reasons for vagrancy to occur like migrants disperse normally but they end up flying unusually long distances during their migration; flying in the wrong direction during migration; Migrants being blown off by the course of wind; Migration overshoots, flying in the proper direction but well beyond their actual destination; Mirror-image migration, where birds migrate at a mirror angle to their normal migrating direction; Reverse-direction migration, where birds migrate in the opposite direction to their normal migratory direction.

The probability of arrival of a vagrant is in such a way that they are expected to see more in the islands that are close to continents than that in remote islands. Even the season and weather also involve a lot, as some vagrant species of birds arrive in numbers every year and some are very rare like once in decades, contradictorily, some birds occur both in autumn and spring. According to Robbins (1980) and McLaren et al. (2006), body size of birds is certainly relevant to the vagrancy, not only because longer sustainability of larger birds even with scarcity of food than the smaller ones, but also because they are more likely to survive a tough journey and are more conspicuous after their arrival as a vagrant. It is harder to identify pelagic seabirds because they majorly depend on birds that are being sighted and recorded from land or from the dead ones which will be washed up on the shoreline (as they sometimes end up in storm accidentally). Larger-sized vagrants such as ducks and geese recordings are way more than the smaller-sized ones like warblers (Figs. 16.7 and 16.8).

Fig. 16.7 Lesser white-fronted goose *Anser erythropus* (to the left) and Greater white-fronted goose *Anser albifrons* (to the right) both in a single frame. One individual from both the species as Vagrants at Hadinaru Lake, India for the first time ever in 2023. (Photo: Arjun Shridhar)

Fig. 16.8 Great white pelican *Pelecanus onocrotalus* recorded in Kukkarhalli Lake, Karnataka, India in 2012. (Photo: Saptha Girish)

16.6 Changes in Migratory Behaviour

Bird migration is one such kind of study in which, one directly visualizes if there are any changes that are going on in any of its aspects like its pattern, location, etc. This is also a reason why bird migratory behavioural studies have attracted so many people. Recently, there are increasingly many changes in bird migration and migratory behaviour in relation to climate change and other different human activities and impacts. As birds and their migration have been recorded and studied for so many years, birds have become one of the best indicators of environmental change. Migratory bird species are the ones highly vulnerable and susceptible to the effects of environmental changes than non-migratory ones because migratory birds' route is highly influenced by the climatic conditions of not just one location such as for wintering, migration, and breeding.

There are some changes in the habits of birds' migration that are significantly evident not only in the tropics but all over the world. Almost all the changes happening to the migratory habits of birds are linked to the changes that are happening in their food availability, erratic changes in climatic conditions, etc.

Birds are evidently adjusting their migratory behaviour and habits without the need for altering or modifying the genetic control mechanism. However, every part of migration process indeed has a genetic component to it, as a result, some behavioural changes happening in migration may reflect genetic changes which are induced by natural selection, and it is very hard to visualize the genetic changes that are happening in the wild. So, the main impactful changes that are happening are:

- Changing from migratory bird to sedentary bird.
- Changing from a sedentary bird to a migratory bird.
- Altering migrating routes.
- Changes in migrating direction.
- Changes in migration timing.

16.7 Declining Migrants

People have always looked forward to the return of all kinds of bird migrants like ducks, geese, flycatchers, cuckoos, swallows, etc. According to the recent observation, there is a gradual decline in bird migrants. This is not a thing that is only occurring in tropics but happening throughout the world for various reasons such as habitat destruction, forest fragmentation, change in food availability, urbanization of forest grasslands and agricultural lands and change in land usage.

16.8 Genetics of Migration in Birds

The study of the genetical aspect of bird migration is challenging because of the level of constraints it has. Basically, migratory behaviour itself is a complex process and is difficult to properly define and manipulate. Migratory species are considered non-model organisms, and there is a lacuna of genetic tools of it. To identify the genes involved and understand the variability of migratory genetic architecture to get to know about evolutionary aspects, the important things are: (1) To define, distinguish, describe, explain, and quantify, the traits of the phenotypes. (2) Disentangle genetic and plastic effects underlying the process of migration (3) To understand the consequences of fitness associated with the phenotypes of migration (Liedvogel et al. 2011).

If a bird has evolved any adaptive traits by natural selection, at least once there existed or there presently exists a genetic variation between and within the individuals. Genetic variability is widely present in natural and wild populations than the domestic ones. Studies on migratory traits that are inherited have revealed that there is a considerable amount of additive genetic variation for single traits, but migration requires group of traits. The complex phenomenon of migration includes group of morphological, sensory, and physiological traits that are inherited as the components of "Migratory gene package" (which includes information about the directionality of migration, physiological adaptations that are necessary for the successful migratory journey like fat reserves, timing, moulting, etc.).

Most of the genetic understanding of migratory traits is based on the cross-breeding and selection experiments carried out on blackcaps, and the results were that the migratory birds have a considerable amount of genetic variation and that they are "linked" or "correlated genetically" and they are inherited components as migratory gene package. This indicated that there can be change in the suite of traits by selection and this change can happen very rapidly (Pulido and Berthold 1998, 2003, 2010).

Genetical studies on bird migration were carried out for the first time by Margaret Nice 1933, 1937. It was on Song Sparrows *Melospiza melodia*, about the inheritance of migratoriness and sedentariness. The most striking examples of genetic variation study within natural populations come from the work of Berthold. He and his group studied migratory behaviour in some of the European warblers and succeeded in breeding Eurasian blackcaps (*Sylvia atricapilla*) easily in captivity when compared with others. In 2003, he studied the phenomenon of migratory restlessness in different populations of blackcaps. These birds breed during summer in the Canary Islands of Europe and fly in winter to Africa. They hand-reared nestling blackcaps from four different populations where each one from a different area (Canary Islands, Southern France, Southern Germany, and Southern Finland) and they were kept in aviaries away from the adults. After some time, the birds' natural migration time came which is autumn, in which they would migrate to Africa. During that time the birds showed migratory restlessness, fluttering, and jumping all around the aviaries. There was also a correlation between how far the birds would have travelled in their migration and the degree of their migratory restlessness (intensity and duration of migratory restlessness). Birds from the Canary Islands are close to Africa and birds from Finland and Germany are far and also would have to migrate more, and in hand-reared birds, ones which were from Canary Island showed less migratory restlessness than the birds from Finland and Germany. The genetic basis for these differences in the migratory restlessness was clinched by cross-breeding the birds of different populations. Those Hybrids showed intermediate levels of migratory restlessncss than the parents.

16.8.1 Genetics of Migration Dispersal

Dispersal is basically leaving from one place to reach another. It is mainly reported as a process with low or no heritability. The decision of dispersal is mainly triggered by the surrounding environmental and social factors rather than the genetic ones but in some birds, there is a report of high heritability (Hansson et al. 2003; Pasinelli et al. 2004). But it appears to be that natal dispersal is linked with exploratory behaviour of birds, which has been shown to have some kind of genetic basis in great tits *Parus major* (Dingemanse et al. 2003). Even, there have been some findings in which there is an association between the polymorphism and exploratory behaviour in the gene encoding dopamine receptor D4 (DRD4, GenBank accession no: DQ006801.1, DQ006802.1) of at least one great tit *Parus major* Population (Fidler et al. 2007; Korsten et al. 2010). Recently, the DRD4 genes of blackcaps (HM196845) were tested and found no association with either the distance of migration or restlessness (Mueller et al. 2011). So, the question of the presence or absence of heritability and to what extent the migration is governed by genes is yet to be answered and needs more research in that aspect.

16.8.2 Genetics of Migration Propensity

This is basically about birds considering that, "Should I stay, or should I migrate?". Recent research on blackcaps examined polymorphism in six candidate genes which were identified previously for the traits of behaviour that might be linked with migration and found a significant correlation in the allelic variation in the locus (ADCYAP1 gene) of one candidate; Higher migration is seen in the ones with longer alleles (Mueller et al. 2011). This association was not only seen within the population but also across the population. Further research is needed for the confirmation of the same in other passerines.

16.8.3 Genetics in Migration Directionality

Directionality has a major role in bird migration. It could be fatal for a bird if it migrates in the wrong direction. Directionality and dispersal of migration are totally contrasting. The present knowledge that we have on the genetical basis of directional information of migration is by conducting two types of experiments; they are cross-breeding (Berthold 1992; Helbig 1994) and displacement experiments (Perdeck 1958; Chernetsov et al. 2008).

The results of blackcaps cross-breeding experiment suggested that they are inherited with migratory directionality as a single component of the migratory gene package (Berthold 1999). However, these genetically determined migration directionality is highly vulnerable towards the tiny evolutionary changes (i.e., microevolution) that often happen at the population level (Berthold 1992). According to Rolshausen et al. 2009, (an experiment which was done on blackcaps) there is an association between the morphological changes like wing and beak shape and new migratory routes of the birds.

16.8.4 Genetics of Migration Timing

This is basically the inherited package that will prepare the bird with necessary information about the timing of the migratory departure, stopover timings, etc. (Åkesson and Hedenström 2007; Pulido et al. 2001). The gene responsible for the timings of the activity is *Clock*, which is a central and molecularly regarded component of the animal's circadian clock (O'Malley and Banks 2008; O'Malley et al. 2010).

In a study carried out on blue tits *Cyanistes caeruleus,* conveyed a latitudinal cline in the variability of the *Clock* gene AY376853. There is a chance that this might manifest a generalized adaptation in birds towards the ecological factors that are correlated with latitude (Johnsen et al. 2007). The result from this study also showcases a possible connection between the bird's migration timing and breeding in reproductive isolation between sympatric populations (Friesen et al. 2007). The complete inherited traits of migration can be called a 'magic trait.' To accomplish and achieve in this field, there is a need to conduct research that has next-generation sequencing tools with advanced technology to quantify the phenotype of bird migration.

16.9 Evolution of Migration

Migration is an adaptive response to the seasonal changes and environment, that allows the animal to migrate to take complete advantage of the spatial variation occurring in the seasonal fluctuations of the resources needed (Gauthreaux 1982; Rappole 1995; Berthold 1996). Evolution of any adaptive trait by the means of natural selection conveys that there exists or at least there has been once existed in the past—genetic variation between the individuals. Bird migration is highly liable, developed and evolved independently in many avian lineages.

Although, bird migration requires a lot of adaptations in morphology, behaviour, and physiology, it would be not so tough to see how this migration process has evolved, because the intermediate stages or transitional stages still occur. For example, in a bird population of the same species, some birds don't migrate at all, some birds do short-distance migration, some do short-distance migration and yet cover long distances, and some more do long-distance migration.

Migratory behaviour has been studied in an evolutionary and phylogenetic context (Outlaw et al. 2003; Outlaw and Voelker 2006; Winger et al. 2012; Zink 2002, 2011). For example, Outlaw and Voelker (2006) examined the migratory behaviour in the motacillidae family of birds, like White wagtail *Motacilla alba*, golden pipit *Tmetothylacus tenellus,* and yellow-throated longclaw *Macronyx croceus*. One of the hypotheses that they were about to examine was the "Evolutionary precursor" model of migration, which said that migration is more likely to occur in those species of birds that live in edge or open habitats (non-buffered areas) rather than the ones which live in forests (buffered area; Chesser and Levey 1998; Levey and Stiles 1992). The logic that goes here is that, the edge and open habitat show greater seasonal variation in terms of availability of food resources than that of forest habitats and this variation might make the bird species that are living in edge and open habitats select migration. Their analysis did, however, find one ecological variable associated with the process of migration and that is species living in higher altitudes are more likely to migrate than species living in lower altitudes.

Later, using a set of different methods and techniques, Peter Berthold compared 18 different bird species' migratory activity in both laboratory and wild ones (Berthold 1990) where he found a strong correlation in the migratory activity of both groups of birds which suggested that the timing mechanism of migration is under genetical control. As this just was not satisfactory, Pulido and his team (which included Berthold) did so by designing a new selection experiment on autumn migration behaviour by choosing German blackcap bird *Sylvia atricapilla* (Pulido et al. 2001). What they did was, they bought blackcaps from the field and measured their restlessness during the autumn migratory season. The caged birds showed extreme and intense activity and movement during migratory restlessness like wing flapping and moving all around the cage. Again in 1994, they captured 40 blackcaps and put them in cage. Ten birds with the latest migratory restlessness were chosen and were left to mate, producing 26 offspring. From this, four breeding pairs with late migratory restless were taken and were allowed to mate. These produced 14 second-generation birds in a line selected for late migratory activity. It was found that in last two generations, migratory restlessness activity was delayed by an average of 7.65 days they calculated a heritability of 0.72 for this behaviour. By considering this and other findings, it suggests that variance in migratory patterns may be completely additive due to genetic variance in migratory behaviour rather than a type of phenotypical response.

16.10 Techniques to Study Bird Migration

Bird migration is one of the most fascinating and interesting topics to study and research. Up until the recent century, bird migration knowledge was just based on seasonal changes happening in bird distributions. Birds in large numbers were seen flying but didn't know where it was going or coming from. Later, researchers and observers increased and also started the usage of technology to understand bird migration.

In the earliest days, migratory birds were studied in laboratories by keeping them inside cages. Growth and progress in the research field of bird migration started to break through in the early twentieth century when they introduced numbered and labelled metal leg rings to the birds by which they could be individually marked and reported. By the mid-twentieth century, radar was in use to study bird migration which enabled the researchers to track day by day movement of the journey of the birds. Recently, researchers have developed methods that enable them to record bird movement and migration without the need for re-trapping like feather dying, Neck collars, geotagging, etc.

16.10.1 Visual Observation

Visual encountering technique is one of the most widely used methods to understand bird migration, it is indeed one of the oldest methods. Nevertheless, this will enable us to get to know about migratory season and direction determined by the bird well but height and speed at which a bird is travelling will be tough in this technique.

Some birds migrate majorly during the day (Diurnal migrants) and some by night (Nocturnal migrants). Diurnal migrants can be easily seen, identified, and counted by observers on the ground just through their eyes. But, in contrast, understanding the migration of nocturnal migrants is a lot tougher because of increased constraints like very little visibility, and toughness to get the photographs and direction. Aids can be used in order to minimize errors in the data such as binoculars, telescopes, bird guides. and books, still photographic cameras, etc. Using a telescope with 40X magnification, an estimated 50% of the birds flying at 1.5 km distance from the observer could be detected, reducing to zero at 3.5 km (Liechti et al. 1995).

The fact that bird migration is highly prominent near offshore and seacoast, many research programmes would keep trained observers in lighthouse for weeks during the nineteenth century. Be it day or night, a proper scheduled and systematic observation will definitely give some indication of the number and the species composition of birds migrating within the range, seasonal timings of the birds, direction of the migration and its relationship with the weather and climate (Fig. 16.9).

16.10.2 Ringing

Ringing is a scientific procedure in which a bird's leg will be put on with a metal ring which will help in understanding bird migration and movement. Bird metal leg ring was first produced by Hans Christian Cornelius Mortensen in 1899. That ring had information like number and address. This method spread all over the world and became a mainstay in the field of bird migration studies.

Fig. 16.9 A Bird watcher doing visual observation using binoculars

A bird leg ring is made up of light and tough metal, which will be placed around the legs of the nestlings or adults. It should be neither tight nor loose on the birds' legs. The size of the ring varies from bird to bird depending on the birds' size. A ring carries information like a unique number engraved on it, individual identification, address. Therefore, the ringed birds would provide information of longevity of the species and also the movement pattern.

A systematic bird migration study majorly attempting to ring the bird was done in the 1950s by Bombay Natural History Society (BNHS) in India. Back in 1926, a ruler Dhar State (presently in Madhya Pradesh, India) used hand-crafted rings on birds some of which were recovered from as far as eastern and northern USSR. By 2000s, BNHS had ringed 400,000 birds which included more than 40 different species of birds all over India, with about 1% being recovered.

Salim Ali had recorded a ringed grey wagtail *Motacilla cinerea* returning from its breeding grounds in Himalayas to the same exact lawn which was 2000 km away in Greater Bombay on almost the same dates in September for 5 years in a row (Kothari 2007) (Figs. 16.10 and 16.11).

This leg ring led to the discovery of a new technique for recording bird migration and that is bird collaring. As the name says, it is just a collar which will be put on a bird's neck which also contains the same kind of information as a bird leg ring does. For example, in Fig. 16.12. we can see a collar on the bar-headed goose's neck with a label as 'F88.' This is a green colour banded F88 bird which was captured and marked by the Wildlife Conservation and Science Centre in Mongolia on July 13th 2019 at the Salt Lake near Tsakhir soum in province Arkhangai of Central Mongolia. The capture location was N48.126314 E99.143451. This particular collared bird is an adult male, which weighed about 2815 g. It was also marked aluminium band X000542 in its tarsus region and track tag KE1916 (wscc.org.mn).

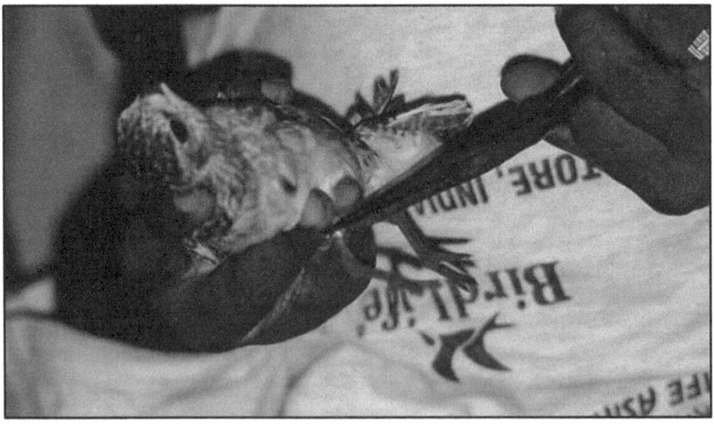

Fig. 16.10 Leg ring being fitted on a Pacific Golden Plover's *Pluvialis fulva* leg in 2001 by BNHS. (Photo: Thejaswi Shivanand)

Fig. 16.11 Curlew sandpiper *Calidris ferruginea* with a leg-ring spotted in Mumbai, India which was tagged in Mumbai itself during December 2014 and February 2015. (Photo: Raju Kasambe)

Fig. 16.12 Bar-headed goose *Anser indicus* (First from top) with Bird Collar attached. (Photo: Shylajesha)

16.10.3 Radar-Based Technique

The usage of radar in bird studies began somewhere in 1950s. A radar emits radio waves in short pulses, and it will record the echoes which come back after hitting the target. The usage of radar in bird migration studies revolutionized the whole aspect. No matter at what speed, light condition, time, altitude and weather a bird is migrating in, we can collect the data by using this radar technique. It will provide information about seasonal migration, migration timing, speed, direction, and altitude of the flight (for discussion of procedures, see Schmaljohann et al. 2008).

But the main disadvantage of this is the cost. It is highly expensive to use radar technique. The equipment is itself very expensive. There should be always trained personnel who will operate and maintain the complete instrument. Other drawbacks are that the birds flying below the radar horizon are usually missed, and sometimes backscatter from the ground can blur the images that appear on the screen.

16.10.4 Radio Transmitter Technique

This is a device which a bird wears like harness or a rucksack with a small protruding aerial. The harness is made of biodegradable material so that it will be worn off after a certain period of time. This will have a transmitter which is called the platform transmitter terminals (PTTs). This will give the data of the bird to which it has been attached. By this, we can get the data and can now follow the bird individually and automatically with the help of satellite. This is basically based on the satellites which are continually circling our globe, it will detect the signals coming from these transmitters anywhere on earth. Then, those satellites will transmit the information to a ground station, and from there it will be sent to the researcher. The accuracy of each and every reading is completely known, so all other less reliable ones can be discarded if not needed. This method is very expensive, but it will provide the best and most accurate data available on the movements of individual migratory birds. PTTs are now available which are literally only 5 g in weight, and it is expected that further miniaturization will be coming very soon.

The usage of this will enable us to monitor all the birds no matter their size day by day on their long migratory journeys. In the beginning, there were individual battery-powered PTTS which used to last only for a few weeks or months. But in 1995, solar-powered PTTs became available, which are currently being used in most cases throughout the world. This will last for years together, and we can also follow the birds for successive years of their migratory journey. This method has opened a plethora of ideas for research and migration studies have advanced in new directions. Through this, we are enabled to get all the latest migratory routes, progress, stopover locations, duration, speed, etc.

Later on, in the 2000s, other types of sensors and data storage devices were invented and developed. Geolocation systems (GLS) are a type of sensor which is based on the continual measurements by photosensors of the light intensity, which can later be used to calculate the position of the bird geographically at different dates (latitude from day length and longitude from the time of dawn and dusk). In contrast, the Global positioning system (GPS) will receive the data required for the calculation of the bird position at a regular interval (the intervals can be pre-set like every 30 min) through a network of satellites. In the early stage, birds were used to be recaptured to recover the tag and to collect data from it. But now as the GLS and GPS are linked to satellite receivers, it will allow us to download the stored data directly from bird by sitting on ground and also without recapturing the bird. In past couple of years, researchers have added additional sensors to this so that extra variables can also be measured like temperature, heart rate of birds, and mortality, but it will keep on adding weight to the device and bird.

The latest development is the use of passive integrated transponders (PIT tags). This is very similar to our microchips which will be fitted to our pets like dogs and cats. It can either be attached to the leg ring or can be implanted under the skin of the bird. It will have a unique code and these codes can be read automatically by receivers and antenna placed near its visiting or nesting sites or any other sites which are being regularly visited by the bird. This method has been used to record the return dates of individual common terns to their nesting colony (Dittman and Becker 2003), for example, and to identify individual Peregrines at nest sites, thus providing information on survival and site fidelity (Smith and McGrady 2009).

Some of the drawbacks of this technique are that the cost of a single unit itself is very expensive, it needs some subsequent servicing, weight of the device is also a major issue (it should not be more than 3–4% of the bird's weight) which is very challenging (Fig. 16.13).

16.10.5 Isotopes and Other Internal Markers

There are many Biochemical substances and markers which are present naturally inside the tissues of migratory birds. These things help to provide information on its movement pattern on a broader scale. Stable isotopes have already proved that they are useful, especially in these kinds of studies. Isotopes of the same element will differ in the number of neutrons in each atom, and therefore in their atomic mass, which can be measured using a mass spectrometer. Several stable isotopes like carbon, hydrogen, nitrogen and other, vary across different geographical regions or also between the habitat of migratory birds and food types (Hobson 1999). For example, the ratio of hydrogen to its isotope deuterium will vary consistently across Europe, in the west it will be deuterium rich and in the east its deuterium depleted.

Fig. 16.13 Graylag goose *Anser anser* fitted with a GPS tracker on its neck. (Photo: Sandeep Gabhane)

Similarly, the isotope ratio of carbon and nitrogen will differ between the terrestrial and aquatic ecosystems.

Birds will absorb isotopes from the food which they consume and deposit them in their body tissues, which will give the isotope signature. As they consume food from one place and migrate to another place the isotopes deposited in their tissues show either the region from where the food was consumed, habitat form where the food occurred or the type of the food, depending on the element which gets analysed. For a short timescale, blood plasma is considered and for a long timescale, keratinous tissues like feathers will be used. So, basically just catching a bird and plucking one of its feathers is more than enough to give us the required information on a very broad scale about its migration.

Lately, this is being used to assess information like breeding location, wintering locations, previous breeding and wintering location, passage times of different populations migrating through the same location, whether the eggs produced by it using the food of breeding location or wintering location and many more aspects of migration ecology are being studies using isotope analyses. Trace elements present in the feathers have also been used to indicate their geographical origins. This is possible because the trace elements present in the feather vary from location to location.

Many other researchers have also tried and attempted the use of DNA in distinguishing different populations of a species from among the mixture of various populations often found at the major migration and wintering sites. However, only in very few species individuals from different populations will be convincingly separated in this manner (examples are in Wennerberg 2001; Smith et al. 2005). Within most of the species, the DNA differences between regional populations are too tiny to be of use in this respect, at least for the methods which are being used presently.

Acknowledgement I wish to thank Dr. A K Chakravarthy, for this wonderful opportunity and for his valuable input, which added the weightage to the chapter scientifically. I am also very obliged to Mr. Shivaprakash Adavanne, for the addition of information. I do not know how much I owe to the generosity of Mr. Karim Iliya, Dr. Raju Kasambe, Dr. Sandeep Gabhane, Mr. Thejaswi Shivanand, Ms. Vijayalakshmi Rao, Mr. Shivaprakash Adavanne, Mr. Shylajesha S, Mr. Saptha Girish MK, Mr. Arjun Shridhar, Mr. Harsha K M, who provided their photographs which made the lines in the chapter more colourful, illustrative and enliven, moreover added life to it. Last but not least, I thank my parents, family, and friends.

References

Adkins B (2017) Animal behaviour. Library Press, New York

Åkesson S, Hedenström A (2007) How migrants get there: migratory performance and orientation. BioScience 57:123–133

Alerstam T, Hedenström A, Åkesson S (2003) Long-distance migration: evolution and determinants. Oikos 103:247–260

Berthold P (1990) Wegzugbeginn und Einsetzen der zugunruhe bei 19 Vogelpopulationen: Eine vergleichende untersuchung. In: van den Elzen R, Schuchmann K, Schmidt-Koenig K (eds) Current topics in avian biology: Proceedings of the International Centennial Meeting of the Deutsche Ornithologen Gesellschaft, pp 217–222

Berthold P (1992) Rapid microevolution of migratory behavior in a wild bird species. Nature 360:668–670

Berthold P (1996) Control of bird migration. Chapman and Hall, London

Berthold P (1999) A comprehensive theory for the evolution, control and adaptability of avian migration. Ostrich 70:1–11

Berthold P (2001) Bird migration, 2nd edn. Oxford University Press, Oxford

Bingman VP, Cheng K (2005) Mechanisms of animal global navigation: comparative perspective and enduring challenges. Ethol Ecol Evol 17:295–318

Bingman V, Jechura T, Kahn MC (2006) Behavioral and neural mechanisms of homing and migration in birds. Animal Spatial Cognition: Comparative, Neural, and Computational Approaches, [On-line]. Available: https://www.pigeon.osy.tufts.edu/asc/bingman

Bowlin MS, Bisson IA, Shamoun BJ, Reichard JD, Sapir N, Marra PP, Kunz TH, Wilcove DS, Hedenström A, Guglielmo CC, Åkesson S, Romanowsky M, Wikelski M (2010) Grand challenges in migration biology. Integr Comp Biol 50(3):261–279

Brines ML, Gould JL (1982) Skylight polarization patterns and animal orientation. J Exp Biol 96:69–91

Bruderer B (1999) Three decades of tracking radar-studies on bird migration in Europe and the Middle East. In: Leshem Y, Mandelik Y, Shamoun-Baranes J (eds) Migrating birds know no boundaries. International Centre for the Study of Bird Migration, Latron, Israel, pp 107–141

Bruderer B, Boldt A (2001) Flight characteristics of birds: 1. Radar measurements of speeds. Ibis 143:178–204

Chernetsov N, Kishkinev D, Mouritsen H (2008) A long-distance avian migrant compensates for longitudinal displacement during spring migration. Curr Biol 18:188–190

Chesser RT, Levey DJ (1998) Austral migrants and the evolution of migration in new world birds: diet, habitat, and migration revisited. Am Nat 152:311–319

Cristol DA, Reynolds EB, Leclerc JE, Donner AH, Farabaugh CS, Ziegenfus CW (2003) Migratory dark-eyed juncos, Junco hyemalis, have better spatial memory and denser hippocampal neurons than non-migratory conspecifics. Anim Behav 66:317–328

Devaraju TM, Rao N, Hajeera BB (2016) Diversity and distribution of avian Fauna in Cubbon Park, Bengaluru, international journal of research in environmental. Science 2(5):24–27

Dingemanse NJ, Both C, Noordwijk AJV, Rutten A, Drent PJ (2003) Natal dispersal and personalities in great tits (*Parus major*). Proc R Soc B 270:741–747

Dingle H (2006) Animal migration: is there a common migratory syndrome? J Ornithol 147:212–220

Dingle H, Drake VA (2007) What is migration? BioScience 57:113–121

Dittman T, Becker PH (2003) Sex, age, experience, and condition as factors affecting arrival date in prospecting common terns *sterna hirundo*. Anim Behav 65:981–986

Egevang C, Stenhouse IJ, Phillips RA, Petersen A, Fox JW, Silk JRD (2010) Tracking of Arctic terns *Sterna paradisaea* reveals longest animal migration. PNAS 107(5):2078–2081

Fidler AE, Oers KV, Drent PJ, Kuhn S, Mueller JC, Kempenaers B (2007) Drd4 gene polymorphisms are associated with personality variation in a passerine bird. Proc R Soc B 274:1685–1691

Friesen VL, Smith AL, Diaz EG, Bolton M, Furness RW, Solis JG, Monteiro LR (2007) Sympatric speciation by allochrony in a seabird. Proc Natl Acad Sci U S A 104:18589–18594

Gauthreaux SA Jr (1982) The ecology and evolution of avian migration systems. Avian Biol 6:93–168

Gill RE Jr, Piersma T, Hufford G, Servranckx R, Riegen A (2005) Crossing the ultimate ecological barrier: evidence for an 11 000-km-long nonstop flight from Alaska to New Zealand and eastern Australia by bar-tailed godwits. Condor 107:1–20

Goodenough J, McGuire B, Jakob E (2010) Perspectives on animal behavior. Wiley, Hoboken, NJ

Gould JL (1982a) Ethology: the mechanisms and evolution of behavior. W. W. Norton, New York

Gould JL (1982b) The map sense of pigeons. Nature 296:205–211

Gould JL, Gould CG (2012) Nature's compass: the mystery of animal migration. Princeton University Press, Princeton, NJ

Graham RS, Hawkes LA, Frappell PB, Butler PJ, Bishop CM, Milsom WK (2015) How Bar-headed geese Fly over the Himalayas. J Physiol 30:107–115
Green M, Alerstam T (2000) Flight speeds and climb rates of Brent geese: mass-dependent differences between spring and autumn migration. J Avian Biol 31:215–225
Hagstrum JT (2000) Infrasound and the avian navigational map. J Exp Biol 203:1103–1111
Hansson B, Bensch S, Hasselquist D (2003) Heritability of dispersal in the great reed warbler. Ecol Lett 6(4):290–294
Helbig AJ (1994) Genetic basis and evolutionary change of migratory directions in a European passerine migrant Sylvia atricapilla. Ostrich 65:151–159
Hobson KA (1999) Tracing origins and migration of wildlife using stable isotopes: a review. Oecologia 120:314–326
Holberton R, Dufty AM Jr (2005) Hormones and variation in life history strategies of migratory and nonmigratory birds. In: Greenberg R, Marra P (eds) Birds of two worlds. Johns Hopkins University Press, Baltimore, pp 290–302
Holberton R, Parrish JD, Wingfield JC (1996) Modulation of the adrenocortical stress response in neotropical migrants during autumn migration. Auk 113:558–564
Johnsen A, Fidler AE, Kuhn S, Carter KL, Hoffmann A, Barr IR, Biard C, Charmantier A, Eens M, Korsten P, Siitari H, Tomiuk J, Kempenaers B (2007) Avian clock gene polymorphism: evidence for a latitudinal cline in allele frequencies. Mol Ecol 16:4867–4880
Kennedy JS (1966) Nervous coordination of instincts. Cambridge Research 2: 29–32
Ketterson ED, Nolan V Jr (1983) The evolution of differential bird migration. Curr Ornithol 1:357–402
Koenig KS, Schlichte H (1972) Homing in pigeons with impaired vision. Proc Natl Acad Sci 69:2446–2447
Korsten P, Mueller JC, Hermannstadter C, Bouwman KM, Dingemanse NJ, Drent PJ, Liedvogel M, Matthysen E, Oers KV, Overveld TV, Patrick SC, Quinn JL, Sheldon BC, Tinbergen JM, Kempenaers B (2010) Association between DRD4 gene polymorphism and personality variation in great tits: a test across four wild populations. Mol Ecol 19:832–843
Kothari A (2007) Birds in our lives. University Press, Hyderabad
Levey DJ, Stiles FG (1992) Evolutionary precursors of long-distance migration: resource availability and movement patterns in neotropical birds. Am Nat 140:447–476
Liedvogel M, Åkesson S, Bensch S (2011) The genetics of migration on the move, Trends in ecology and evolution, Elsevier 26(11):561–569
Liechti F, Bruderer B, Paproth H (1995) Quantification of nocturnal bird migration by moonwatching: comparison with radar and infrared observations. J Field Ornithol 66:457–468
Lovette IJ, Fitzpatrick JW (2016) Handbook of bird biology, 3rd edn. Cornell University, Ithaca, NY
Mackinnon J, Verkuil YI, Murray N (2012) IUCN situation analysis on east and southeast Asian intertidal habitats, with particular reference to the Yellow Sea (including the Bohai Sea). In: Occasional paper of the IUCN species survival commission no. 47. IUCN, Gland and Cambridge, pp ii–70
McCaffery BJ (2008) On scimitar wings. Birding 40(5):50–59
McLaren IA, Lees AC, Field C, Collins KJ (2006) Origins and characteristics of Nearctic land birds in Britain and Ireland in autumn: a statistical analysis. Ibis 148:707–726
Mueller JC, Pulido F, Kempenaers B (2011) Identification of a gene associated with avian migratory behaviour. Proc R Soc B 278:2848–2857
Newton I (2008) The migration ecology of birds. Academic Press, London
Newton I (2010) Bird migration. HarperCollins, New York, NY
O'Malley KG, Banks MA (2008) A latitudinal cline in the Chinook salmon (*Oncorhynchus tshawytscha*) clock gene: evidence for selection on PolyQ length variants. Proc R Soc B 275:2813–2821
O'Malley KG, Ford MJ, Hard JJ (2010) Clock polymorphism in Pacific salmon: evidence for variable selection along a latitudinal gradient. Proc R Soc B 277:3703–3714
Outlaw DC, Voelker G (2006) Phylogenetic tests of hypotheses for the evolution of avian migration: a case study using the Motacillidae. Auk 123:455–466

Outlaw DC, Voelker G, Mila B, Girman DJ (2003) Evolution of long-distance migration in and historical biogeography of Catharus thrushes: a molecular phylogenetic approach. Auk 120:299–310

Perdeck AC (1967) Orientation of starlings after displacement to Spain. Ardea 55(3-4):194–202

Pasinelli G, Schiegg K, Walters JR (2004) Genetic and environmental influences on natal dispersal distance in a resident bird species. Am Nat 164:660–669

Pennycuick CJ (1969) The mechanics of bird migration. Ibis 111:525–556

Pennycuick CJ (1972) Soaring behaviour and performance of some east African birds observed from a motor glider. Ibis 114:178–218

Pennycuick CJ (1975) Mechanics of flight. In: Farner DS, King JR (eds) Avian biology, vol 5. Academic Press, London, pp 1–75

Pennycuick CJ (2008) Modelling the flying bird. Academic Press, London

Perdeck AC (1958) Two types of orientation in migrating starlings Sturnus vulgaris L. and chaffinches Fringilla coelebs L., as revealed by displacement experiments. Ardea 46:1–37

Piersma T, Tris PJ, Mouritsen H, Bauchinger U, Bairlein F (2005) Is there a "migratory syndrome" common to all migrant birds? Ann N Y Acad Sci 1046:282–293

Pulido F (2007) The genetics and evolution of avian migration. BioScience 57(2):165–174

Pulido F, Berthold P (1998) The microevolution of migratory behaviour in the blackcap: effects of genetic covariances on evolutionary trajectories. Biol Conserv Fauna 102:206–211

Pulido F, Berthold P (2003) Quantitative genetic analysis of migratory behavior. In: Berthold P et al (eds) Avian migration. Springer, Berlin, pp 53–77

Pulido F, Berthold P (2010) Current selection for lower migratory activity will drive the evolution of residency in a migratory bird population. Proc Natl Acad Sci U S A 107:7341–7346

Pulido F, Berthold P, Mohr G, Querner U (2001) Heritability of the timing of autumn migration in a natural bird population. Proc R Soc Lond Ser B 268:953–959

Rappole JH (1995) The ecology of migrant birds: a neotropical perspective. Smithsonian Institution Press, Washington, DC

Rappole JH (2013) The avian migrant, the biology of bird migration. Columbia University Press, New York, NY

Rayner JMV (1990) The mechanics of flight and bird migration performance. In: Gwinner E (ed) Bird migration: physiology and ecophysiology. SpringerVerlag, Berlin, pp 283–299

Robbins CS (1980) Predictions of future Nearctic land bird vagrants to Europe. Br Birds 73:448–457

Rolshausen G, Segelbacher G, Hodson KA (2009) Contemporary evolution of reproductive isolation and phenotypic divergence in sympatry along a migratory divide. Curr Biol 19:2097–2101

Ronacher B (2008) Path integration as the basic navigation mechanism of the desert and *Cataglypbis fortis* (Forel 1902) (hymenoptera: Formicidae). Myrmecol News 11:53–62

Sapir N, Horvitz N, Wikelski M, Avissar R, Mahrer Y, Nathan R (2011) Migration by soaring or flapping: numerical atmospheric simulations reveal that turbulence kinetic energy dictates bee-eater flight mode. Proc R Soc B 278:3380–3386

Schmaljohann H, Liechti F, Bachler E, Steuri T, Bruderer B (2008) Quantification of bird migration by radar a detection probability problem. Ibis 150:342–355

Scott GR, Milsom WK (2007) Control of breathing and adaptation to high altitude in the bar-headed goose. Am Physiol Soc 239:379–391

Smith GD, McGrady MJ (2009) Using passive integrated transponder (PIT) tags to better understand Peregrine falcon population in South Scotland and Northeast England. In: Sielicki J, Mizera T (eds) Peregrine falcon populations—status and perspectives in the 21st century. European Peregrine Falcon Working Group, Turul, Warsaw, pp 399–414

Smith TB, Clegg SM, Kimura M, Ruegg K, Milá B, Lovette I (2005) Molecular genetic approaches to linking breeding and overwintering areas in five neotropical migrant passerines. In: Greenberg R, Marra PP (eds) Birds of two worlds. Smithsonian Institution, Washington, DC, pp 222–234

Wennerberg L (2001) Breeding origin and migratory pattern of dunlin (*Calidris alpina*) revealed by mitochondrial DNA analysis. Mol Ecol 10:1111–1120

Wiltschko W, Wiltschko R (2005) Magnetic orientation and magnetoreception in birds and other animals. J Comp Physiol A 191:675–693

Wiltschko W, Wiltschko R (2007) Magnetoreception in bird: two receptors for two different tasks. J Ornithol 148:61–76

Winger BM, Lovette IJ, Winkler DW (2012) Ancestry and evolution of seasonal migration in the Parulidae. Proc R Soc B 279:610–618

Yasukawa K, Martínez TZ (2014) Animal behavior: how and why animals do the things they do. Praeger, Westport, CT

Zink RM (2002) Towards a framework for understanding the evolution of avian migration. J Avian Biol 33:433–436

Zink RM (2011) The evolution of avian migration. Biol J Linn Soc 104:237–250

17 Urban Birds and Adaptive Behaviours

Manjula Menon

Abstract

Urbanization is one of the important causes of biodiversity loss that modifies natural habitats and alters the ecosystem processes and functions. Major anthropogenic drivers causing biodiversity loss include habitat loss, climate change, and invasion of non-native species. Different species exhibit differential responses to urbanization wherein a few species thrive well and evolve and others seldom adapt to changing environments and perish. Few urban exploiters thrive well in the urban settings, resulting in a dynamic process of colonization, thus contributing to biotic homogenization that may negatively impact biodiversity. Urban sprawl also affects the floral and faunal communities by modifying the vegetation composition, resource availability, species migration, density, and structure. Urban birds have shown wide modifications in their breeding, nesting and foraging patterns, thus exhibiting adaptive responses to urbanization. Urbanization and habitat fragmentation reduce the breeding success of specialized species thriving in urban habitats. Bird specialists with narrow guild requirements were found to decrease with increasing urbanization. Bird abundance followed a reverse trend depicting the dominance of a few synanthropic species.

Keywords

Urbanization · Anthropogenic · Birds · Taxonomic homogenization · Communities

---

M. Menon (✉)
Department of Environmental Science and Management, Bharathidasan University, Tiruchirappalli, Tamil Nadu, India

S. Sridhara et al. (eds.), *Animal Behavior in the Tropics*,
https://doi.org/10.1007/978-981-97-7324-4_17

17.1 Introduction

Urbanization is one of the important causes of biodiversity loss that modifies natural habitats and alters the ecosystem processes (Marzluff et al. 2001). Urbanization is also one of the fundamental characteristics that evolved cities that provide refuge and innovative habitats for different species of wildlife. Rapid urbanization has altered the ecological niches (Wood et al. 2021), migration patterns, contamination of soil, air, and water (Roberts et al. 2009) and led to biodiversity loss (Seto et al. 2012). Different species exhibit differential responses to urbanization wherein a few species thrive well and evolve and a few others seldom adapt to changing environments and perish. Those species that lose their natural habitats migrate to cities and adapt themselves to novel environments, overcoming the pressures of predation, competition, thermal fluctuations, alteration in the supply of food and climatic conditions (Powell et al. 2015; Reynolds et al. 2019). Urbanization has large-scale perturbations on communities and ecosystems worldwide. Such perturbations replace natural habitats and alter microhabitat conditions, social structure, species movements and migratory patterns. Fragmentation of forest habitats often leads to the disappearance of forest-dwelling species and the understory insectivores (Zuluaga and Rodewald 2015). Factors associated with urbanization may also exhibit land-cover changes having a considerable impact on temperature and precipitation, thus influencing the microhabitat of the species. Fragmented habitats alter resource availability and species interactions in a community, thus resulting in biotic homogenization and biodiversity loss (Evans and Gawlik 2020). Urban sprawl also affects the floral and faunal communities by modifying vegetation composition, species migration, trophic dynamics, density, and structure (Donnelly and Marzluff 2006). Urbanization also reduces the floral richness and the abundance of pollinators; however, they demonstrate resilience to the effects of urbanization through phenotypic plasticity and evolutionary traits (Ruas et al. 2022).

Birds are highly diverse and are indicators of habitat quality. Birds are exemplar models to study the effects of urbanization. Birds are the most successful urban adapters providing a range of ecosystem services and a few disservices. However, during the last century, many common species of birds have declined from the urban settings, impacting the processes and functioning of the ecosystems (Barton et al. 2020). The benefits urban habitats provide birds are milder climates, higher detectability of food and adaptive nesting sites for species well accustomed to the dynamics of urban environments (Jimenez-Penuela et al. 2019). The loss of common birds from the urban settings may be attributed to the fragmentation of their habitats, thus, exposing exposing them to threatening environments (Mbiba et al. 2021).

Fragmentation of habitats leads to microhabitat variations (McKinney 2002), thus increasing edge effects on species by exposing them to altered biotic and abiotic conditions (Pereira et al. 2004). In the urban environment, interspecific competition and predation are important in determining avian habitat selection and community organization in addition to food and nesting sites (Kurucz et al. 2021). Though urban areas are recognized as potential areas of biodiversity, however, species competition and limitations of resources may expose birds to threats. The ability to exploit dynamic resources by birds helps them in successfully colonizing the urban environments. Successful urban bird invaders tend to have large clutch size, large brain size, and generalist feeding behaviour (Callaghan et al. 2019). Some urban species of birds have increased range size while in few others, there is a decrease in range size (Catterall et al. 1998). In another scenario, urbanization may also expose species to infection (Giraudeau et al. 2014), which may affect the reproduction and fitness of bird populations (Marzal et al. 2005; Marzal et al. 2008).

The Grinnellian niche concept states that the niche of a species is determined by its habitat and the accompanying behavioural adaptations that help the species to thrive in that environment (Grinnell 1917). Among the guild types, urbanization provides more favourable habitats to aerial and ground foraging insectivores (O'Connell et al. 2000). Urban exploiters compared to the rural species are comparatively more aggressive, bolder, and less neophobic (Vincze et al. 2021). The complex novel environments in cities accelerate certain selective pressures that predominantly alter the genetic structure of the species, behaviour, and morphology (Shochat et al. 2006). In a study in Seattle, USA, the human population and their cultural differences showed discouraging behaviour towards American Crows and European Starlings, which inturn influenced bird behaviour (Clucas and Marzluff 2012). Urbanization has profound impact on species sensitive to habitat modifications, such as short-distance migrants, granivores, ground-nesting species, and ground-gleaners (Lakatos et al. 2022). Urban ecosystems filter those species that are smaller in body mass and have lower habitat specificity (Tai et al. 2022). These ecosystems also depict a lower seasonal variation of bird communities and this pattern could be associated with fewer occurrences of migratory species and less seasonal variations of resident birds (Leveau 2018). A recent study has shown that climate change-associated environmental changes have modified the migratory habits of birds occupying the urban landscapes compared to rural habitats (Tryjanowski et al. 2013). Urbanization also affects bird migration, wherein urban adapters fail to migrate from their breeding range or wintering stations due to the availability of surplus resources in the urban areas (Bonnet-Lebrun et al. 2020).

The diversity of bird species declines with increasing urbanization (Jokimaki and Suhonen 1993; Menon et al. 2016). A more generalized conclusion of urbanization is that it provides dominance of a few species, and long-term homogenization of diversity (Kurucz et al. 2021). Similarly, bird communities often change with urban development (Kluza et al. 2000), favouring fewer species over others, thus increasing the abundance of a few species in the urban environment (Kondratyeva et al. 2020).

Urban landscapes are often dominated by a few abundant non-native species (Barton et al. 2020), that modify habitats with successional changes in vegetation. Tree-dominated matrixes in the urban landscape play a crucial role in connecting fragmented habitat patches and non-vegetative areas that may affect the dispersal ability of the species (Zuluaga and Rodewald 2015). Large-sized birds were found absent from the city centres as they are affected by human presence and noise, both of which are detrimental factors for the initiation of longer flights (Morelli et al. 2018). An interesting study by Leveau (2019) has shown that urbanization induces homogenization of bird colours and sizes in avian species. Urbanization also negatively affects the colour diversity of birds. Several species of nocturnal migratory birds were affected by collision with buildings, including species of conservation concern as they often get disoriented by artificial lighting (Longcore and Rich 2004).

Bird species diversity and richness were found to decrease with increasing urbanization (Clergeau et al. 2006; Menon et al. 2016). Similarly, bird specialists with narrow guild requirements were also found to decrease with increasing urbanization (Devictor et al. 2007). Bird abundance followed a reverse trend (Grimm et al. 2008; Menon et al. 2016), depicting the overall dominance of a few synanthropic species facilitating biotic homogenization (Clergeau et al. 2006). Though there are several studies on the effects of urbanization on birds, however, these studies lack an understanding on the effects of urban dynamics on interacting species, their communities and social systems.

17.2 Urbanization and the Nesting Patterns in Birds

Urban birds have shown wide modifications in nesting patterns and nest designs in response to ecological challenges posed by biotic homogenization, nest predation, sexual selection and ecological traps in the urban settings. Several studies have reported modifications in nest design, and nest composition along a gradient of urbanization and changes in the nesting patterns within conspecific species (Reynolds et al. 2019). The choice of urban structures by urban adapted species has widened their survivorship in urban settings. Among birds, passerines are most successful in building nests in artificial structures (Lambrechts et al. 2017). Urban birds usually nest in a wide variety of sites such as factories, shops, houses, rice mills, under bridges, and abandoned residential buildings where they use a variety of structures like crevices, eaves of houses, rooftops, cloth, fur, metal pipes, vertical pipes, fences, and wires (Fig. 17.1). Birds often select materials that are locally available and those that resemble natural vegetation. Anthropogenic structures in urban areas though provide favourable nesting and breeding sites, however, may also function as ecological traps, which may induce fear in birds and may induce fear in birds and may reduce their fitness (Schlaepfer et al. 2002). The use of such anthropogenic materials also has negative impacts on nestling success and fitness. A study by Lee et al. (2015) has shown an increase in the presence of anthropogenic materials the urban nests and a decrease in natural materials like leaves, stems, and feathers (Fig. 17.2). The study also found that older individuals incorporated more anthropogenic materials than younger ones. The use of cigarette butts as an ectoparasite repellent, paper strips as water absorbent, and wires to strengthen the nest are the most widely used by urban adapters (Jagiello et al. 2022). A study has demonstrated that the use of anthropogenic materials had a strong negative impact on the reproductive success of Blue tits. In contrast, no such negative impacts were recorded in great tits (Jagiello et al. 2022). Urbanization also affects the nest design in birds due to fewer native species and composition of the nest changes along an urbanization gradient (Reynolds et al. 2019).

Fig. 17.1 Manmade structures used by urban birds for nesting

Fig. 17.2 The plastic wires act as ecological trap for birds

With the available resources, birds mimic their natural nests in the urban settings and safeguard them by incorporating bright-coloured materials that could be closely associated with territorial defence or as an indicator to alarm the predators. Several species of birds in the urban settings prefer a lighter nest; the classic example is the house sparrow. The house sparrows construct a dense nest of hay at sites where food availability is abundant and a lighter nest with few strings of hay in the urban settings, where procuring nesting materials may be arduous (Fig. 17.3). A lighter nest in the urban may help the species adapt to thermal fluctuations or could be a strategy to make the nest less visible to predators. Such observations were also recorded by Hanmer et al. (2017) and Biddle et al. (2016). Thus, the potential for biotic differentiation after environmental filtering in the urban ecosystems was based on species-specific traits and their adaptations, that would potentially differ across cities and seasons. In a highly urbanized metropolitan city like Bangalore, urban structures have facilitated the success of *Columba livia* (blue rock pigeon). However, high-rise buildings were less preferred by *Corvus splendens* (house crow), and *Corvus macrorhynchos* (large-billed crow) and in recent years, have shown a significant decline in their population. However, a reverse trend was observed in less urbanized cities where populations of House crows and Large-billed crows dominated the urban matrix.

Urbanized birds are usually generalist consumers. However, urban landscapes with dense vegetation provide suitable habitats for a variety of cavity-nesting birds and the abandoned nest would be subsequently used by other fauna. It has been observed that cavity-nesting birds and those species that nest on buildings aggressively compete with native species for nesting sites (Barton et al. 2020). Hence,

Fig. 17.3 Sack ropes used by Sparrows for nest construction due to its resemblance to hay

cavity-nesters are successful urban exploiters, as this trait is exhibited by the most successful breeders in the urban environment. A study by Kurucz et al. (2021) found that shrub-nesting in urban habitats would be safer, probably due to lower predation pressure and favourable nesting sites in cities. In cavity-nesting species like the woodpeckers, abundance may be limited by the availability of canopy cover, and standing dead trees for foraging, roosting, and nesting (Tomasevic and Marzluff 2020). Thus, alterations in their population may affect the food web dynamics in an ecosystem. Inhabited buildings provide suitable nest sites for *Lonchura punctulata* (scaly-breasted munia), *Pycnonotus cafer* (red-vented bulbul), *Columba livia* (blue rock pigeon), *Passer domesticus* (house sparrow), *Acridotheres tristis* (common myna), *Corvus splendens* (house crow), *Corvus macrorhynchos* (large-billed crow), *Cinnyris asiaticus* (purple sunbird), and *Leptocoma zeylonica* (purple-rumped sun bird), These birds successfully breed in urban settings and are able to exploit the urban environment due to their wider geographic range and large population sizes (De Laet and Summers-Smith 2007). Soldatini et al. (2008), reported that the use of rooftops by urban gulls has resulted in a widespread increase in population over the past few decades, resulting in the outbreak of pathogens (Coulson et al. 1983). The adaptive nesting strategy adopted by *Falco peregrinus* (peregrine falcon), recorded better productivity (Gahbauer et al. 2015), as artificial structures safeguard against predators and subsequently higher abundance. However, open-cup nesters, that prefer densely vegetated areas for nesting were found to be negatively associated with urbanization (Mathe and Batary 2015).

A study by Vincze et al. (2021) has shown that there was lesser detection of natural nests and higher detection of artificial nests by predators with increasing urbanization. It has also been shown that nests placed on non-native vegetation, may act as ecological traps, exposing species to predators (Rodewald et al. 2010). Wetland birds such as storks have shown adaptive responses to changes in nesting patterns in response to alteration of habitat structure and decrease in food availability (Frederick et al. 2009). In a city like Delhi native bird species diversity declined with an increase in the density of exotic plants (Khera et al. 2009), thus an absence of densely vegetated sites for nesting. Nesting in proximity to human population also induces behavioural changes in urban bird populations affecting their breeding biology, foraging behaviour and frequency of nest visits for feeding the nestlings.

17.3 Urbanization Influences Breeding Birds

Breeding is the most interesting and phenomenal phase in the lifecycle of a bird (Daan and Tinbergen 1997). Optimal environmental conditions play an important role in the breeding success of birds and thus are important aspects in studying population and demographic processes (Thompson 2007). Urbanisation often induces novel environmental pressures on birds, leading to selective adaptations. Urbanization and habitat fragmentation also reduce the breeding success of species thriving in the urban habitats (Reynolds et al. 2019).

The breeding success of urban birds depends on the availability of green patches, how connected the fragmented landscapes are, and how birds adapt to human presence (Jokimaki et al. 2020). Urban birds well adapted to urban environments, defend smaller territories due to sufficient resources available in these habitats (Juarez et al. 2020). A study by Lan et al. (2021), found that urbanization influences the bird nesting period, the selection of nesting materials and the choice of nest sites in breeding birds. Studies have also recorded early nest initiation for urban raptors and species that visit urban feeders (Kettel et al. 2018). Breeding birds in cities have diverse diets with few species breeding on the ground (Jokimaki et al. 2016), and a few others on trees or urban structures. The onset of breeding in birds tends to be earlier in urban matrixes than the rural ones due to increased food availability (Chamberlain et al. 2009). However, clutch sizes were recorded smaller and nestling conditions were unsatisfactory in urban habitats (Rose et al. 2017). A study by Weaving et al. (2016), showed that *Podargus strigoides* (tawny frogmouth), an endemic nocturnal bird species of Australia, synchronizes breeding period to coincide with peak environmental conditions, a strategy well adapted to maximize reproductive potential. A study in central Argentina, found that during the breeding season, the diversity of bird plumage colour increased along the rural-urban gradient (Leveau 2022).

European birds such as *Falco tinnunculus* (common kestrel), recorded higher populations with an increase in human density and built-up area, whereas species like *Hirundo rustica* (barn swallow) and *Muscicapa striata* (spotted flycatcher) decreased with an increase in the proportion of built-up areas (Jokimaki et al. 2018). A study by Branston et al. (2021), has shown that food availability may be a constraining factor for urban bird nestlings as the availability of insects would be lower in the urban areas. It was recorded that the natural selection process for urban tits that were early breeders and had large clutch sizes was less favourable when compared to non-urban tits (Branston et al. 2021). The changes in day length in urban environments because of artificial lighting and increased temperature have also affected urban birds (Dominoni et al. 2020). In the urban environments, if a male song is unheard by a female, then the male has to physically fight out invaders to attract the females (Slabbekoorn and den Boer-Visser 2006). Once the territories are established during the breeding season, birds select suitable places for nesting, food procurement and mate attraction (Woltmann and Sherry 2011). A study by Goodale et al. (2015), have shown that the loss of a few individuals from mixed-species flocks due to habitat alteration, could reduce the fitness of many other species. Urbanization has also altered the stress hormone levels and reproductive hormone levels in birds living in urbanized locations (Patankar et al. 2021).

The availability of resources alters the breeding phenology of species occupying the urban environments (Branston et al. 2021). Urban birds were found to have smaller clutches, early laying of eggs, and lower productivity than rural birds (Sepp et al. 2018; Branston et al. 2021). However, a study by Wawrzyniak et al. (2015) on blue tits found a seasonal decline in clutch size in urban habitats compared to the forest. The lower clutch size in urban areas may be attributed to the non-availability of high-quality food and increased competition during the breeding season between the conspecifics thus affecting clutch size, fitness and breeding success of birds (Chamberlain et al. 2009).

A study by Mills et al. (1989), found that vegetation factors showed more variance in breeding bird density than housing density in southwestern desert habitats. Similarly, seasonal breeding performance differs in African raptors across an urbanization gradient (Rose et al. 2017). Urbanization also affects the breeding patterns of migratory birds where surplus resources in the urban during the breeding season prevent birds from migrating (Bonnet-Lebrun et al. 2020). The most adapted bird to human habitations, the *Passer domesticus* has shown wide variations in breeding patterns. The interesting visualizations represent the male claiming its nest site and defending its territory. Often a male hops around the selected mate, flicking the wings and the tail, followed by attracting the female to its nest by carrying an exceptionally long string of hay in its beak. A study by Chamberlain et al. (2009), found that most urban birds are characterized by slightly greater annual productivity. The effect of urbanization is such that passerines that are more sexually dichromatic are less likely to occur in the urban areas (Iglesias-Carrasco et al. 2019). Urbanization also drives changes in acoustic communication, thus affecting the breeding success of birds (Marin-Gomez and MacGregor-Fors 2021). It was also found that the average size of tree patches and the characteristics of street trees functioned as buffers to reduce the negative effects of noise on bird communities, especially during the breeding season (Pena et al. 2017). The noises from anthropogenic activities mask bird songs and their transmission resulting in lower mate attraction, territorial defence and subsequently low reproductive success. The challenge is to understand how urban birds modify the song and call structure to avoid bird call masking (Patankar et al. 2021). A classic example is that of the European robins (*Erithacus rubecula*), adapted to nocturnal singing to avoid song masking (Leveau 2018). Higher temperatures in the urban settings have been shown to correspond to earlier dawn chorus onsets (Hasan and Badri 2016), and advanced breeding periods (Dominoni et al. 2020).

17.4 Urbanization Affects Foraging Strategies in Birds

Urbanization is a rapidly growing threat to biodiversity. So, it is necessary to understand how biodiversity is linked to ecosystem processes and services. Urbanization impacts the foraging activities and habitat use of birds at various levels of urban development. Species that exhibit high plasticity in their foraging behaviours take advantage of urban centres as novel sites for foraging, which subsequently helps them to occupy a wide array of habitats (Shaffer et al. 2017). Urban environments often have complex trophic dynamics (Shochat et al. 2006). Birds provide important ecosystem services through pollination and the absence of pollinators in an ecosystem depicts the vulnerability of the species to habitat fragmentation. Pollinators and seed dispersers link fragmented habitats, maintaining the gene flow and contributing to the resilience of the ecosystems (Pauw and Louw 2012). Urban-adapted species exhibit certain traits and adaptive responses inorder to adjust to the pressures of urbanization (Evans et al. 2011). Urbanization also results in the alteration of resources (Chamberlain et al. 2009), that may signal species to adapt to optimal foraging strategies. Their ability to exploit novel resources in the urban environments, facilitated by an abundance of food sources such as discarded food, alternate feeding sites, and road kills have provided birds an easy refuge in the urban environments (Corsini et al. 2022). Successful urban exploiters tend to have larger clutch sizes, larger brain, and a generalist feeding behaviour (Callaghan et al. 2019). Novel food opportunities in a limited space open new ventures for bird resource utilization. A study by Evans and Gawlik (2020), in *Mycteria americana* (Wood Stork), found that during suboptimal conditions, the stork broadened its diet to include anthropogenic food items, resulting in increased productivity. Such responses are often exhibited by wetland birds that exhibit a strong phylogenetic trait towards urban tolerance. Though patches in the forest do not mimic the vegetation in the urban environment however, few species are more acclimatized to urban environments and prefer non-native vegetation and thus, successfully survive by prioritizing their feeding strategies. A biodiversity hotspot at the Cape of Africa found that urbanization reduces the functional diversity of the nectarivore guild (Pauw and Louw 2012).

Flocking is said to affect the foraging efficiency of birds by providing improved foraging strategies and increased predator avoidance (Sridhar and Sankar 2008). Kurucz et al. (2021), found that the consumption of food by birds inhabiting the urban matrix was independent of the level of urbanization. Similarly, food resources in urban areas were found to be low in protein, thus compromising avian reproductive success (Jimenez-Penuela et al. 2019). During winter, birds procure food at the onset of dawn and the time at which bird forages is determined by the reserves left over the previous day and also by their capacity to capture prey in low light intensities (Ockendon et al. 2009). Leveau (2019), has found that colour and life-history traits of birds were interrelated in their responses to urbanization with birds of highly urbanized city centres sharing similar guilds and nesting sites. Fuirst et al. (2018), on *Larus argentatus* (Herring gulls), showed that urbanized landscapes impacted the foraging behaviours of generalist seabirds however, did not substantially influence the microbial diversity at the foraging sites.

Menon and Rangaswamy (2020) showed that *Dicrurus macrocercus* (black drongo) opted for optimal foraging strategies that resulted in bird-farmer mutualism. Farming activities in the rice fields disturb the beetles actively feeding on the crops and are easily preyed upon by the birds. Farming activities in the field, provide signalling strategy for optimal resource utilization in species like *D. macrocercus*. The classical niche theory states that both biotic and abiotic factors synergistically shape species distributions in an ecosystem, which is the result of spatiotemporal heterogeneity (Chisholm and Pacala 2010). The hypothesis stated that vegetative factors and modified urban elements at multiple spatial scales influence patterns of avian abundance, distribution, and resource utilization (Kirk et al. 2011; Menon and Rangaswamy 2020). Hensley et al. (2019), recorded that during a particular season, species dissimilarity and diet guild composition were higher between cities than in regional species pools. Range shifts in bird species and communities have been recorded as an impact of climate change. Even the impact on migratory birds has been documented involving changes in arrival and departure, changes in routes and travel distances, brood size and alterations in breeding behaviour depending on the resources available at the destination (Trautmann 2018).

17.5 Conclusion

Modern structural designs in urban planning should be incorporated to facilitate nesting by birds in the avian urban environments. The concept of shelter belts needs to be introduced in urban governance and planning as these spaces would ensure a safe refuge for urban birds and their survival. Such green corridors will ease the damage caused by habitat fragmentation and will provide a safe space for bird movement and migration. Introducing selective floral species, would attract birds to the urban landscape. To maintain biodiversity within the corridors, species distribution, bird community structure and their breeding success needs to be monitored. Native vegetation should be maintained in cities to prevent taxonomic homogenization. Increasing shelter belts of native species would create edge effects, that would protect species sensitive to urbanization and facilitate the movement of birds across habitat corridors. Landscapes should be designed based on bird communities visiting the nearest forested patches and by integrating citizen science initiatives to facilitate the restoration of landscapes.

Acknowledgement The author thanks the University Grants Commission, Government of India, for financial assistance and support for the research work.

References

Barton JH, Morris K, Meritt D, Magle S, LaMontagne JM (2020) Does urbanization influence population trends of cavity-nesting birds and their relationship with European starlings? Acta Oecol 108:103636

Biddle LE, Goodman AM, Deeming DC (2016) Geographical effects on the mass and dimensions of finch (Fringillidae, Passeriformes) and thrush (Turdidae, Passeriformes) nests. Avian Biol Res 9:13–21

Bonnet-Lebrun A-S, Manica A, Rodrigues ASL (2020) Effects of urbanization on bird migration. Biol Conserv 244:108423

Branston CJ, Capilla-Lasheras P, Pollock CJ et al (2021) Urbanisation weakens selection on the timing of breeding and clutch size in blue tits but not in great tits. Behav Ecol Sociobiol 75:155

Callaghan CT, Major RE, Wilshire JH, Martin JM et al (2019) Generalists are the most urban-tolerant of birds: a phylogenetically controlled analysis of ecological and life history traits using a novel continuous measure of bird responses to urbanization. Oikos 128:845–858

Catterall CP, Kingston MB, Park K, Sewell S (1998) Deforestation, urbanisation and seasonality: interacting effects on a regional bird assemblage. Biol Conserv 84(1):65–81

Chamberlain DE, Cannon AR, Toms MP, Leech DI, Hatchwell BJ, Gaston KJ (2009) Avian productivity in urban landscapes: a review and meta-analysis. Ibis 151:1–18

Chisholm RA, Pacala SW (2010) Niche and neutral models predict asymptotically equivalent species abundance distributions in high-diversity ecological communities. Proc Natl Acad Sci U S A 107:15821–15825

Clergeau P, Croci S, Jokimaki J, Kaisanlahti-Jokimaki ML, Dinetti M (2006) Avifauna homogenisation by urbanisation: analysis at different European latitudes. Biol Conserv 127:336–344
Clucas B, Marzluff J (2012) Attitudes and actions toward birds in urban areas: human cultural differences influence bird behavior. Auk 129(1):8–16
Corsini M, Leanza P, Rodewald AD, Sudyka J, Dhondt AA, Szulkin M (2022) Coping with novelty across an urban mosaic: provisioning latency increases closer to roads and is associated with species-specific reproductive success in two urban adapters. Sci Total Environ 847:157450
Coulson JC, Butterfeld J, Thomas C (1983) The herring gull *Larus argentatus* as a likely transmitting agent of Salmonella montevideo to sheep and cattle. J Hyg (Lond) 91:437–443
Daan S, Tinbergen JM (1997) Adaptation of life histories. In: J.R. Krebs, & N.B. Davies (Eds.), Behavioural Ecology: An evolutionary approach (pp. 311–333). London:Blackwell.
De Laet J, Summers-Smith J (2007) The status of the urban house sparrow *Passer domesticus* in north-western Europe: a review. J Ornithol 148(2):S275–S278
Devictor V, Julliard R, Couvet D, Lee A, Jiguet F (2007) Functional homogenization effect of urbanization on bird communities. Conserv Biol 21:741–751
Dominoni DM, Jensen JK, de Jong M, Visser ME, Spoelstra K (2020) Artificial light at night, in interaction with spring temperature, modulates timing of reproduction in a passerine bird. Ecol Appl 30(3):e02062
Donnelly R, Marzluff JM (2006) Relative importance of habitat quantity, structure, and spatial pattern to birds in urbanizing environments. Urban Ecosyst 9:99–117
Evans BA, Gawlik DE (2020) Urban food subsidies reduce natural food limitations and reproductive costs for a wetland bird. Sci Rep 10:14021
Evans KL, Chamberlain DE, Hatchwell BJ, Gregory RD, Gaston KJ (2011) What makes an urban bird? Glob Change Biol 17(1):32–44
Frederick P, Gawlik DE, Ogden JC, Cook MI, Lusk M (2009) The White ibis and Wood stork as indicators for restoration of the everglades ecosystem. Ecol Indic 9(6):S83–S95
Fuirst M, Veit RR, Hahn M, Dheilly N, Thorne LH (2018) Effects of urbanization on the foraging ecology and microbiota of the generalist seabird *Larus argentatus*. PLoS One 13(12):e0209200
Gahbauer MA, Bird DM, Clark KE, French T, Brauning DW, Mcmorris FA (2015) Productivity, mortality, and management of urban peregrine falcons in northeastern North America. J Wildl Manag 79:10–19
Giraudeau M, Mousel M, Earl S, McGraw K (2014) Parasites in the city: degree of urbanization predicts poxvirus and coccidian infections in house finches (*Haemorhous mexicanus*). PLoS One 9(2):e86747
Goodale E, Ding P, Liu X et al (2015) The structure of mixed-species bird flocks, and their response to anthropogenic disturbance, with special reference to East Asia. Avian Res 6:14
Grimm NB, Faeth SH, Golubiewski NE, Redman CL, Wu JG, Bai XM, Briggs JM (2008) Global change and the ecology of cities. Science 319:756–760
Grinnell J (1917) The Niche-Relationships of the California Thrasher. The Auk 34(4):427–433
Hanmer HJ, Thomas RL, Beswick GJF, Collins BP, Fellowes MDE (2017) Use of anthropogenic material affects bird nest arthropod community structure: influence of urbanisation, and consequences for ectoparasites and fledging success. J Ornithol 158:1045–1059
Hasan NM, Badri M (2016) Effect of ambient temperature on dawn chorus of house sparrows. Environ Ecol Res 4(3):161–168
Hensley CB, Trisos CH, Warren PS, MacFarland J, Blumenshine S, Reece J, Katti M (2019) Effects of urbanization on native bird species in three southwestern US cities. Front Ecol Evol 7:71
Iglesias-Carrasco M, Duchene DA, Head ML, Moller AP, Cain K (2019) Sex in the city: sexual selection and urban colonization in passerines. Biol Lett 15:20190257
Jagiello Z, Corsini M, Dylewski L, Ibanez-Alamo JD, Szulkin M (2022) The extended avian urban phenotype: anthropogenic solid waste pollution, nest design and fitness. Sci Total Environ 838:156034
Jimenez-Penuela J, Ferraguti M, la Puente JM, Soriguer R, Figuerola J (2019) Urbanization and blood parasite infections affect the body condition of wild birds. Sci Total Environ 651(2):3015–3022

Jokimaki J, Suhonen J (1993) Effects of urbanization on the breeding bird species richness in Finland : a biogeographical comparison. Ornis Fennica 70:71–77
Jokimaki J, Suhonen J, Kaisanlahti-Jokimaki ML, Carbo-Ramirez P (2016) Effects of urbanization on breeding birds in European towns: impacts of species traits. Urban Ecosyst 19:1565–1577
Jokimaki J, Suhonen J, Jokimaki MLK (2018) Urban core areas are important for species conservation: a European-level analysis of breeding bird species. Landsc Urban Plan 178:73–81
Jokimaki J, Suhonen J, Benedetti Y, Diaz M, Kaisanlahti-Jokimaki ML et al (2020) Land-sharing vs. land-sparing urban development modulate predator prey interactions in Europe. Ecol Appl 30(3):e02049
Juarez R, Chacon-Madrigal E, Sandoval L (2020) Urbanization has opposite effects on the territory size of two passerine birds. Avian Res 11:11
Kettel EF, Gentle LK, Quinn JL et al (2018) The breeding performance of raptors in urban landscapes: a review and meta-analysis. J Ornithol 159:1–18
Khera N, Mehta V, Sabata BC (2009) Interrelationship of birds and habitat features in urban greenspaces in Delhi, India. Urban For Urban Green 8:187–196
Kirk D, Lindsay K, Brook R (2011) Risk of agricultural practices and habitat change to farmland birds. Avian Conserv Ecol 6:5
Kluza DA, Griffin CR, DeGraaf RM (2000) Housing developments in rural New England: effects on forest birds. Anim Conserv 3:15–26
Kondratyeva A, Knapp S, Durka W, Kuhn I, Vallet J et al (2020) Urbanization effects on biodiversity revealed by a two-scale analysis of species functional uniqueness vs. redundancy. Front Ecol Evol 8:73
Kurucz K, Purger JJ, Batary P (2021) Urbanization shapes bird communities and nest survival, but not their food quantity. Glob Ecol Conserv 26:e01475
Lakatos T, Chamberlain DE, Garamszegi LZ, Batary P (2022) No place for ground-dwellers in cities: a meta-analysis on bird functional traits. Glob Ecol Conserv 38:e02217
Lambrechts MM, Charmantier A, Demeyrier V, Lucas A et al (2017) Nest design in a changing world: great tit *Parus major* nests from a Mediterranean city environment as a case study. Urban Ecosyst 20:1181–1190
Lan F, Ma X, Lu J, Li Y, Chai R, Li X, Luo Y, Zhang Y, Ye Z, Fu C, Bao W, Li L, Xing X (2021) Effects of urbanization on bird nesting: a review. Biodiv Sci 29(11):1539–1553
Lee K, Jang YC, Hong S, Lee J, Kwon In-Ki (2015) Plastic marine debris used as nesting materials of the endangered species Black-faced Spoonbill *Platalea minor* decreases by conservation activities. J Korean Soc Mar Environ Energy 18(1):45–49
Leveau LM (2018) Urbanization, environmental stabilization and temporal persistence of bird species: a view from Latin America. PeerJ 6:e6056
Leveau LM (2019) Urbanization induces bird color homogenization. Landsc Urban Plan 192:103645
Leveau LM (2022) Does spring bring colors? Seasonal bird plumage color variation along urban-rural gradients. Perspect Ecol Conserv 20(2):103–110
Longcore T, Rich C (2004) Ecological light pollution. Front Ecol Environ 2(4):191–198
Marin-Gomez OH, MacGregor-Fors I (2021) A global synthesis of the impacts of urbanization on bird dawn choruses. Ibis 163(4):1133–1154
Marzal A, de Lope F, Navarro C, Moller AP (2005) Malarial parasites decrease reproductive success: an experimental study in a passerine bird. Oecologia 142(4):541–545
Marzal A, Bensch S, Reviriego M, Balbontin J, De Lope F (2008) Effects of malaria double infection in birds: one plus one is not two. J Evol Biol 21(4):979–987
Marzluff JM, Bowman R, Donnely R (eds) (2001) Avian ecology and conservation in an urbanizing world. Kluwer Academic, Boston
Mathe O, Batary P (2015) Insectivorous and open-cup nester bird species suffer the most from urbanization. Bird Study 62:78–86
Mbiba M, Mazhude C, Fabricius C, Fritz H, Muvengwi J (2021) Bird species assemblages differ, while functional richness is maintained across an urban landscape. Landsc Urban Plan 212:104094

McKinney ML (2002) Urbanization, biodiversity, and conservation. Bioscience 52:883–890
Menon M, Rangaswamy M (2020) Dynamics of farmland birds in traditional agroecosystems: the value of countryside elements in avian conservation. Proc Zool Soc 73:352–361
Menon M, Devi MP, Rangaswamy M (2016) Avifaunal richness and abundance along an urban rural gradient with emphasis on vegetative and anthropogenic attributes in Tiruchirappalli, India. Landsc Res 41(1):131–148
Mills GS, Dunning JB, Bates JM (1989) Effects of urbanization on breeding bird community structure in southwestern desert habitats. Condor 91(2):416–428
Morelli F, Mikula P, Benedetti Y, Bussière R, Jerzak L, Tryjanowski P (2018) Escape behaviour of birds in urban parks and cemeteries across Europe: evidence of behavioural adaptation to human activity. Sci Total Environ 631–632:803–810
O'Connell TJ, Jackson LE, Brooks RP (2000) Bird guilds as indicators of ecological condition in the central Appalachians. Ecol Appl 10(6):1706–1721
Ockendon N, Davis SE, Miyar T, Toms MP (2009) Urbanization and time of arrival of common birds at garden feeding stations. Bird Study 56(3):405–410
Patankar S, Jambhekar R, Suryawanshi K, Nagendra H (2021) Which traits influence bird survival in the city? A review. Land 10:92
Pauw A, Louw K (2012) Urbanization drives a reduction in functional diversity in a guild of nectar-feeding birds. Ecol Soc 17(2):27
Pena JCC, Martello F, Ribeiro MC, Armitage RA, Young RJ, Rodrigues M (2017) Street trees reduce the negative effects of urbanization on birds. PLoS One 12(3):e0174484
Pereira HM, Daily GC, Roughgarden J (2004) A framework for assessing the relative vulnerability of species to land-use change. Ecol Appl 14:730–742
Powell LL, Wolfe JD, Johnson EI, Hines JE, Nichols JD, Stouffer PC (2015) Heterogeneous movement of insectivorous Amazonian birds through primary and secondary forest: a case study using multistate models with radio telemetry data. Biol Conserv 188:100–108
Reynolds JS, Ibanez-Alamo JD, Sumasgutner P et al (2019) Urbanisation and nest building in birds: a review of threats and opportunities. J Ornithol 160:841–860
Roberts AD, Prince SD, Jantz CA, Goetz SJ (2009) Effects of projected future urban land cover on nitrogen and phosphorus runoff to Chesapeake Bay. Ecol Eng 35:1758–1772
Rodewald AD, Shustack DP, Hitchcock LE (2010) Exotic shrubs as ephemeral ecological traps for nesting birds. Biol Invasions 12:33–39
Rose S, Sumasgutner P, Koeslag A, Amar A (2017) Does seasonal decline in breeding performance differ for an African raptor across an urbanization gradient? Front Ecol Evol 5:47
Ruas RB, Costa LMS, Bered F (2022) Urbanization driving changes in plant species and communities-a global view. Glob Ecol Conserv 38:e02243
Schlaepfer MA, Runge MC, Sherman PW (2002) Ecological and evolutionary traps. Trends Ecol Evol 17:474–480
Sepp T, McGraw K, Kaasik A, Giraudeau M (2018) A review of urban impacts on avian life-history evolution: does city living lead to slower pace of life? Glob Chang Biol 24(4):1452–1469
Seto KC, Gueneralp B, Hutyra LR (2012) Global forecasts of urban expansion to 2030 and direct impacts on biodiversity and carbon pools. Proc Natl Acad Sci U S A 109:16083–16088
Shaffer SA, Cockerham S, Warzybok P, Bradley RW, Jahncke J, Clatterbuck CA et al (2017) Population level plasticity in foraging behavior of western gulls (*Larus occidentalis*). Mov Ecol 5(1):27
Shochat E, Warren PS, Faeth SH, McIntyre NE, Hope D (2006) From patterns to emerging processes in mechanistic urban ecology. Trends Ecol Evol 21(4):186–191
Slabbekoorn H, den Boer-Visser A (2006) Cities change the songs of birds. Curr Biol 16(23):2326–2331
Soldatini C, Albores-Barajas YV, Mainardi D, Monaghan P (2008) Roof nesting by gulls for better or worse? Ital J Zool 75:295–303
Sridhar H, Sankar K (2008) Effects of habitat degradation on mixed-species bird flocks in Indian rain forests. J Trop Ecol 24(2):135–147

Tai D, Chen C, Song Y, Tan X, Yang X, Wang Y (2022) Ecological traits and landscape characteristics predicting bird sensitivity to urbanization in city parks. Basic Appl Ecol 58:110–120
Thompson FR (2007) Factors affecting nest predation on forest songbirds in North America. Ibis 149(s2):98–109
Tomasevic JA, Marzluff JM (2020) Roosting, reproduction, and survivorship of pileated woodpeckers (*Dryocopus pileatus*) in a suburban setting. Avian Conserv Ecol 15(1):13
Trautmann S (2018) Climate change impacts on birds. In: Tietze DT (ed) Bird species: how they arise, modify and vanish. Fascinating life sciences series. Springer, Cham, 266pp. https://doi.org/10.1007/978-3-319-91689-7_12
Tryjanowski P, Sparks TH, Kuzniak S, Czechowski P, Jerzak L (2013) Bird migration advances more strongly in urban environments. PLoS One 8(5):e63482
Vincze E, Bokony V, Garamszegi LZ, Seress G et al (2021) Consistency and plasticity of risk-taking behaviour towards humans at the nest in urban and forest Great tits, *Parus major*. Anim Behav 179:161–172
Wawrzyniak J, Kalinski A, Glądalski M, Banbura M, Markowski M, Skwarska J et al (2015) Long-term variation in laying date and clutch size of the great tit *Parus major* in central Poland: a comparison between urban parkland and deciduous forest. Ardeola 62:311–322
Weaving M, John W, Bronwyn I, Anthony R, Raylene C (2016) Adaptation to urban environments promotes high reproductive success in the tawny frogmouth (*Podargus strigoides*), an endemic nocturnal bird species. Landsc Urban Plan 150:87–95
Woltmann S, Sherry TW (2011) High apparent annual survival and stable territory dynamics of Chestnut-backed Antbird (*Myrmeciza exsul*) in a large Costa Rican rain forest preserve. Wilson J Ornithol 123(1):15–23
Wood ZT, Palkovacs EP, Olsen BJ, Kinnison MT (2021) The importance of eco-evolutionary potential in the Anthropocene. Bioscience 71(8):805–819
Zuluaga GJC, Rodewald AD (2015) Response of mixed-species flocks to habitat alteration and deforestation in the Andes. Biol Conserv 188:72–81

18 Behavioural Ecology of Larger Bandicoot Rat, *Bandicota indica* (Bechstein), Burrow Ecology, Social and Feeding Behaviour

Shakunthala Sridhara

Abstract

In India, the Larger Bandicoot Rat, *Bandicota indica*, the Lesser Bandicoot Rat, *Bandicota bengalensis* and species of *Rattus* are pests on agricultural crops, in godowns. The ecology of the lesser bandicoot rat in Calcutta. Bombay Natural History Society and the John Hopkins University Centre, Calcutta, 1968 and poultry (Chopra, Rodents in Indian agriculture. Scientific Publishers, Jodhpur, pp 309–330, 1992; Parshad, Integr Pest Manag Rev 4:97–126, 1999; Sridhara, Vertebrate pest management: the Indian scenario. Scientific Publishers, Jodhpur, pp 205–226, 2006). In Malaysia, *B. indica* caused 30% damage to fruit bunches of oil palm. Bandicoots dig simple to complex burrows on the bunds, inside paddy fields, along the boundary of vegetable crops, on the slopes of ponds and other water bodies including sewage canals. The species is extremely aggressive towards same and small mammals when confronted. Much like other rodents, it exhibits two behavioural adaptations to poison baiting viz. Neophobia and Bait Shyness towards acute poisons. It is a vector/reservoir of several viral, bacterial, and Rickettsial diseases and hosts several species of dangerous helminth parasites. In spite of such damage to agriculture and human/animal health, in-depth studies on its behavioural ecology are needed. Although detailed data on its burrow ecology and feeding behaviour are available, information on population dynamics, breeding, and reproduction, as well as social behaviour are scanty.

Keywords

Bandicota indica · *Bandicota bengalensis* · *Rattus* species · Crop damage · Burrow ecology · Feeding behaviour · Neophobia · Bait Shyness · Reservoir/vector

S. Sridhara (✉)
Department of Zoology, University of Agricultural Sciences, Bengaluru, Karnataka, India

S. Sridhara et al. (eds.), *Animal Behavior in the Tropics*,
https://doi.org/10.1007/978-981-97-7324-4_18

18.1 Introduction

The Larger bandicoot rat, *Bandicota indica* (Bechstein) is widely spread and is a common pest in much of South Asia (India, Bangladesh, Pakistan) and Southeast Asia (Myanmar, Malaysia, Thailand, Vietnam, Sumatra, Laos, Southern China, Hongkong, Taiwan, Nepal and Sri Lanka (Fig. 18.1). It is the largest rat on the Indian subcontinent, with a body and head length of 300–400 mm and an equally long tail; this is twice as large as the black rat or common house rat. The juvenile larger bandicoot weighs 300 g heavier than the other big field rat, *Bandicota bengalensis* while a fully grown rat weighs 500–1500 g and more (Prater 1980). This large size immediately distinguishes the bandicoot from other rats. Its pelage is darker compared to the lighter greyish of lesser bandicoot rats. *B. indica* has a robust body, a rounded head, large rounded or oval ears, and a short, broad muzzle. Its long and naked scaly tail is like other rats and mice. When exited larger bandicoots stand on their hindfeet with the piles of long hairs on their dorsal side becoming erect and grunt threateningly. However, except for some detailed investigations on their pest status (Spillet 1968; Poche et al. 1982; Islam et al. 1993; Singleton and Petch 1994; Sridhara et al. 2010), studies on ecology, behaviour, population dynamics and reproductive biology are scanty. One possible reason could be its extremely aggressive behaviour deterring researchers. When faced suddenly by humans or other small mammals like dogs, larger bandicoots make a grunting noise and the guard hairs on the back stand erect making the rat appear big, dangerous, and extremely threatening. It is said that hungry rats attack animals larger than them and may even gnaw the feet of captive elephants!

The larger bandicoot rat is essentially a parasite on man, living mostly in or around human dwellings. It causes significant damage to ground and floorings because of its burrowing habits; it also digs tunnels through bricks and masonry. Like most other rats it is omnivorous feeding on household refuse, grain, paddy, vegetables, insects, molluscs and eggs, and occasionally attacks poultry (Chakraborty and Chakraborty 1999; Parshad 1999; Sridhara 2006).

Fig. 18.1 The larger bandicoot, *Banicota indica*: Piloerection (Sridhara et al. 2010)

18.2 Species Association

In India, as a commensal *B. indica* is associated with *Bandicota bengalensis* and *Rattus rattus* (Seal and Banerjee 1969; Deoras and Pradhan 1977; Chakraborty 1985). However, Sridhara et al. (2010) observed *B. indica* as the sole species in poultry sheds in villages around Bengaluru. However, she noticed a clear-cut replacement of hitherto dominant commensal, *Rattus rattus* by *B. bengalensis* and *B. indica* in Bengaluru. This contrasts with less than 1% of *B. indica* in other urban areas of India namely Chennai, Mumbai, and Kolkata (Spillet 1968; Deoras 1966; Seal and Banerjee 1969). Although the species was abundant in Chennai during nineteenth century (Jerdan 1874; Llyod 1909), not a single species was seen in that city in 1983 (Save Grain Campaign 1983). This decline of *B. indica* in these cities was attributed to the reclamation of tanks and marshy lands in and around cities, coupled with introduction of underground drainage system in place of open drains and adaptation of the aggressive, field rat, *B. bengalensis* to a commensal way of life, competing strongly with larger bandicoots. Contrastingly *B. indica* successfully colonized bunds of newly introduced Indira Gandhi canal in Rajasthan, Haryana, and Punjab (Saxena and Chandna 1990). The delta islands of West Bengal characterized by huge marshy areas harboured dense populations of *B. indica* (62%) followed by a far lesser proportion of *B. bengalensis* (15%; Chakraborty and Chakraborty 1991a, b).

Of the three species of genus *Bandicota* Gray, two of them, *Bandicota savilei* and *Bandicota indica* coexist in mainland Thailand, Cambodia, Laos, and Vietnam (Marshall 1977; Corbet and Hill 1992; Musser and Brothers 1994; Lunde and Son 2001; Alpin et al. 2003a, b; Wilson and Reeder 2005). In most areas, the population densities of *B. indica* were quite low, although high densities might occur locally (Alpin et al. 2003b; Singleton 2003).

B. indica is not a very active animal, rather it is sedentary in habit, not moving far from its burrow especially if it lives near marshy areas. It follows definite routes to find food i.e., predetermined 'runs' are made from burrows to food sources. The 'runs' are usually concealed below bushes and other foliage and are rarely exposed. Such runs were found radiating from the burrow entrance towards rice plots in Thailand (Marshall 1977). A lone study on its home range revealed that it varied from 304 to 749 m^2 (Wang and Wang 2001). In sugarcane fields, response to sugarcane harvest differed amongst individual bandicoots. Some moved to unharvested fields, others stayed deep inside their burrows and some were crushed by the harvester (Wang and Wang 2001).

18.3 Space Technologies to Map Distribution of *B. indica*

Space technologies have been used increasingly for assessing the risk of infection by vector-borne diseases, providing tools for mapping the distribution of vectors. As rodent-borne diseases have been a growing concern in Thailand since the emergence of leptospirosis in 1998, a study was carried out to determine the hazard related to the potential presence of the main rodent vector for leptospirosis namely, *B. indica*, widely distributed in that country using satellites (Herbreteau et al. 2005). This species is found throughout Southeast Asia and immigrated recently into Thailand with the development of agriculture. Greater Bandicoot rats are now major pests in agricultural and urban settings (Nowak and Pardisiao 1991) and are involved in the transmission of diseases, especially leptospirosis. It is a major field rat (74.5% of all rodents trapped), occurring at far lesser densities around (21%) or inside houses (4.5%). The index assessing its potential presence, Bi-index, in each land use class takes values from 0 to 0.283 with the maximum for the "mixed agriculture" class, meaning a high probability of finding *B. indica* there. Using the map of land use and this index, a map of potential presence of *B. indica* was created for Nakhon Pathom province in Thailand. The first area comprising 27.6% of land with grasslands, plantations and wetlands had low presence of *B. indica.* The second area (35.7%) comprised urban and suburban areas with had higher density of *B. indica* due to human inhabitation which provides food for rodents. The last class of 36.6% represented agricultural fields (crops or paddy) and some heterogeneous transitional areas between agricultural and urban classes. These areas are often bare lands or mixed agriculture fields where rats are likely to find refuge and food as they reside proximately to larger agricultural fields.

18.4 Burrow Ecology

Bandicoots are strictly ground-dwelling, territorial animals. Piles of granular forms of fresh earth resembling large molehills at the entrance of the burrow are a good indication of their presence. Burrows of bandicoots are found on the bunds of paddy fields, inside vegetable gardens. Crop fields and poorly constructed houses with stacked stored grains, inside and outside godowns, along the banks and slopes of ponds, marshy areas and by the sides of open drains (Spillet 1968; Chakraborty and Chakraborty 1999; Sridhara et al. 2010).

The number of active burrows is indicative of the population density and relative surface activity levels of the rodents (Burhanuddin and Hafidzi 2018). Burrows can be on the bunds (elevation separating two adjacent field units) or along the boundary of vegetable fields, parks and gardens. Live burrows of *B. indica* are easily identified by the presence of fresh wet globules of soil around the burrow openings. Also visible around the burrow openings are its footprints, tail marks and soft faecal matter. If the rat is living close to water bodies, freshly broken shells of crabs and molluscs may also be seen around the burrow openings. Arjunwadkar and Gadgil (1974) reported 2–16 burrow systems of *B. indica* organized into a colony. Such burrow systems were found adjacent to grain storage godowns and mud houses near Pune, India. Within the colony, burrows were aligned next to each other. They considered each colony to be a discrete entity. However, no such colonies were found by Chakraborty and Chakraborty in West Bengal in India (Chakraborty and Chakraborty 1999). Marshall (1977) asserts that "Actually what looks like a colony with mounds, holes, runways and crop

Fig. 18.2 Burrows of *B. indica* in a poultry shed of layers (Sridhara et al. 2010)

Fig. 18.3 Burrow of *B. indica* below the steps leading to a godown (Sridhara et al. 2010)

predation turns out upon trapping to be that of just one individual or one family." Burrows were observed throughout the year along the ponds, open drains, or other water resources having some aquatic weeds, fishes, and molluscan population (Chakraborty and Chakraborty 1991a, b). They inhabited burrows on land surface from November to May but shifted to bunds bordering water bodies from June to December. In vegetable fields, burrows occurred along the boundary with their numbers fluctuating with food availability (Chakraborty and Chakraborty 1999). Poultry sheds with mud flooring (Fig. 18.2) and occasionally boundaries of godowns have permanent burrows (Fig. 18.3). Pattern and size of the burrow vary from habitat to habitat and are categorized as under:

18.4.1 Ponds and Marshy Areas

Burrows were found throughout the year in these habitats Usually, ponds having thick vegetation along the edges harboured a large number of burrows. The burrows along the slope of ponds and marshy areas were simple. Each burrow system had a few openings (1–3) on the slope of the pond and up to 11 on the surface adjacent to the pond. Often one of the burrow openings was below the level of the ground (Chakraborty and Chakraborty 1999).

When the opening of simple burrows leads to a short, unbranched, curved tunnel descending up to two strata of soil, it becomes a burrow system. The maximum length of such tunnels was 22 cm and their diameter ranged 6–14 cm. Usually, bandicoots do not live in such burrows. But their wet floor, fresh markings of feet and tail, and presence of broken molluscan shells indicate that such burrows are used as temporary shelter and as feeding sites (Chakraborty and Chakraborty 1999). Such burrows as they age, descend to a depth of 30–103 cm. Most of the tunnels and their branches ended blindly below the surface with one or two enlarging at the terminal. This enlarged end served as a brood chamber while one dead end was used to hoard food. Brood chamber was found in 34% of the burrows. Occurring at a depth of 40–90 cm it was lined with soft material like grasses, leaves and roots of aquatic weeds. The main opening was elliptical in shape, 12 cm in diameter and smooth in outline. Other openings were mostly roundish, 6–19 cm in diameter with irregular or smooth outlines. Openings with irregular openings are fresh ones. The diameter of tunnels ranged from 6 to 24 cm except at the bulged dead end (Chakraborty and Chakraborty 1999).

18.4.2 Agricultural Fields

18.4.2.1 On the Bunds

Live burrows were found on highlands adjacent to cultivated paddy fields from June to December. Such burrow systems were open on one side of the bund (Fig. 18.4). Some of them remained underwater when the field was irrigated (paddy). Each opening was connected to the central tunnel lying horizontally at a depth of

Fig. 18.4 Burrow on a bund in agricultural fields

30–60 cm which ended blindly on either side. One end of the tunnel enlarged slightly, was padded with finely cut roots and of paddy leaves, and grasses and served as a brood chamber. Usually, burrow openings were clean but when fields become dry, globules of earth were found around the burrow openings (Chakraborty and Chakraborty 1999).

18.4.2.2 Crop Fields

The frequency of occurrence of burrows of Larger bandicoot rats inside the paddy field was very low in Karnataka, south India. Chakraborty and Chakraborty (1999) observed live burrows only from late November to early May coinciding with ripening and harvest and post-harvest stages of paddy. During November–December, burrow systems were very simple with one or two openings. If there were two tunnels they were joined by a horizontal tunnel or they both got connected at their extremities to form a 'V' joint. From January onwards the tunnels became elaborate with many vertical branches getting interconnected forming a network of tunnels below the ground. The increased number of vertical tunnels led to several more burrow openings on the surface. Such tunnels extended up to 198 cm in length and their maximum depth was 72 cm (Chakraborty and Chakraborty 1999).

A survey on paddy plots in Kedah which is considered as the granary of rice in Malaysia showed a small increase in density based on 29% and 31% active burrows in 2018 and 2019 respectively. Over, 30% proportion of active burrows indicated that the level of infestation remained substantial (Burhanuddin and Hafidzi 2018).

18.4.2.3 Vegetable Fields

Burrows were seen along the bushes/bunds that formed the boundary of vegetable plots. The burrow had a single opening leading to a tunnel with smooth walls ending in a blind, slightly enlarged end. Another burrow system had three openings. From each opening, an oblique tunnel of 30–40 cm diameter descended downwards. Two such tunnels after extending to about 90 and 112 cm horizontally ended blindly into a slightly enlarged brood chamber. The third tunnel after running about 133 cm opened inside the vegetable field. Interestingly the burrow openings along the bushy border were simple openings, the ones that opened inside the vegetable fields were covered with loose soil. There was no brood chamber or signs of hoarding but partially eaten potatoes were seen. Such borrows were seen in the fields of sweet potato, potato, pumpkin, *colocasia*, etc. (Chakraborty and Chakraborty 1999).

18.4.3 Mud Houses and Poorly Constructed Buildings

Live burrows were found throughout the year inside and outside of mud houses and in poorly constructed godowns storing food grains. Each burrow system had 3–19 openings. The tunnels from the burrow entrance descended to a depth of 65 cm and were interconnected.

More often bandicoots excavated tunnels below the floor inside buildings/houses and opened outside beneath the walls. Such tunnels measured 120 cm in length, whose diameter ranged 4–13 cm and that of the interiors of the tunnel was about 5–13 cm. Opening outside of walls was a huge pile of the earth but those inside were plain openings. There were also simple holes outside used for quick entry and exit. There was no hoarding of food grains but a meagre amount of food grains, bits of old clothes and other soft, household articles were found in different parts of the burrows (Chakraborty and Chakraborty 1999; Sridhara et al. 2010).

18.4.4 Open Drains/Canal Bunds

Burrow systems of *B. indica* were found along the slopes of open drains throughout the year. Each burrow system had several openings whose diameter ranged from 4 to 7 cm. Tunnel diameter varied from 5 to 10 cm while their length ranged 30–40 cm. The burrow openings were covered with fine globules of earth during dry season. The central tunnel also gave rise to blindly ending branches but situated in an upward direction. Most burrow systems had brood chambers lined with smooth soil, pieces of paper, bits of old cloth, dry leaves and roots (Chakraborty and Chakraborty 1999).

18.5 Behaviour

Behavioural studies on *B. indica* are limited except for some scant reports on social dominance, neophobia, aversion learning, movement, and migration. Dominance by a single male was observed by Sridhara (1987) and Chakraborty (1992). The initial response of *B. indica* on encountering other animals including humans is highly aggressive. Standing on its hindfeet, with piloerection i.e., the hair on the body becoming erect, emitting threatening grunts, growls and explosive snarls, the rat induces fear in humans and pets including big dogs. The caged rats exhibit such behaviour only during initial days of caging, Once acclimated to caged conditions, they become docile and can be tamed (Blanford 1881; Marshall 1977; Sridhara 1987).

18.5.1 Social Dominance

Low intra-specific aggression was reported amongst *B. indica* (Spillet 1968). The males displayed characteristic behavioural postures when confronted with another male (Sridhara 1987). The dominant male slowly approached the submissive male, with hair becoming erect (piloerection), uttered threatening noises chasing the submissive male (Fig. 18.5). The later rat displayed a crouching, submissive posture i.e., standing on a hind foot with belly exposed and making squeaky noises. While the dominant rat moved around the cage casually, the subordinate sat in a corner. Sometimes the submissive rat is so frightened it jumps into the air inside the observation chamber. If both the males are equally dominant, chasing by one rat presumably to ascertain dominance while the less dominant tries to escape, but the ensuing

Fig. 18.5 Dominant *B. indica* (right) chasing (Sridhara et al. 2010)

fight establishes one of them as dominant (Chakraborty 1992; Sridhara et al. 2010). When two or more *B. indica* were kept in the same cage and offered food in a tray, the dominant male always guarded the food container and scared the submissive ones, not allowing them to feed. However, this behaviour was temporary as all three rats learnt to tolerate each other and shared the food within a few days (Chakraborty 1992).

18.6 Food Preferences and Feeding Behaviour

B. indica is omnivorous and exhibits versatile feeding behaviour. It feeds on grains, vegetables, and domestic refuse and is a pest in poultry and a variety of crop fields. It has a keen sense of smell and hearing and this enables it to locate insect larvae below the ground. Its snout is used to find food and the forepaws are used for digging burrows and to retrieve food. It eats the insect pests that cause damage in lawns and keeps down the populations of cockroaches and spiders. *B. indica* is an excellent swimmer and is capable of diving to retrieve prey items from the sediment at the bottom. This enables it to exploit a wide variety of aquatic and terrestrial foods such as molluscs, crustaceans, water lily fruit, water hyacinths, insects and earthworms.

18.6.1 Stomach Analysis

Investigations of stomach contents of *B. indica* at monthly intervals revealed the presence of both plant and animal matter indicating consumption of diverse foods. Molluscs, fishes, insects, earthworms, and grasses viz. *Colocacia, Derris* sp. were consumed throughout the year while quantitatively more crabs and molluscs were eaten during monsoon. Paddy and potato were found during post-monsoon. Stomach analysis revealed consumption of paddy as the maximum (9.0%) followed by potato tubers (6%), grasses (4.3%), snail, *Pila globosa* (3.8%), the bivalve, *Lamellidens marginalis* (3.0%), monitor, *Varanus* sp. (1%) and frog, *Rana* sp. (1%) (Chakraborty and Chakraborty 1999).

18.6.2 Feeding Behaviour

Studies conducted at Sundarbans and Calcutta indicated molluscs and crabs as the main food during pre-monsoon and monsoon. Occasionally fish and prawns were also consumed. Occasionally bandicoots preyed on ground-nesting birds and attacked freshwater snakes (Nandi 1984; Chakraborty and Chakraborty 1999) and chicks reared in open poultry sheds (Sridhara et al. 2010).

During harvest season, larger bandicoot rats regularly visited the adjacent paddy fields and often settled by excavating new burrows. Damage is severe at tillering stage during monsoon. When grain formation sets in damage is seen mostly at the centre of the fields and near the bund. Their number increased dramatically during ripening, harvesting, and booting stages of rice (November–December) in Myanmar (San and Sein 2019). Amongst vegetables potato, cucurbits, gourd, and musk melon were damaged significantly (Chakraborty and Chakraborty 1982).

Shukor et al. (2018) found *B. indica* consuming around six times (67.65 g) more fruitlets of oil palm than *Sundamys muelleri* (10.80 g) indicating that the species caused higher damage even with a smaller population. The rat highly preferred young oil palm meristem as a food source causing severe damage to immature and young plantings. The authors calculate that one adult greater bandicoot rat can destroy daily two immature, 3-year-old palms. Further, *B. indica* preferred the young oil palm fruitlets at the early ripening stage over the ripe ones (Hormaza et al. 2012), where the kernel has already turned solid but the outer shell of the kernel has not yet hardened (Forero et al. 2012).

18.6.3 Food Preferences Under Caged Conditions

Laboratory investigations on the food preferences of *B. indica* were carried out in three types of tests namely single-choice, double-choice and multiple-choice tests by Chakraborty and Chakraborty (1999) and Sridhara and Srihari (1983).

18.6.3.1 In Single-Choice Tests

In single-choice tests comprising *Pila globosa*, rice, wheat, green gram and Bengal gram, the snail was found to be the most preferred. Amongst food grains and pulses, rice was consumed more than other grains.

18.6.3.2 Paired Choice

When five food items namely *P. globosa*, rice, wheat, green gram and Bengal gram in a total of ten paired combinations were evaluated, a simple preference for *P. globusa* was observed in all the four dyadic tests, rice in three out of four, Bengal gram in two out of four, green gram in one out of four, while wheat was not at all preferred (Chakraborty and Chakraborty 1999).

18.6.4 Oil Preference

Paired and multiple-choice tests involving different oils mixed with cereal baits revealed the preferential order of coconut oil > ground nut oil > sesame oil > mustard oil in Calcutta (Chakraborty and Chakraborty 1999).

18.6.5 Multiple-Choice Tests

The food preference of Larger bandicoots in caged conditions was determined by offering multiple choices of cereals and pulses were studied by Sridhara and Srihari (1983). The order of preference of intake of cereals was rice > jowar > wheat > maize. Amongst pulses, the green gram was preferred the most to cowpea and field bean. Whole grain and broken forms of rice were consumed more than the flour form. When different oils mixed with rice at 10% were offered, the preferential intake was ground nut oil > gingili oil > coconut oil. The addition of 2% salt and 2% sugar to the bait did not affect food preference.

In studies in Malaysia, when offered a choice of six types of food i.e. sugarcane, ripe oil palm fruitlet, young oil palm fruitlet, young oil palm meristem, rice grains and corn to *B. indica* in the laboratory, sugarcane was the most preferred food with an average consumption of 78.99 g/day. The result was not surprising as the rats were from a field where sugarcane was cultivated before it was replaced by oil palm (Shukor et al. 2018).

18.7 Neophobia and Bait Shyness

18.7.1 Neophobia

B. indica, akin to other species of rats is highly suspicious of new foods and new objects avoiding them till it judges them to be harmless. It avoids unfamiliar objects even if it is a container with familiar food but in a different form. This was evident when the rat consumed more food from a familiar food box compared to food from unfamiliar box when both were offered simultaneously. Similarly, intact apple snails were preferred to flesh of apple snails as these rats eat normally flesh from captured live snails and thus are familiar only with intact forms of snails (Chakraborty and Chakraborty 1999).

This avoidance of unfamiliar food hampers bait-based poisoning of rodents. To overcome this avoidance, prebaiting, i.e. offering plain bait without poison for 2–3 days before poisoning is advised.

18.7.2 Bait Shyness and Acute Rodenticides

B. indica exhibited aversion towards poison-associated foods, a phenomenon called bait shyness. When bandicoots were exposed to two acute rodenticides namely sub-lethal doses 'vacor' and zinc phosphide (both at 0.05% in the cereal bait), they learnt to avoid baits associated with these poisons. Before exposure to poison, the rice bait was consumed more than the jowar bait. But when sub-lethal doses (0.05%) were added to rice bait, there was a dramatic reduction in the consumption of rice bait but a concurrent increase in the intake of jowar bait. Later, when the rats were given a choice of rice bait and jowar bait without poisons, the developed persisted aversion for up to 30 days for 'vacor' exposed rats and up to 60 days in rats that consumed sub-lethal doses of zinc phosphide (Sridhara and Srihari 1980). The practical implication of these results is to advise the farmers if they wished to use zinc phosphide to kill the larger bandicoots again, to use it 60 days after the first poisoning.

18.8 Economic Importance

The economic importance of *B. indica* lies in the damage it causes to agricultural crops and equally important is its role as a vector/reservoir of several dangerous zoonotic diseases transmitting several harmful helminth parasites to humans.

18.8.1 Damage to Crops

Larger bandicoot rat is a major pest on many crop fields like paddy (Bira et al. 2013; Borah and Bora 2012), wheat (Singal and Pasahan 1993; Hussain et al. 1992; Sheiker and Malhi 1983; Burhanuddin and Hafidzi 2018), sugarcane (Tariq et al. 2017), coconut (Yashoda 1979; Bhat and Sujatha 1986), Sorghum, groundnut (Hussain et al. 1992) and oil palm (Shukor et al. 2018). In deepwater rice in Bangladesh, the two bandicoot species, *Bandicota indica* and *Bandcota bengalensis* caused 5–10% yield losses in many fields (Catling and Yasin 1981; Ahmed et al. 1986). In Malaysia, some farmers claimed damage of up to 50% per planting season (Burhanuddin and Hafidzi 2018). In Thailand, a 1976–1978 damage survey estimated that eight rodent species, including *B. indica* had caused losses of 6.9% (Weis 1981).

Reports of yield losses to various crops by *B. indica* in India is mind-boggling. Spillet (1968) quantified rodent losses in the godowns of Calcutta, which was a cumulative estimate of losses caused by three major species of rodent pests of stored foods in India namely *B. indica*, *B. bengalensis* and *R. rattus*. Chakraborty and Chakraborty (1982) reported that actual crop losses were not significant but physical destruction to tillers was much more serious affecting crop yields, a fact which was further supported by Agarwal and Chakraborty (1985). *B. indica* caused damage to sugarcane (Srivastava 1982), coconut seedlings (Bhat 1992), cardamom

(Srihari and Chakravarthy 1992), uprooting of rubber seedlings (Jayratnam 1980), cocoa (Bhat and Sujatha 1986), poultry (Chopra 1992; Sridhara et al. 2010) and foundation of houses, and godowns (Sridhara 2006).

18.8.2 Disease Transmission

Rodents are reservoirs of several zoonotic diseases which include viral (Hantavirus, arenavirus), bacterial (Leptospirosis, plague), Ricketssial and are host to several harmful parasitic helminths, nematodes and cestodes. The diseases are transmitted directly or through contamination of food with their urine or faeces (Leptospirosis), or indirectly such as fleas biting infected rodents first and then human beings (Plague) or even by airborne transmission (Hantaviruses). *B. indica* and *Rattus exulans* were found to be the primary potential reservoirs/vectors of Hantavirus in Thailand (Elwell et al. 1985; Nitatpattana et al. 2000). Lung tissue samples from four *B. indica* confirmed positive for the Hantaviral N-antigen (Hugot et al. 2009). Studies in Thailand have shown that *B. indica* in the North-Eastern part of the country are seropositive for *Leptospira autumnalis* and *L. javanica.* In Myanmar, one specimen of *B. indica* (out of four examined) was positive for murine typhus.

Two cestodes (*Raillietina* sp., *Hymenolepis diminuta*), one nematode (*Notocotylus* sp.), ten nematodes (*Trichuris muris, Eucoleus* sp., *Aonchotheca* sp., *Ganguleterakis spumosa, Syphacia muris, Physaloptera* sp., *Protospiura* sp., *Mastophorus* sp., *Cyclodontostomum purvisi, Trichostrongylidae* sp.) and one *Pentastomida* sp., were found in *B. indica* captured from Northern and Northeast Thailand (Chaisiri et al. 2012). Six helminth species that potentially infect humans namely Raillietina sp., *Hymenolepsis nana, H. diminuta, Echinostoma ilocanum, E. malayanum* and *Angiostrongylus cantonens*is infected *B. indica, B. savilei* and *Leopoldamys edwardsi* in Laos (Sithay et al. 2020).

18.9 Conclusion

Most studies on damage caused by *B. indica* also take into consideration losses caused by other rodent pest species including *B. bengalensis* in South Asia and several *rattus* species in Southeast Asia as they occupy the same habitat. A simulated damage study in deepwater rice in Bangladesh revealed yield loss of 5–10% in many fields, with 33–83% of fields attacked by two bandicoot species and three other rodent species (Catling and Yasin 1981), and at 2–12% (mean = 6%) yield loss per hectare in 12 districts (Ahmed et al. 1986). At the national level combined losses to rice and wheat caused by rats were estimated at 20 million MT (US$ 25 million) in Bangladesh. In Thailand, a 1976–1978 damage survey (Weis 1981) estimated that eight rodent species, including *B. indica*, had caused losses of 6.9% damage. Losses to oil palms in Thailand have been estimated at 6–36%, with *B. indica* and *Rattus losea* damaging palms that are less than 4 years old and *R. tiomanicus* and

R. diardii damaging older palms (Singleton and Petch 1994). In Myanmar, rats cause extensive damage to groundnuts at certain times of the year (Weis 1981).

Gratz (1994) reported the importance of *B. indica* as a vector of rat lungworm, a nematode (*Angiostrongylus cantonensis*) that is widespread in Southeast Asia. Both humans and rodents can act as the primary host for the lungworm (Singleton and Petch 1994). In Myanmar, one specimen of *B. indica* (out of four examined) was positive for murine typhus as measured by indirect fluorescent antibody tests; however, this specimen was negative for plague (Weis 1981). Studies in Thailand have shown that *B. indica* in the North-Eastern part of the country was seropositive for *Leptospira autumnalis* and *L. javanica*.

Studies in Thailand were focused on public health impact of larger bandicoots, its pestiferous activities were studied on rice in Malaysia and Myanmar and in oil palm fruits in Malaysia. In Indian villages where the poor live, in ill-constructed structures larger bandicoots flourish. They are also abundant in small-scale poultry farms mostly again poor in structure. Although they do not predate on birds or eggs, considerable damage is caused to the feed made available to free-ranging birds.

References

Agarwal VC, Chakraborty S (1985) Rodent population and their damage in godowns and paddy fields. Non-insect pests and predators, New Delhi, pp 192–198

Ahmed SA, Mian MY, Haque ME, Brooks JE (1986) Rat damage characteristics and assessment methods in deep water rice. Bangladesh J Sci Res 4(2):131–146

Alpin KP, Frost NP, Tuan LP, Hung NM (2003a) Identification of rodents of the genus Bandicota in Vietnam and Cambodia. In: Singleton GR, Hinds LA, Krebs CJ, Spratt DM (eds) Rats, mice and people: rodent biology and management. Australian Centre for International Agriculture, ACIAR Monograph No. 96, Canberra, pp 531–535

Alpin KP, Brown PR, Jacobs J, Krebs CJ, Singleton GR (2003b) Field methods for rodent studies in Asia and the Indo-Pacific. Australian Centre for International Agriculture, Monograph No. 100, Canberra, p 136

Arjunwadkar AV, Gadgil M (1974) Burrowing habits of the greater bandicoot rat, *Bandicota indica*. J Bombay Nat Hist Soc 34:937–946

Bhat K (1992) Production losses due to rodents and counter strategies. In: Ishwar Prakash Ghosh PK (ed) Rodents in Indian agriculture. Scientific Publishers, Jodhpur, pp 279–288

Bhat K, Sujatha A (1986) A study of the species of small mammals in mixed coconut-cocoa gardens of Kerala. Planter Malaysia 62:428–432

Bira A, Ahmed SKZ, Anantharaj A, Thripati RS, Dam Roy S (2013) Occurrence of bandicoot rat, *Bandicota bengalensis* in paddy fields of Bay Islands, India. Ann Plant Prot Sci 22(1):190–239

Blanford WT (ed) (1881) Fauna of British India: mammalia. Taylor and Francis, London

Borah RK, Bora DK (2012) Incidence and damage of *Bandicota bengalensis* in rice based cropping system in upper Brahmaputra valley of Assam. Ann Plant Prot Sci 20(2):380–382

Burhanuddin M, Hafidzi HM (2018) Preliminary study of bandicoot rat population in paddy field, Kedah. Int J Agric For Plant 8:70–74. ISSN 2462-1757 201970

Catling HD, Yasin Y (1981) Rat damage in deep water rice in Bangladesh. Bangladesh J Zool 9(1):27–36

Chaisiri K, Chaeychomsri W, Sirunta-WIneti J, Ribas A, Herbreteau V, Morand S (2012) Diversity of gastrointestinal helminths among murid rodents from northern and northeastern Thailand. Southeast Asian J Trop Med Public Health 43(1):21–28

Chakraborty R (1985) Eco-toxicology of the larger bandicoot rat Bandicota indica (Bechstein) [Rodentia Muridae]. Ph.D. thesis University of Calcutta, Calcutta, India

Chakraborty R (1992) The larger bandicoot rat, *Bandicota indica* (Bechstein). In: Ishwar Prakash Ghosh PK (ed) Rodents in Indian agriculture. Scientific Publishers, Jodhpur, pp 193–210

Chakraborty R, Chakraborty S (1982) Food habits of larger bandicoot rat, *Bandicota indica* (Bechstein) in the fields of West Bengal during rainy season. Rod Newsl 6:27193–27210

Chakraborty R, Chakraborty S (1991a) Taxonomic review of the genus *Bandicota* gray and its species with a note on the intraspecific geographical variation in the larger bandicoot rat, *Bandicota indica* (Bechstein) [Mammalia; Rodentia]. Rec Zool Surv India 88(1):87–99

Chakraborty R, Chakraborty S (1991b) Burrow system of the larger bandicoot rat, *Bandicota indica* (Bechstein) [Rodentia: Muridae] in West Bengal. Rec Zool Surv India 89(1–4):105–122

Chakraborty R, Chakraborty S (1999) Feeding behaviour of the larger bandicoot rat, *Bandicota indica* (Bechstein) [Rodentia: Muridae]. Rec Zool Surv India 97(part 2):45–72

Chopra L (1992) Production losses due to rodents and control strategies: poultry farms. In: Ishwar Prakash Ghosh PK (ed) Rodents in Indian agriculture. Scientific Publishers, Jodhpur, pp 309–330

Corbet GB, Hill JE (1992) Mammals of the Indo Malayan region. A systematic review. Oxford University Press, Oxford, p 488

Deoras PJ (1966) Significance of probable changes of rat population in Bombay. In: Harris KL (ed) Proc Indian rodent symposium (1966), Calcutta, pp 58–68

Deoras PJ, Pradhan MS (1977) Observations on the bandicoot rats from Goregaon-Malad in Bombay. In: Proc. All India rodent seminar (1975), Ahmadabad, India, pp 61–69

Elwell MR, Ward GS, Tingpalapong M, LeDuc JW (1985) Serologic evidence of Hanta-like virus in rodents and man in Thailand. S East Asian T Trop Med Pub Health 16:349–354

Forero DC, Hormaza P, Romero HM (2012) Phenological growth stages of African oil palm (Elaeis guineensis). Ann Appl Biol 160:56–65

Gratz NG (1994) Rodents as carriers of disease. In: Buckle AP, Smith RH (eds) Rodent pests and their control. CAB International, Oxford

Herbreteau V, Gonzalez J, Andrianasolo H, Kittayapong P, Hugot J (2005) Mapping the potential distribution of Bandicota indica, vector of Zoonoses in Thailand, by use of remote sensing and geographic information systems (a case of Nakhon Pathom Province). Nat Hist J Chulalongkorn Univ 5:61–67

Hormaza P, Fuquen EM, Romero HM (2012) Phenology of the oil palm: interspecific hybrid, Elaeis oleifera × Elaeis guineensis. Sci Agric 69:275–280

Hugot JP, Plyusnina A, Herbreteau V, Nemoriv K, Laakkonen J, Lunddkuist A, Supputamongkol Y, Henttonen H, Plyusnin A (2009) Genetic analysis of Thailand Hantavirus in *Bandicota indica* trapped in Thailand. http://www.VirologyJ.com/3/1/72

Hussain I, Ahmad MM, Brooks JE (1992) Body size, sex ratio and reproduction of *Bandicota bengalensis* in northern Punjab, Pakistan. Pak Zool 24(3):193–198

Islam Z, Morton RG, Jupp BP (1993) The effects of rat damage on deep water rice yields in Bangladesh. Int J Pest Manag 39(2):250–254

Jayratnam K (1980) Pests in rubber plantations. In: Pillai PNR (ed) Handbook of natural rubber production in India. St. Joseph's Press, Trivendrum, pp 315–322

Jerdan RE (1874) Mammals of India. Wyman and Sons, London

Llyod RE (1909) The races of Indian rats. Rec Indian Mus 3:1–100

Lunde DP, Son NT (2001) An identification guide to the rodents of Vietnam. Center for Biodiversity and Conservation at the American Museum of Natural History, New York, p 56

Marshall JT (1977) Rats and mice in mammals of Thailand. In: Lekagal B, McNeely AJ (eds) The association for the conservation of wildlife, Bangkok, pp 397–487

Musser GG, Brothers EM (1994) Identification of bandicoot rat from Thailand (Bandicota, Muridae, Rodentia). American Museum Novitates No 3110

Nandi NC (1984) Bandicoot rat seizing a snake. J Bombay Nat Hist Soc 81:183

Nitatpattana N, Chauvancy G, Dardaine J, Poblap T, Jumronsawat K, Tangka-nakul W, Poonsuksombat D, Yoksan S, Gonzalez JP (2000) Serological study of Hantavirus in the rodent

population of Nakhon Pathom and Nakhon Ratchasima Province Thailand. Southeast Asian J Trop Med Pub Health 31:277
Nowak RM, Pardisiao JL (1991) Walker's mammals of the world, 5th edn. The John Hopkins University Press, Baltimore and London, p 1629
Parshad VR (1999) Rodent control in India. Integr Pest Manag Rev 4:97–126
Poche RM, Mian MY, Haque ME, Sultana P (1982) Rodent damage and burrowing characteristics in Bangladesh wheat fields. J Wildl Manag 46(1):139–147
Prater SH (1980) The book of Indian animals. Bombay Natural History Society, Bombay
San AY, Sein MM (2019) Population structure and trap success of *Bandicota indica* (Bechstein, 1800) in the four allocated study sites of Magway region in Myanmar. Int J Avian Wildl Biol 4(4):102–104. https://doi.org/10.15406/ijawb.2019.04.00160
Save Grain and Campaign (1983) National rodent control campaign, Government of India, Ministry of Food and Civil Supplies, Chennai. Rod Newsl 7:23–24
Saxena Y, Chandna S (1990) Prevalence of Greater Bandicoot rat, *Bandicota indica* in border areas of Haryana, Punjab and Rajasthan. Rod Newsl 14:2–3
Seal SC, Banerjee RN (1969) Changing pattern of rodent population in Calcutta and Howrah. In: Harris KL (ed) Indian Rodent Symp (1966), Calcutta, India, pp 69–83
Sheiker C, Malhi CS (1983) Territorial and hoarding behaviour in *Bandicota* sp and *Mus* spp. of Garwal Himalayas. Proc Indian Natl Sci Acad B 49(4):332–335
Shukor MI, Abidin CMRZ, Hamid NH (2018) Greater bandicoot rat, Bandicota indica infestation in oil palm plantation and its management. Planter 94(1110):100–110
Singal RK, Pasahan SC (1993) Exploration of the status of the lesser bandicoot rat, Bandicota bengalensis Gray in wheat crop. Ann Arid Zone 32(4):41–244
Singleton GR (2003) Impact of rodents on rice production in Asia. USA: IRRI Discussion Paper Series No. 43. International Rice Research Institute, Los Banos
Singleton GR, Petch DA 1994 A review of the biology and management of rodent pests in Southeast Asia. ACIAR Technical Report No. 30. p 65
Sithay P, Thongseesuksai T, Chanthavong S, Savongsy O, Khaminsou N, Boonmars T, Laummaunwai P (2020) Zoonotic Helminthiases in Rodents (*Bandicota indica*, *Bandicota savilei*, and *Leopoldamys edwardsi*) from Vientiane Capital, Lao PDR. Am J Trop Med Hyg 103(6):2323–2327. https://doi.org/10.4269/ajtmh.20-0778. Epub 2020 Sep 17. PMID: 32959774; PMCID: PMC7695065
Spillet JJ (1968) The ecology of the lesser bandicoot rat in Calcutta. Bombay Natural History Society and the John Hopkins University Centre, Calcutta
Sridhara S (1987) Social interactions in the larger bandicoot rat, *Bandicota indica*. In: Proc National Sem. Anim. Behav, Bangalore, India
Sridhara S (2006) Rodent pest management in poultry and other animal facilities. In: Vertebrate pest management: the Indian scenario. Scientific Publishers, Jodhpur, pp 205–226
Sridhara S, Srihari K (1980) Bait shyness towards zinc phosphide and 'vacor' in larger bandicoot rat, *Bandicota indica* (Bechstein). Indian J Exp Biol 18(9):1029–1031
Sridhara S, Srihari K (1983) Food preference studies of the larger bandicoot rat, *Bandicota indica* (Bechstein). Proc Indian Acad Sci 92:43–48
Sridhara S, Chaitanya N, Raveendra Babu T, Tripathi RS (2010) The larger bandicoot rat, *Bandicota indica* (Bechstein). Tech Bull No 18. All India Network Project on Rodent Control (Indian Council of Agricultural Research) Jodhpur, p 40
Srihari K, Chakravarthy AK (1992) Cardamom. In: Ishwar Prakash Ghosh PK (ed) Rodents in Indian agriculture. Scientific Publishers, Jodhpur, pp 289–308
Srivastava DC (1982) Sugarcane. In: Ishwar Prakash Ghosh PK (ed) Rodents in Indian agriculture. Scientific Publishers, Jodhpur, pp 231–248
Tariq SA, Pervez A, Zafar H, Ali R (2017) Population dynamics of rodent pests in sugarcane, (*Saccharum officinarum* L.) in lower Sindh, Pakistan. Int J Biol Biotech 14(2):283–288
Wang YP, Wang ZT (2001) Telemetry-tracked range and movements of Bandicota indica and Rattus losea in sugarcane field. Report of the Taiwan Sugar Research Institute 174:34

Weis N (ed) (1981) Rodent pests and their control. GTZ Report. German Agency for Technical Cooperation, 1A-1, Eschborn

Wilson DE, Reeder DM (eds) (2005) Mammal species of the world a taxonomic and geographic reference, 3rd edn. Johns Hopkins University Press, p 2142. ISBN: 1-800-537-5487 or (410) 516–6900. http://www.press.jhu.edu

Yashoda LU (1979) Damage to coconuts in a nursery plantation by *Bandicota bengalensis*. Int J Pest Manag 25(2):147–149

Seasonal Reproductive Strategies of Indian Palm Squirrel, *Funambulus pennanti*: Modulation by Climate, Pineal Gland and Melatonin

19

Chandana Haldar, Sameer Gupta, Rakesh Verma, and Vartika Malik

Abstract

Rhythmic environmental changes caused by planetary movements are responsible for the evolution of life on Earth. Environmental and climatic variations prepare the animals to adapt themselves to the changing conditions accordingly. These changes in the environment ensure the most essential events occur optimally just once in a year, e.g. seasonal reproduction, hibernation, fur-colour change, etc., or at a particular time of day, e.g. sleep-wake, activity-rest cycle, etc. Seasonality in reproduction is an essential component of the life cycle for the perpetuation of species and is also the primary physiological phenomenon that is frequently and repeatedly being studied. It is among the major components of the "complex web" of seasonal physiological adjustments that permit organisms to protect their progeny and maintain a positive energy balance throughout the fluctuating ambient seasonal temperature and food availability in challenging environmental settings.

Keywords

Indian palm squirrel · Reproduction · Climate · Pineal gland

19.1 Introduction

Mammals, the most successful creature of the animal kingdom began to populate the earth around 250 million years ago when it comprised of a single landmass called the supercontinent Pangaea. Some parts of the landmass experienced robust changes in the climate and thereby the food availability which formed the basis of

C. Haldar (✉) · S. Gupta · R. Verma · V. Malik
Pineal Research Lab, Department of Zoology, Banaras Hindu University, Varanasi, India

S. Sridhara et al. (eds.), *Animal Behavior in the Tropics*,
https://doi.org/10.1007/978-981-97-7324-4_19

year-round-variation in survival and reproductive strategies of the early evolved mammals (Crowley 1994). These seasonal physiological variations helped animals to cope with the environmental fluctuation in food and thereby energy availability and to adjust their reproduction to the most favourable time of the year or sometimes throughout the year. In this way, seasonality in reproduction of mammals evolved as a consequence of selection pressure for survival (Bronson 1989). Therefore, the survival and perpetuation of the species with the environment and their adaptive mechanisms were referred to as "Seasonal Reproduction". This temporally orientated reproduction became a tool for natural selection and has played a decisive role in the evolution of seasonality in mammals (Prendergast et al. 2009). On the basis of the breeding pattern, mammals can be categorised into three basic classes:

1. **Seasonal Breeders**
   Animal species, that successfully breed once only during a favourable time of the year, e.g. *Mesocricetus auratus* (Golden hamster), *Funambulus pennanti* (Indian palm squirrel), *Mus terricolor* (Indian pigmy field mice).
   (a) **Long Day Breeders** are those seasonally breeding species that breed under long photoperiodic of nature (summer) or under artificial long light conditions, e.g. ferrets, minks, hedgehogs, white-footed mice, raccoons and mares.
   (b) **Short-Day Breeders** are those animals that breed under short photoperiodic of nature (winter) when the length of the day gets shortened, e.g. sheep, deer and goats are well-known short-day breeders.
2. **Aseasonal or Continuous Breeders:** Species that breed successfully throughout the year, i.e. humans (*Homo sapiens*).
3. **Opportunistic Breeder**
   Animal species that successfully breed depending upon the immediate climatic and nutritional conditions are called "Opportunistic breeder". when the optimum conditions are transient, this strategy of reproduction is effective only for species that can produce offspring rapidly and therefore can complete lactation/weaning, etc. before the end of favourable conditions (Heideman and Bronson 1992). Best example is the tree kangaroo (*Dendrolagus* sp.).

19.2 Geophysical Changes and Biological Rhythms

Geophysical and biological rhythms are universal in nature and were known even to the ancient people who recognised the diurnal/annual variations of the light and dark cycle according to season that is accompanied by autumnal changes in the foliage of trees, migration of birds and period of activity and rest. Almost all the physiological functions of living organisms are rhythmic in nature. Behavioural patterns of most of the animals exhibit a rhythmic variation in accordance with the environmental variables. The circa-annual and circadian variation in temperature and photoperiod are among the most striking examples of a temperate zone that regulates reproduction, immunity and other physiological functions. The annual cycle of rainfall, humidity consequently with the cycle of food availability are important variables for tropical inhabitants where the photoperiod has only a few hours difference in comparison to the temperate zone.

Animals adapt to their environment by various means like migration, hibernation and especially managing reproduction. A meticulous study of the seasonal variation in the reproductive pattern of the male Indian tropical palm squirrel, *F. pennanti* establishes that unlike the temperate zone mammal—a small variation (circa 1 h 40 min to 2 h 30 min) during summer and winter photoperiod can be analysed by the Indian tropical animals via the retinohypothalamic tract and pineal gland (Pevet and Challet 2011; Haldar et al. 2006). The variation in the circulatory melatonin by the pineal gland is responsible for the regulation of seasonal rhythms in reproduction. The minor changes in photoperiod/temperature in summer/winter as well as rainfall/humidity in the tropics influence the levels of circulatory steroids (gonadal/adrenal) and pineal melatonin significantly (Ahmad and Haldar 2009). However small, the changes in season/environment induce stress affecting the adrenal steroidogenesis and can significantly compromise the reproduction and immunity in tropical squirrel *F. pennanti* (Mukherjee et al. 2014; Ahmad and Haldar 2009). Our studies establish that the pineal gland is involved in mediating the environmental cues such as photoperiod, rainfall/humidity, temperature and availability of food with reproduction to maintain the life/perpetuation of this arboreal rodent. But, the abiotic factors of nature can induce stress and can suppress immunity in a reproductive phase-dependent manner thereby compromising their survival (Ahmad and Haldar 2010a, b).

19.3 Seasonality and Seasonal Reproduction

Marked variations in environmental conditions are being experienced by most of the animal species throughout the year. For higher reproductive success, animals restrict their breeding to a particular time of year (season) when all the required conditions are favourable, i.e. abundance of food and low thermoregulatory requirements for successful rearing of the offspring, etc. (Baker 1938; Bronson 1989). Animals can retard their breeding schedule during unfavourable conditions otherwise they had to invest greater energy towards individual survival (e.g. physiological maintenance, body temperature regulation or immunity; Demas and Nelson 1998; Demas 2004; Ahmad and Haldar 2010b). Mammalian species respond to the variation in annual day-length by the adaptive alterations in their behaviour and many physiological processes in anticipation of the upcoming season (Reiter 1975, 1993). The activation and de-activation of the reproductive cycle in seasonal breeders is one of the most significant examples of a photo-periodically controlled phenomenon (Reiter 1980, 1981; Reiter et al. 2013).

Reproduction is an extremely complex and energy-demanding process. Its physiological cost is primarily borne by the female (Wade and Schneider 1992; Schneider 2004). However, males of many species use considerable energy to mate as well as to prevent other males from mating with partners. Hence, the mating of many species is restricted to a specific mating season as an adaptation to cope with this high energy cost (Prendergast 2005; Simonneaux et al. 2013). Moreover, the precise timing of the reproduction minimises maternal "cost-to-benefit" ratio thereby ensuring better survival of the offspring (Dardente 2012). The onset of the breeding season has specific physiological underpinnings and is associated with the activation of the hypothalamo-pituitary-gonadal axis in male individuals (Aleandri et al. 1996; Greives et al. 2008).

19.4 Factors Regulating Seasonal Reproduction

The synchronisation of the HPG axis to specific environmental cues is of prime importance in male reproduction in many species (Chemineau et al. 2008). The control of seasonal reproduction is an intricate interplay of both exogenous and endogenous factors (Bronson 1989, 2009). The exogenous variables involved in the control of reproduction are mostly environmental and other factors associated with it like photoperiod (length of the day), temperature, water content of the air (humidity), rain, food, social cues, predation pressure, etc. The environmental variables affecting seasonal reproduction are categorised into:

19.4.1 Ultimate Factors

Ultimate factors are those environmental variables that exert selection pressure to restrict breeding activity to a particular season or a time of the year when it has to be most successful. Reproductive timings are important under selection pressure because any errors can result in wasted effort for reproduction and can compromise survival (Baker 1938; Reiter 1975). The ultimate factors include mainly the dietary (food availability) and climatic factors (rainfall and temperature). Predator pressure and inter-specific competition can also be regarded as ultimate factors.

19.4.2 Proximate Factors

The immediate environmental cues that control the annual rhythm by regulating physiological processes are known as proximate factors. The proximate factors include photoperiod, temperature, humidity, rainfall, etc. The proximate factor will only be considered as a reliable signal of seasonal events when it consistently retains its stability over a longer period of time (Hoffmann 1973; Steger et al. 1985). The most accurate and reliable proximate factor is the amount of environmental light and dark hours, i.e. photoperiod (Baker and Ranson 1932).

19.5 Photoperiod: The Core Proximate Factor

The variations in photoperiod are the most accurate and reliable natural measure for determining the season. Thus, it acts as the priming factor for reproduction (Goldman 2001; Zucker 2001; Gwinner 2003). Photoperiod is the most "noise-free" annual predictor of seasons in mammals (Turek and Campbell 1979). Since seasonal changes in climate and food availability can be predicted on the basis of changing photoperiod, it is the most reliable predictor of upcoming environmental conditions (Bradshaw and Holzapfel 2007). Photoperiod is one of the most extensively studied extero-receptive factors in the conditions of temperate and tropical zone countries (Gwinner 1981). Photoperiodism refers to the ability of an organism to access day-length information as a predictive signal to time their seasonal events (Goldman 2001). Environmental day length is an important cue in initiating physiological and developmental functions typically irrevocable and it culminates in a future period (Bradshaw and Holzapfel 2007).

The first demonstration of the role of environmental day length as a detectable signal for the control of reproductive function in mammals was found in a colony of field voles where reproduction became continuous following exposure to 15 h of light; however, exposure to 9 h of light hampered the reproductive processes (Baker and Ranson 1932). Thereafter, the photoperiod has been reported to affect the reproductive process in almost all orders of mammals. The influence of day-length/photoperiod on reproductive physiology has been thoroughly studied in sheep (Thwaites 1965), Syrian hamsters (Reiter 1991; Steger et al. 1985) and Siberian hamsters (Hoffmann 1973). In Syrian and Siberian hamsters, short-day exposure resulted in a marked decline in concentration of peripheral gonadotropins [luteinising hormone (LH), follicle-stimulating hormone (FSH)] and testosterone, accompanied with testicular regression and cessation of spermatogenic activity within 8–12 weeks, whereas maintenance of hamsters under long day condition testicular involution (Hoffmann 1973; Steger et al. 1985). In sheep, exposure to short days induces HPG axis and activates reproduction whereas exposure to long days leads to reproductive inhibition (Chemineau et al. 2010).

To understand photoperiodism, both the Syrian and Siberian hamsters were extensively used as models, as their neuroendocrine axis is sensitive to the changing environmental day-length and they can be easily maintained in the laboratory conditions. Several other species of wild rodents such as the short-tailed field vole, *Microtus agrestis* (Spears and Clarke 1986), four-striped grass mouse, *Rhabdomys pumilio* (Jackson and Bernard 1999), white-footed mouse, *Peromyscus leucopus* (Young et al. 2000), Mongolian gerbils (Karkas et al. 2005), the tete veld rats (*Aethomys ineptus*), Namaqua rock mice *Aethomys namaquensis* (Muteka et al. 2006) and Indian palm squirrel, *Funambulus pennant* (Ahmad and Haldar 2010a), were also investigated to elucidate the effects of photoperiodic changes on seasonal reproduction.

19.5.1 Critical Photoperiod: The Duration Matters

The minimum duration of light exposure that supports the reproduction of a seasonal animal is called the "critical photoperiod" (Elliott 1976). Critical photoperiod differs significantly among animals living at different altitudes and latitudes (Bradshaw and Holzapfel 2001) and is not a fixed variable. Adult Syrian hamsters undergo gonadal (testicular) regression when exposed to <12.5 h of light per day (Elliott 1981). However, Siberian hamsters, living at higher latitudes display a critical photoperiod of 13 h of light per day for gonadal maturation, compared to Syrian hamsters (Hoffmann 1982). Manipulation of the photoperiodic history of individual rodents can also affect critical photoperiod. The critical photoperiod in Syrian hamster is independent of the previous light regime however, in Siberian hamster the critical photoperiod always appears to be dependent on the photoperiodic history (Pevet 1988).

19.5.2 Photo-Refractoriness: A Disobedience to Photoperiod

The term photo-refractoriness refers to the loss of responsiveness to photoperiod. In seasonally breeding animals, continuous/longer exposure to the typical duration of photoperiod induces photo-refractoriness, i.e. spontaneous reversion of behaviour and body physiology and to a state related to the opposite photoperiod (Lincoln et al. 2005). The photo-refractoriness is particularly noted during the spring and autumn equinoxes when seasonally breeding species of mammals have completely different physiology and despite the varying photoperiodic length the associated melatonin signal, remains the same at both times of the year (Freeman and Zucker 2001). In Syrian hamsters, prolonged exposure to short photoperiod (SD; 10–12 weeks) results in gonadal regression like the winter phenotype and then it is followed by a recovery to long photoperiod (LD: summer) type gonadal physiology within a period of 12–20 week without any change in the prevailing seasonal photoperiod (Pevet 1988). The SD-induced refractory response allows the animals in nature to reboot reproductive physiology in anticipation of spring without any stimulus of a long photoperiod (Reiter 1972). Photo-refractoriness in hamsters is terminated by exposure to a long photoperiod of about 10–15 weeks. Following such light exposure, the animals may again react favourably to short photoperiods (Pevet 1988).

19.6 Regulation of Seasonal Reproduction

Reproduction is a highly complex and energy-consuming physiological process that is being driven by the hypothalamic-pituitary-gonadal (HPG) axis. Substantial research demonstrates that photoperiod regulates the seasonal reproductive activity through modulation of pineal function that ultimately alters GnRH release by the hypothalamus (Roy and Belsham 2002) and gonadotropin (FSH/LH) secretion from the pituitary (Filippa et al. 2005). Maintenance of reproduction in mammals depends upon the activation of the HPG axis. Pulsatile release of GnRH from the hypothalamus into the hypophysial-portal systems triggers the release of gonadotropins (LH/FSH) from the pituitary which in turn, stimulates the production of gonadal steroids for the induction of spermatogenesis (Levine 2003; Bliss et al. 2010). Gonadal steroids activate sex differences and also feedback to the brain, for improvised sexual behaviour and manage GnRH and gonadotropin secretion. The HPG axis experiences episodic activation/deactivation throughout the lifetime. The modification of the HPG axis to regulate reproductive function can be influenced by photoperiod at multiple levels (Arendt 1998) and this photoperiodic message is conveyed by the chronobiotic neurohormone melatonin.

19.6.1 Melatonin: The Regulator of Seasonal Endocrine Orchestra

Melatonin, chemically known as *N*-acetyl-5-methoxytryptamine, was initially isolated and identified in the 1950s as the hormone of the neuroendocrine pineal organ (Lerner et al. 1958). Its name is indicative of its skin-lightening attribute noted primarily in fishes and amphibians as melatonin's first identified function.

The hormone melatonin received considerable interest when it was noted to regulate/reset circadian/diurnal rhythms (Redman et al. 1983; Armstrong et al. 1986). In photoperiodic animals, it also measures the day length (photoperiodic time measurement, PTM), an environmental signal for timing seasonal reproduction, metabolism and behaviour (Tamarkin et al. 1985; Arendt 1986; Reiter 1991). The presence of melatonin was noted in all vertebrates and is ubiquitously found in various taxa including, e.g. unicellular eukaryotes, plants and even bacteria (Hardeland and Fuhrberg 1996; Hardeland and Poeggeler 2008; Hardeland et al. 2007; Paredes et al. 2009) indicating that over time melatonin, has gained regulatory control over numerous additional functions during the process of evolution (Hardeland et al. 1995; Tan et al. 2010). Exceptional multiplicity in the actions of melatonin has been noted because of its integrative and orchestrating role by which it became remarkable amongst many other important signalling molecules.

19.6.2 Neuroendocrine Regulation of Seasonal Reproduction

In seasonally breeding animals, photoperiod conventionally modulates the reproductive activity to a particular time of the year or season to ensure that the progeny is born at the most favourable time of year to facilitate their survival (Reiter 1972; Goldman 2001; Reiter et al. 2009). Seasonal breeding in mammals is regulated by a distinctive photo-neuroendocrine circuit, via modulation of neuro hormone melatonin (Dardente 2012). In most of the mammals, seasonal reproduction is modulated by circulatory melatonin levels. The cellular mechanism and anatomical substrate through which melatonin modulates seasonal sexual activity are slowly being unravelled (Reiter and Sharma 2021). Recent reports indicate that hypothalamic

GnRH synthesis requires the stimulatory input from upstream hypothalamic areas to the HPG axis (Simonneaux et al. 2012) and kisspeptin (Kp) and its receptor GPR54 plays a critical role in activating these brain areas and thus controls seasonal reproductive activity (Revel et al. 2008).

19.6.2.1 Annual Cycle of Light and Dark and Variation in Circulatory Melatonin Level

The biosynthetic pathway of melatonin synthesis and the enzymes involved was for the first time described by Axelrod and co-workers (1960). The variety of modalities of light stimulus (intensity, wavelength, time and duration) is important for determining the amplitude of its effect on the circadian and neuroendocrine reproductive system in mammals (Aschoff 1981b; Cardinali et al. 1972; Takahashi et al. 1984; Brainard et al. 1983, 1986) via the circulatory melatonin (Binkley 1990; Reiter 1991). Melatonin, the chemical signal of dark period, helps the animals to make seasonal adjustments not only for reproduction but also for various other important functions such as regulation of immunity, metabolism, energy/stress/oxidative balances in order to help the animals to cope with changing environmental conditions (Figs. 19.1 and 19.2).

19.6.2.2 Melatonin, Photoperiodism and Seasonal Reproduction

The duration of nocturnal pineal melatonin production is responsible for the adjustment of reproductive processes in mammals. As the environmental light-dark condition changes with the earth's rotation, so does the melatonin rhythm. Therefore, photoperiod is considered a key priming factor for reproduction (Goldman 2001; Gwinner 2003) as changes in climate and other resources are predicted with changing photoperiod. The term "photoperiodism" is thus an organism's ability to utilise daylight information to time seasonal events in their life history. Endogenous timekeepers of the hypothalamus-hypophyseal-epiphyseal system are attached to the

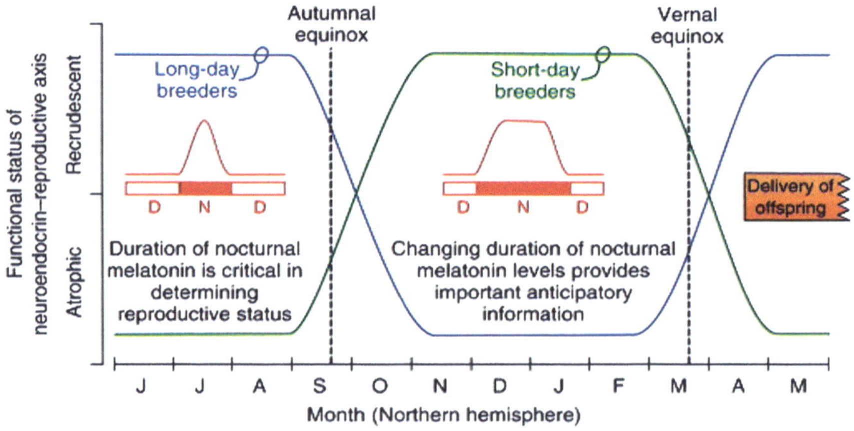

Fig. 19.1 Melatonin in regulation of seasonal reproduction. (Adapted from Reiter et al. 2018)

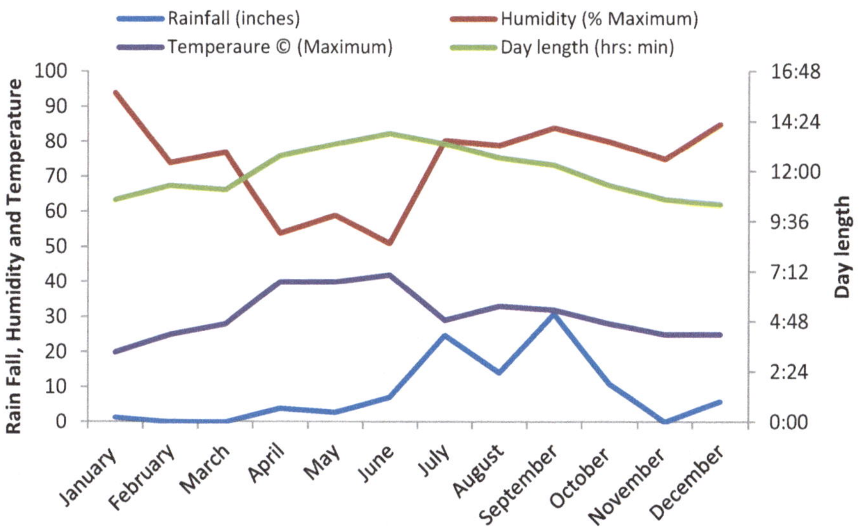

Fig. 19.2 Annual variation in rainfall, humidity, temperature and day length. (Source: Redrawn from PhD Thesis, Prof. Haldar's Lab., BHU)

photoperiod sensing centre of the retina and SCN, but the nature and length of the endogenous rhythm could be different.

19.7 Seasonal Variations in Indian Tropics

Timed environmental inputs are the most important abiotic factors that regulate mammalian reproductive physiology. Photoperiod being a proximate factor occupies the highest position in the hierarchy of environmental signals due to its ability to affect the HPG axis via retina-epiphysis-hypophysis connection. Other abiotic factors like temperature, humidity, rainfall, etc. create general stress for the squirrels (Haldar and Saxena 1988; Haldar et al. 2001). Like movement restrictions or constraints as they are hyperactive rodents and have less food availability during rainy seasons. It also creates general stress for the squirrels resulting in high circulatory corticosterone that leads to untimely reproductive regression and low immunity (Haldar et al. 2001, 2006). Such a hypothesis was experimentally examined in the Indian tropical plam squirrel, *F. pennanti*, a field rodent and also a pest of field products as enemy for farmers that commonly destroys food grain crops, fruiting plants and flowering.

In the tropical zone (Varanasi-Lat. 25°18′ North; Long. 83°1′ East) the duration of day-length in summer (June 13 h 50 min) and winter (Dec. is 10 h 26 min) comparatively have less differences than the temperate zone, while the annual rainfall cycle (July—24.72′′and humidity in %—Max. 80.34; Min. 67.5; Temp. in Jan. Max 20 °C and min 9° and in June 46° max and 22°min C; no rainfall) and the total period of solar insolation during monsoon is having a significant impact in regulation of seasonal reproductive physiology. Hence, the processes have evolved specific reproductively active and inactive phases (Haldar et al. 2001, 2006; Lahiri and Haldar 2009) (Figs. 19.2 and 19.3).

Being arboreal in habit squirrels are found in abundance in the trees close to human habitation and farms. Relative food constituent of North India varies with the season and so the food availability for the squirrels varies with different seasons of the years, such as grains, nuts, seeds and fruits in the spring and early summer, while in winter when the temperature starts decreasing, (but not the photoperiod) and harvesting are over, the food grains and seeds, etc. are not so abundantly available hence, squirrels decrease their daily activity and undergoes very short hibernation Dec.–Jan. for ~45–55 days. It can very well be noted by their disappearance from the fields. Unlike the northern palm squirrel, the palm squirrel of the southern part of India where variations in seasons are not so marked, the south Indian palm squirrel *Funambulus palmarum* breeds throughout the year.

Some animals are circa-annual species defined by their continuous annual cycles of reproduction but under persisting environmental conditions (Dardente et al. 2019). Photoperiodic species on the other hand have endogenous rhythms that are represented only during half of their life cycle. Seasonal physiological responses are guided by endogenous timekeepers and are synchronised by the annual environmental light/dark cycle, a mechanism called photoperiodic entrainment. Nocturnal melatonin synthesis provides an endogenous hormonal representation of the changes in the length

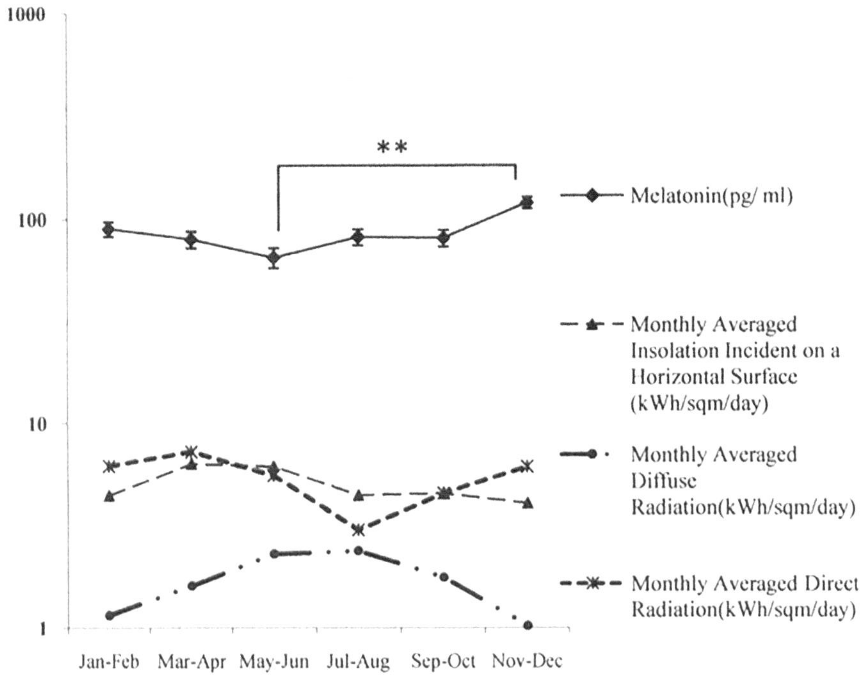

Fig. 19.3 Solar insolation during monsoon and seasonal variation in melatonin level *F. pennanti*. (Source: Redrawn from PhD Thesis, Prof. Haldar's Lab., BHU)

of the external photoperiod. Repeated daily exposure to light schedule and concomitant changes in the peripheral levels of melatonin is a critical marker for the photoperiodic changes in seasonal reproductive and metabolic processes (Dardente et al. 2014).

19.8 Seasonal Variation in Photoperiod and Reproduction of *F. pennanti*

Photoperiod is the most reliable environmental cue that regulates reproductive functions in the seasonal breeders via the neuroendocrine gland pineal and its secretory product, melatonin. The pineal gland and melatonin are the main mediators of the environmental photic signals to the body physiology of most of the seasonal breeder, which then brings about necessary behavioural, metabolic and reproductive changes via the HPG and HPA axes for adaptation to the changing environments. All seasonal breeders of temperate/tropical zones understand the external environmental condition by slight/prolonged changes in photoperiodic or day-length and then they adjust their physiology accordingly.

The sexual cycle of *F. pennanti* is essentially denoted by a short duration of quiescence prevailing in nature by the end of October. Sexual quiescence is depicted by an arrest of spermatogenesis and a decrease in the weight and activity of the accessory sex organs by the autumn. The sexual recrudescence of the squirrels starts at the beginning of the year (January). In March–April, a small but clear decrease in sexual activity especially evident at the level of the seminal vesicle and prostate weight was observed without an arrest in spermatogenesis (Srivastava and Haldar 1986). July onwards a decrease in sexual capabilities appears to have consequences on the breeding cycle of the squirrels. This period of the year is also characterised by frequent rainfall. It is difficult for the squirrel to roam around in search of food when the ground is wet and/or is filled with water as well as trees are extremely wet. This condition certainly takes part in the induction of sexual quiescence and is a strong abiotic factor-dependent phenomenon. (Fig. 19.4a, b: (a) Seasonal variation in body weight, (b) testis and accessory sex organ weight).

Our data on the regulation of reproduction in the squirrels by factors other than light may compel the reproduction biologist to re-think about their proposed theory on "environmental light as the primary indicator in the regulation of circadian rhythms or seasonal cycles of reproduction". We found factors such as neuroendocrine responses due to other endocrine glands like thyroid and adrenal and their hormones/functions could influence reproduction. The daily metabolism and seasonal stress such as extreme winter or rainfall (Aschoff 1981a; Binkley 1990; Reiter 1991) may also be the factors of influence and concern.

The exact pathways through which animals may anticipate the seasonal changes are still being deciphered. By now, it is well established that circadian pacemakers harboured in the hypothalamic suprachiasmatic nucleus (SCN) and the pineal gland are two separate and indispensable components of the translating system that senses and integrates the environmental information. Photoperiodic information is conveyed by the pineal gland to the body via daily pattern (amplitude and duration) of pineal melatonin synthesis. Seasonal variation in environmental photoperiod profoundly

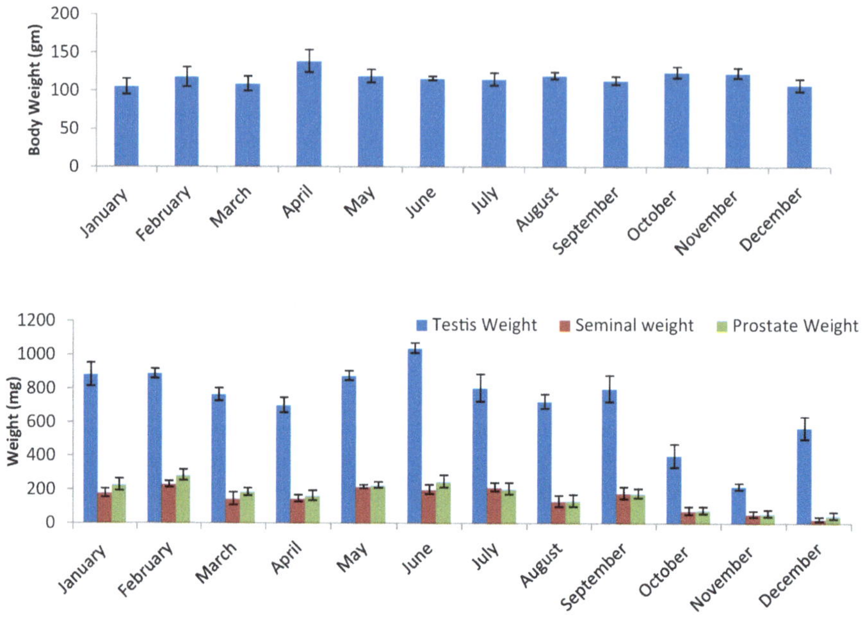

Fig. 19.4 Seasonal variations in (**a**) body and (**b**) testis, seminal and prostate gland weight of Indian palm squirrel, *F. pennanti*. (Source: Redrawn from Ph.D. Thesis, Prof. Haldar's Lab, BHU)

alters the onset timing and amplitude of melatonin peak. In mammalian species, the duration of the nocturnal melatonin is directly associated with the duration of dark period. The accumulating data suggests that seasonal regulation of melatonin synthesis/release is complex and is governed by photoperiod as well as a feedback mechanism involving melatonin itself. The melatonin-mediated synchronisation of the neuroendocrine activities (seasonal physiology and behaviour) with the prevailing environmental conditions are far more important for the animals living in nature. In the absence of the synchronising signal of melatonin, animals may become out of phase with the environment, a situation that might be risky for their survival.

In several physiological systems, the duration of melatonin secretion drives the other required metabolic changes. Thus, melatonin can act as a key regulator in mediating the redistribution of energy resources among the important physiological processes that help to maximise fitness and survival. We provided evidence for the understanding of the mechanisms underlying the neuroendocrine-immune interactions and suggested that the complex interactions among these systems get regulated at all the levels of physiology (Ahmad and Haldar 2009, 2010b).

19.9 Variations in Circulatory Melatonin and Reproduction of *F. pennanti*

Normally, melatonin at the physiological level is known to have a negative effect on reproductive status via the hypothalamo-pituitary-gonadal axis. The most important mechanism of melatonin action is via its membrane receptor pathway and is reported to act through its two high-affinity membrane receptors MT1 & MT2 to modulate number of physiological functions. Further, unlike other melatonin receptors for which to date controversy exists, MT1 has been shown to enjoy a prominent expression. The receptors, MT1 & MT2, were selected to explain the photoperiodic regulation of reproduction by melatonin, gonadal steroid and androgen receptor of a tropical seasonal breeder *F. pennanti*. An intricate relationship between MT1 and androgen receptor—AR at the translational level elucidated exact mechanism behind the trade-off relationship between immunity and reproduction. Immunity is another seasonally affected physiological phenomenon of the animals that has an intricate relation with reproduction. Healthy reproduction is only possible if the animals are healthy.

19.10 Influence of Solar Insolation and Experimental Photoperiod on Expression of Melatonin Receptor Subtype MT1 in Indian Palm Squirrel *F. pennanti*

The effects of natural solar insolation (due to the dense clouds for longer the day during monsoon reducing the intensity of the light drastically) for days and weeks (Fig. 19.3) and experimental photoperiodic conditions on circulatory melatonin level and expression of MT1 receptor on gonads of *F. pennanti* was noted in relation to testicular function. We noted high pineal secretory activity resulting in higher peripheral melatonin levels with a decreased testicular melatonin receptor expression in *F. pennanti* during short days of winter months. Moreover, constant exposure to light mimicked an "enhanced summer" like condition/effect for the testicular physiology and vice versa was observed when exposed to constant dark conditions.

During annual study of environmental effect on the reproduction of *F. pennanti,* a slight increase in melatonin levels during monsoon months was noted when the duration of the photoperiod was less and the solar insolation reaching the earth's surface was significantly reduced in terms of direct radiations. Not only the time period of natural sunlight but also the intensity of sunlight expressed as solar insolation affected the circulatory level of melatonin and its receptor expression on the testes of this wild field tropical rodent (Lahiri and Haldar 2009). We noted that an increase in circulatory level of melatonin as well as evening melatonin injections decreased receptor subtype MT1 expression in the testes at both transcriptional as well as translational levels, reflecting autoregulatory downregulation of melatonin receptors. So, in *F. pennanti,* melatonin may be suggested as a molecular messenger of photoperiodic signals that operate via MT1 receptor. Such an expression of melatonin and its receptor was also noted for the spleen of *F. pennanti* and hence these results suggest a similar photoperiodic regulation/adaptation for immune function may exist in animals residing in the tropical zone (Lahiri and Haldar 2009; Ahmad and Haldar 2009, 2010b).

To everything there is a season and a time for every purpose under heaven

References

Ahmad R, Haldar C (2009) Photoperiod-testicular-immune interaction in a seasonal breeder Indian palm squirrel *Funambulus pennanti* during the reproductively inactive and active phases. J Neuroendocrinol 21:2–9

Ahmad R, Haldar C (2010a) Photoperiodic regulation of MT1 and MT2 melatonin receptor expression in spleen and thymus of a tropical rodent *Funambulus pennanti* during reproductively active and inactive phases. Chronobiol Int 27:446–462

Ahmad R, Haldar C (2010b) Melatonin and androgen receptor expression interplay modulates cell mediated immunity in tropical rodent *Funambulus pennanti*. Scand J Immunol 71:420–430

Aleandri V, Spina V, Morini A (1996) The pineal gland and reproduction. Hum Reprod Update 2:225–235

Arendt J (1986) Role of the pineal gland and melatonin in seasonal reproductive function in mammals. Oxf Rev Reprod Biol 8:266–320

Arendt J (1998) Melatonin and the pineal gland: influence on mammalian seasonal and circadian physiology. Rev Reprod 3:13–22

Armstrong SM, Cassone VM, Chesworth MJ, Redman JR, Short RV (1986) Synchronization of mammalian circadian rhythms by melatonin. J Neural Transm Suppl 21:375–394

Aschoff J (ed) (1981a) Handbook of behavioral neurobiology: biological rhythms, vol 4. Plenum Press, New York, pp 1–563

Aschoff J (1981b) Free running and entrained circadian rhythms. In: Aschoff J (ed) Handbook of behavioral neurobiology: biological rhythms, vol 4. Plenum Press, New York, pp 81–92

Baker JR (1938) The relationship of breeding seasons. In: de Beer GR (ed) Evolution, essays of aspects of evolutionary biology. Oxford University Press, Oxford, pp 161–177

Baker JR, Ranson RM (1932) Factors affecting the breeding of the field mouse (*Microtus agrestis*) part 1, Light. Proc R Soc Lond B Biol Sci 110:3–13

Binkley S (1990) The clock work sparrow: time, clocks and calendars in biological organisms. Prentice Hall, Englewood, pp 1–259

Bliss SP, Navratil AM, Xie J, Roberson MS (2010) GnRH signaling, the gonadotrope and endocrine control of fertility. Front Neuroendocrinol 31:322–340

Bradshaw WE, Holzapfel CM (2001) Phenotypic evolution and the genetic architecture underlying photoperiodic time measurement. J Insect Physiol 47:809–820

Bradshaw WE, Holzapfel CM (2007) Evolution of animal photoperiodism. Annu Rev Ecol Evol Syst 38:1–25

Brainard GC, Richardson BA, King TS, Matthews SA, Reiter RJ (1983) The suppression of pineal melatonin content and N-acetyltransferase activity by different light irradiances in the Syrian hamster. Endocrinology 113:293–296

Brainard GC, Vaughan MK, Reiter RJ (1986) Effect of light irradiance and wavelength on the Syrian hamster reproductive system. Endocrinology 119:648–654

Bronson FH (1989) Mammalian reproductive biology. Univ Chicago Press, Chicago, p 325

Bronson FH (2009) Climate change and seasonal reproduction in mammals. Philos Trans R Soc B 364:3331–3340

Cardinali DP, Lynch HJ, Wurtman RJ (1972) Binding of melatonin to human and rat plasma proteins. Endocrinology 91(5):1213

Chemineau P, Guillaume D, Migaud M, Thiéry JC, Pellicer-Rubio MT, Malpaux B (2008) Seasonality of reproduction in mammals: intimate regulatory mechanisms and practical implications. Reprod Domest Anim 2:40–47

Chemineau P, Bodin L, Migaud M, Thiéry JC, Malpaux B (2010) Neuroendocrine and genetic control of seasonal reproduction in sheep and goats. Reprod Domest Anim 45(Suppl 3):42–49

Crowley TJ (1994) Pangean climates. In: Klein GD (ed) Pangea: paleoclimate, tectonics and sedimentation during the accretion, zenith and breakup of a supercontinent, vol 288. Geological Society of America, pp 25–55

Dardente H (2012) Melatonin dependent timing of seasonal reproduction by the pars tuberalis: pivotal roles for long day lengths and thyroid hormones. J Neuroendocrinol 24:249–266

Dardente H, Hazlerigg DG, Ebling FJ (2014) Thyroid hormone and seasonal rhythmicity. Front Endocrinol (Lausanne) 5:19
Demas GE (2004) The energetics of immunity: a neuroendocrine link between energy balance and immune function. Horm Behav 45:173–780
Demas GE, Nelson RJ (1998) Social, but not photoperiodic, influences on reproductive function in male *Peromyscus aztecus*. Biol Reprod 58:385–389
Elliott JA (1976) Circadian rhythms and photoperiodic time measurement in mammals. Fed Proc 35:2339–2346
Elliott JA (1981) Circadian rhythms, entrainment and photoperiodism in the Syrian hamster. In: Follett BK, Follett DE, Wright J (eds) Biological clocks in seasonal reproductive cycles. Bristol, pp 203–217
Filippa V, Penissi A, Mohamed F (2005) Seasonal variations of gonadotropins in the pars distalis male viscacha pituitary. Effect of chronic melatonin treatment. Eur J Histochem 49:291–300
Freeman DA, Zucker I (2001) Refractoriness to melatonin occurs independently at multiple brain sites in Siberian hamster. Proc Natl Acad Sci U S A 98:6447–6452
Goldman BD (2001) Mammalian photoperiodic system: formal properties and neuroendocrine mechanisms of photoperiodic time measurement. J Biol Rhythm 16:283–301
Greives TJ, Kriegsfeld LJ, Bentley GE, Tsutsui K, Demas GE (2008) Recent advances in reproductive neuroendocrinology: a role for RFamide peptides in seasonal reproduction? Proc Biol Sci 275:1943–1951
Gwinner E (1981) Annual rhythms: perspective. In: Aschoff J (ed) Handbook of behavioral neurobiology: biological rhythms, vol 4. Plenum Press, New York, pp 381–389
Gwinner E (2003) Circannual rhythms in birds. Curr Opin Neurobiol 13:770–778
Haldar C, Saxena N (1988) Pineal gland and humidity effects on testicular function of the Indian palm squirrel, Funambulus pennanti. J Pineal Res 5:411–418
Haldar C, Singh R, Guchhait P (2001) Relationship between the annual rhythms in melatonin and immune system status in the tropical palm squirrel, *Funambulus pennanti*. Chronobiol Int 18(1):61–69
Haldar C, Sharma S, Singh SS (2006) Reproductive phase dependent circadian variation of plasma melatonin, testosterone, thyroxine and corticosterone in Indian palm squirrel, Funambulus pennanti. Biol Rhythms Res 37(1):1–10
Hardeland R, Fuhrberg B (1996) Ubiquitous melatonin-presence and effects in unicells, plants and animals. Trends Comp Biochem Physiol 2:25–45
Hardeland R, Poeggeler B (2008) Melatonin beyond its classical functions. Open Physiol J 1:1–23
Hardeland R, Balzer I, Poeggeler B, Fuhrberg B, Uría H, Behrmann G, Wolf R, Meyer TJ, Reiter RJ (1995) On the primary functions of melatonin in evolution: mediation of photoperiodic signals in a unicell, photooxidation, and scavenging of free radicals. J Pineal Res 18:104–111
Hardeland R, Pandi-Perumal SR, Poeggeler B (2007) Melatonin in plants-focus on a vertebrate night hormone with cytoprotective properties. Funct Plant Sci Biotechnol 1:32–45
Heideman PD, Bronson FH (1992) A pseudo-seasonal reproductive strategy in a tropical rodent, *Peromyscus nudipes*. J Reprod Fertil 95:57–67
Hoffmann K (1973) The influence of photoperiod and melatonin on testis size, body weight, and pelage colour in the Djungurian hamster (*Phodopus sungorus*). J Comp Physiol 95:267–282
Hoffmann K (1982) The critical photoperiod in the Djungarian hamster *Phodopus sungorus*. In: Aschoff J, Daan S, Groos G (eds) Vertebrate circadian systems. Springer, Heidelberg, pp 297–304
Jackson C, Bernard RTF (1999) Short day length alone does not inhibit spermatogenesis in the seasonally breeding four-striped field mouse (*Rhabdomys pumilio*). Biol Reprod 60:1320–1323
Karkas A, Camasari C, Serin E, Gunduz B (2005) Effect of photoperiod and food availability on growth, leptin, sexual maturation and maintenance in the Mongolian gerbils (*Meriones unquiculatus*). Zool Sci 22:665–670
Lahiri S, Haldar C (2009) Response of melatonin receptor MT1 in spleen of a tropical Indian rodent, *Funambulus pennanti*, to natural solar insolation and different photoperiodic conditions. Chronobiol Int 26:1559–1574

Lerner AB, Case JD, Takahashi Y, Lee TH, Mori N (1958) Isolation of melatonin, pineal factor that lightens melanocytes. J Am Chem Soc 80:2587–2589
Levine JE (2003) Gonadotropin-releasing hormone (GnRH). In: Helen LH, Anthony WN (eds) Encyclopedia of hormones. Academic Press, New York, pp 157–165
Lincoln GA, Johnston JD, Andersson H, Wagner G, Hazlerigg DG (2005) Photorefractoriness in mammals: dissociating a seasonal timer from the circadian-based photoperiod response. Endocrinology 146:3782–3790
Mukherjee A, Haldar C, Vishwas DK (2014) Melatonin prevents dexamethasone induced testicular oxidative stress and germ cell apoptosis in golden hamster, *Mesocricetus auratus*. J Androl 47(8):920–931
Muteka SP, Chimimba CT, Bennett NC (2006) Reproductive seasonality in *Aethomys namaquensis* (Rodentia: Muridae) from Southern Africa. J Mammal 8:67–74
Paredes SD, Korkmaz A, Manchester LC, Tan DX, Reiter RJ (2009) Phytomelatonin: a review. J Exp Bot 60:57–69
Pevet P (1988) The role of the pineal gland in the photoperiodic control of reproduction in different hamster species. Reprod Nutr Dev 28:443–458
Pevet P, Challet E (2011) Melatonin: both master clock output and internal time-giver in the circadian clocks network. J Physiol Paris 105:170–182
Prendergast BJ (2005) Internalization of seasonal time. Horm Behav 48:503–511
Prendergast BJ, Pyter LM, Galang J, Kay LM (2009) Reproductive responses to photoperiod persist in olfactory bulbectomized Siberian hamsters (*Phodopus sungorus*). Behav Brain Res 198:159–164
Redman J, Armstrong S, Ng KT (1983) Free-running activity rhythms in the rat: entrainment by melatonin. Science 219:1089–1090
Reiter RJ (1972) Evidence for refractoriness of the pituitary-gonadal axis to the pineal gland in golden hamsters and its possible implications in annual reproductive rhythms. Anat Rec 173:365–372
Reiter RJ (1975) The pineal gland and seasonal reproductive adjustments. Int J Biometeorol 19:282–288
Reiter RJ (1980) Photoperiod: its importance as an impeller of pineal and seasonal reproductive rhythms. Int J Biometeorol 24:57–63
Reiter RJ (1981) Pineal control of reproduction. Prog Clin Biol Res 59:349–355
Reiter RJ (1991) Neuroendocrine effects of light. Int J Biometeorol 35:169–175
Reiter RJ (1993) The melatonin rhythm: both a clock and a calendar. Experientia 49:654–664
Reiter RJ, Sharma R (2021) Central and peripheral actions of melatonin on reproduction in seasonal and continuous breeding mammals. Gen Comp Endocrinol 1(300):113620
Reiter RJ, Tan DX, Manchester LC, Paredes SD, Mayo JC, Sainz RM (2009) Melatonin and reproduction revisited. Biol Reprod 81:445–456
Reiter RJ, Rosales-Corral SA, Manchester LC, Tan DX (2013) Peripheral reproductive organ health and melatonin: ready for prime time. Int J Mol Sci 14:7231–7272
Reiter RJ, Tan DX, Sharma R (2018) Historical perspective and evaluation of the mechanisms by which melatonin mediates seasonal reproduction in mammals. Melatonin Res 1(1):59–77
Revel FG, Saboureau M, Pevet P, Simonneaux V, Mikkelsen JD (2008) RFamide-related peptide gene is a melatonin-driven photoperiodic gene. Endocrinology 149:902–912
Roy D, Belsham DD (2002) Melatonin receptor activation regulates GnRH gene expression and secretion in GT1-7 GnRH neurons. Signal transduction mechanisms. J Biol Chem 277:251–258
Schneider JE (2004) Energy balance and reproduction. Physiol Behav 81:289–317
Simonneaux V, Bur I, Ancel C, Ansel L, Klosen P (2012) A kiss for daily and seasonal reproduction. Prog Brain Res 199:423–437
Simonneaux V, Ancel C, Poirel VJ, Gauer F (2013) Kisspeptins and RFRP-3 act in concert to synchronize rodent reproduction with seasons. Front Neurosci 7:22
Spears N, Clarke JR (1986) Effect of male presence and of photoperiod on the sexual maturation of the field vole (*Microtus agrestis*). J Reprod Fert 78:231–238

Steger RW, Matt K, Bartke A (1985) Neuroendocrine regulation of seasonal reproductive activity in the male golden hamster. Neurosci Biobehav Rev 9:191–201

Takahashi JS, Decoursey PJ, Bauman L, Menaker M (1984) Spectral sensitivity of a novel photoreceptive system mediating entrainment of mammalian circadian rhythms. Nature 308:186–188

Tamarkin L, Baird CJ, Almeida OF (1985) Melatonin: a coordinating signal for mammalian reproduction? Science 227:714–720

Tan DX, Hardeland R, Manchester LC, Paredes SD, Korkmaz A, Sainz RM, Mayo JC, Fuentes-Broto L, Reiter RJ (2010) The changing biological roles of melatonin during evolution: from an antioxidant to signals of darkness, sexual selection and fitness. Biol Rev Camb Philos Soc 85:607–623

Thwaites CJ (1965) Photoperiodic control of breeding activity in the Southdown ewe with particular reference to the effects of an equatorial light regime. J Agric Sci Cambr 65:57–65

Turek FW, Campbell CS (1979) Photoperiodic regulation of neuroendocrine-gonadal activity. Biol Reprod 20:32–50

Wade GN, Schneider JE (1992) Metabolic fuels and reproduction in female mammals. Neurosci Biobehav Rev 16:235–272

Young KA, Zirkin BR, Nelson RJ (2000) Testicular regression in response to food restriction and short photoperiod in white-footed mice (*Peromyscus leucopus*) is mediated by apoptosis. Biol Reprod 62:347–354

Zucker I (2001) Circannual rhythms. In: Takahashi JS, Turek FW, Moore RY (eds) Handbook of behavioral neurobiology: circadian clocks, vol 12. Kluwer Academic/Plenum Publishers, New York, pp 509–527

Behaviour and Ecology of Swamp Deer (*Rucervus duvaucelii duvaucelii*) in Captivity in India

20

Parvez, Akshay Kumar Chakravarthy, and Subhadra Barik

Abstract

Experiments were conducted on the behaviour of swamp deer (*Rucervus duvaucelii duvaucelii*) at the National Zoological Park (NZP) New Delhi, India, in 2020. Three species of swamp deer with restricted range are endemic to India. Swamp deer is an obligate grassland-dwelling endemic cervid and the most extinction-prone mega herbivore in the Indian subcontinent. The species is vulnerable and differs from other members of the deer family (Corvidae) having antlers with three tines. A health chart and disease history have been observed, and health monitoring was conducted during the study period based on behavioural observations of 16 (5M:11F) swamp deers (*R. duvaucelii duvaucelii*) at Beat No. 16, NZP, Delhi. Tuberculosis and pulmonary tuberculosis illnesses were identified to harm swamp deer population in the zoological park.

Keywords

Swamp deer · Captivity · Behaviour · Ecology

Parvez (✉)
Society for Science and Technology Applications (SSTA), Bengaluru, Karnataka, India

A. Kumar Chakravarthy
Division of Entomology and Nematology, Indian Institute of Horticultural Research, Bengaluru, Karnataka, India

S. Barik
Sálim Ali Centre for Ornithology and Natural History, Coimbatore, Tamil Nadu, India

S. Sridhara et al. (eds.), *Animal Behavior in the Tropics*,
https://doi.org/10.1007/978-981-97-7324-4_20

20.1 Introduction

Swamp deer (*Rucervus duvaucelii duvaucelii*; Corvidae; Artiodactyla) is one of the three living species of swamp deer in the world. The name of this species has its origin in the Indian language Hindi, *Barasingha*, which means '12-antlered deer' and also refers to the habitat preferred by the species. The swamp deer (*R. duvaucelli duvaucelli*) name commemorates the French naturalist Alfred Duvaucel. A swamp deer male is called a stag, a female is called a doe and newborn is called a fawn. Their coat colour is chestnut brown on the back, fading to a lighter brown on the sides, creamy white on the belly and on the inside of limbs and underside of the tail. Their throat, insides of ears and chin are white. The adult males have darker coats than females with conspicuously large antlers (grow up to 1 m in length) covered by velvet in summer except during their breeding season. This subspecies (*duvaucelli*) of swamp deer has splayed hooves because of its swampy habitat.

These are highly sensitive species having social, terrestrial, nocturnal and herbivorous (polyphagous) mammals. Swamp deer is the only deer species with one breeding season in a year, that's why it is also known as monoestrous (females coming into the heat only once in year). Swamp deer are normally found in herds of similar gender and age, herds have about 10–20 individuals (Ahmad and Khan 2014). Generally female leads the herd followed by males in the rear. Their population is decreasing and is listed as a vulnerable species due to habitat destruction (Primack 2006) as a result of deforestation (Sadadev et al. 2021) followed by draining of swamps and marshes for agricultural purposes. Currently, these mammals are hunted for their horns used in cutlery handles, pistol butts, buttons and trophies to decorate houses (Dhami et al. 2021; Khulal et al. 2021). It is also used in traditional oriental medicines. Also, they are threatened by diseases (Regmi et al. 2022) from domestic cattle. On the other hand, a considerable number of mortalities among this species occur in captivity. Given the limited distribution and marginalized habitat status, pose significant conservation and management challenges for the remaining fragmented population and in north, north-east and central India. So, a swamp deer-centric conservation approach is required to save these animals in human-dominated habitats.

According to Holloway (1973) in 1965, there were 11 populations of the already declining northern race of swamp deer in Uttar Pradesh, India. But by 1992, eight of these disappeared through mainly habitat destruction. Nandy et al. (2012) identified the habitat for swamp deer in the Jhilmil Jheel Conservation Reserve in Uttarakhand, India. During the early twentieth century, the swamp deer existed in the Indo-Gangetic plains (Duckworth et al. 2013). However, according to Karanth et al. (2010), among all the Indian herbivores, swamp deer has the highest extinction probability. The above studies highlight the importance of the riverine-grasslands system of northern India where a considerable amount of biodiversity co-exists with a very high human population. Therefore, there is a pressing need to focus on population enumeration and grassland management, particularly found in conservation areas such as the Ganga and Sharda blocks.

A captive breeding programme, important for the conservation of the species, requires detailed knowledge about the behavioural pattern of the species. Available published data on swamp deer is primarily from Jhilmil Jheel Conservation Reserve, Haridwar, Uttrakhand, India; *Feed and feeding habitat of swamp deer* by R. Tewari and G.S. Rawat, and Dudhwa National Park, Uttar Pradesh, India; by *Kaleem Ahmed and Jamal A. Khan* on seasonal activity pattern of swamp deer. Keeping all points in view, an attempt was made to prepare a standard ethogram for swamp deer in the present paper based on the observations in captivity at National Zoological Park (NZP), New Delhi, India.

Daily the concentration of food, including mesh (concentrated) (2 kg), kutti (4 kg), gram soaked (500 g), tree fodder (500 g), cucumber (summer)/carrot (winter) (100 g), and amla green (winter) (50 g), was served to an individual.

20.2 Distribution

It is the largest grassland-dwelling endemic cervid of India and Nepal and its population is declining and is found in the fragmented habitat. Swamp deer is distributed in central India (*R. duvaucelii branderi*), northern and north-eastern India (*R. duvaucelii anjitsinhji*) and southern Nepal (Fig. 20.1).

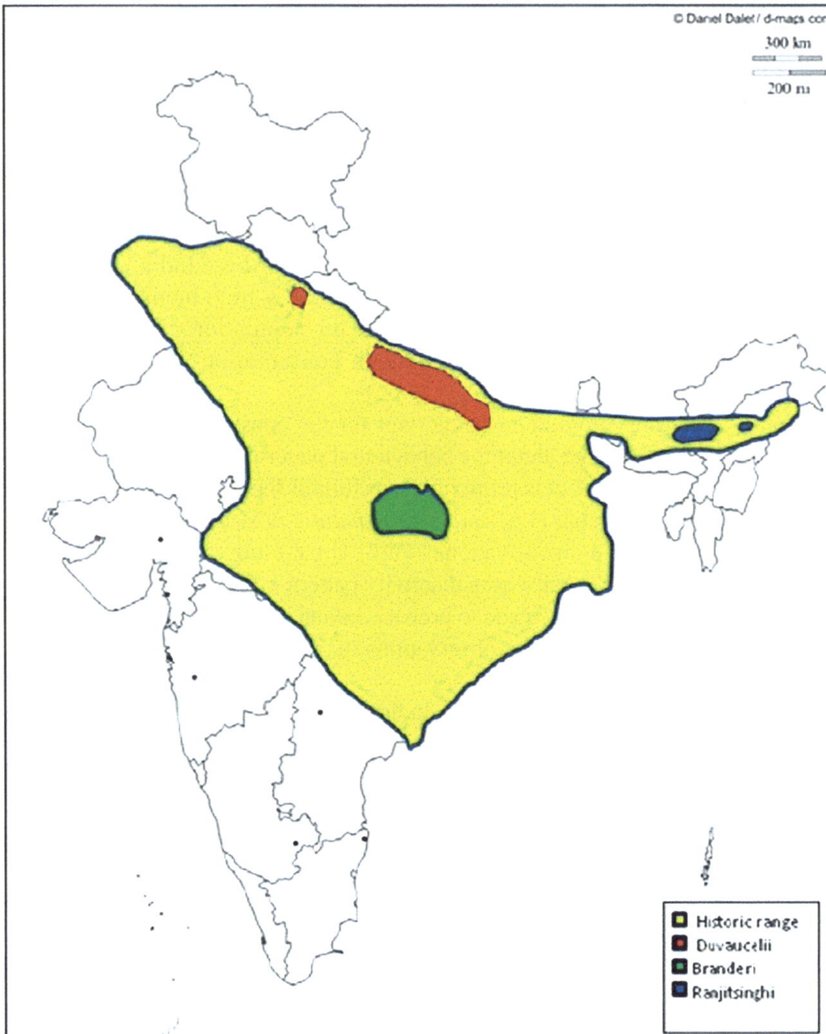

Fig. 20.1 Distribution of swamp deer. (Source: Tewari and Rawat 2013)

20.3 Breeding

The primary purpose of captive breeding of animals is to secure the survival rate of endangered species and maintain in controlled environment to create a sizeable, stable and healthy population in order to avoid extinction, and to reintroduce species back into their natural habitats when conditions allow. Captive breeding plans help to ensure large, healthy and genetically diverse populations that otherwise would not exist. National Zoological Park plays a vital role in captive breeding to bolster an existing population enough to allow the species to be reintroduced into the wild in one or more areas. The role of NZP in the captive breeding of swamp deer (*R. duvaucelli duvaucelli)* is quite impressive. Swamp deer are seasonal breeders, they are monoestrous and monotocous in nature (Qureshi et al. 2004). Stag is capable of reproducing at the age of 2–3 years and is capable of reproducing at the age of 4 years onwards. They breed in NZP from mid-October to early January, with a gestation period of 240–280 days. Rutting starts around mid-September, while shedding of antlers begins mid-August, and infighting occurs during mating to show the dominance in the herd.

Swamp deer is a highly endangered species in the urge of extinction listed as a Schedule-I species under the Wildlife (Protection) Amendment Act, 2022 and vulnerable in the International Union for Conservation of Nature (IUCN) red list and Appendix-I of the Convention on International Trade of Endangered Species (CITES). Swamp deer in captive breeding are not mated randomly. Stags and does are crossed in such a way as to preserve maximum genetic variability and to obtain a demographically stable population. For this reason, breeding programmes use pedigree records or the so-called studbooks books as a management tool.

These records are a complete inventory of the captive animals of the species included in the breeding programme wherever they are in the world. These books give important data about their dates of birth and death, individual identification of sire and dam and each individual's location throughout their life. Individual identification is done by assigning a 'studbook number' at birth that acts as the identity of each species and allows it to be identified up until its death (Table 20.1 and Figs. 20.2 and 20.3).

Table 20.1 List of participatory zoos in swamp deer conservation breeding

Alipore Zoological Garden, Kolkata	Nawab Wajid Ali Shah Zoological Garden, Lucknow
Arignar Anna Zoological Park, Chennai	Nehru Zoological Park, Hyderabad
Indira Gandhi Zoological Park, Visakhapatnam	Sakkarbaug zoo, Junagadh
Kanan Pendari zoo, Raipur	Sanjay Gandhi Biological Park, Patna
Kanpur Zoological Park, Kanpur	Sri Chamarajendra zoological gardens, Mysore
Mahendra Chaudhary Zoological Park, Punjab	Sri Venkateswara Zoological Park, Tirupati
Nandankanan Zoological Park, Bhubaneswar	Thiruvananthapuram Zoological Park, Thiruvananthapuram

Fig. 20.2 Chasing for mating

Fig. 20.3 Resting pregnant female deer

20.4 Behaviour and Ecology

The average time spent in each behavioural activity by male and female is depicted in Fig. 20.4. These observations were recorded at Beat No.16, NZP. August to November 2019–2020. Of the 16 activities recorded, the animals spent maximum time resting, followed by feeding, standing and walking (Fig. 20.4).

When the time budgeting was considered on the adult sex basis, it was found that females were more active than males. The females spent more time feeding and standing and walking while the males spent most of the time resting (Fig. 20.5).

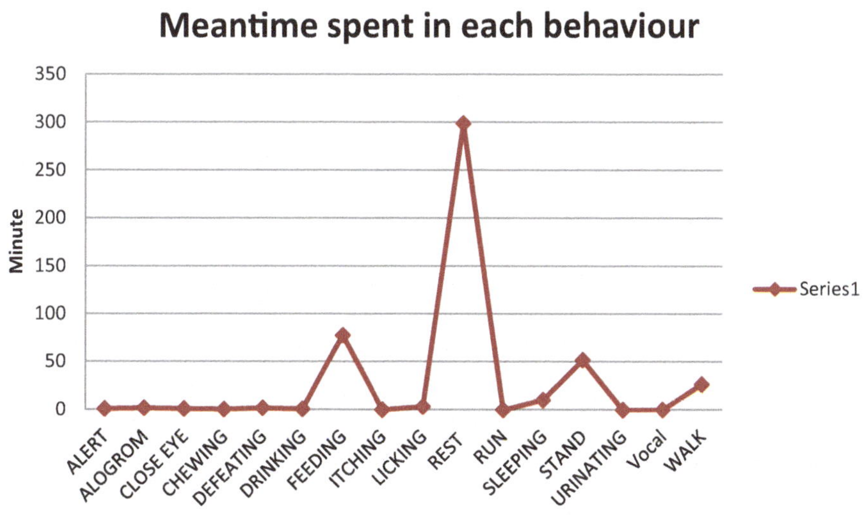

Fig. 20.4 Mean time spent by the animal for each behavioural trait

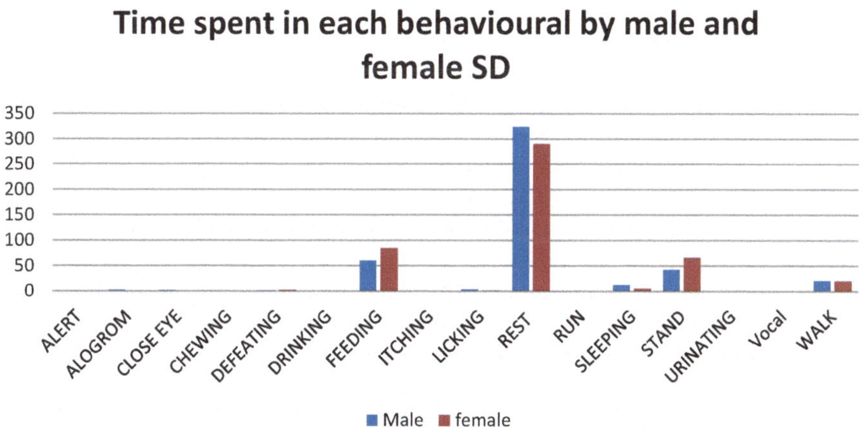

Fig. 20.5 Time spent for each behavioural trait by male and female swamp dear

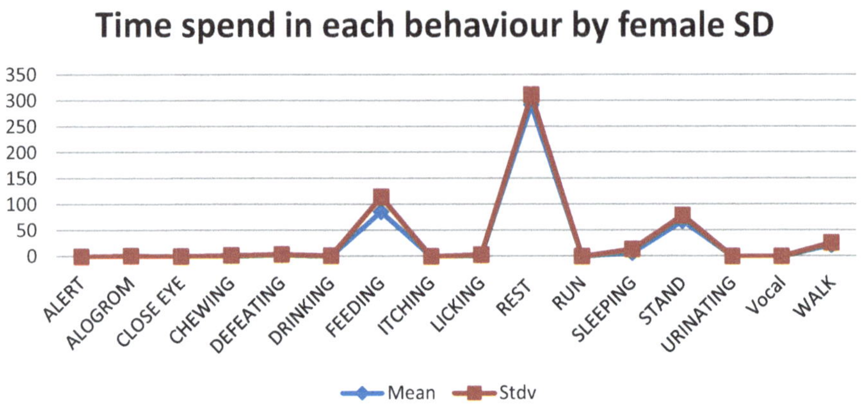

Fig. 20.6 Time spent for each behavioural trait by female swamp deer

The detailed analysis of time budgeting of activities of females revealed that maximum time was spent resting followed by feeding and standing. This is obvious because to feed, the animal has to make movements (Figs. 20.6 and 20.7).

Similar observations were recorded in Dhudhwa National Park (DNP), Uttar Pradesh, India (Ahmad and Khan 2014). The DNP represents a highly diverse and productive deciduous forest dominated by Sal (*Shorea robusta*) trees. The workers recorded observations on swamp deer herd from sunrise to sunset. When analyzing the time allocation for various activities, it was noted that resting constituted 63.77% of their time, while feeding accounted for 24.70%. Interestingly, regardless of the season—be it winter or summer—resting emerged as the predominant activity for the deer. The workers also found that there were significant seasonal differences in the time spent on feeding. The reduction in feeding time in winter was compensated by an increase in resting time. This suggests that the herd spent more time feeding during the night. Schaller (1967) also recorded a herd of 20 SD grazing throughout the night. The increased summer season feeding time was attributed to the high energy demand of pregnant females in the population. Herds of SD maximize their feeding activity around sunset (Martin 1977).

Observations on the diurnal activity pattern showed a peak in resting around 10:00 in the morning and 03:00 in the noon. Subsequent to these peaks, there was an increase in feeding and other activities.

The behavioral activities of swamp deer were consolidated into seven major categories: Defensive, Explorative, Locomotory, Stationary, Maintenance, Reproductive & Social, and others (Table 20.2). It was found that SD spent maximum time resting, the next activity in which maximum time was spent was feeding and the next order of activity was standing, walking and sleeping. There were slight differences in the time spent for each category by male and female confirming the above observation (Table 20.2 and Fig. 20.4).

Scanty information exists about swamp deer. Although studbook of swamp deer has maintained this species and its breeding in captive have been recorded in 15

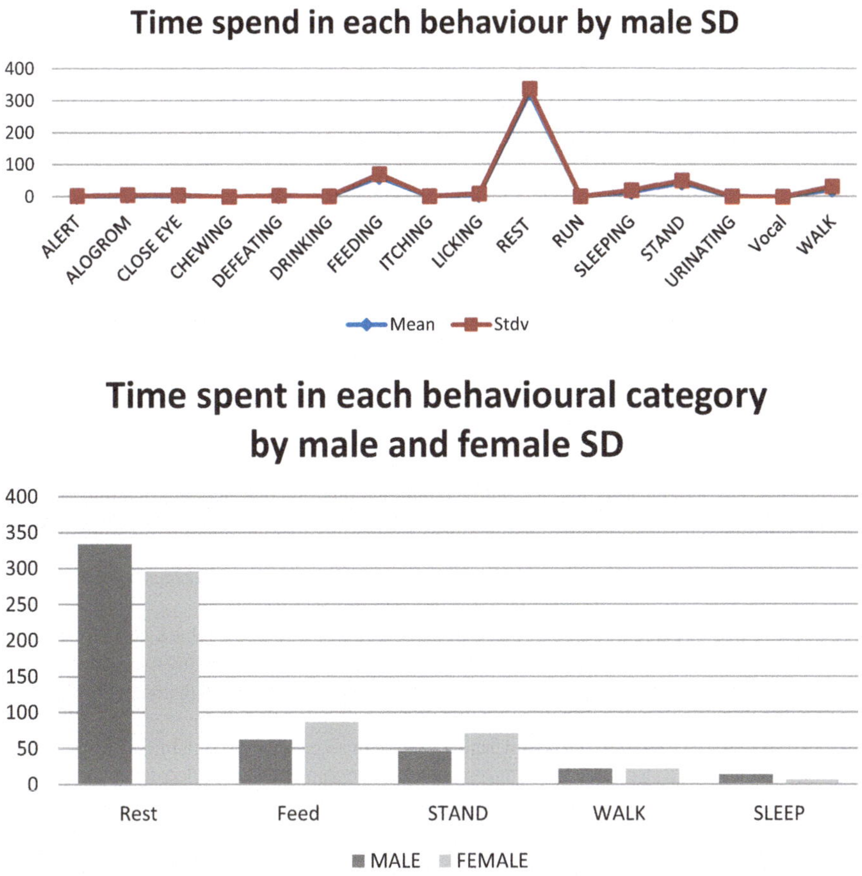

Fig. 20.7 Time spent for each behavioural trait by male swamp deer

zoos, including National Zoological Park, New Delhi (Table 20.1). At NZP, relevant mating pair selection is based on their courtship behaviour observed between opposite sex, when swamp deer was housed in an open enclosure. Preliminary, the mating pair of swamp deer approach each other selectively, and sniffing and licking each other with naso-genital or naso-nasal contact (Figs. 20.8 and 20.9), followed by chasing and mounting attempts by the stags when mutual understanding is established. Then the doe remains standing in a respective place and attains a specific posture (to increase the gap between hind limbs and bend hind limbs slightly); the male approaches the female from the backside and clasps her from the top. Afterwards, the male moves its body towards the neck of the female and the female positions its tail for insertion of penis into the female's genitalia for 20–30 s. Subsequently, the female swamp deer gets out of the male's grip. Proper monitoring of the mating event is carried out by direct observation. Swamp deer are well known and close with their keepers, who cooperate in the identification of each individual

Table. 20.2 Behavioural description of swamp deer activities

Category	Behaviour	Description
Defensive	Alert	A brief immobile posture; head and ears point to the direction of a sound source
	Defensive	Deer bends its head down, positioning it between the hind limbs
	Vocalization	Vocal cords, through its throat and out of its mouth
Explorative	Sniffing	Sniffing genitalia by naso-genital or naso-nasal contact
	Bipedal stand	Stand on bipedal for attacking or feeding by using their forelimbs
Locomotory	Running	Quadrupedal movement faster than walk
	Walking	Normal locomotion from one site to another
Stationary	Resting	Lying on ground or sitting with open or blinking eyes
	Sleeping	Lying on ground with closed eyes attain a steady posture
Maintenance	Bathing	Lying or standing on ground in water
	Defecating	Refer to elimination of scats pellets in a quadrupedal standing position with hind limbs slightly bent towards ground
	Drinking	Head below the level of ground and consume water by lapping movement of the tongue
	Feeding	Lowering of head below to level of back and ingestion of feed by frequent in and out movement of its tongue
	Grooming	Licking and itching one another by in and out moment of tongue
	Urinating	Usually urinate in a quadrupedal standing position the posture merely the same as when defecate
	Territorial marking	Marking territory by urinating on soil substrate and tree trunk

(continued)

Table. 20.2 (continued)

Category	Behaviour	Description
Reproductive and social	Alogrooming	Licking or rubbing another individual
	Approaching	Climbing and walking acts towards other individuals followed by nosonasal or nosogenital sniffing
	Licking	Licking or cleaning itself
	Copulation	Throat fully enlarged stag points at the doe's back, grasps his tail and stag inserts his genitalia in doe's genitalia
	Courtship	Opposite sex individuals attract to involve in displaying their desire to copulate
	Chasing	Walk or run for female to mount
	Mounting	Grasping doe's back from rear or side by stag hooves, stag's fully enlarges his throat over the doe's back and remains parallel to her
	Mating	Male climbs up the female's back from rear or side with his claws of forelimbs grasping female's body
	Retreat	Uplift female from resting posture by male's head and hooves
Others	Urine and semen spraying	Release semen and spray urine by stag mainly during heating
	Infighting	Head down towards ground against another stag to compete to claim the territory with the best doe for mating

and to facilitate proper captive breeding, research and management. At National Zoological Park, swamp deer are housed in a 4903 m^2 naturalistic enclosure (Fig. 20.8; Table 20.1), where can find a chance to interact with the opposite sex individuals.

During heat, male deer antlers start shedding around mid-August to September; afterwards, males squirt urine on their ventral side, rub their antlers in the ground and vocalize rutting calls to attract females. Such behaviour is performed by males as a sexual display to attract the opposite sex. On behalf of the abovementioned courtship behaviour, females select the male for copulation. The stag approaches doe by showing their dominancy in the herd. A quadrupedal chasing behaviour may be observed when stag tries to mount with the doe. Mounting may occur during dorsoventral body posture. After termination of copulation, the mating pair stands at the same place for about 1–2 h. Females show sluggishness for 1–2 days and males remain with females in these days (Fig. 20.9).

Swamp deer did not show any special behavioural changes during pregnancy. However, to assess the pregnancy of swamp deer is not difficult because management can find out by observing the increasing size of the belly towards the ground. Births have been reported from August to late October. At birth, a fawn weighed 2.5 kg and measured length at shoulder 12–13 inches and length 18–20 inches (head to tail). Usually, a single fawn is produced. In the overall birth of swamp deer recorded at NZP, the litter size is always one fawn. Sankaran (1989)

Fig. 20.8 Nasogenital contact

Fig. 20.9 Insertion

Table 20.3 General supplements and drugs provided by veterinary

Drugs/supplement	Remarks
Probiotics (live microbial feed supplement)	In feeding for animal health, and are considered health and growth stimulator
Liver tonic	Made of good quality vitamins and minerals. Helps in liver regeneration. It contains silymarin to prevent the effect of toxins
Mineral mixture	It contains all the essential minerals. It helps increases milk production in animals. Improves reproductive efficiency
Supplement	Ca^{+}, vitamins, iron, etc.

Fig. 20.10 Swamp deer fawn

and Qureshi et al. (1995) reported that the gestation period in swamp deer ranges from 240 to 250 days with a reproductive rate of 20–45 fawns per 100 hinds. The fawn used to stay at the bottom of trees/bushes or wooden logs on the bed of paddy straw (*Praal*) made by the mother. The mother introduced the fawn to the herd and different parts of the enclosure. Fawn follows her mother for suckling after 1–2 h of birth. The mother is very protective towards her fawn. When disturbed, the female shows aggressive behaviour and calls to protect her fawn. The fawn is independent at 3–4 months of age. The period of maternal care is estimated about 3–5 months in swamp deer (Table 20.3 and Figs. 20.10 and 20.11).

Fig. 20.11 Fighting

20.5 List of Known Causes of Death of Swamp Deer from 1962 to 2020

Swamp deer suffer from various causes, such as traumatic shock, pneumonia, transportation, capture stress, capture myopathy, acute impaction, septicaemia, tuberculosis (TB), acute gastroenteritis, acute tympanitis shock, fracture, asphyxia, shell shock, compound fracture, multiple infection shock, acute enteritis, multiple infection and salinity, pulmonary tuberculosis, cirrhosis liver and old age.

From 2003 to 2007, the number of swamp deer was reduced from 16 to 3 and several causes of death were introduced during this period; tuberculosis and pulmonary tuberculosis were highly responsible for adult swamp deer's death, and the time veterinarian effectively stopped the medicine for treating animals. After 2007, population of swamp deer started increasing again due to good fawn survival rate, and the number again increased to 16 till 2014, including one swamp deer (♂) received from the Chhatbir Zoo, Punjab on 22 June 2013.

20.6 Future Conservation Plan

Swamp deer (*R. duvaucelli duvaucelli*) is on the verge of extinction; our goal is to preserve the species in captivity with good facilities and increase the number of population. Afterwards, reintroduce the species into its natural habitat so that it can help in maintaining the ecosystem. Also, we need to focus on marking the population by tagging; it helps in future conservation breeding and easily detects each and every individual and provides facilitative conditions, if needed. Genetic study of each individual is required to collect detailed information about the species, to know about genetic disorders in the population, and to reduce the rate of inbreeding in animals which could affect the abnormalities in fawning. In future, animal exchange programmes will continue to account for genetic variability in species.

The existing enclosure of swamp deer is 4903 m^2 with only higher plant vegetation (Table 20.4). We need to introduce grass species and bushes for grazing (*Anethum graveolens, Saccharum spontaneum, Imperata cylindrica, Narenga porphyrocoma, Phragmites karka, Oryza rufipogon*). Add a thick layer of soil in moats with water to grow aquatic plants (*Hydrilla* species). The size of night house and

Table 20.4 Enclosure and future plan for conservation of swamp deer

Existing enclosure (overview)

Future plan (overview)

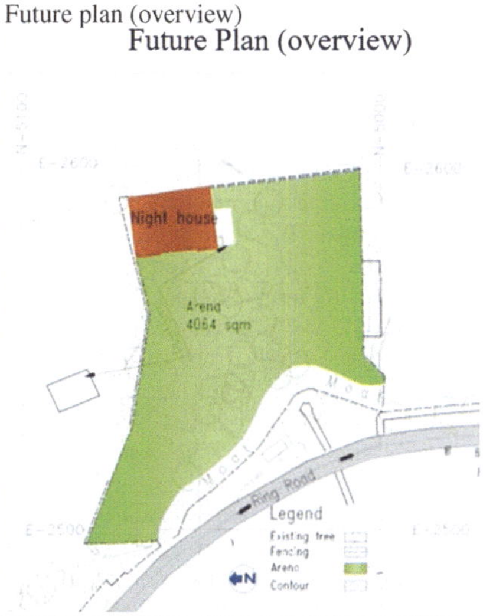

Current vegetation (*existing arena*)	Current vegetation (*future plan*)
Neem tree (*Azadirachta indica*)	Neem tree(*Azadirachta indica*)
Kabuli Kikar tree (*Prosopis juliflora*)	Kabuli Kikar tree (*Prosopis juliflora*)
Desi Kikar tree (*Acacia nilotica*)	Pilkhan tree (*Ficus virens*)
Pilkhan tree (*Ficus virens*)	Pipal tree (*Ficus religiosa*)
Pipal tree (*Ficus religiosa*)	Doob grass (*Cynodon dactylon*)
Gular tree (*Ficus racemosa*)	
Khajoor tree(*Phoenix dactylifera*)	

kraal will be modified in future and few open cubicles will also be implemented in their enclosures to reduce animal stress in case of disease or wound. Digital surveillance will also be implemented on both sides (krall and arena) for the well-planned studies.

References

Ahmad K, Khan AJ (2014) Seasonal activity pattern of swamp deer (Rucervusduvauceliiduvaucelii) in Dudhwa National Park, Uttar Pradesh, India. J Appl Nat Sci 6(1):246–253

Dhami B, Timilsina S, Adhikari A, Neupane B, Chhetri NB, Sharma A, Paudel A, Miya MS, Sharma B, Chhetri A (2021) Research trends, conservation issues and approaches for the endangered red panda (Ailurus fulgens): a systematic review of literatures across their home-range. J Anim Divers 3:57–68

Duckworth JW, Samba Kumar N, Pokheral CP, Sagar Baral H, Timmins RJ (2013) Rucervus duvaucelii. The IUCN Red List of threatened species. Version 2014.3. www.iucnredlist.org. Accessed 11 May 2015

Holloway CW (1973) Swamp deer in Uttar Pradesh. Oryx 12:41–48

Karanth KK, Nichols JD, Karanth KU, Hines JE, Christensen NL Jr (2010) The shrinking ark: patterns of large mammal extinctions in India. Proc R Soc B Biol Sci 277(1690):1971–1979

Khulal R, Neupane B, Dhami B, Regmi S, Tiwari GP, Parajuli M (2021) Habitat use and conservation threats to wild water buffalo Bubalus arnee (mammalia: Artiodactyla: bovidae) in Koshi Tappu wildlife reserve, Nepal. J Threat Taxa 13:19714

Martin C (1977) Status and ecology of barasingha (Cervus duvauceli branderi) in Kanha National Park (India). J Bombay Nat Hist Soc 74(1):60–132

Nandy S, Kushwaha SPS, Gaur P (2012) Identification of Swamp Deer (Cervus duvauceli duvauceli Cuvier) Potential Habitat in Jhilmil Jheel Conservation Reserve, Uttarakhand. India Using Multi-Criter Anal Environ Manag 49:902–914. https://doi.org/10.1007/s00267-012-9826-5

Primack RB (2006) Essentials of conservation biology. Sinauer Associates, Sunderland

Qureshi Q, Singh VP, Sankaran R, Seal US, Walker S, Molur S (1995) Population and habitat viability assessment workshop (P.H.V.A.) for Barasingha (cervus duvauceli duvauceli). Wildlife Institute of India, Dehradun, 31 pp

Qureshi Q, Sawarkar VB, Rahmani AR, Mathur PK (2004) Swamp deer or Barasingha (Cervus duvauceli Cuvier, 1823). ENVIS Bull 7:181–192

Regmi S, Neupane B, Dhami B, Gautam D, Panthi S, Poudel M (2022) Niche breadth and overlap of spotted deer and domestic cattle with swamp deer in tropical region of Nepal. Ecol Process 11:1–10

Sadadev BM, Silwal T, Dhami B, Thapa N, Neupane V, Rana A, Singh HB (2021) Do grassland burning practices affect the distribution of the hispid hare, Caprolagus hispidus (Pearson, 1839)? A study at the Shuklaphanta National Park, Nepal. J Anim Divers 3:86–92

Sankaran R (1989) Status of the swamp deer (Cervus duvauceli duvauceli) in Dudhwa National Park (1988–1989). Bombay Natural History Society Technical Report, No. 14, 25 pp

Schaller GB (1967) The deer and the tiger. A study of wildlife in India. The University of Chicago Press

Tewari R, Rawat GS (2013) Assessment of swamp deer habitat in and around Jhilmil Jheel Conservation Reserve, Haridwar, Uttarakhand, India. Int J Conserv Sci 4(2):243–249

21 Behavioural Activity Patterns of Indian Spotted Chevrotain *Moschiola indica* (Mammalia: Tragulidae) in Captivity

Parvez and Rajesh Kumar Mohapatra

Abstract

The present study examined the activity patterns of the eight males and three females of Indian spotted chevrotain (*Moschiola indica*) from 4 October 2019 to 5 November 2019 between 08:00 and 17:00 h at the Nandankanan Zoological Park, Odisha, India. Behavioural observations of 252 h were recorded using instantaneous sampling with 2-min intervals. A total of 15 behaviours were recorded and clubbed into four distinct behavioural categories, such as active, inactive, maintenance and social/reproductive. Analysis of targeted behaviours revealed that captive Indian spotted chevrotains spent 59% of their daytime being inactive. High inactivity in daytime and increased activity between 16:00 and 17:00 h indicate their nocturnal habits. However, increased activity between 10:00 and 11:00 h correlates to feed delivery time in their enclosure.

Keywords

Behavioural activity budget · Captive condition · *Moschiola indica* · Nandankanan · Zoological Park · Odisha · India

Parvez (✉)
Environmental Management & Policy Research Institute, Bengaluru, Karnataka, India

Ballygunge Science College, University Of Calcutta, Kolkata, West Bengal, India

R. K. Mohapatra
Nandankanan Zoological Park, Bhubaneswar, Odisha, India

S. Sridhara et al. (eds.), *Animal Behavior in the Tropics*,
https://doi.org/10.1007/978-981-97-7324-4_21

21.1 Introduction

Indian spotted chevrotains or mouse deer (ISC), a deer without having antlers in males! In fact, they are not true deer, they are small ungulates belonging to the family Tragulidae and order Cetartiodactyla. It is named for its resemblance to a mouse in its face and body, and its toes are similar to those of a deer. This creature has remained unchanged throughout 25–30 million years of evolution, hence it's known as 'living fossil'. Scent glands associated with other ruminant families are absent in mouse deer and a chin gland is used for marking. They weigh between 2–4 kg and have a shoulder height of 25–30 cm. The mouse deer possesses a three-chambered simple stomach. By selecting a highly digestible diet, it enables rapid fermentation and efficient digestion with swift passage through the gut. They are usually solitary and the home range of a male may overlap that of several females. Their home ranges are small in size and range from 13–24 hectares. The species Moschiola indica is a nocturnal and shy animal, rapidly moving into dense cover on being disturbed (Raman 2004). They can be found in evergreen, semi-evergreen and deciduous forests, as well as in grassland. They are distributed across India and Nepal (Duckworth and Timmins 2015). Poaching and hunting in the wild for their meat and fur and habitat degradation accounts for their population decline in wild (Sarvani et al. 2018). To ensure their survival from these antropogenic threats they are protected as Schedule I species of Wildlife Protection Act, 1972; however, the IUCN Red List (2015) based on an assessment by Duckworth and Timmins (2015) considers it a species of Least Concern (LC). Conservation breeding plays a key role in rehabilitation and supplementation to their wild population, along with in-situ conservation strategies. In this context, the Central Zoo Authority, New Delhi initiated conservation breeding programmes to support species conservation. Nehru Zoological Park, Hyderabad is the coordinating zoo and Nandankanan Zoological Park, Bhubaneswar is a participatory zoo for the conservation breeding programme of the said species. It is worth mentioning that, a detailed knowledge about the behaviour and biology of species is essential for success of such breeding programme. This pilot study aimed to record behaviours of Indian spotted chevrotain at the conservation breeding facility of Nandankanan Zoological Park, Bhubaneswar to prepare an ethogram and quantify the time spent in different behavioural activities and individual variation, if any.

21.2 Methodology

21.2.1 Animal Housing and Husbandry

Nandankanan Zoological Park, Bhubaneswar is a premier large zoo of India. The zoo has successfully bred many endangered animals Indian pangolin, gharials, ratel, tiger, mouse deer etc. The study was carried out at Enclosure No. 37C, housing 11 Indian spotted chevrotain, including 8 Males and 3 Females. This naturalistic

Fig. 21.1 A view of Indian spotted chevrotain's enclosure

enclosure has its top covered but with proper ventilation, visual barriers and provision of sun shade mosaic. This soil-substrate has palm trees bamboo, grasses and some additional artificial hideouts (Fig. 21.1). The zoo has the experience of maintaining and breeding Indian spotted chevrotain since 1971. This pilot study was conducted from 4 October 2019 to 5 November 2019.

21.3 Identification

Each mouse deer has a unique stripe pattern and spots on both sides of the body (Fig. 21.2) (Kumbhar et al. 2013). The male mouse deer have canines protruding out of maxilla. The female mouse deer are bigger in size than males, and even the way of urinating is different in both the sexes as excretion parts are located differently. We used these ques for identification of these Indian spotted chevrotains during the study.

21.4 Data Collection

Behavioral data was recorded for 11 Indian spotted chevrotains using a data sheet during the zoo opening hours only i.e. from 08:00h to 17:00h 6 days a week except Monday, the zoo closure day using instantaneous behavioural sampling at 2 min intervals (Mohapatra et al. 2021; Altmann 1997). Based on the initial observations

Fig. 21.2 Figure showing unique patches of spots and stripes on right and left sides of a Indian spotted chevrotain

an ethogram was created and each ISC has given a unique ID for observing individual's behavior, the ID was ISC1-ISC11. Observed behaviours of with their respective time were documented in a data sheet. Observation where animals were out of sight was not included in the analysis. Frequencies of behaviours were summed on a daily basis for each behaviour and behavioural category. These frequencies were converted to the proportion of time spent on different behaviour (percentage of time spent). The percentage of time spent in different behavioural activities was compared with respect to their sex using a t-test.

21.5 Results

A total of 15 different behavioural events were recorded and described as an ethogram (Table 21.1). All the behaviours were grouped into the four different behavioural categories. These categories were active, inactive, maintenance and social behaviours. Inactivity was the dominant behavioural category followed by active, maintenance and social respectively. Resting, sleeping and yawning were the behaviours clubbed in inactive category, in which mouse deer spent more time in resting 318.1 ± 15.3 minute (m), followed by sleeping at 12.9 ± 6.5 m and yawning at 0.9 ± 0.7 m. Therefore, walking at 90.8 ± 14.4 m was the dominant behaviour in an inactive category, followed by standing at 22.5 ± 8.5 m, alert at 62.2 ± 1.3 m and running at 3.5 ± 1.4 m. Feeding behaviour was found dominant with an average time spent of 64.4 ± 17.3 m, in maintenance behaviour followed by defecating at 1.4 ± 0.6 m, urinating at 0.8 ± 0.5 m, and drinking behaviour at 0.1 ± 0.1 m, respectively. Afterwards, licking 13.7 ± 2.2 m behaviour was found dominant in social behavioural categories, then allogrooming, 4.0 ± 1.9 m and then matting 0.6 ± 1.0 m, respectively. Indian spotted chevrotain were found inactive with an average time (330.23 ± 8.59) m followed by their activity (123.92 ± 7.84) m, maintenance (67.47 ± 3.97) m and social behaviours (18.33 ± 0.74) m (Fig. 21.3). The mean percentage of time spent (± standard deviation) for the different behaviours by individual Indian spotted chevrotain is shown in Table 21.2. The first peak in active

Table 21.1 Ethogram showing behavioural description

Category	Behaviour	Description
Active	Alert	Interrupting any activity it was engaged in
	Running	Quadrupedal movement faster than walk
	Standing	A steady posture on feet
	Walking	Normal locomotion from one site to another
Inactive	Resting	Lying on ground or sitting with open or blinking eyes
	Sleeping	Lying on ground with closed eyes attain a steady posture
	Yawning	Stretching body posture after rest or sleep
Maintenance	Defecating	Refer to elimination of scat pellets in a quadrupedal standing position with hind limbs slightly bent towards ground
	Drinking	Head below the level of ground and consuming water by lapping movement of the tongue
	Feeding	Lowering of head below to level of back and ingestion of feed by frequent in and out movement of its tongue
	Urinating	Usually urinate in a quadrupedal standing position, the posture merely the same as when defecate
Social/ reproductive	Alogrooming	Licking or rubbing another individual
	Licking	Licking or cleaning itself
	Chasing	Walk or run for female to mount
	Mating	Male climbs up the female's back from rear or side with his claws of forelimbs grasping female's body

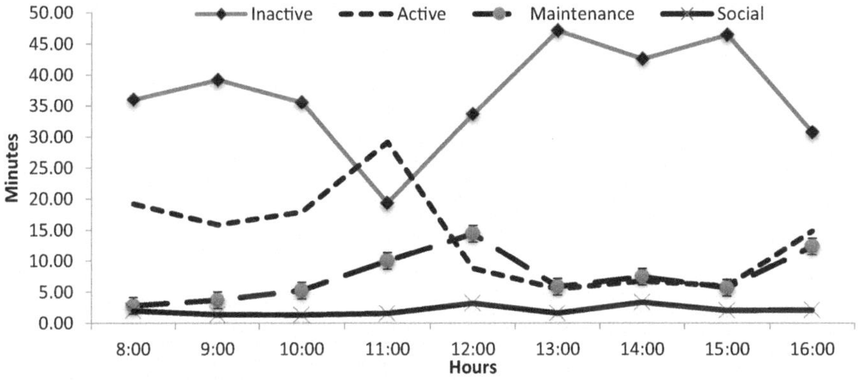

Fig. 21.3 Mean time spent in active, inactive, maintenance and social behavioural categories

behaviour was observed between 10:00 and 11:00 h and the second peak between 16:00 and 17:00 h. The stationary state was observed most of the time during the study with a peak between 13:00 and 15:00 h (Table 21.3, Fig. 21.3). Generally, feeding time between 11:00 and 12:00 h makes the chevrotain highly active for feeding, walking and searching for food. A drop in resting behaviour was observed between 11:00 and 12:00 h and 16:00 and 17:00 h this is because they used to spent their time in feeding and walking. Reproductive behaviour was observed for

Table 21.2 Mean time spent (± standard deviation) for the different behaviours by individual Indian spotted chevrotain

Behaviours	ISC1	ISC2	ISC3	ISC4	ISC5	ISC6	ISC7	ISC8	ISC9	ISC10	11
Resting	355.60 ± 43.08	367.50 ± 29.63	307.80 ± 41.39	346 ± 43.59	327.30 ± 34.77	341.90 ± 36.61	346.70 ± 52.14	291.90 ± 31.23	279.70 ± 31.2	286.00 ± 50.72	248.80 ± 99.62
Standing	22.60 ± 5.21	21.10 ± 15.15	27.50 ± 18.89	23.5 ± 9.15	16.70 ± 10.52	22.20 ± 12.15	19.00 ± 5.82	22.80 ± 9.43	19.40 ± 5.73	23.10 ± 8.71	29.90 ± 10.63
Walk	64.80 ± 4.72	59.50 ± 20.17	89.30 ± 28.05	76.2 ± 22.49	83.30 ± 28.35	73.10 ± 25.37	81.00 ± 34.44	109.20 ± 28.21	129.50 ± 22.14	114.60 ± 28.44	117.80 ± 42.11
Feeding	45.20 ± 0.43	47.60 ± 12.98	65.30 ± 15.75	49.8 ± 23.29	61.20 ± 24.42	49.30 ± 19.08	54.50 ± 29.39	73.50 ± 20.49	78.80 ± 22.87	72.40 ± 32.12	110.60 ± 61.89
Licking	13.00 ± 5.25	17.40 ± 8.1	21.10 ± 6.03	13.2 ± 5.78	11.70 ± 5.59	13.50 ± 6.04	15.30 ± 5.77	11.70 ± 4.01	8.60 ± 3.95	15.40 ± 10.22	9.60 ± 5.17
Alert	4.30 ± 2.99	6.70 ± 2.99	7.10 ± 4.74	6.8 ± 4.22	5.70 ± 2.84	6.00 ± 2.51	5.30 ± 3.26	9.20 ± 5.28	3.00 ± 2.93	7.40 ± 4.21	7.00 ± 3.97
Sleeping	17.10 ± 2.98	11.90 ± 8.88	10.80 ± 5.89	15.8 ± 8.43	20.60 ± 11.94	17.10 ± 14.95	10.70 ± 9.06	13.40 ± 8.41	11.40 ± 5.66	8.60 ± 9.53	4.70 ± 4.11
Running	5.20 ± 3.39	3.60 ± 2.11	2.30 ± 1.34	4.5 ± 3.6	2.70 ± 3.57	5.70 ± 3.13	2.40 ± 1.39	3.10 ± 2.71	3.80 ± 1.28	3.50 ± 2.32	1.20 ± 1.5
Allogrooming	8.60 ± 5.62	1.20 ± 1	3.50 ± 4.14	1.6 ± 2.21	5.00 ± 5.04	7.70 ± 6.65	1.80 ± 2.5	4.20 ± 6.92	0.50 ± 0.88	6.10 ± 4.17	3.80 ± 3.3
Defaecating	2.00 ± 1.45	0.90 ± 1.02	1.70 ± 1.75	1.4 ± 1.31	3.30 ± 2.17	1.90 ± 2.19	1.30 ± 1.97	0.50 ± 1.27	1.30 ± 1.49	0.40 ± 0.82	1.20 ± 1.36
Yawning	0.10 ± 0.44	1.30 ± 1.49	2.40 ± 2.01	0.1 ± 0.44	2.00 ± 1.58	0.20 ± 0.61	1.10 ± 1.02	0.30 ± 0.97	2.00 ± 1.71	0.00 ± 0	0.20 ± 0.61
Urinating	1.40 ± 0.94	1.30 ± 1.17	0.50 ± 0.88	0.7 ± 0.97	0.50 ± 0.88	1.20 ± 1.36	0.90 ± 1.37	0.20 ± 0.61	0.60 ± 0.94	0.80 ± 1.19	0.50 ± 0.88
Mounting	0	0	0	0	0	0	0	0	1.10 ± 1.77	1.70 ± 3.13	3.90 ± 7.12
Drinking	0.10 ± 0.44	0	0.70 ± 1.17	0.1 ± 0.44	0	0.20 ± 0.61	0	0	0	0	0.10 ± 0.44

Table 21.3 Percentage of time spent (mean ± standard deviation) in active, inactive, maintenance and social behaviours

Time	Behavioural categories			
	Inactive	Active	Maintenance	Social
08:00–08:59	36.01 ± 20.62	19.22 ± 5.98	2.81 ± 1.37	1.96 ± 0.30
09:00–09:59	39.14 ± 21.03	15.85 ± 5.71	3.68 ± 1.69	1.33 ± 0.46
10:00–10:59	35.53 ± 19.3	17.92 ± 6.39	5.24 ± 2.53	1.31 ± 0.47
11:00–11:59	19.32 ± 10.9	29.11 ± 11.25	10.01 ± 4.85	1.56 ± 0.52
12:00–12:59	33.61 ± 17.7	8.80 ± 2.53	14.42 ± 7.05	3.17 ± 1.49
13:00–13:59	47.08 ± 25.44	5.48 ± 1.37	5.83 ± 2.88	1.58 ± 0.85
14:00–14:59	42.49 ± 22.39	6.74 ± 1.66	7.42 ± 3.59	3.33 ± 1.56
15:00–15:59	46.35 ± 24.81	5.95 ± 1.36	5.68 ± 2.65	2.02 ± 0.90
16:00–17:00	30.70 ± 17.3	14.85 ± 5.13	12.38 ± 5.59	2.07 ± 0.60

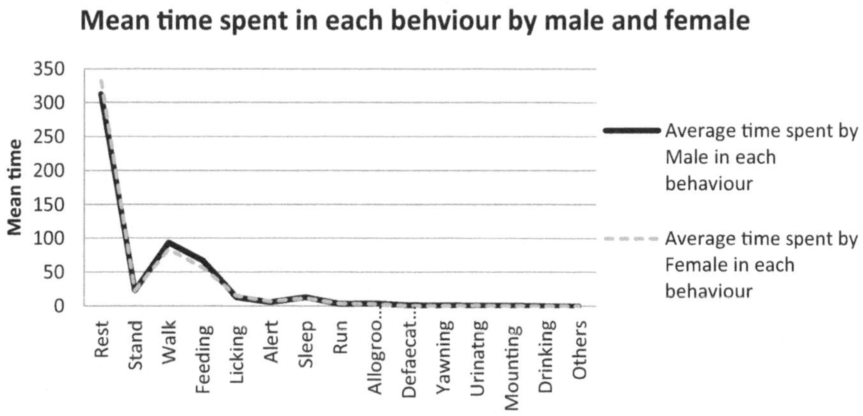

Fig. 21.4 Mean time spent in each behaviour by male and female

4–5 days during the whole study period; the observed behaviours with their respective behavioural categories were briefly described in an ethogram.

Comparison of males and females using t-test found no significant difference in male (8 No.) and female's (3 No.) behavioural activities except their resting behaviour which is insignificant between both the sex (Figs. 21.4 and 21.5).

There is no significant difference found in behavioural activities of male and female mouse deer, and the *P* value was found greater than 0.05 for each behaviour so the null-hypothesis is accepted.

21.6 Discussion

All individuals were observed extremely healthy and active at the facilities. There is an ample amount of sunlight, space and water provided in the enclosure for animals. There are many species of the Tragulidae family that were recorded as being

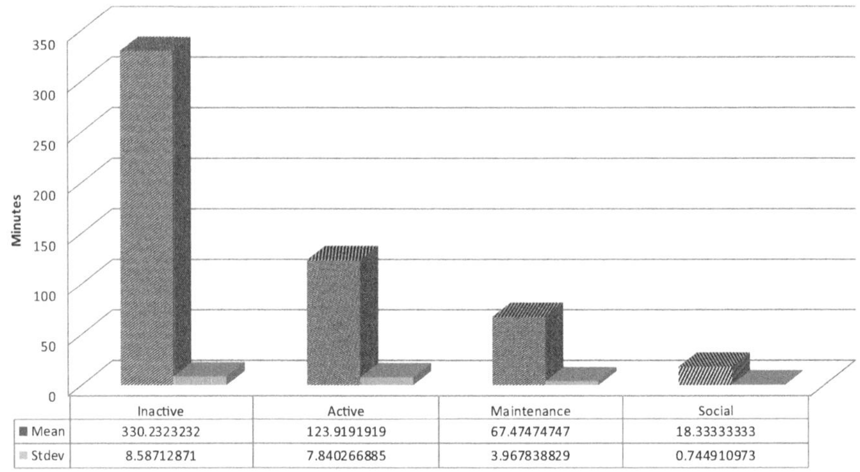

Fig. 21.5 Mean time spent in each behavioural category

nocturnal (Barrette 1987; Cranbrook 1987; Dubost 1975, 1978, 1984; Eisenberg and McKay 1974; Nowak 1991) except *Tragulus javanicus* (Matsubayashi and Bosi 2003). However, Dubost (1978) reported in his research that African mouse deer were active during the night. Few studies on *Moschiola indica* have been documented to date by Acharjyo et al. (2005) studied sex ratio, weight and length of young one at birth, litter size (one only) and breeding season/month of mouse deer at NKZP round the year except April and July, whereas birth season of Delhi Zoo recorded in all the month except October, November and December (Sankhala and Desai 1969), and Ahmedabad zoo recorded birth year-round except December (Arora 2002), therefore ISC breeding season in captivity is throughout the year and locality specific. Groves and Meijaard (2005) described interspecific variation in *Moschiola* sp. from India and Sri Lanka based on conventional taxonomy, Baral et al. (2009) discussed the current assumptive distribution of ISC in Nepal's lowland, Sridhara et al. (2013) studied the role predator in influencing habitat use of ISC and the threats they are facing due to habitat fragmentation, Swapna and Sandeep (2013) addressed the housing condition, breeding season, and average gestation period (162 days) of ISC at Nehru Zoological Park, Ramesh et al. (2013) found mouse deer occurrence positively associated with moist bamboo brakes during dry season, and Kamalakannan and Manna (2013) have strengthen the identification of ISC using Tricho-taxonomic, Kumar and Umapathy (2023) reviewed the success of captive breeding and rewilding *M. indica* by highlighting reproductive monitoring, genetic studies and demographic management, behavioral studies and future perspective). Hence, none of them have recorded ISC behavior systematically. This is the first study on the behaviour of *M. indica* in captivity.

Mating behaviour has been briefly recorded for the greater mouse deer (*Tragulus napu*). During courtship, this species uses a combination of chemical and vocal cues. Males use their inter-mandibular glands to mark females by rubbing them while mating. Rhythmic vocalizations are often exchanged before mounting (Acharjyo et al.

2005). A similar courtship activity was performed during a study in *M. indica*. Though, chasing, rubbing, mounting and sniffing female reproductive organs were common behaviours before mating was observed between 16:00 and 17:00 h. Mouse deer were mostly found active during feed time (11:00 and 13:00 h); during this time, they used to perform walking, running, standing, alert and maintenance behaviour though feeding, drinking, defecating and urinating were mostly in the evening. In animal kingdom, dominance is the sign of being dominant and being first to mate among all males. Perhaps most of the cases of fighting have been reported from night time. As there is no confined time for reproduction in mouse deer, though females were capable of mating even after an hour of offspring's births; moreover, males were sighted for mounting during suckling also (Fig. 21.6).

Captive mouse deer of NKZP live on a diet of fresh vegetables and fruits with lactate (10–15 mL) for better growth. Indian chevrotain plays an important ecological role as seed dispersers, and they are the main prey for dholes (*Cuon alpines*) and leopards (*Panthera pardus*) in the wild. Predation by birds of prey, e.g. forest eagle owls (*Bubo nipalensis*) has also been reported (Ramesh et al. 2013; Sridhara et al. 2013).

Members of the Tragulidae family were recorded as nocturnal (Barrette 1987; Cranbrook 1987; Dubost 1975, 1978, 1984; Eisenberg and McKay 1974; Nowak 1991), except for *Tragulus javanicus* (Matsubayashi and Bosi 2003). Earlier studies also recorded *Moschiola indica* being nocturnal and timid, which disappears in dense vegetation in slightest disturbance. Male lives a solitary live except during the mating season (Swapna and Sandeep 2013). Acharjyo et al. (2005) reported breeding of mouse deer at Nandankanan Zoological zoo round the year, except in April and July. In contrast, birth records in Delhi Zoo were observed in all months, except in October, November and December (Sankhala and Desai 1969). Similarly, Ahmedabad zoo recorded births year-round, except in December (Arora 2002). Therefore, ISC breeding season in captivity is throughout the year and locality specific. All these zoos have reported a litter size of one; however, Hyderabad zoo recorded an exceptional case where a female gave birth to two fawns (Swapna and

Fig. 21.6 Male try to mate when female nourish her offspring

Sandeep 2013). Notably, there is no confined breeding season for mouse deer, allowing females to mate shortly after giving birth, sometimes even within an hour. Males have been observed attempting to mate during the suckling period as well (Fig. 21.6). There were instances of infighting recorded between males (Nandankanan Zoological Park records). The current study reported what the Indian spotted chevrotains do during the day time? The study observed that they spent a significant amount of time in inactivity or resting behaviour (Fig. 21.7). Earlier studies also noted day resting behaviour of mouse deer often in dense undergrowth (Matsubayashi et al. 2003; Barrette 1987; Nowak 1991). The active behaviours are also mostly oriented to food or feeding time. Swapna and Sandeep (2013) reported the average daily diet of ISC consists of apple 50 g, banana 100 g, carrot 50 g, sweet potato 50 g, mixed grains (black gram, horse gram and green gram) 50 g, lucerne 100 g, hay 500 g, Acalypha/Ficus 500 g per adult individual. At Nandankanan Zoological Park, it consists of ripe banana 250 g, apple 50 g, carrot 125 g, beans 125 g, pumpkin 50 g, lady finger 100 g, green sag 50 g and in season (sweet potato 100 g [October–December], Momordica 50 g [July–September]) per adult a day. Nandankanan Zoological Park has been breeding mouse deer since 1971 and a participating zoo in the conservation breeding programme of mouse deer since 2009. There have been 36 births of mouse deer in captivity since 1971, including 23 births under the conservation breeding programme, i.e., after 2009. The zoo has sent mouse deer to the coordinating zoo Nehru Zoological Park, Hyderabad, in 2013 and 2016 and Sri Chamarajendra Zoological Garden, Mysore, in 2018 and received mouse deer from Nehru Zoological Park, Hyderabad, in 2016 to infuse new blood line and increase heterozygosity. Ramesh et al. (2013) found that the occurrence of ISC was positively associated with moist bamboo brakes during the dry season. In this study, bamboo clumps present in the enclosure were found as a preferred area for resting and hiding place to rear fawns.

Fig. 21.7 Mouse deer resting at NKZP

Before giving birth to a young one, females used to rest longer than the usual time and avoid running. After this study, additional observations were made just after the birth of a young one. A young one can walk within an hour of its birth. On average, the size of a young one is 26.5–30.5 cm (length) and 14.0–16.0 cm (shoulder height) with an approximate weight of 288–382 g (Acharjyo and Mishra 1981). Primarily, 1–2 weeks offspring depend on suckling and then start chewing small green leaves. Then after the age of 40–45 days, they can start feeding on soft fruits and vegetables.

21.7 Conclusion

The present study describing 15 different behaviour patterns clustered in four behavioural categories provides an initial framework for understanding the behavioural biology repertoires of Indian spotted chevrotain. Further study on behavioural biology of Indian spotted chevrotain is required to facilitate both ex situ and in situ conservation of the species. Since a comprehensive knowledge of behavioural biology of the Indian spotted chevrotain is essential for its management in captivity, the description of this chapter may also be useful in improving the husbandry and management of captive Indian spotted chevrotain and designing future conservation efforts. If cage size increases, the group should be isolated to avoid breeding, which may result in a genetic disorder. Night study is also required in zoos as well as in wild areas by using infrared and camera trapping methods for their nocturnal way of living. Further, a birth chart of each and every individual should be made with monthly monitoring by using the above identification technique, which may help researchers to develop a monograph about the species.

Acknowledgement I am thankful to Mr. Sashi Pual, Principal Chief Conservator of Forest, Odisha, for permitting (Memo No. 8425) the research at Nandankanan Zoological Park and Wildlife Sanctuary, Odisha. I am thankful to Anjan Kumar Das (Deputy Director, NKZP) for allowing me and thankful to Milan Panda (Education Officer, NKZP) for the support. Authors are thankful to Dr. Debabrat Mohapatra (Senior Veterinarian) and Dr. Santosh (Veterinarian) for their guidance and valuable suggestions. At North Orissa University, I am thankful to Dr. Susmita kar, Khan Ashaharraza and Subhadra Barik for their comments on observations noted. The authors are extremely thankful to zookeeper Mr. Praful Bag and the ground staff for sharing knowledge on mouse deer.

Conflicts of Interests None of them have a conflict of interest.

References

Acharjyo LN, Mishra CG (1981) Notes on weight and size at birth of eight species of Indian wild ungulates in captivity. J Bombay Nat Hist Soc 78(2):373–375

Acharjyo L, Mahapatra M, Sinha S (2005) On some aspects of reproductive pattern of the Indian chevrotain or mouse-deer (*Tragulus meminna*) in captivity. ZOO's Print 6(1):10

Altmann J (1997) Observational study on behavior sampling methods. Behaviour 49(1):227–267

Arora BM (2002) Reproduction in wild mammalian and Conservation. Association of Indian Zoo and wildlife Veterinarians, Bareilly (Uttar Pradesh), pp 28–29

Baral HS, Shah KB, Duckworth JW (2009) A clarification of the status of Indian chevrotain *Moschiola indica* in Nepal. Vertebr Zool 59(2):197–200

Barrette C (1987) The comparative behaviour and ecology of chevrotains, musk deer, and morphologically conservative deer. In: Wemmer CW (ed) Biology and management of the Cervidae. Smithsonian Institution Press, Washington DC, pp 200–213

Cranbrook E (1987) Riches of the wild: land mammals of South-East Asia. Oxford University Press, Singapore

Dubost G (1975) Le comportement du chevrotain african, Hyemoschus aquaticus Ogiby (Artiodactyla, Ruminantia). Zeitschrift fu¨r Tierpsychologie 37:449–501

Dubost G (1978) Un aperc¸u sur l'e´cologie du chevrotain african Hyemoschus aquaticus Ogiby, Artiodactyle Tragulide´. Mammalia 42:1–62

Dubost G (1984) The encyclopedia of animals. Andromeda Oxford Ltd., Oxford

Duckworth JW, Timmins R (2015) Moschiola indica. The IUCN Red List of Threatened Species 2015:e.T136585A61979067. https://doi.org/10.2305/IUCN.UK.2015-2.RLTS.T136585A61979067.en. Accessed on 14 September 2024

Eisenberg JF, Mckay GM (1974) Comparison of ungulate adaptations in the New World and Old World tropical forests with special reference to Ceylon and the rainforests of Central America. In: Geist V, Walther F (eds) The behavior of ungulates and its relation to management. International Union for Conservation of Nature, Morges, pp 585–602

Groves CP, Meijaard E (2005) Interspecific variation in *Moschiola*, the Indian chevrotain. Raffles Bull Zool 12(1):413–421

Kamalakannan M, De JK, Manna CK (2013) Identification of dorsal guard hairs surface structure of Indian chevrotain Moschiola indica Gray, 1852 (Tragulidae: Artiodactyla: Mammalia). Biolife 1(3):155–158

Kumar V, Umapathy G (2023) From captivity to conservation success: a reviewon the mouse deer breeding program and its implications for biodiversity preservation. J Endocrinol Reprod 27(3):179–185. https://doi.org/10.18311/jer/2023/34990

Kumbhar A, Prabhu CL, Yogesh JK, Francies P, Patwardhan G, Qureshi Q, Jhala Y (2013) Application of photographic capture-recapture sampling for estimating abundance of Indian mouse deer *Moschiola indica*. World J Zool 8(4):388–391

Matsubayashi H, Bosi E (2003) Activity and habitat use of lesser mouse-deer (*Tragulus javanicus*). J Mammal 84(1):234–242

Mohapatra RK, Panda S, Khan P (2021) Behavioural study of zoo animals. Indian Zoo Year Book 9(1):125–134

Nowak RM (1991) Walker's mammals of the world, vol II, 5th edn. The John Hopkins University Press, Baltimore

Raman TRS (2004) Mouse deer (Moschiola meminna Erxleben, 1777). ENVIS Bull 7:131–140

Ramesh T, Kalle R, Sankar K, Qureshi Q (2013) Dry season factors determining habitat use and distribution of mouse deer (*Moschiola indica*) in the Western Ghats. Eur J Wildl Res 59:271–280

Sankhala KS, Desai JH (1969) Reproductive pattern of some Indian mammals. Cheetal 12(l):114–119

Sarvani RK, Parmar DR, Tabasum W, Thota N, Sreenivas A, Gaur A (2018) Characterization of the complete mitogenome of Indian mouse deer, *Moschiola indica* (Artiodactyla: Tragulidae) and its evolutionary significance. Sci Rep 8(1):1–12

Sridhara S, Edgaonkar A, Kumar A (2013) Understorey structure and refuges from predators influence habitat use by a small ungulate, the Indian chevrotain (*Moschiola indica*) in Western Ghats, India. Ecol Res 28:427–433

Swapna P, Sandeep M (2013) Captive breeding of Indian spotted chevrotain or mouse-deer (Tragulus meminna) in Nehru Zoological Park, Hyderabad. Indian Zoo Year Book 8(1):26–32

Behaviour and Breeding of Himalayan Goral (*Naemorhaedus goral*) in Captivity

22

Parvez, Subhadra Barik, and Akshay Kumar Chakravarthy

Abstract

The Himalayan Goral (*Naemorhaedus goral*) is endemic to the Himalayan region embracing India, Nepal, Pakistan and China. It is a near-threatened and not much-researched species. It is a herbivore closely related to Serows. The experiments conducted at Delhi Zoo, India, revealed that gorals preferred a mixed vegetarian diet. Their breeding season in the zoo is throughout the year but generally between September and December. Gorals live 8–15 years in the wild and in captivity for 17 years. In captivity, male establishes dominance through aggressive behaviour and eventually in battle with competitive males. The courtship and mating include flank stork, flehmen, gamboling, territorial and maternal behaviour. Conservation plans for the goral include providing a naturalistic enclosure like ridges and cliffs on mountainous tracts and conservation of natural habitats in the Himalayan region.

Keywords

Goral · Himalayan region · Captive breeding · Behaviour

Parvez (✉)
Ballygunge Science College, University of Calcutta, Kolkata, West Bengal, India

Environmental Management and Policy Research Institute, Bengaluru, Karnataka, India

S. Barik
Sálim Ali Centre for Ornithology and Natural History, Coimbatore, Tamil Nadu, India

A. Kumar Chakravarthy
Division of Entomology and Nematology, Indian Institute of Horticultural Research, Bengaluru, Karnataka, India

S. Sridhara et al. (eds.), *Animal Behavior in the Tropics*,
https://doi.org/10.1007/978-981-97-7324-4_22

22.1 Introduction

The Himalayan goral (*Naemorhaedus goral*, Bovidae, Artiodactyla) is a goat antelope often found on rocky hillsides and high altitude areas with an elevation of 1000–2500 m or 4200 m on sleeper slopes with low vegetation (Johnsingh 1992; Hoffmann et al. 2008; Srivastava and Kumar 2018). The Himalayan goral is the smallest member of the Rupicaprini tribe of goat antelopes. They are grazing animals and feed on grasses, green leaves, buds of flowers and fruits in their natural habitat. Interestingly, Valdez (2011) reported six species exist under the genus *Naemorhedus*. However, the International Union for Conservation of Nature (IUCN) still recognizes only the four species, i.e. Himalayan goral (*Naemorhedus goral*), Chinese goral (*N. caudatus* and *N. griseus*) and Red goral (*N. bailey*) (Duckworth and MacKinnon 2008; Duckworth et al. 2008a, b). One more interesting thing is that IUCN listed Himalayan goral (*N. goral*) as Near Threatened under the list of the Red List of Threatened Taxa 2020 and categorized it under *Appendix I* of Convention on International Trade in Endangered Species (CITES) (Hemley 1994; Wijnstekers 2011). However, the Wild Life (Protection) Act, Amendment 2022 listed *N. goral* under *Schedule I* species. Goral populations in the Himalayas are on the urge of extinction by experiencing habitat degradation, fragmentation, changes in land use patterns, illegal poaching and human disturbances (Yang et al. 2013; Challender et al. 2015). The number of *Naemorhedus* species has been questionable (Shukla et al. 2018; Mori et al. 2019), where Valdez (2011) reported six species under the *Naemophedus* genus. On the other hand, Li et al. (2020) described five species. Even though, recent trends in phylogenetics using Markov Chain Monte Carlo (MCMC) revealed six species (Joshi et al. 2020).

The Himalayan goral is 95–130 cm (37–51 in.) in length and weighs 35–42 kg (77–93 lb). It has a grey or grey-brown coat with tan legs, lighter patches on its throat and a single dark stripe along its spine. Males have short manes on their necks. Both males and females have short and backward-facing horns that can grow up to 18 cm (7.1 in.) in length. The male goral has longer horns, is more divergent and is thicker at the base in comparison to the female's horn. Besides this, the ridges of the male's horns are more prominent than the ridges of the female's horns. The Himalayan goral has strong long legs and broad, heavy hooves well adapted for climbing and jumping. The Himalayan goral's back is slightly arched. Himalayan goral has a curved face, with a distinct nasal bone and eyes set close together. Himalayan gorals have large ears that measure 10–14 cm in length.

Gorals occur in the Himalayan region embracing India, Pakistan, Nepal, Bhutan, Myanmar and China where it is endemic (Fig. 22.1).

In addition to certain peculiarities in the form of the skull, gorals are chiefly distinguished from the closely related serows (Capricornis species) in that they do not possess preorbital glands below their eyes, nor corresponding depressions in their skulls. The Himalayan gorals are social animals living in small groups of 4–12 individuals. Older males often prefer living solitary life. They are diurnal and active in the early morning to late evening (Crepuscular). Litter size is always one with a gestation period of 6 months.

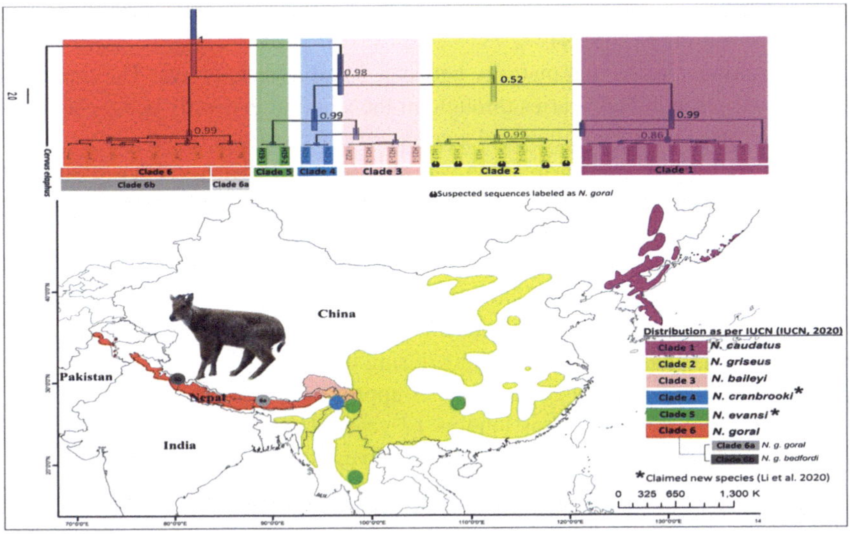

Fig. 22.1 Phylogenetic relationship of different goral species using the partial fragment of mitochondrial cytochrome b gene. The four colours represent the distribution ranges of four goral species as per IUCN (Accessed June 2020: Duckworth and Mackinnon 2008; Duckworth et al. 2008a, b), and two additional dots of sky blue and green represent the distribution of two newly claimed species by and Li et al. (2020) from the Northern Myanmar and Southern China. Above the clades posterior probability is shown. *Cervus elaphus* (NC_007704.2) was used as an outgroup for analysis. (Source: Joshi et al. 2020)

The food composition in the zoo was mesh (concentrated with the wheat ban, Bengal gram, maize, (crushed), barley (crushed), oats, groundnuts oil cake, turmeric powder and common salt) (2 kg), Kutti (Green fooder) ficus (4 kg), gram soaked (500 g), tree fodder (500 g), cucumber (summer)/carrot (winter) (100 g), amala green (winter) (50 g).

22.2 Breeding

Studies were conducted on goral breeding at National Zoological Park (NZP), From November 2020 to January 2021. Observations were recorded through a hideout behind a tree with a pair of 8 * 30 binoculars (Nikon). The observations revealed that the animals were very active in the early morning and late evening. At NZP, gorals usually spend feeding, grooming, making movement and resting during midday on the rocky cliffs and inside the walled enclosures. But when the weather is cloudy, the goral were found active throughout the day. Female gorals and their fawn, including yearlings, form small herds of 2–14 individuals. Adult males were found solitary except during the breeding season. A herd's territory generally encompasses approximately 0.4 km^2 (100 acres). During the mating season, adult males occupied and marked territories of about 0.2 km^2 (22–25 ha). Unlike other herding animals, gorals rely more heavily on their acute eyesight than on their sense of smell or hearing (TNECN).

The goral reaches sexual maturity between 2 and 3 years of age. The timing of the mating season in NZP varies throughout the year but generally occurs between September and December. The goral's gestation period lasts 6 months, after which females give birth to one fawn. Fawn was able to stand after just a few hours and begin following their mothers on the same day. The period of maternal care of goral's fawn is estimated about 7–8 months. Mothers are very protective towards their fawns, and after 1 or 2 h, the mother introduced his fawn to the herd and all sides of the enclosure. Fawns follow their mother for suckling and in the starting days mostly prefer to rest inside the small houses of enclosure or side of the walls. Fawns were reported 4–5 times suckling during the day.

The life span of the Himalayan goral is approximately 8–15 years in the wild. But, in captivity, gorals lived as long as 17 years. The Himalayan goral's natural predators include leopards, lynx, tigers, wolves, wild dogs and martens. The Himalayan goral is a sensitive and shy animal, but here in captivity they are familiar with their keepers and animal attendants. The Himalayan gorals have an excellent camouflage coat providing the goral in its natural habitat, so the goral's first reaction when confronted by potential danger is to freeze, in hopes of being overlooked. If the goral does happen to be spotted, it emits a hissing or sneeze-like alarm call by repeatedly tossing limbs into the ground and fleeing. The Himalayan gorals are excellent spry and agilely move across rocky areas. The Himalayan gorals are adapted to and prefer red rocks and cliff areas so that they can escape from predators.

22.3 Behaviour

The Himalayan gorals (*Nemorhaedus goral*) are polygynous. Dominant males of a herd have the right to mate with all females in their range during the breeding season. They mate during November and December in the wild, but at NZP, they bred throughout the year. The behaviours were observed and recorded by using Altmann (1974) behavioural random sampling.

The dominance is established through alarming or threatening displays and lastly by battle with other males of a herd. When animals fight, they attempt to wound their opponent with small, backward-curving horns instead of engaging in butting. Heptner et al. (1989) wrote that the Himalayan goral shows a subdued rutting behaviour pairing off from September to November. The female goral never forms a harem, but the adult male defends their territories.

When the male courts the female, he starts approaching the female by walking and shaking his head and with body movements, chases and walks close to her with croup touching. Afterwards, head-down behaviour was performed by the male to drive the female from other males.

Flank stroke is another behaviour performed by male gorals. It consists of softly caressing the female's body backside, the male continuously chasing and walking close to her. Gamboling is another behaviour by males to show dominance by often running and jumping playfully near females.

Flehmen behaviour is a behaviour in which a male bends back its upper lip, showing its front teeth, inhales, normally with the nostrils closed and then frequently maintains this position for a number of seconds. The term flehmen comes from the German verb flehmen, which means to flash one's top teeth. Its root is the Upper Saxon flemmen, which means "to appear spiteful." Karl Max Schneider, the Leipzig zoo director and an expert in big cats kept in captivity, coined the term in 1930.

Chasing behaviour

Flank stroke

Naso-nasal contact

Naso-genital contact

Mating

Suckling

Marking dominance display is yet another behaviour exhibited by gorals during courtship. It comprises the head being rubbed gently on vegetation or ground by both sexes, simultaneously having naso-nasal and naso-genital contacts engaged by both sexes.

Table 22.1 General supplements and drugs provided under captivity

Drugs/supplement	Remarks
Probiotics (live microbial feed supplement)	Help in feeding for animal health. They are considered as health and growth stimulators
Liver tonic	Made of good quality vitamins and minerals. Helps in liver regeneration. It contains Silymarin to prevent the effect of toxins
Mineral mixture	It contains all the essential minerals. It helps in increasing milk production in animals. Improves reproductive efficiency
Supplement	Ca^+, vitamins, iron, etc.

The courtship in gorals eventually terminates in mounting behaviour common in most of the animals. Males during mating mount on the females from the back. When the female stands with a wide gap between hind limbs, males slightly adjust the female's tail and clasp her from the top. Afterwards, insertion of the male penis into the female's genitalia occurs, during which snorting soft whistles are uttered by courting males, and harsher panting sounds (Panting) are vocalized during copulation. The panting of Himalayan goral can be heard up to 80–500 m away (Lovari and Apollonio 1994). NZP played an impressive role in Himalayan goral captive breeding and helped in conserving this species from 23 April 1963, and now the population size is 20 (5M:12F:3U) (Table 22.1).

Existing enclosure

Existing enclosure overview

Generally, Goral's deaths tended to occur due to a range of factors, encompassing traumatic events, infectious diseases, complications with various organs, and the natural process of aging. This includes a variety of conditions such as pneumonia, tuberculosis, and issues leading to organ failure. Factors like injuries, exhaustion, and specific diseases such as cirrhosis of the liver and septicaemia also played significant roles. The culmination of these health challenges, including complications from birth and the effects of old age, contributed to the mortality observed in these instances.

22.4 Conservation Plan

For the Himalayan goral (*Nemorhaedus goral*), the conservation plan is to maintain the conservation strategy and boost the population to re-wilding the species into its natural habitat so that it can help maintain the ecosystem and the natural resources. In future, plans are ahead to establish a conservation centre for Himalayan goral because animals prefer facilitated naturalistic enclosures like in Indian zoos. The natural habitat of Himalayan goral in zoos includes cliffs and ridges for climbing and rocky lodges for resting. The surroundings of the enclosure are very similar to their natural habitat. In future, the size of the enclosure should be larger for their growing population. Construction of one or more rocky cliffs and animal houses is in progress for upgraded enclosures. It will be a part of the enrichment drill. Exchange programmes will also be a part of the future plan for genetic variability in species. It helps to avoid inbreeding problems.

Table 22.2 Trees in enclosures for gorals

Vegetation (*previous enclosure*)	Current vegetation (*existing enclosure*)
Neem tree (*Azadirachta indica*)	Neem tree (*Azadirachta indica*) outside arena
Kabuli Kikar tree (*Prosopis juliflora*)	Pilkhan tree (*Ficus virens*)
Pilkhan tree (*Ficus virens*)	Pipal tree (*Ficus religiosa*)
Pipal tree (*Ficus religiosa*)	

The existing enclosure of Himalayan goral (*Nemorphaedus goral*) has been constructed to provide cliff-type habitat. The enclosure has a stone step to reach the top with a cubical inside. Himalayan goral looks to be in hills when it climbs to the top of the cliff. It is a massive structure of stone masonry that merges with the background of the Purana Quila.

The arena has enough space for the movement of animals, with a number of trees (Table 22.2) and a moat on the northern side. Visitors can see the gorals through a chain link fence from four sides. But in future, it will be advisable to block one or two sides of the arena so animals feel safe and comfortable. Arena fulfils Central Zoo Authority (CZA) norms of containment requirements. The area of the arena is 1333 m^2.

Acknowledgement The authors are thankful to Mr Ramesh Kumar Pandey, Director NZP, for his encouragement and support during the study. They are also indebted to Dr. Manoj Kumar, Biologist at NZP, for guidance and to Mr. Riaz Ahmad Khan, curator at NZP, for technical support. The assistance of ground staff and veterinarian at NZP is gratefully acknowledged.

References

Altmann J (1974) Observational study of behavior: sampling methods. Behaviour 49(3–4):227–266

Challender DWS, Harrop SR, Macmillan DC (2015) Understanding markets to conserve trade threatened species in CITES. Biol Conserv 187:249–259

Duckworth J, Mackinnon J (2008) *Naemorhedus goral*. Red List of endangered species. Accessed 14 Jun 2016

Duckworth JW, MacKinnon J, Tsytsulina K (2008a) *Naemorhedus caudatus*. IUCN Red List threat species 2008 eT14295A4429742.

Duckworth JW, Steinmetz R, Chaiyarat R (2008b) *Naemorhedus griseus*. IUCN Red List threat species 2008 eT14303A4430834

Hemley G (1994) International wildlife trade: a CITES sourcebook. World Wildlife Fund/Island Press, Washington, DC

Heptner VG, Nasimovich AA, Bannikov AG, Hoffman RS (1989) Mammals of the Soviet Union, vol 1. EJ Brill, Leiden

Hoffmann RS, Lunde D, MacKinnon JR, Wilson DE, Wozencraft WC (2008) In: Smith AT, Xie Y (eds) A guide to the mammals of China. Princeton University Press, Princeton

Johnsingh AJT (1992) Prey selection in three large sympatric carnivores in Bandipur insects or lion hunting buffalo, depends on the quality and quantity of its diet time tested strategies against several ecological constraints such as prey size. Mammalia 56:517–527.

Joshi BD, Singh VK, Singh H, Singh A, Singh SK, Chandra K et al (2020) New insights of zoogeographical distribution of Himalayan goral (Naemorhedus goral) from Indian Himalayan region. bioRxiv, 2020-08

Li G, Sun N, Swa K, Zhang M, Lwin YH, Quan RC (2020) Phylogenetic reassessment of gorals with new evidence from northern Myanmar reveals five distinct species. Mammal Rev 50(4):325–330

Lovari S, Apollonio M (1994) On the rutting behaviour of the Himalayan goral Nemorhaedus goral (Hardwicke, 1825). J Ethol 12(1):25–34

Mori E, Nerva L, Lovari S (2019) Reclassification of the serows and gorals: the end of a neverending story? Mamm Rev 49:256–262

Shukla MA, Joshi BD, Kumar VP, Kumar S, Mehta AK (2018) Forensic examination of goral samples seized under wildlife offence cases in India using DNA based techniques. J Environ Biosci 32:313–317

Srivastava T, Kumar A (2018) Seasonal habitat use in three species of wild ungulates in Sikkim Himalaya. Mamm Biol 88:100–106

Valdez R (2011) Subfamily Caprinae. In: Wilson D, Mittermeier R (eds) Handbook of the mammals of the world, vol 2. Hoofed mammals. Lynx Editions, Barcelona, pp 742–759

Wijnstekers W (2011) The evolution of CITES, 9th edn. CIC—International Council for Game and Wildlife Conservation, Budapest

Yang C, Xiang C, Qi W, Xia S, Tu F, Zhang X et al (2013) Phylogenetic analyses and improved resolution of the family Bovidae based on complete mitochondrial genomes. Biochem Syst Ecol 48:136–143

Wintering Diet Selectivity of Sloth Bear, *Melursus ursinus* Shaw, 1791 in Dry Deciduous Ecosystem of Gujarat State, India

23

Vasantkumar Rabari, Arzoo Malik, and Nishith Dharaiya

Abstract

The sloth bear, *Melursus ursinus* Shaw, is omnivorous and myrmecophagous in food habits, consuming insects and large amounts of plant matter, particularly fruits. The feeding ecology and diet composition of *Melursus ursinus* was studied in the Balaram-Ambaji Wildlife Sanctuary located in the western part of the Aravali Mountain range, India. Scats were collected from the entire study area. Moreover, 118 quadrat plots of 10 m^2 were conducted to record the presence of food resources for *Melursus ursinus* during the cold season. We examined 64 scats, which contained 62% plant material and 33% animal material, while 5% of scats contained both plant and animal material. The selectivity index showed that *Melursus ursinus* consumed *Zizyphus* sp. to a greater proportion than their availability, with less preference for *Cassia fistula*, *Grewia flavescens* and *Ficus* sp. The findings of the study suggest that plantations of wintering fruit species (mainly *Zizyphus* sp.) and protection of food plants in the forest may restrict *Melursus ursinus* movements outside the forest and help mitigate human-bear conflicts.

Keywords

Selectivity index · Scat analysis · *Melursus ursinus* · Food availability · Sloth bear · Myrmecophagy

V. Rabari (✉) · A. Malik
Wildlife and Conservation Biology Research Lab, Department of Life Sciences, Hemchandracharya North Gujarat University, Patan, Gujarat, India

N. Dharaiya
Wildlife and Conservation Biology Research Lab, Department of Life Sciences, Hemchandracharya North Gujarat University, Patan, Gujarat, India

Centre of Excellence for Wildlife and Conservation Studies, Bhakta Kavi Narsinh Mehta University, Junagadh, Gujarat, India

S. Sridhara et al. (eds.), *Animal Behavior in the Tropics*,
https://doi.org/10.1007/978-981-97-7324-4_23

23.1 Introduction

Dietary preferences influence the habitat selection of a species, as nutrition plays an important role in the growth and development of the animal (Mewada et al. 2019). Habitat destruction and fragmentation majorly affect food availability in the forest, forcing *Melursus ursinus* towards human-dominated areas in search of food (Dharaiya and Ratnayeke 2009). According to Rajpurohit and Krausman (2000), food has a greater impact on habitat quality, which plays a significant role in the mitigation of conflict by reducing bear movements near human settlements in search of food. *M. ursinus* is unique among all the ursids (Fig. 23.2b, c) by their myrmecophagous feeding habit, i.e. adapted for feeding on insects and fruits (Joshi et al. 1997). It is also found that variations in the diet composition of *M. ursinus* appeared to be based on the availability of fruiting species and colonies of termite mounds (Laurie and Seidensticker 1977; Rabari and Dharaiya 2022).

Studies on the feeding ecology of *M. ursinus* showed that bears consumed both animal and plant matter in their regular diet (Sukhadiya et al. 2013). Many studies have found that plant matter is an important component of *M. ursinus* diets (Schaller 1967; Sukhadiya et al. 2013; Mewada 2015; Philip et al. 2022; Rabari and Dharaiya 2022). In contrast, Joshi et al. (1997), Ratnayeke et al. (2007), Ramesh et al. (2009) and Khanal and Thapa (2014) observed an inclination towards animal matter in the diet. These studies suggest that the dietary composition of *M. ursinus* is dependent on food availability in their habitat that labels them as opportunistic feeders. But, when both types of food (fruits and insects) are available, *M. ursinus* feeds on fruits to meet nutrition requirements due to the bulk of availability and easier access to fruits than insects (Palei et al. 2020). Similarly, selective feeding has been observed in black bears based on nutrition requirements (Noyce et al. 1997). It is also found that fleshy fruits are rich in sugar, which fulfils instant energy requirements, and excess sugar can be converted and stored as tissue fat for further utilization (Palei et al. 2020). A high number of *M. ursinus* attacks were reported in winter from the North Gujarat region (Garcia et al. 2016), suggesting that bears may be moving around human settlements in search of food, increasing the possibilities of human-*bear* conflicts. As a result, *M. ursinus*'s dietary preference for fruits is notable and warrants further investigation during the winter season. Therefore, the current investigation was aimed to understand the food preference of sloth bear during winter season in dry deciduous ecosystem of Gujarat, India.

23.2 Material and Methods

23.2.1 Study Area

The study was conducted in the Balaram-Ambaji Wildlife Sanctuary, located in the northern part of Gujarat state in Banaskantha district, western India, during the winter (December 2019 to March 2020) (Figs. 23.1 and 23.2a). Geographically, the sanctuary is located between 24°10′ to 24°21′N latitude and 72°38′ to 73°00 E longitude. The topography of the area mainly falls in the foothills of the Aravalli hill range, which constitutes the southern boundary of the expanding Thar Desert. Two major rivers of north Gujarat, Banas and Sabarmati, irrigate the study area, which makes the sanctuary rich in floral and faunal diversity. Balaram-Ambaji Wildlife Sanctuary falls in the semi-arid zone, where the climate is mostly subtropical (Rabari et al. 2022b). The climate can be divided into three seasons based on seasonal variations in humidity, temperature and precipitation: summer (March to June), winter (November to February) and monsoon (July to October). The sanctuary's temperature ranges from 7 °C in the winter (January) to 44 °C in the summer (June). The rainfall is comparatively very low, with an average annual rainfall of 600–750 mm (Rabari et al. 2023). Among the mammalian species, the sanctuary is well known for the *Melursus ursinus* and *Panthera pardus* Linnaeus (Rabari et al. 2022a). There are more than 483 plant species recorded in the sanctuary (Patel and Varshney 2018). The dominant plant species are *Cassia fistula*, *Zizyphus* sp., *Ficus*

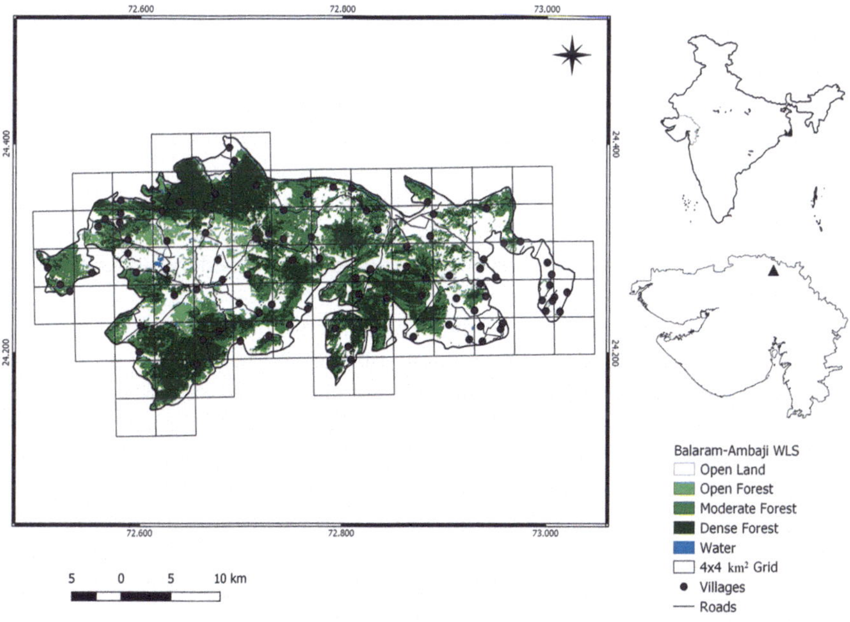

Fig. 23.1 Map of the study area, Balaram-Ambaji Wildlife Sanctuary, Gujarat, India

Fig. 23.2 (**a**) Landscape of the study area, and (**b** and **c**) photographs of sloth bears *Melursus ursinus* Shaw, 1791

sp., *Butea monosperma*, *Acacia senegal*, *Diospyros melanoxylon*, *Mallotus philippensis*, *Mangifera indica*, *Mitragyna parviflora*, *Tectona grandis*, *Wrightia tinctoria*, *Syzygium cumuni*, *Thespesia lampas* and *Adina cordifolia* (Patel and Varshney 2018).

23.2.2 Data Collection

To evaluate the food availability, the entire study area was divided into a grid with an equal size of 4 km^2 (Sharma et al. 2022). The size of the grid was kept consistent with the approximate size of the home range of *M. ursinus*, as described by Joshi et al. (1995), and also to minimize the effect of spatial autocorrelation. During fieldwork, each grid was surveyed by laying a 10 × 10 m^2 quadrat to record the availability of fruiting species for *Melursus ursinus* food as described by Sukhadiya et al. (2013) and Mewada (2015).

The nocturnal foraging habit of the bear primarily does not permit adequate data to be gathered based on observations of their feeding behaviour. Feeding ecology is mainly studied by scat analysis, one of the widely used techniques to study the diet of large carnivores. Scat analysis is also described as one of the best available methods for studying the food habits of *M. ursinus* (Dharaiya and Ratnayeke 2009; Mewada 2015). In the present study, the same methodology was used to identify the food composition in the *bear* scats collected from the study area. During the survey, scats of the bear have been identified and collected with collection details such as date, place and GPS location. The bear scats can be distinguished from other mammalian scats by their shape, size and undigested food (seeds, bee wax, ant heads and insect body parts). All the collected scats were maintained at the laboratory for further analysis. In the laboratory, scats were washed in water and sun-dried. Later on, each component was sorted into different Petri dishes. Animal matter was classified as black ants, honey bee wax, termites, weaver ants and other insect body parts. The plant seeds were sorted and identified to the species by comparing them with seeds of winter fruits found in the study area.

23.2.3 Data Analysis

Based on the presence of seeds in the scat and abundant fruiting species found in quadrates, a selectivity index (SI) was calculated to understand food preference among available fruiting species in the study area. The SI was calculated for each fruiting plant recorded in both *M. ursinus*'s scats and quadrates (*Zizyphus* sp., *Grewia flavescens*, *Ficus* sp. and *Cassia Fistula* were found in scats). This method has been previously used for the food preferences of mammals including bears. The SI was calculated using the following formula (Kays 1999):

$$S = E / A,$$

where "E" represents the use of each plant species as defined as the overall amount of the plant species found in scats, and "A" is the availability of a particular species in forests, defined as the frequency of plant species in quadrats. Selectivity indices >1.0 indicated that a species was eaten more than expected given its availability, while a value of index <1.0 indicated that the species was eaten less than expected. The logarithm of the index is used to produce symmetrical scales. Moreover, the Chi-square (X^2) test was employed to check the significance of food consumption and availability in the area.

23.3 Results

A total of 64 bear scats were analyzed during the study period. Here, the scat composition was mainly classified into three different categories: plant matter (62%), animal matter (33%) and both (5%) (Fig. 23.3). In the occurrence of different food items in scats, the frequency of *Zizyphus* sp. was found to be the highest (56.25%), followed by black ants (32.81%), honey wax (10.93%), *Ficus* sp. (7.81%), *Cassia fistula* (6.25%), termites (6.25%), *Diospyros melanoxylon* (3.12%), *Grewia flavescens* (3.12%), insect body parts (3.12%), *Zea mays* (1.56%), weaver ants (1.56%) and unidentified plant (1.56%) and unidentified animal material (1.56%) (Fig. 23.4).

Forty-three woody plant species were listed from 118 quadrates throughout the sanctuary. Out of which 11 were identified as food plants by *M. ursinus* in fruiting

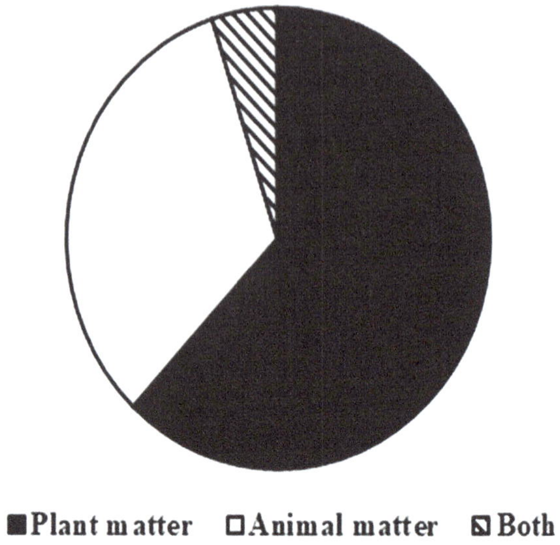

Fig. 23.3 Food components with respect to plant and animal matters found in the *Melursus ursinus* scat during the study period

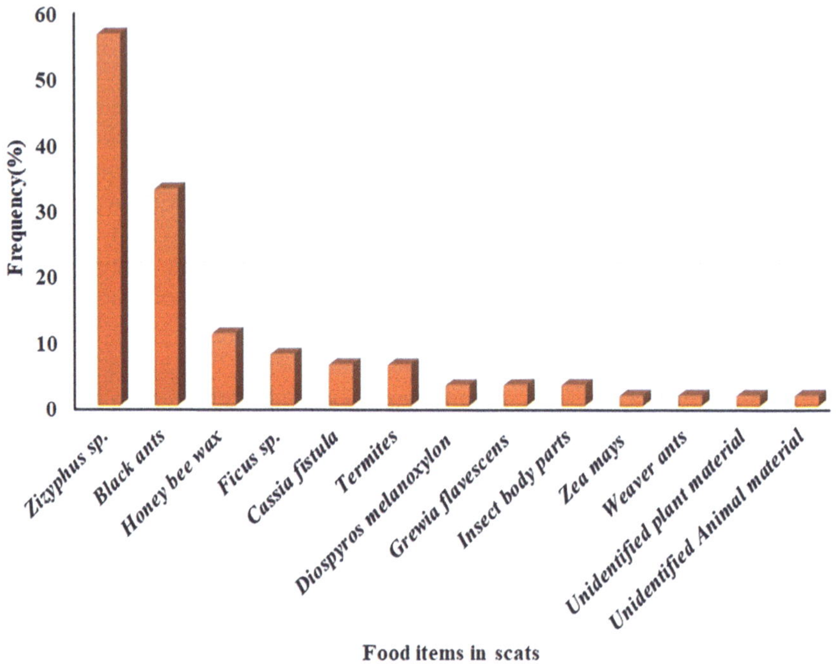

Fig. 23.4 Occurrence of different food items in *Melursus ursinus* scats ($n = 64$)

or non-fruiting stages based on available literature (Mewada and Dharaiya 2010; Sukhadiya et al. 2013; Mewada 2015) (Table 23.1). Among all the 11 food species, *Cassia fistula*, *Ficus benghalensis*, *Aegle marmelos*, *Grewia flavescens* and *Zizyphus*

Table 23.1 A list of plant species available as food for *Melursus ursinus* in the study area

No	Common name	Scientific name	Fruiting status in winter
1	Timru	*Diospyros melanoxylon*	Non-fruiting
2	Umbiyo	*Miliusa tomentosa*	Non-fruiting
3	Khajuri	*Phoenix sylvestris*	Non-fruiting
4	Jambu	*Syzygium cumini*	Non-fruiting
5	Mahudo	*Madhuca indica*	Non-fruiting
6	Peepal	*Ficus religiosa*	Non-fruiting
7	Garmalo	*Cassia fistula*	Fruiting
8	Vad	*Ficus benghalensis*	Fruiting
9	Bili	*Aegle marmelos*	Fruiting
10	Trambat	*Grewia flavescens*	Fruiting
11	Bor	*Zizyphus* sp.	Fruiting

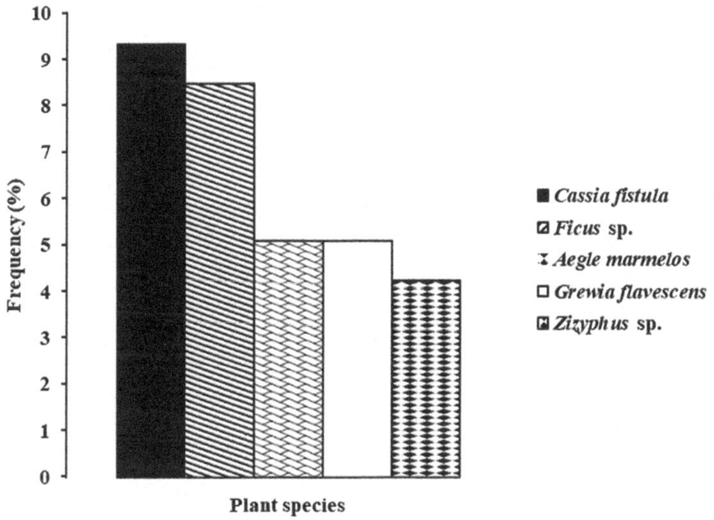

Fig. 23.5 Frequency of fruiting species available in winter as food for bears in the study area

sp. are reported as available fruiting species during the study period (winter) (Fig. 23.5), showing a high abundance of *Cassia fistula* (F = 9.32%), followed by *Ficus* sp. (F = 8.47%), *Aegle marmelos* and *Grewia flavescens* (f = 5.08%). The abundance of *Zizyphus* sp. (f = 4.23%) was lower compared to other plant foods available in Balaram-Ambaji Wildlife Sanctuary.

To understand the preferred fruit of the sloth bears, the SI was calculated based on the availability of fruiting species in quadrates and consumption during winter in Balaram-Ambaji Wildlife Sanctuary. The selectivity index of *Zizyphus* sp. is (=1), followed by *Grewia flavescens* (=0.14), *Ficus* sp. (= −0.16) and *Cassia fistula* (= −0.3) (Fig. 23.6). As an availability of food in forests, *Cassia fistula* was found dominant, while *Zizyphus* sp. is less. However, *Zizyphus* sp. consumption

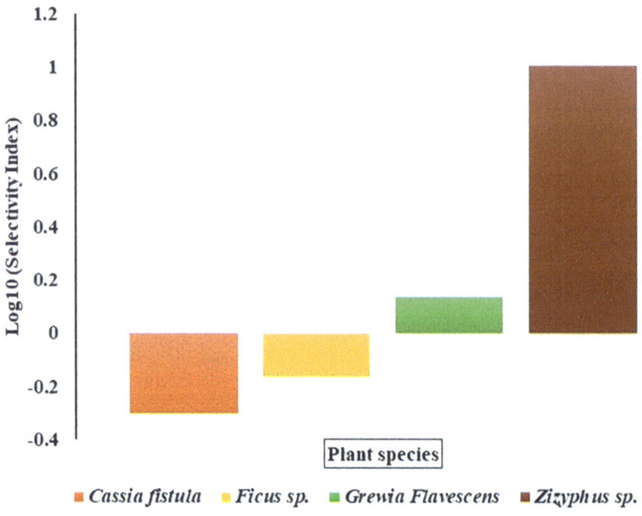

Fig. 23.6 Selectivity index for top food species consumed by bears in winter

(availability in scat) was observed more frequently in *M. ursinus* scat, despite the fact that other species such as *Cassia fistula* and *Ficus benghalensis* were in the fruiting stage and more abundant in the area. This might indicate an inclination of the *Melursus ursinus*'s dietary habit towards *Ziziphus* sp. This is also supported by the chi-square test, which suggests no significant association between food availability and consumption ($X^2 = 2.43$, df = 2).

23.4 Discussion

The research focuses primarily on the availability of fruiting species in forests and their consumption in order to better understand the preference of food species by sloth bears. This study provides adequate information about the food preferences of *M. ursinus* during the winter when both plant and animal matter were plentiful. It was found that *Melursus ursinus* consumes more plant-based food than animal matter during winter. The same was reported by Schaller (1967), Sukhadiya et al. (2013), Mewada (2015) and Philip et al. (2022). Contrastly, an inclination towards animal matter is observed in the diet of *Melursus ursinus* by Joshi et al. (1997), Ratnayeke et al. (2007), Ramesh et al. (2009) and Khanal and Thapa (2014).

Zizyphus sp. was the most preferred food with a high index of selectivity compared to others. The same species was also recorded with a high index of relative importance by Mewada (2015), revealing the importance of *Zizyphus* sp. as a food of *Melursus ursinus*. The strong preference of *Zizyphus* sp. over other fruiting species possibly revealed that the nutrient intake of this fruit may compensate cost of time spent foraging for it. According to Garcia et al. (2016), high number of

Melursus ursinus attacks have been reported during the winter in the study area. This suggested that *Melursus ursinus* may be travelling towards human settlements in search of food, which may raise the possibility of conflict between humans and *Melursus ursinus*. Consequently, the present study recommends plantations of *Zizyphus* sp. should be encouraged in the dry deciduous forests as one of the key factors that may help to reduce the movement of bears out of protected areas, especially in winter to reduce human-bear conflicts.

23.5 Conclusion

The present study revealed that *Melursus ursinus* is reported to feed more on plant-based food than animal food during the winter season in the Balaram-Ambaji Wildlife Sanctuary in Gujarat, India. It also suggests strong selectivity towards *Zizyphus* sp. among other fruiting plant species, which may show a high nutrition value of this plant species that can be used for better management practices to improve habitat quality for the bears in the tropics, world over.

23.6 Recommendations

The present study highlighted the preference of *Melursus ursinus* towards the fruiting species, especially *Zizyphus* sp. Plantation of such fruiting species in the sanctuary may restrict the movement of *Melursus ursinus* outside the protected area, which may be helpful to mitigate the human-bear conflict in the study area.

23.7 Research Gap

- There are gaps in information regarding the nutritional value of different food items of the sloth bears.
- A year-long study could provide more robust information on *Melursus ursinus* food preferences and selectivity during different seasons.
- A holistic understanding of *Melursus ursinus* ecology in different landscapes by comparing the nutritional value of food and selectivity by *M. ursinus* can be the backbone of the conservation and future management of *M. ursinus* habitat. To ensure the long-term survival of sloth bear populations and promote conservation strategies that account for both ecological and socio-economic factors are essential.

Acknowledgements The authors are thankful to the Chief Wildlife Warden and Principal Chief Conservator of Forests, Gujarat. We are also thankful to the rangers and forest field staff for their support and logistics during the fieldwork. The authors thank Mr. Pratik Desai for his technical support. Our heartfelt thanks to the Wildlife and Conservation Biology Research Lab and Department of Life Science, HNG University—Patan for providing lab facilities.

References

Dharaiya N, Ratnayeke S (2009) Escalating human–sloth bear conflicts in North Gujarat: a tough time to encourage support for bear conservation. Int Bear News 18(3):12–14

Garcia KC, Joshi HM, Dharaiya N (2016) Assessment of human–sloth bear conflicts in North Gujarat, India. Ursus 27(1):5–10. https://doi.org/10.2192/ursus-d-15-00012.1

Joshi AR, Garshelis DL, Smith JL (1995) Home ranges of sloth bears in Nepal: implications for conservation. J Wildl Manag 59:204–214

Joshi AR, Garshelis DL, Smith JL (1997) Seasonal and habitat-related diets of sloth bears in Nepal. J Mammal 78(2):584–597. https://doi.org/10.2307/1382910

Kays RW (1999) Food preferences of kinkajous (*Potos flaws*): a frugivorous carnivore. J Mammal 80(2):589–599

Khanal S, Thapa TB (2014) Feeding ecology of sloth bears in Chitwan National Park, Nepal. J Sci Technol 9(2):118–122

Laurie A, Seidensticker J (1977) Behavioural ecology of the sloth bear (*Melursus ursinus*). J Zool 182(2):187–204. https://doi.org/10.1111/j.1469-7998.1977.tb04155.x

Mewada TP (2015) Index of relative importance of the dietary proportions of sloth bear (*Melursus ursinus*) in semi-arid region. Not Sci Biol 7(3):281–288

Mewada T, Dharaiya N (2010) Seasonal dietary composition of sloth bear (*Melursus ursinus*) in the reserve forest of Vijaynagar, North Gujarat, India. Tiger Paper 37(2):8–13

Mewada TP, Tiwari UL, Kotia A (2019) Fruiting species influence the seasonal use of the habitat by sloth bear in and around Balaram Ambaji wildlife sanctuary, Gujarat, India. Not Sci Biol 11(2):183–190

Noyce KV, Kannowski PB, Riggs MR (1997) Black bears as ant-eaters: seasonal associations between bear myrmecophagy and ant ecology in north-central Minnesota. Can J Zool 75(10):1671–1686. https://doi.org/10.1139/z97-794

Palei HS, Debata S, Sahu HK (2020) Diet of sloth bear in an agroforest landscape in eastern India. Agrofor Syst 94:269–279. https://doi.org/10.1007/s10457-019-00389-1

Patel RH, Varshney AV (2018) Floristic diversity of forest of Balaram Ambaji Wildlife Sanctuary, Banaskantha District, Gujarat. J Agroecol Nat Resour Manag 5(3):132–134

Philip R, Bhatnagar C, Koli VK (2022) Feeding ecology of the vulnerable sloth bear (*Melursus ursinus*) in and around Mount Abu wildlife sanctuary, Rajasthan, India. Int J Environ Stud 79(4):651–661

Rabari V, Dharaiya N (2022) A systematic review on the feeding ecology of sloth bear Melursus ursinus Shaw, 1791 in its distribution range in the Indian subcontinent. J Threat Taxa 14(12):22329–22336

Rabari VM, Malik A, Dharaiya N (2022a) A preliminary study on relative abundance of wild mammals based on camera trap in Balaram-Ambaji Wildlife Sanctuary, Gujarat State, India. J Anim Divers 4(3):53–61. https://doi.org/10.52547/JAD.2022.4.3.7

Rabari VM, Malik A, Dharaiya N (2022b) First camera-trap record of rusty-spotted cat *Prionailurus rubiginosus* (I. Geoffroy Saint-Hilaire, 1831) from Balaram-Ambaji Wildlife Sanctuary, Gujarat, India. J New Biol Rep 11(1):7–9

Rabari V, Chaudhary S, Desai P, Dharaiya N (2023) A unique observation of masturbation in a male common langur (*Semnopithecus entellus*) in Jessore sloth bear sanctuary, Gujarat, India. Mun Ent Zool 18(1):617–619

Rajpurohit KS, Krausman PR (2000) Human-sloth-bear conflicts in Madhya Pradesh, India. Wildl Soc Bull 28:393–399

Ramesh T, Sankar K, Qureshi Q (2009) Additional notes on the diet of sloth bear *Melursus ursinus* in Mudumalai Tiger Reserve as shown by scat analysis. J Bombay Nat Hist Soc 106(2):204–206

Ratnayeke S, Van Manen FT, Padmalal UKGK (2007) Home ranges and habitat use of sloth bears *Melursus ursinus inornatus* in Wasgomuwa National Park, Sri Lanka. Wildl Biol 13(3):272–284

Schaller GB (1967) The deer and the tiger: a study of wildlife in India. University of Chicago Press, Chicago

Sharma B, Joshi R, Sathyakumar S (2022) Habitat suitability modelling of *Melursus ursinus* (Shaw, 1791) (Mammalia: Carnivora) in the Chitwan National Park, Nepal. J Anim Divers 4(3):31–43. https://doi.org/10.52547/JAD.2022.4.3.5

Sukhadiya D, Joshi JU, Dharaiya N (2013) Feeding ecology and habitat use of sloth bear (*Melursus ursinus*) in Jassore Wildlife Sanctuary, Gujarat, India. Indian J Ecol 40(1):14–18

Social Organisation of Asian Elephants: Findings from the Kabini Elephant Population, Southern India

24

Chowdari Jabili, Ankana Sanyal, Divya Choudhary, and T. N. C. Vidya

Abstract

Findings on the social organisation of Asian elephants from a long-term study of individually identified elephants in Nagarahole and Bandipur National Parks (Kabini elephant population), southern India, are summarised here. First, female social structure, genetic relatedness and associations of females, dominance relationships and tests of socioecological theory, relating food resources and agonism within and between groups, are described. We then mention the work on differences in female social structure and group sizes in the presence and absence of calves and move on to the development of calf behaviour and interactions between calves and female conspecifics in the context of allomothering. Finally, we summarise male social organisation, describing the social structure of adults and subadult males, dominance relationships between males, the effect of musth on associations and dominance, and subadult male associations and dispersal.

Keywords

Asian elephant · Kabini · Nagarahole and Bandipur National Parks · Social organisation · Dominance relationships

C. Jabili · A. Sanyal · D. Choudhary · T. N. C. Vidya (✉)
Animal Behaviour and Sociogenetics Laboratory, Evolutionary and Organismal Biology Unit, Jawaharlal Nehru Centre for Advanced Scientific Research, Bengaluru, Karnataka, India
e-mail: tncvidya@jncasr.ac.in

S. Sridhara et al. (eds.), *Animal Behavior in the Tropics*,
https://doi.org/10.1007/978-981-97-7324-4_24

24.1 Introduction

Studying the interactions and relationships amongst individuals is crucial to understanding the social organisation and behavioural ecology of social animals. As highlighted in a recent editorial (Salguero-Gómez et al. 2022), 'Time is of the essence in ecology'; thus, programmes that monitor populations of identified individuals over long periods are valuable in obtaining such understanding (see examples in Clutton-Brock and Sheldon 2010; de Silva 2016; Sheldon et al. 2022). Here, we summarise the findings from a long-term study of an Asian elephant (*Elephas maximus*) population in southern India. Much has been written about certain aspects of free-ranging Asian elephants—feeding ecology, elephant-human interactions, conservation strategies, population sizes and trends (McKay 1973; Olivier 1978; Sukumar 1989, 2003; Baskaran et al. 1995, 2010; Armbruster et al. 1999; Williams et al. 2001; Bist 2002; Leimgruber et al. 2003; Campos-Arceiz and Blake 2011; Goswami et al. 2014; Fernando et al. 2020), and, to a lesser extent and more recently, population genetic structure, phylogeography and stress hormones, among others (Vidya et al. 2005, 2009; Vidya 2016, Pokharel et al. 2017, 2019; Vijayakrishnan et al. 2018). However, studies on Asian elephant social organisation are somewhat lacking (but see McKay 1973; Sukumar 1989; Fernando and Lande 2000; de Silva et al. 2011, 2017). This is not entirely surprising: Asian elephants largely occupy forested habitats unlike their African savannah cousin and are, therefore, much more difficult to identify or follow. While indirect methods might provide some insights (Vidya and Sukumar 2005; Ahlering et al. 2011; Chakraborty et al. 2014), radio tracking elephants in forested habitats (Baskaran et al. 1995; Fernando et al. 2008; Abram et al. 2022) or opportunistically observing identified individuals in more open areas (de Silva et al. 2011, 2017; Nandini et al. 2017, 2018; Keerthipriya et al. 2020, 2021; Gautam and Vidya 2023) is often necessary to obtain information on social organisation. There have been a few programmes monitoring individually identified Asian elephants, such as in Uda Walawe (de Silva et al. 2013) and Minneriya and Kaudulla National Parks, Sri Lanka, and in Mudumalai Tiger Reserve (Arivazhagan and Sukumar 2005) in India, apart from Kabini (Vidya et al. 2014). While Athira and Vidya (2021) compared the social systems of elephants across populations and species, in this chapter, we largely restrict ourselves to an overview of some of the findings from a single population—the Kabini elephant population.

24.2 The Asian Elephant

The Asian elephant (*Elephas maximus*) is one of the three extant species of the order Proboscidea, the other two species being the African savannah elephant (*Loxodonta africana*) and the African forest elephant (*Loxodonta cyclotis*). The Asian elephant is endangered (Williams et al. 2020), with its current range reduced to less than 6% of its historical range (Sukumar 2003, p. 45) and its global population size in the wild is estimated at 48,300–51,700 individuals (Williams et al. 2020). Although distributed across 13 countries, about 60% of the wild Asian elephants (approximately 29,900 individuals) are found in India, distributed across four regions:

southern India, northeastern India, east-central India and northern India (Baskaran et al. 2011). Elephants inhabit a wide range of forest types, from wet evergreen forests to moist and dry deciduous forests to dry thorn forests (Sukumar 2003; Williams et al. 2020). They also have large annual home ranges of several hundred square kilometres, although this may vary based on habitat type and the availability of space (Baskaran et al. 1995; Fernando et al. 2008; Alfred et al. 2012; Wilson et al. 2020, 2021; Abram et al. 2022). Asian elephants are megaherbivores (weighing 2500–4500 kg) with visible body-size sexual dimorphism (Sukumar et al. 1988; Mumby et al. 2015). They spend most of their waking hours feeding on a variety of plants, depending upon the habitat and season (Vancuylenberg 1977; Sukumar 1989; Baskaran et al. 2010; English et al. 2014), and their large ranges and foraging habits make them important seed dispersers (Campos-Arceiz et al. 2008; Campos-Arceiz and Blake 2011).

The Asian elephant is a social species, with females and their offspring forming matrilineal social units and pubertal males dispersing from their natal clans (Sukumar 1989). They have high cognitive abilities (Plotnik et al. 2011) and use olfactory, acoustic and tactile communication, with vision probably being limited to shorter ranges (see Sukumar 2003). They are long-lived (living up to 60–70 years), polygynous, and slow reproducing (often with an interbirth interval of 4 years or more (Sukumar 2003; de Silva et al. 2013)). Females conceive from the age of about 10 years or older (Sukumar 2003; de Silva et al. 2013; Kabini Elephant Project unpublished data). Males may be physiologically able to reproduce when they are about 15 years old but compete for mates only when they are older (Sukumar 2003).

Although Asian elephants do not have any natural predators (calves may be preyed upon by tigers), the species faces enormous threats from loss, fragmentation and degradation of habitats, as well as the ensuing human-elephant conflict over space and food (Johnsingh et al. 1990; Leimgruber et al. 2003; Williams et al. 2020). Poaching of males for ivory (only males, and not all males, carry tusks) resulting in extremely skewed adult sex ratios, and the historical capture of elephants for warfare, historical sport hunting and control killing have been other threats (Sukumar 2003).

24.3 Study Area and Population

The Kabini Elephant Project was initiated in early 2009 to carry out long-term monitoring of identified Asian elephants in southern India. This is one of the few such programmes worldwide of monitoring hundreds of identified elephants and the only one of this scale in India. The study area includes Nagarahole National Park and Tiger Reserve and Bandipur National Park and Tiger Reserve (Fig. 24.1). These parks lie in the Nilgiris-Eastern Ghats landscape in southern India, which harbours the single largest population of Asian elephants, with over 8800 individuals (Baskaran et al. 2011). The parks cover a range of habitat types, from evergreen and moist deciduous forests in the west to dry deciduous forests over large areas, to dry thorn forests in the east (Pascal 1986, 1988). The average annual rainfall ranges from about 625 mm in the west to 1500 mm in the east, with a dry season from

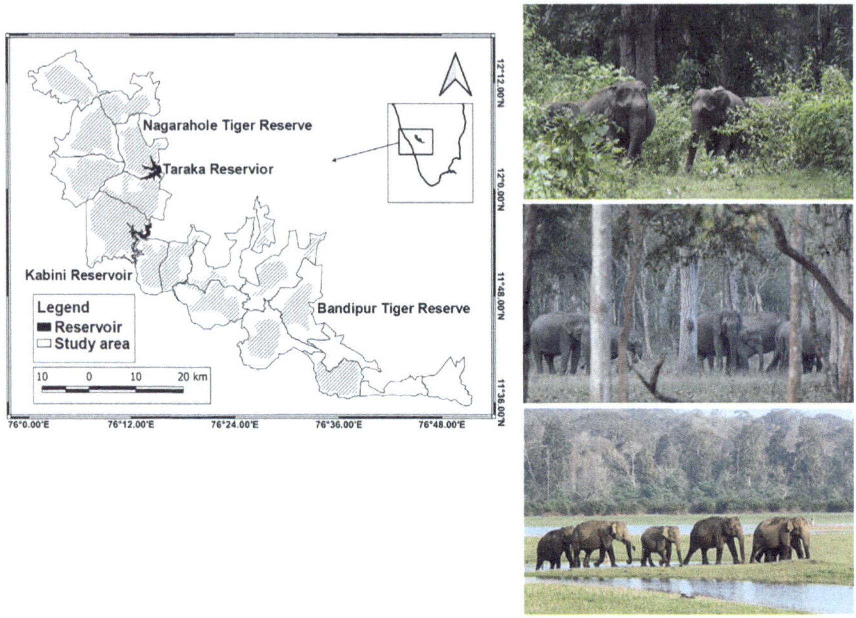

Fig. 24.1 Map of Nagarahole and Bandipur National Parks, areas sampled (in grey hatching) and views of the forest in the wet season (top) and dry season (middle), and the grassland habitat (bottom)

about December to early June and a wet season from June to November (Nandini et al. 2017; the area receives rainfall from the SW and NE monsoons, but as the forest remains wet in between, the entire duration is considered the wet season). The two parks are separated by the perennial river, Kabini. The Beechanahalli Dam built on this river in the early 1970s led to the creation of the Kabini reservoir, the backwaters of which recede during the dry season and form an open grassland, attracting elephants (Fig. 24.1). This open area has allowed identification and observation of elephants. During the wet season, the backwaters submerge the grassland and elephants are more dispersed in the forests on the two sides. Nagarahole and Bandipur National Parks have high elephant densities (~2 elephants/km^2; Baskaran et al. 2011). The adult sex ratio of elephants is about 1 male:4.9 females (Gupta et al. 2016) due to historical poaching. We refer to the elephants using the study area centred around the Kabini reservoir as the Kabini elephant population for convenience. This is not geographically separated from other areas and is not a distinct population in terms of gene flow.

We have been carrying out stratified sampling, with the areas around the backwaters sampled more intensively than the forests on either side. Sampling has been carried out from about 6:30 AM to about 5:30 or 6:30 PM based on daylight hours and field permits. When elephants are encountered, they are aged and sexed, and the group compositions are noted down after individually identifying animals (Vidya et al. 2014). Identification is carried out based on a combination of natural physical characteristics, such as ear, tail, tusk and other features (Fig. 24.2; Vidya et al. 2014).

Fig. 24.2 Individual identification of Asian elephants using natural physical characteristics. Individual ID sheets of a female (top left) and a male (top right), and right and left ears of four adult females of a clan (below)

24.4 Female Elephant Social Organisation and Structure

It has long been known that Asian elephants form matrilineal societies comprising adult females and dependent offspring of both sexes (McKay 1973; Sukumar 1989). However, while Asian and African savannah elephants were initially thought to have similar social organisation, with a multitiered fission-fusion society comprising family groups, bond groups and clans (Eltringham 1982; Moss and Poole 1983; Sukumar 1989), Fernando and Lande (2000) suggested that the social organisation of the former was restricted to family groups. A fission-fusion society is one in which group sizes and compositions change over short time periods, often in response to spatiotemporally varying resources (Kummer 1971; Dunbar 1988; Connor et al. 2000; Asensio et al. 2009). Thus, family groups might fuse temporarily into bond groups, which in turn might fuse temporarily into clans, which might then fission into the component bond groups or family groups. The subsequent work of de Silva et al. (2011), again in Sri Lanka, showed that female Asian elephants had more extended associations than those within the family group. However, the extent and strength of associations were much reduced compared to the African savannah elephant (de Silva and Wittemyer 2012). Studies in the Kabini population showed a social organisation between that of an individual-based fission-fusion society and a flexible-nested multi-level fission-fusion society (Nandini et al. 2018; Fig. 24.3).

A female group (equivalent to a 'party' in the primate literature) is a set of adult females and their associated young ones (present during a sighting in the field) showing affiliative interactions or coordinated movement and within about 50–100 m of each other (Nandini et al. 2018). Based on field data collected on identified adult females from 2009 to 2014, associations between different female dyads were quantified using the association index (AI) (Ginsberg and Young 1992; AI_{AB} = the number of times individual A and B were seen together (N_{AB}) divided by the total number of times either was seen). These AI values were used to construct a female social network, with the nodes representing individual females, the edges representing associations between females and edge thicknesses representing AI values (see Fig. 24.3). This network was highly non-random and was partitioned into communities (a set of nodes with more connections within the set than across

Fig. 24.3 Illustration of female social organisation inference: (**a**) female groups seen at different points in time, of different sizes and compositions. Each female is coloured differently. Sighting data on group compositions are used to construct an association index (AI) matrix (**b**), and AI values are used as edge weights in a social network (**c**). AI values in the matrix are depicted here in the form of darker shades representing higher AI values. Hierarchical community detection in the social network allows one to obtain clans—the most inclusive communities—and first-level communities within clans. There are three clans in the network shown, and one of the clans has four first-level communities within the clan (flexible-nested fission-fusion society), while the other two smaller clans do not have such sub-communities (individual-based fission-fusion society). The clan and first-level communities are drawn on to the illustration of the groups seen in the field (from a) in order to show how females associate within these communities (**d**)

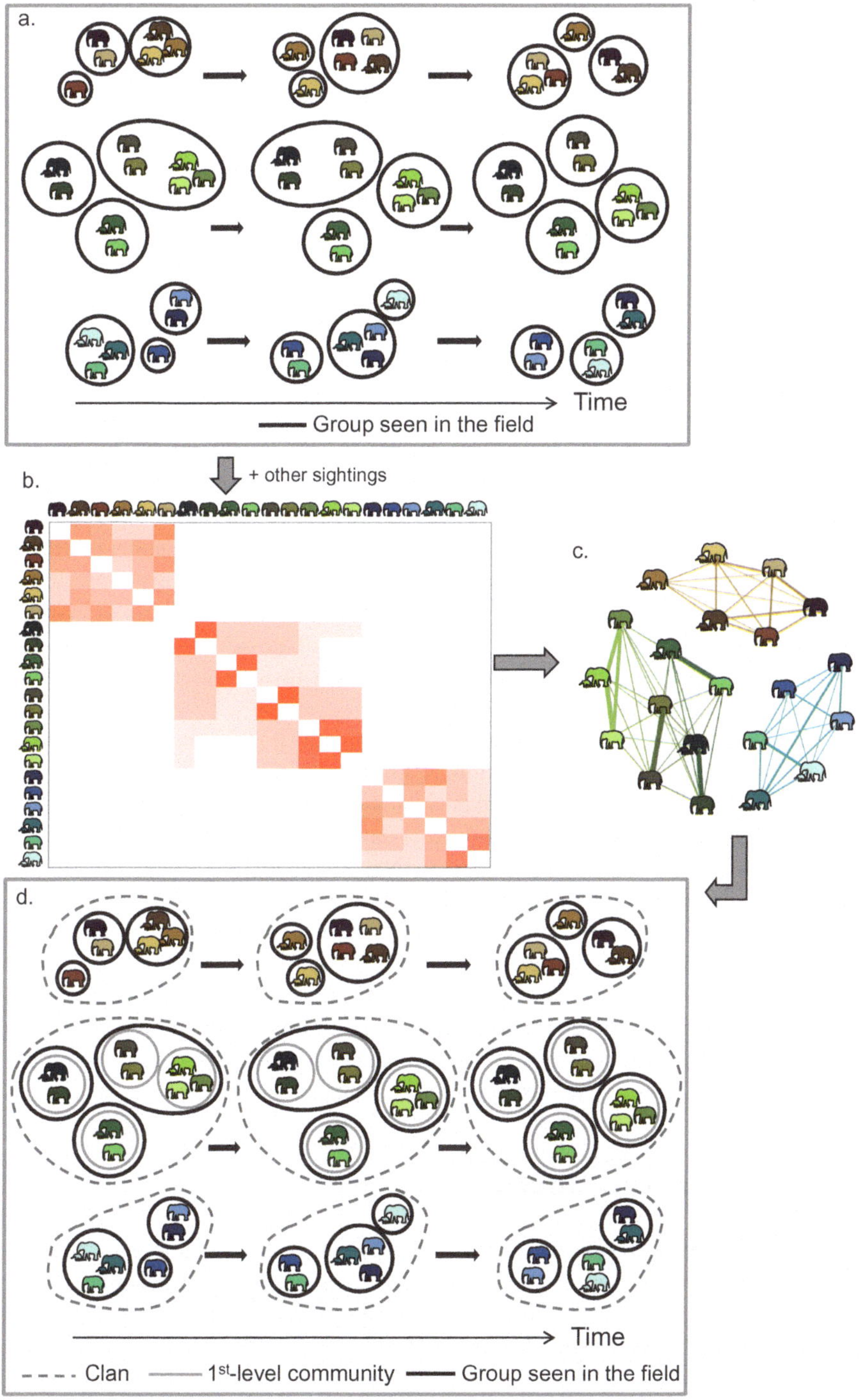
a.
Time
Group seen in the field
b.
+ other sightings
c.
d.
Time
Clan
1st-level community
Group seen in the field

sets) using the statistic modularity (see Nandini et al. 2018). Communities represented the most inclusive level of female social organisation and were termed clans. Associations between females of different clans were rare.

Unlike species with fixed social groups, female elephants live in a fission-fusion society, in which group sizes and compositions change over time. Thus, all the females of a clan are not always seen together, and subsets of the clan are observed as groups (parties) at different points in time. Based on hierarchical community detection within clans (Louvain method: Blondel et al. 2008), it was possible to find out whether there were sub-communities within clans. If individual females joined or left groups independent of other females in the clan, no distinct sub-communities would be found within clans, and this would be termed an individual-based fission-fusion society. On the other hand, if sets of females within the clan associated together more often than with others, there could be sub-communities, giving rise to multi-level societies, which in turn would be flexible or strictly nested communities within communities (van Schaik 1999; Grüter and Zinner 2004). Nandini et al. (2018) found no sub-structuring within 38% of the clans and sub-structuring in other, mostly larger, clans. The sub-communities were called first-level communities, partly analogous to the family groups of African savannah elephants. (First-level and second-level communities were found in one clan, which meant two levels of sub-structuring within that clan.) Therefore, the Kabini population was inferred to have a female social organisation in between that of an individual-based fission-fusion society and a flexible-nested multilevel fission-fusion society (Nandini et al. 2018; see Athira and Vidya 2021; see Fig. 24.3).

A comparison of social organisation in the Kabini population (using a 500-m distance cutoff to match the definition of groups in other studies) with the Uda Walawe population and the Samburu African savannah elephant population found the first to have similar group sizes to Uda Walawe and to have network statistics in between those of the two other populations (Nandini et al. 2018). This comparative study also showed that, although the African savannah elephant had a greater extent of sociality (more associates, stronger associations and a social structure between a flexible-nested and a strictly nested multilevel fission-fusion society), there were basic underlying similarities in the social organisation of the Kabini and Samburu populations, with the differences possibly emanating from the larger average group sizes seen in Samburu compared to Kabini (Nandini et al. 2018). More such long-term studies are necessary to further compare social structures.

24.5 Female Social Structure

Animals that show fission-fusion dynamics often change their group sizes in response to spatiotemporally varying resource availability, with groups splitting when resources are low or spread out or increase travel costs, and fusing when resources support large groups (Kummer 1971; Dunbar 1988; Chapman et al. 1995; Wittemyer et al. 2005; see Aureli et al. 2008; Smith et al. 2008; Asensio et al. 2009). As mentioned above, a group (party) is the number of individuals seen in a sighting, which can vary over time in a fission-fusion society and is often a subset of the most inclusive community (clan in the Asian elephant). Therefore, it was natural to look at the variability in group size and social structure across seasons, with the expectation that group sizes might be larger and associations stronger during the season with greater resource availability. While food availability in the forest would be higher during the wet than in the dry season, it would be more clumped around the Kabini reservoir during the dry season. The latter, along with water availability, might increase group size during the dry season. Nandini et al. (2017) found that group sizes were, in fact, larger overall in the dry season than in the wet season. The percentage of non-zero AI values was higher in the dry season than in the wet season due to a greater number of associations with rarer individuals in the dry season. When only common individuals were considered, a greater number of stronger connections was found in the wet season. However, when group sizes and associations of individual clans were compared across seasons, there was no consistently significant difference in group size or association patterns across seasons (Nandini et al. 2017). Thus, the overall difference at the population level likely arose from the over-representation of some clans, highlighting the need to consider individually identified animals in such studies.

An important finding from this work was that average group sizes were similar across clans of different sizes (clans could have different numbers of adult females depending on how many females were born into the clan and survived). Although one might have expected larger clans to associate in the form of larger groups and vice versa, most of the larger clans were also found in the form of relatively small groups, thus lowering the within-clan average AI (Fig. 24.4) and increasing the time required to meet all the clan-mates (Nandini et al. 2017). Thus, there seemed to be a restriction on female group size in this population, possibly due to food limitation in larger groups (Nandini et al. 2017). Since the small group sizes did not give rise to small, fixed, groups, but instead to groups that showed changes in group compositions, there were possibly benefits from interacting with other clan-mates. It was

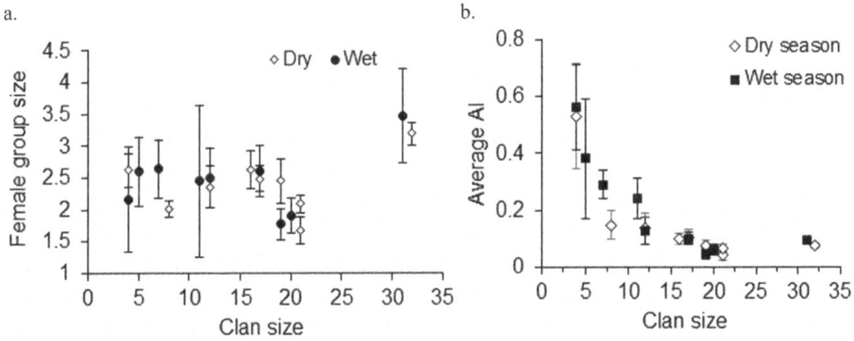

Fig. 24.4 (**a**) Average (± 1.96 SE) female group sizes (number of adult females) seen in nine common clans of different sizes during the dry and wet seasons, and (**b**) average (± 1.96 SE) Association Index (AI) within clans of different sizes during the dry and wet seasons. Graphs based on data from Nandini et al. (2017)

suggested that allomothering might be one such benefit. Thus, fission-fusion dynamics in this population was not related to seasonality in resources, unlike that seen in several species, and possibly helped females meet clan-mates under conditions of restricted group size (Nandini et al. 2017).

24.6 Genetic Relatedness and Female Associations

Since female group sizes were generally small (2–3 adult females) compared to clan sizes in the Kabini population, it was of interest to find out how females chose clan-mates with whom to group. One possibility was that females preferred to associate with relatives. First- and second-order relatives had been found to associate together in the neighbouring Mudumalai Wildlife Sanctuary (Vidya and Sukumar 2005), but clans and sub-communities had not been identified in that study. Therefore, Nandini (2016) examined associations and genetic relatedness amongst females in different clans in Kabini to find out (i) whether clans were assemblages of related females, (ii) whether first-level communities (see Sect. 24.4 above) within clans were composed of close relatives and (iii) whether females associated most with their relatives. Fresh dung samples were collected as the source of DNA from identified adult females upon observed defecation and samples were stored in ethanol. DNA extracted from the dung was used to genotype individuals using 14 microsatellite loci. The dataset included 174 identified adult females.

The (average ± SD) genetic relatedness (Queller and Goodnight 1989) between adult females within clans was significantly greater than zero ($R = 0.079 \pm 0.044$, $N = 15$ focal clans), indicating overall relatedness, but not high relatedness (Nandini 2016). Many clans did not have a relatedness value significantly greater than zero. However, first-level communities within-clans had closely related females ($R = 0.215 \pm 0.082$, $N = 20$ first-level communities), suggesting that such structuring might be based on relatedness. While the overall correlation between genetic relatedness and associations (AI) was weak at the population level and was seen in only about half the clans at the clan level, adult females were significantly more closely related to their top associate (i.e. associate with the highest AI; $R = 0.254 \pm 0.046$, $N = 124$ dyads from 15 focal clans), possibly as first- or second-order relatives, than to the average associate (Nandini 2016). They were also more closely related to their second associate, but not to their third associate, compared to the average associate. The average relatedness between female dyads with association index greater than 0.5 was 0.336 ± 0.116 ($N = 21$ dyads from 10 focal clans). It is possible that associations of such highly related individuals might help with the nurturing and/or protection of young ones (allomothering), leading to possible indirect fitness benefits. However, there were also unrelated females within clans and whether advantages during between-clan contests lead to direct benefits of clan membership needs to be examined (Nandini 2016). Low relatedness may ensue from increased clan size, multiple females in a clan dying and the remaining not being very related, or different males siring offspring with females in a clan. The last female survivor of a diminished clan may also eventually join another clan (Keerthipriya and Vidya 2021), resulting in lowered relatedness.

24.7 Dominance Relationships Among Females Within and Between Clans

Competition for food is often a cost of group living and can shape social organisation and structure (Wrangham 1980; Watts 1985; van Schaik 1989; Janson and Goldsmith 1995). Different kinds of dominance structures are found in animal societies, ranging from egalitarianism (with competitors having similar access to the limiting resource) to despotism with a clear linear dominance hierarchy (with clearly predictable winners and losers) (Vehrencamp 1983; van Schaik 1989). In the latter scenario, a higher rank is expected to provide priority of access to resources and hence reproductive success (Janson 1985; van Schaik 1989; Hodge et al. 2008). It was of interest to examine dominance relationships in the Asian elephant because of three disparate predictions. Since elephants are mixed-feeders, with grass being a large component of forage in a neighbouring forest (Baskaran et al. 2010) as well as in the Kabini grassland, and food resources are likely widely distributed and of low quality, an egalitarian society with low within-group and between-group agonism (aggressive/submissive behaviours) could be expected. Here, within-group and between-group indicate within-clan and between-clan agonism. This is because, while clans could split into groups in the fission-fusion society, we very rarely saw agonistic interactions between groups belonging to the same clan (Gautam and Vidya 2023). On the other hand, since there were close relatives within clans, it was plausible that the Kabini population would show nepotistic dominance relationships, with females supporting relatives (van Schaik 1989; Sterck et al. 1997). Third, African savannah elephants were known to show strong expression of dominance, with an age/size-based hierarchy. The matriarch (oldest female) was the most dominant individual within family groups (Archie et al. 2006) and also influenced the ranks of non-matriarch females in between-group contests (Wittemyer and Getz 2007).

Therefore, behavioural data were collected on adult females in Kabini and agonistic interactions between them were recorded. Based on a few hundred agonistic interactions observed within clans during 2009–2012, an average rate of independent agonistic interactions involving adult females of about one interaction per hour was found (Nandini 2016; Gautam and Vidya 2023). There was an effect of age on winning, with winners being older than losers in over three-fourths of the agonistic interactions between adult females (Nandini 2016). In the Amboseli African savannah elephant population, older adult females won 94% of the agonistic interactions (Archie et al. 2006). Thus, there was a greater proportion of age-reversed dominance and weaker dominance structure within clans in the Asian elephant in Kabini, as also seen in the Uda Walawe Asian elephant population (de Silva et al. 2017; although females from across social units were included in that analysis). Moreover, in Kabini, the matriarch was not the single most dominant female within clans, nor was she the most central animal in the clan's social network (Nandini 2016). Behaviourally distinct matriarchs were not seen in Uda Walawe either (de Silva et al. 2017). Matriarchs are considered to be important repositories of social information in the African savannah elephant (McComb et al. 2011) and have the highest ranks as mentioned above. Thus, sociality and decision-making in the Asian elephant may be less centralised, perhaps partly due to smaller group sizes. The effect of nepotism on dominance has not yet been tested explicitly by examining genetic relatedness and dominance relationships.

There was an effect of age on winning even when adult females from different clans were engaged in agonistic interactions (Nandini 2016). Nandini (2016) also examined between-clan interactions at the level of the clan itself, deeming a clan to have won against another clan if the other clan moved away from the feeding area. There was a clear winner in only about half the clan-level between-clan agonistic encounters. When there was a clear winner, the winning group was larger in size, and the winning group's oldest female present during the encounter was older than the losing group's oldest female present (Nandini 2016). Overall, the Kabini female elephant population seemed to show a regime of weaker within-clan contest competition and stronger between-clan contest with an advantage to larger groups in between-clan contests (Nandini 2016).

24.8 Dominance Relationships and Access to Food Within Clans

To understand the foraging consequences of dominance relationships within clans, data on grass abundance from 1-m × 1-m quadrats were collected at individual feeding sites and from control sites 3 m (one adult female elephant body length) away. If there was a strongly expressed linear dominance hierarchy, one would expect a correspondingly steep rank-related skew in access to feeding opportunities, with dominants obtaining more access and subdominants having less access. On the other hand, if the dominance hierarchy was weak, such rank-related skew might not be found (Janson and van Schaik 1988; Koenig 2002). Despite an ordering by age, significant linear dominance hierarchies were not found within clans in Kabini (Nandini 2016). The directional consistency index (which quantifies the extent of directionality in a dominance relationship; i.e. whether one individual of a dyad is consistently dominant over another) was also not very high within clans (mean = 0.740, SD = 0.069, $N = 8$ clans) (Gautam 2019). Thus, the dominance relationships were weak. Grass biomass (mean = 731.66 g/m^2, 1.96 SE = 38.252, $N = 152$ feeding sites), grass cover and grass height were all significantly higher at the individual feeding sites examined than at their respective control sites (mean biomass = 626.88 g/m^2, 1.96 SE = 39.841), indicating better quality of feeding sites selected by individuals than the sites nearby, but there were large contributions of random effects, and the effect sizes were low (Gautam 2019). In keeping with the expectation based on a weak dominance hierarchy, there was no significant effect of the dominance status of individuals (using z-transformed modified David's scores) on the grass abundance at their feeding sites (Gautam 2019).

24.9 Effect of Food Abundance and Distribution on Within-Clan and Between-Clan Agonism

Female social organisation is thought to have been shaped by resource-risk distributions (Wrangham 1980; van Schaik 1989; Janson and Goldsmith 1995). The Ecological Framework of Female Social Relationships (EMFSR; Wrangham 1980; van Schaik 1989; Isbell 1991; Sterck et al. 1997), a major part of socioecological theory (Crook and Gartlan 1966; Terborgh and Janson 1986), relates the abundance and distribution of food resources and predation to the evolution of various female social structures through food competition. When food availability is limited and highly dispersed within groups, scramble competition is expected within groups, whereas when food is available in small clumps or patches that can be usurped by individuals, contest competition is expected within groups (Resident-Nepotistic Society) (Janson 1985; van Schaik 1989; Sterck et al. 1997). Similarly, contest competition is expected between groups if high-quality food patches are large and can be monopolised by a group to the exclusion of other groups (Resident-Egalitarian Society) (Wrangham 1980; Isbell 1991; Sterck et al. 1997). Potentially strong contest competition between and within groups is expected to lead to the amelioration of within-group contests via the dominants becoming more tolerant of the subordinates to retain the advantage of large group size in between-group contests (Resident-Nepotistic-Tolerant Society; van Schaik 1989; Sterck et al. 1997).

The first tests of the EMFSR in the Asian elephant were carried out in the Kabini elephant population by quantifying food abundance, food distribution, competitor density and agonistic interactions within and between clans. Quantification of grass abundance in the Kabini backwaters and Nagarahole forest habitat showed that the grass biomass in the former was about three times that of the latter (backwaters, 2015: mean = 704.78 g/m^2, $N = 95$ plot clusters; 2016: mean = 583.46 g/m^2, $N = 96$ plot clusters; forest, 2013: mean = 202.24 g/m^2, $N = 40$ plots), indicating that the Kabini backwaters were a large, resource-rich habitat (Gautam and Vidya 2023). The rate of agonistic interactions between adult females between clans (mean = 2.514 interactions/female/h) was over twice the rate of agonistic interactions between adult females within clans (mean = 1.152 interactions/female/h). The behaviours seen were also more aggressive between clans. Overall, 377 independent and 578 non-independent individual-level interactions were recorded during 84 agonistic encounters between clans (Gautam and Vidya 2023), compared to a single between-clan agonistic encounter that had been recorded over a period of a few years in the neighbouring Mudumalai forest (Baskaran et al. 2018). Thus, the food-rich grassland created by the receding backwaters of the Kabini reservoir seemed to enable strong between-clan contests (Gautam and Vidya 2023). While there were clear winners in only about half the between-clan encounters, losing groups lost access to their feeding site, apart from the time invested in the agonism.

The Kabini grassland was divided into zones, and grass biomass was sampled in clusters of plots within these zones. There was variability across these zones and across sampling months, as well as local variability within zones. The duration of between-clan encounters was longer where grass was more abundant, in keeping with socioecological prediction (Gautam and Vidya 2023). However, despite fission-fusion dynamics, within-clan agonism was also frequent (Gautam and Vidya 2023), although hierarchies within clans were not clearly resolved, as described in the sections above. Fission-fusion dynamics had been invoked to explain the lower frequency of agonism and weaker dominance relationships in Uda Walawe, Sri Lanka, compared to those in the Samburu African savannah elephant population (de Silva et al. 2017). In Kabini, within-clan agonism slightly increased with group size to an extent and was not influenced by local food abundance, in contradiction to socioecological expectation, possibly because of the high density of individuals (Gautam and Vidya 2023).

Overall, the work from Kabini showed partial support for socioecological predictions. However, it must also be kept in mind that this is an artificial system created by the construction of a dam on the Kabini river. Thus, anthropogenically created resource-rich habitats may increase contests between groups in species that are otherwise not thought to have high contests (Gautam and Vidya 2023). Additionally, the resultant high elephant density in a resource-rich island amidst resource scarcity might have led to a Red Queen type of scenario with individuals feeding as fast as possible and agonism being high within groups without a strong dominance hierarchy or rank-related skew (Gautam 2019).

24.10 Effects of the Presence of Calves on Female Social Structure

Female Asian elephants have an interbirth interval of about 4 or more years and a long gestation period of approximately 20–22 months (Sukumar 2003; de Silva et al. 2013). The mortality rate of young ones is also often high, and calves may be predated by tigers (Sukumar et al. 1997, Sukumar 2003; Mar et al. 2012, Kabini Elephant Project unpublished data). Therefore, females may show cooperative care, and calves might be the *raison d'être* for female group living (Gadgil and Nair 1984). In order to examine whether there were changes in adult female group size and social structure in the presence of calves, sightings of identified adult females were examined from the long-term data of the Kabini Elephant Project. Group sizes of adult females, as well as experienced group sizes (group size experienced by an adult female in the sightings in which she is seen), were larger in the presence than in the absence of calves (Revathe 2022). This was also the case when examining the presence of all the young individuals up to 2 years old instead of only calves. When association networks of adult females were constructed, networks of females from sightings with at least one calf in the group had higher average AI, degree and clustering coefficient (connectedness of neighbours of a node amongst themselves) and lower path length than networks of females based on sightings with no calves (Revathe 2022). Thus, associations were wider and stronger in the presence of calves than in their absence. Cooperative offspring care might be one of the benefits of maintaining extended associations with other clan-mates, and this is being investigated further.

24.11 Calf Development, Calf-Conspecific Interactions and Allomaternal Care

To find out whether female social structure was different in the presence of calves due to non-mother females taking care of calves, the development of calves and calf-conspecific interactions were studied. Asian elephant calves are precocial but nutritionally dependent on the mother for about 2 years and attain adult body size slowly over many years. A previous study had examined the development of calves in semi-captivity (Nair 1989), and Revathe et al. (2020) studied this in the Kabini population. The development of various categories of behaviours such as feeding-related, resting-related and social behaviours was examined. Calves showed adult-like feeding behaviours from when they were 3–6 months old, and this increased over time, with the activity budgets of calves beginning to approximate that of adults when calves were 1–2 years old; there were further subtle changes until the age of 5–10 years (Revathe et al. 2020). While calves were known to not use their trunk immediately after birth (Nair 1989), early lateralisation of the trunk was found, with trunk skills improving slowly over time (Revathe et al. 2020).

Calves were found to often be trailed by or present near a non-mother female in the group (Gadgil and Nair 1984; Revathe 2022). Such non-mother females who showed greater coordinated movement with the calf than that shown by other group members were termed escorts (Revathe 2022). Observations were carried out to see whether escorts could be considered allomothers (non-mother females who care for conspecific young). It was found that newborn calves (<3 months old) spent more time close to their mothers than infant calves (3 to ≤6 months) did, whereas both spent similar amounts of time with escorts and other females (non-mother, non-escort females in the group) (Revathe 2022). Calves initiated more interactions towards conspecific females than vice versa and also terminated more interactions than conspecifics did. Interactions between calves and escorts were similar to those between calves and mothers and higher than those between calves and other females. Calves even initiated more non-suckling interactions towards their escorts than towards the mother. Similarly, escorts also initiated more interactions towards the calves than the mothers did (Revathe 2022). Escorts were similar to mothers in the proportion of calf-initiated interactions that they terminated and in the proportions of positive and negative responses they showed towards the calves. Thus, escorts and calves seemed to have an important relationship, and escorts could be considered allomothers (Revathe 2022).

Some of the hypotheses to explain allomaternal care have included the learning-to-mother hypothesis, the mother-benefit hypothesis and the infant-benefit hypothesis (Lancaster 1971; Hrdy 1976; Ross and MacLarnon 2000). According to the first, nulliparous, non-lactating females should show allomothering since it provides mothering experience (e.g. Lancaster 1971; Zhang et al. 2018). If the mother-benefit hypothesis holds true, allomothering should decrease the energetic costs of mothering, such as by decreasing calf-mother proximity and interactions so that the mother obtains some foraging freedom (Ross and MacLarnon 2000). It is likely then that the allomothers would be related to the mother. If the infant-benefit hypotheses were held, one would expect an increase in group size in the presence of the allomother in order to decrease predation risk, and young calves might be allomothered more frequently than older calves. Allomothers would likely be related to infants then (e.g. Konrad et al. 2019). In the Kabini population, nulliparous females comprised 67% of all the allomothers, and mothers were older than allomothers in two-thirds of the dyads (Revathe 2022). The frequency of allomothering (number of sightings in which allomothering was seen) was higher in 10–20-year-old females than expected (Revathe 2022). Thus, there was support for the learning-to-mother hypothesis, but allomothering was not solely for learning to mother as some females from older age classes also showed allomothering. Calves with allomothers had fewer social interactions with their mothers and spent a smaller proportion of their time near their mothers than calves without allomothers, suggesting that there might be some maternal benefit, which needs to be further examined in terms of foraging freedom (Revathe 2022). Since group size was significantly higher in the presence than in the absence of allomothers and young calves received more allomaternal care than older calves, there was also some support for the infant-benefit hypothesis (Revathe 2022). It would be interesting to further examine which exact females show allomothering and why.

24.12 Non-musth Adult Male Social Structure and Associations

Male sociality differs from that seen among females in the Asian elephant. As mentioned in the Introduction, females are philopatric, whereas males disperse from their natal clans. Socio-sexual differentiation begins post the juvenile stage (although even juvenile males and females may show differences, as seen in the African savannah elephant; Lee 1986; Lee and Moss 2011), with subadults beginning to associate with non-natal clan members, often other subadult males, occasionally. Adolescent males, about 10 years of age, start moving away from their clans and eventually disperse over a period of a few years, associating with other males and establishing their new home ranges as largely solitary adults (Desai and Johnsingh 1995). Thus, males may experience different social environments and challenges as they grow and show age-specific associations. We describe first, what is known about adult male social structure, and then, subadult male social structure in the Kabini population.

We considered males to be adults when they were at least 15 years old as they were likely to have finished the process of dispersal. They could also enter into musth around that age and probably be physiologically capable of reproducing.

Musth is analogous to rutting in ungulates and characterised by high serum testosterone levels (Jainudeen et al. 1972a, b; see Sect. 24.15). However, males that had just turned adult would likely not be able to reproduce competitively. Therefore, adult males were subdivided into young adult males (15 to <30 years old) and old adult males (≥30 years old), and their patterns of associations were examined (Keerthipriya et al. 2021). Old adult males came into musth regularly and were expected to be reproductively competitive. Since males might show different associations in and out of musth and in the presence and absence of females, their associations were examined separately. Males were considered to associate with each other in female absence if they fed within about 50 m of each other (a distance at which visual signals might be used, although they would likely be able to detect each other using sound or smell at greater distances) in the absence of any female group (Keerthipriya et al. 2021). Males were considered to associate with a female group if they were within about 10 m (easy physical reach to check the reproductive status of females) of the group. Males associating with the same female group at the same time were considered to associate in female presence (Keerthipriya et al. 2021).

Based on sightings of 38 non-musth adult males seen at least on five different days during 2009–2014, adult males were found to spend the majority of their time alone (average = 53.8% of their time), followed by that (29.6%) in mixed-sex groups (associating with female groups), then in all-male groups (12.3% of their time) and in all-male groups with a single adult male (which referred to a group comprising an adult male and one or more subadult and/or juvenile males; 4.4%) (Keerthipriya et al. 2021). Although old males would be reproductively more active, they spent a smaller proportion of their time in mixed-sex groups than young males (Fig. 24.5). Thus, old males spent similar proportions of their time in all-male groups and mixed-sex groups, whereas young males spent a greater proportion of their time in

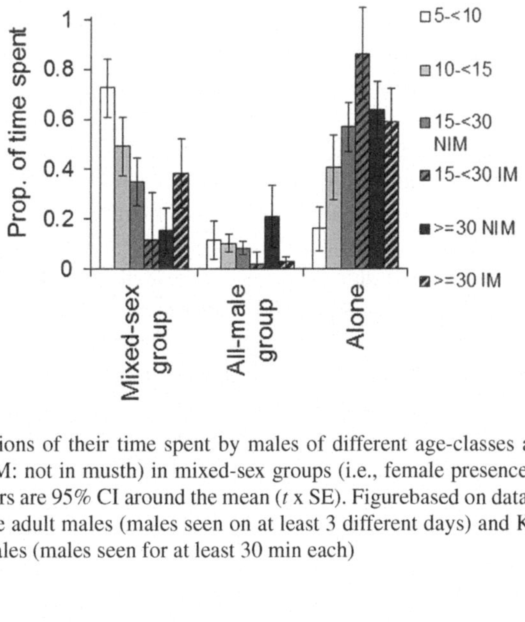

Fig. 24.5 Proportions of their time spent by males of different age-classes and musth statuses (IM: in musth, NIM: not in musth) in mixed-sex groups (i.e., female presence), all-male groups, and alone. Error bars are 95% CI around the mean (*t* x SE). Figurebased on data from Keerthipriya et al. (2020) for the adult males (males seen on at least 3 different days) and Keerthipriya (2018) for the subadult males (males seen for at least 30 min each)

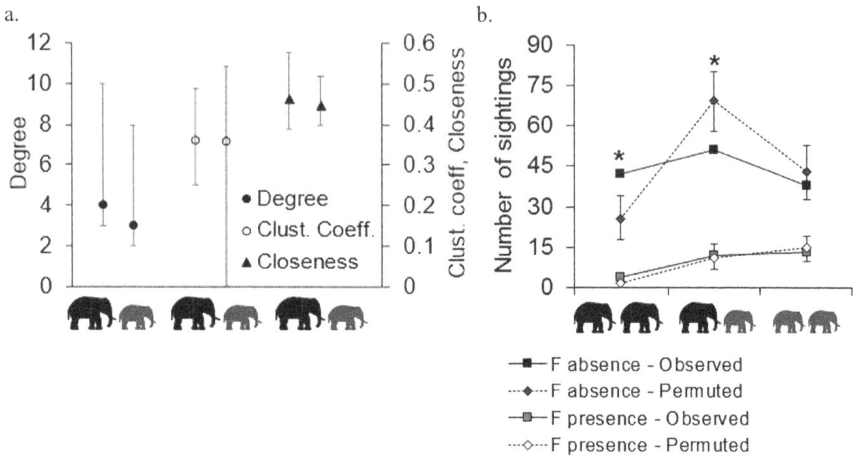

Fig. 24.6 (**a**) Network statistics (degree, clustering coefficient, and closeness centrality) of old adult (30+ years; in black) and young adult (15 to <30 years; in grey) males, and (**b**) the observed and permuted number of sightings in which adult males were with others of the same or different age classes in the presence and absence of females. Median, 25 and 75 percentiles are shown in a. Asterisks mark significant differences in b. Figure based on data from Keerthipriya et al. (2021)

mixed-sex groups than in all-male groups (Keerthipriya et al. 2021). When it came to associations amongst adult males, in the absence of females, old adult males associated with other old adult males more than expected at random, old and young adult males associated less than expected at random, and young adult males associated as expected at random (Keerthipriya et al. 2021, Fig. 24.6). Associations amongst adult males were not different from random in the presence of females (Fig. 24.6), suggesting that they just happened to meet when independently associating with female groups. Keerthipriya et al. (2021) also found that both old and young males were equally likely to initiate association when such a pair was in the vicinity, and young males spent similar proportions of their sightings with old males and with other young males. The number of associates and a measure of how central the animal was in the social network (closeness centrality) were also not different between old and young adult males (Fig. 24.6). Therefore, these findings were unlike those in the African savannah elephant, in which old males had higher centrality in networks and more associates than young males, young males preferred to associate with older males and older males were thought to facilitate social learning (Evans and Harris 2008; Chiyo et al. 2011 Murphy et al. 2020). Thus, non-musth adult male associations in Kabini did not seem to be based primarily on social learning from old males, and there was some support for males forming associations to test the strengths of old males (Keerthipriya et al. 2021). More frequent associations between old males than expected had also been observed in the African savannah elephant (Chiyo et al. 2011).

24.13 Non-musth Adult Male Group Size and Feeding Competition

Adult males in the Kabini population spent a much smaller proportion of their time (~12%) in all-male groups than that seen in the Amboseli African savannah elephant population (~63%; Chiyo et al. 2011). When they formed groups, group sizes of all-male groups in Kabini were also smaller compared to those in Amboseli (Keerthipriya et al. 2021). Group sizes of male Asian elephants have been reported to be small in other populations also (e.g. Daniel et al. 1987). This difference in sociality might be a result of the differences in resource distribution in their respective habitats, as was suggested in the case of female social structure also in Kabini (Nandini et al. 2017). Asian elephants occupy more forested habitats than the African savannah elephant and food resources might be limited within small feeding patches. Analysis of the proportions of sightings spent with different associates in Kabini showed a negative relationship between the number of associates of a male and the mean proportion of sightings that he spent with an associate (Keerthipriya et al. 2021). Thus, this trade-off between the total number of associates and time spent with each associate suggested a restriction on group size in males also.

Asian elephants in a neighbouring forest in the Nilgiris landscape were found to spend over 50% of the daytime hours feeding (Baskaran et al. 2010). Thus, although female social organisation in mammals are generally expected to be influenced by food competition and male social organisation by competition for females, male sociality may also be directly influenced by food competition in megaherbivores. It may, therefore, be important to examine the feeding ecology of males in different social contexts. Males possibly experience an age-graded difference in energy costs due to changing social environments also. Males might be expected to face higher feeding competition than females on account of their larger sizes and might have lower food intake or greater spacing between individuals to maintain their food intake when they are in groups. However, although old males, which are larger, are expected to face a greater constraint than young males, male age was not found to affect the time spent in all-male groups, nor was there a significant relationship between male age and the average group size experienced by the male (Keerthipriya et al. 2021). Feeding rates by males in different scenarios are being examined currently. Sex-based differences in feeding behaviour have been observed in the African savannah elephant, but not due to competition (Stokke and du Toit 2000).

24.14 Dominance Relationships Between Adult Males

Females are a limiting resource for males due to their long inter-birth interval and gestation period. Females come into oestrus once in about 4 months (16 weeks; Hess et al. 1983) if they are not pregnant or lactating and are sexually receptive only for 3–7 days (Sukumar 2003). Therefore, relationships between adult males are expected to be agonistic rather than affiliative, also perhaps explaining the small adult male group sizes mentioned above. However, if this alone explained the difference in adult male group sizes between the African savannah and Asian elephant, one should find greater competition for mates in the latter and either a higher rate of agonism or a more resolved dominance hierarchy and higher skew in reproductive success. When agonistic interactions were examined in the Kabini population, almost all the independent dominance bouts were resolved, with clear winners and losers (Keerthipriya 2018). Dominance networks were sparse, and there was no linear dominance hierarchy. However, there was some orderliness in dominance relationships, with more transitive triads than that expected by chance (Keerthipriya 2018). Age and body condition had effects on the probability of winning, as did musth, although encounters between musth and non-musth males were not common (Keerthipriya 2018). This was also seen in the Kaziranga Asian elephant population, although with a much higher proportion of encounters between musth and non-musth males and a slightly smaller proportion of resolved interactions (Chelliah and Sukumar 2013). Age/size and musth were found to affect the probability of winning in male African savannah elephants also (e.g. Poole 1989). Unfortunately, the rate of agonism between males is not available from any of these populations. Dominance bouts in Kabini were usually observed between non-musth males in the absence of female groups, and associations between adult males, especially old males, seemed to facilitate dominance interactions, suggesting that associations might be useful for testing strengths (Keerthipriya 2018). Dominance interactions were considered to be part of an association between males if there was sparring between males that also fed together, whereas they were considered to not be part of an association if males showed aggressive interactions and moved away without feeding together. Since musth is likely to be important for gaining access to females (see below), whether dominance relationships of non-musth males result in rank-based priority of entering into musth needs to be investigated.

24.15 Musth and Reproduction

Adult males experience an annual rut-like phenomenon called musth. This is characterised by high serum testosterone levels, temporal gland secretion, urine dribbling at times, greater aggression towards other males and greater interest in females (Jainudeen et al. 1972a, b). Individual males may come into musth at different times of the year and reproduction is not restricted to the musth period. Males may come into musth from about 15 years of age, with the duration of musth increasing with age (Keerthipriya et al. 2020; Fig. 24.7). Musth is thought to be a costly and, therefore, honest signal of intent (Poole 1989) as males in musth reduce their feeding time and lose body condition over the duration of musth (see Sukumar 2003). Examining the associations of musth males in Kabini, it was found that old males (≥30 years) in musth spent a slightly higher proportion of their time (compared to non-musth male time) with females (Fig. 24.5) and associated with larger female groups than when they were not in musth. Young males spent a smaller proportion of their musth time with females compared to their non-musth time (Fig. 24.5) and associated with similarly sized female groups (Keerthipriya et al. 2020). Thus, old

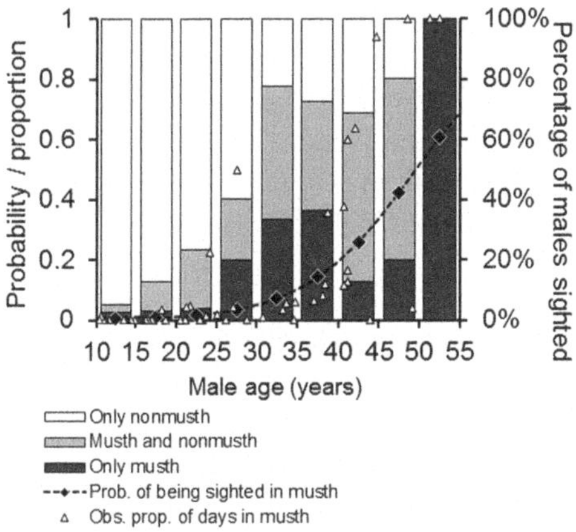

Fig. 24.7 The effect of age on musth. Bars show the percentage of males of different age-classes sighted only when there were not in musth, when they were both in musth and not in musth, and only when they were in musth. The observed proportion of days in which individual males of different ages were sighted in musth are shown as triangles. The dashed line shows the expected probability of being sighted in musth with increasing age based on a logistic regression. Figure based on data fromKeerthipriya et al. (2020)

males gained access to females during musth, whereas young males had reduced access to females during musth, making musth payoffs age-specific (Keerthipriya et al. 2020).

The effect of age on musth. Bars show the percentage of males of different age-classes sighted only when there were not in musth, when they were both in musth and not in musth, and only when they were in musth. The observed proportion of days in which individual males of different ages were sighted in musth are shown as triangles. The dashed line shows the expected probability of being sighted in musth with increasing age based on a logistic regression. Figure based on data from Keerthipriya et al. (2020)

A greater turnover of musth males than non-musth males was seen in Kabini, suggesting different non-musth and musth home ranges, and musth as a roving strategy by old males to gain reproductive opportunities (Keerthipriya et al. 2020). As mentioned above, males in musth also had a greater probability of winning in dominance interactions against non-musth males. It was also seen that musth males engaged in dominance interactions more frequently in female presence than in female absence (Keerthipriya 2018; Chelliah and Sukumar 2013). Thus, musth seemed to be advantageous to old males in both scramble and contest competition for access to females. A comparison of the proportions of sightings of males in musth across populations suggested that the number of very old males (>45 years old) might influence the occurrence of musth (Keerthipriya and Vidya 2019). Whether the access to females provided by musth or dominance relationships based on non-musth contests translates into reproductive success is not yet known and a study to examine the variance in reproductive success of males is underway in

Kabini. Females were found to prefer old musth males over young non-musth males in the Kaziranga population (Chelliah and Sukumar 2015). Sneak mating by young males has been observed in Kabini (Keerthipriya et al. 2020) and Kaziranga (Chelliah and Sukumar 2015). However, the frequency of matings by and reproductive success of young, non-musth males, which associate with female groups to a greater extent than old, non-musth males, is also not yet known.

24.16 Subadult Male Dispersal and Associations

Males disperse from their natal clans when they are about 10 years old, through a gradual process (Desai and Johnsingh 1995). Based on the genetic relatedness of adult males and females within and across locations, dispersal in the Asian elephant has been inferred to be 'locational' (Vidya and Sukumar 2005; Vidya et al. 2005; Ahlering et al. 2011), in which males leave their home ranges and disperse to new locations, as opposed to dispersing from the social group but remaining in the natal home range and temporarily moving outside only to breed (social dispersal; see Isbell and van Vuren 1996). The proximate reason for dispersal may include feeding competition in the natal group as males grow in size, resulting in the males moving away gradually from the centre of the group. Another possibility is that males are evicted by the adult females in the group due to feeding competition, in which case, subadult males may receive increasing dominance from adult females as they grow older. The proximate reason for dispersal is not understood so far in the Asian elephant and a study is ongoing in Kabini, examining feeding rates and dominance received. In Amboseli, it was found that females tolerated natal males but showed aggression towards non-natal dispersing males (Lee et al. 2011). However, in Amboseli, there is presumably some social dispersal at least as inbreeding avoidance was shown by adult males towards close kin in the area (Archie et al. 2007). An understanding of the reasons for dispersal and the factors affecting successful dispersal in males would help link the behavioural ecology of animals to population ecology (see Andreassen et al. 2002).

Since subadult males show a gradual process of dispersal, they may associate with various individuals during that process, gathering social experience, testing strengths with other males and learning male-specific foraging tactics, if any. The associations formed by these subadult males may be based on age or genetic relatedness. Associations based on genetic relatedness are currently being examined in the Kabini population. Subadult males in Kabini were found to spend about three-quarters of their time in mixed-sex groups (with this proportion of time decreasing with increasing age; Fig. 24.5), and they spent most of the remaining time alone (Keerthipriya 2018). Therefore, unlike adult males, who met primarily in the absence of females, subadult males associated amongst themselves to a greater extent in the presence of females. Thus, dispersal would not increase their frequency of meeting subadult males. It would be interesting to look at the identites of associates preferred by dispersing males and the associations of such males with their natal clans and other clans during the process of dispersal. Variation in the age of dispersal and the time taken to complete dispersal would also be interesting to examine.

When subadult males associated in female absence, they associated with their age-class peers (old subadults: 10 to <15 years old; young subadults: 5 to <10 years old) more than expected at random (Keerthipriya 2018). It is possible that this might facilitate testing strength as in adult males. However, since subadult males largely associated in the presence of females, the vast majority of dominance bouts between subadult males occurred in the presence of females (Keerthipriya 2018). Subadult males engaged in more dominance interactions with new opponents compared to adults and also initiated more bouts that they eventually lost compared to adults, suggesting that they might still be learning to assess one another (Keerthipriya 2018). There was an effect of age on the outcome of agonistic interaction among subadult males also. Whether associations amongst subadults are stable over long time periods remains to be studied.

24.17 Looking Ahead

The long-term Kabini Elephant Project has thus provided some understanding of female and male social organisation, but a lot remains to be uncovered. For example, how fission-fusion decisions are made by females and to what extent decisions of grouping, movement and between-clan dominance are individual versus joint decisions are not known. Whether the lack of high centrality of the matriarchs of clans (or the older males in male societies) results in less centralised decision-making and whether they still have important roles as repositories of social information during adverse circumstances such as drought are also unknown. It is also unclear to what extent calves drive female sociality, how exactly groups organise themselves when a calf is born, why certain females become allomothers and what might be the costs of allomothering or increasing group sizes to protect calves. With regard to male sociality, whether food abundance and feeding competition directly affect male associations and subadult male dispersal, how subadult males choose their associates over the period of dispersal, whether dominance relationships between non-musth adult males translate into musth priority, and whether dominance rank and musth translate into reproductive success are interesting questions to address. Some of these areas are currently being investigated while we hope that others will follow.

Acknowledgements We thank JNCASR, DST-SERB and CSIR for funds. Chowdari Jabili was given a Ph.D. fellowship by JNCASR, and Ankana Sanyal and Divya Choudhary were given Ph.D. fellowships by UGC. We thank Hansraj Gautam for his helpful comments on the manuscript.

References

Abram NK, Skara B, Othman N, Ancrenaz M, Mengersen K, Goossens B (2022) Understanding the spatial distribution and hot spots of collared Bornean elephants in a multi-use landscape. Sci Rep 12:12830

Ahlering MA, Hedges S, Johnson A, Tyson M, Schuttler SG, Eggert LS (2011) Genetic diversity, social structure, and conservation value of the elephants of the Nakai Plateau, Lao PDR, based on non-invasive sampling. Conserv Genet 12:413–422

Alfred R, Ahmad AH, Payne J, Williams C, Ambu LN, How PM, Goossens B (2012) Home range and ranging behaviour of Bornean elephant (*Elephas maximus borneensis*) females. PLoS One 7(2):e31400

Andreassen HP, Stenseth NC, Ims RA (2002) Dispersal behaviour and population dynamics of vertebrates. Dispersal Ecol 1:237–256

Archie EA, Morrison TA, Foley CAH, Moss CJ, Alberts SC (2006) Dominance rank relationships among wild female African elephants, *Loxodonta africana*. Anim Beh 71:117-127.

Archie EA, Hollister-Smith JA, Poole JH, Lee PC, Moss CJ, Maldonado JE, Fleischer RC, Alberts SC (2007) Behavioural inbreeding avoidance in wild African elephants. Mol Ecol 16:4138–4148

Arivazhagan C, Sukumar R (2005) Comparative demography of Asian elephant populations (*Elephas maximus*) in southern India. Centre for Ecological Sciences Technical Report No. 106. Indian Institute of Science, Bangalore

Armbruster P, Fernando P, Lande R (1999) Time frames for population viability analysis of species with long generations: an example with Asian elephants. Anim Conserv 2:69–73

Asensio N, Korstjens AH, Aureli F (2009) Fissioning minimizes ranging costs in spider monkeys: a multiple-level approach. Behav Ecol Sociobiol 63:649–659

Athira TK, Vidya TNC (2021) Elephant social systems: what do we know and how have molecular tools helped? J Indian Inst Sci 101(2):257–278

Aureli F, Schaffner CM, Boesch C, Bearder SK, Call J, Chapman CA, Connor R, Di Fiore A, Dunbar RIM, Henzi SP, Holekamp K, Korstjens AH, Layton R, Lee P, Lehmann J, Manson JH, Ramos-Fernández G, Strier KB, van Schaik CP (2008) Fission-fusion dynamics: new research frameworks. Curr Anthropol 49:627–654

Baskaran N, Balasubramanian S, Swaminathan S, Desai AA (1995) Home range of elephants in the Nilgiri Biosphere Reserve, South India. In: Daniel JC, Datye HS (eds) A week with elephants. Bombay Natural History Society, Oxford University Press, Bombay, pp 296–313

Baskaran N, Balasubramanian M, Swaminathan S, Desai AA (2010) Feeding ecology of the Asian elephant *Elephas maximus* Linnaeus in the Nilgiri Biosphere Reserve, southern India. J Bomb Nat Hist Soc 107:3–13

Baskaran N, Varma S, Sar CK, Sukumar R (2011) Current status of Asian elephants in India. Gajah 35:47–54

Baskaran N, Kanakasabai R, Desai AA (2018) Ranging and spacing behaviour of Asian elephant (*Elephas maximus* Linnaeus) in the tropical forests of southern India. In: Sivaperuman C, Venkataraman K (eds) Indian Hotspots. Springer, Singapore, pp 295–315

Bist SS (2002) An overview of elephant conservation in India. Indian For 128:121–136

Blondel VD, Guillaume JL, Lambiotte R, Lefebvre E (2008) Fast unfolding of communities in large networks. J Stat Mech: Theory Exp 10:P10008

Campos-Arceiz A, Blake S (2011) Megagardeners of the forest—the role of elephants in seed dispersal. Acta Oecol 37:542–553

Campos-Arceiz A, Larrinaga AR, Weerasinghe UR, Takatsuki S, Pastorini J, Leimgruber P, Fernando P, Santamaría L (2008) Behaviour rather than diet mediates seasonal differences in seed dispersal by Asian elephants. Ecology 89(10):2684–2691

Chakraborty S, Boominathan D, Desai AA, Vidya TNC (2014) Using genetic analysis to estimate population size, sex ratio, and social organization in an Asian elephant population in conflict with humans in Alur, southern India. Conserv Genet 15:897–907

Chapman CA, Chapman LJ, Wrangham RW (1995) Ecological constraints on group size: an analysis of spider monkey and chimpanzee subgroups. Behav Ecol Sociobiol 36:59–70
Chelliah K, Sukumar R (2013) The role of tusks, musth and body size in male–male competition among Asian elephants, *Elephas maximus*. Anim Behav 86(6):1207–1214
Chelliah K, Sukumar R (2015) Interplay of male traits, male mating strategies and female mate choice in the Asian elephant, *Elephas maximus*. Behaviour 152:1113–1144
Chiyo PI, Archie EA, Hollister-Smith JA, Lee PC, Poole JH, Moss CJ, Alberts SC (2011) Association patterns of African elephants in all-male groups: the role of age and genetic relatedness. Anim Behav 81:1093–1099
Clutton-Brock T, Sheldon BC (2010) The seven ages of *Pan*. Science 327(5970):1207–1208
Connor RC, Wells RS, Mann J, Read AJ (2000) The bottlenose dolphin: social relationships in a fission-fusion society. In: Mann J, Connor RC, Tyack PL, Whitehead H (eds) Cetacean societies: field studies of dolphin and whales. University of Chicago Press, Chicago, pp 91–125
Crook JH, Gartlan JS (1966) Evolution of primate societies. Nature 210(5042):1200–1203
Daniel JC, Desai AA, Sivaganesan N, Datye HS, Rameshkumar S, Baskaran N, Balasubramanian M, Swaminathan S (1987) Ecology of the Asian elephant. Final report. Bombay Natural History Society, Bombay
de Silva S (2016) Need for longitudinal studies of Asian wildlife in the face of crises. Glob Ecol Conserv 6:276–285
de Silva S, Wittemyer G (2012) A comparison of social organisation in Asian elephants and African savannah elephants. Int J Primatol 33:1125–1141
de Silva S, Ranjeewa AD, Kryazhimskiy S (2011) The dynamics of social networks among female Asian elephants. BMC Ecol 11(1):1–6
de Silva S, Webber CE, Weerathunga US, Pushpakumara TV, Weerakoon DK, Wittemyer G (2013) Demographic variables for wild Asian elephants using longitudinal observations. PLoS One 8(12):e82788
de Silva S, Schmid V, Wittemyer G (2017) Fission–fusion processes weaken dominance networks of female Asian elephants in a productive habitat. Behav Ecol 28:243–252
Desai AA, Johnsingh AJT (1995) Social organisation and reproductive strategy of the male Asian elephant (*Elephas maximus*). Abstract. In: Daniel JC, Datye HS (eds) A week with elephants. Bombay Natural History Society and Oxford University Press, Bombay, p 532
Dunbar RIM (1988) Primate social systems: studies in behavioural adaptation. Croom Helm, London
Eltringham SK (1982) Elephants. Blandford Press, Poole, Dorset
English M, Gillespie G, Ancrenaz M, Ismail S, Goossens B, Nathan S, Linklater W (2014) Plant selection and avoidance by the Bornean elephant (*Elephas maximus borneensis*) in tropical forest: does plant recovery rate after herbivory influence food choices? J Trop Ecol 30:371–379
Evans KE, Harris S (2008) Adolescence in male African elephants, *Loxodonta africana* and the importance of sociality. Anim Behav 76:779–787
Fernando P, Lande R (2000) Molecular genetic and behavioural analysis of social organization in the Asian elephant (*Elephas maximus*). Behav Ecol Sociobiol 48:84–91
Fernando P, Wikramanayake ED, Janaka HK, Jayasinghe LKA, Gunawardena M, Kotagama SW, Pastorini J (2008) Ranging behaviour of the Asian elephant in Sri Lanka. Mammal Biol 73(1):2–13
Fernando P, Ekanayaka SKK, Pastorini J (2020) The elephant at the fence: almsman, panhandler, friend or foe? Eur J Wildl Res 66(6):1–3
Gadgil M, Nair VP (1984) Observations on the social behaviour of free ranging groups of tame Asiatic elephants (*Elephas maximus Linn*). Proc Indian Acad Sci Anim Sci 93(3):225–233
Gautam H (2019) Resource availability, within-clan and between-clan agonistic interactions, and dominance relationships amongst female Asian elephants in Nagarahole National Park, Southern India. Ph.D. thesis, Jawaharlal Nehru Centre for Advanced Scientific Research, Bengaluru, India
Gautam H, Vidya TNC (2023) Do food distribution and competitor density affect agonistic behaviour within and between clans ina high fission-fusion species? R Soc Open Sci 10:230990

Ginsberg JR, Young TP (1992) Measuring association between individuals or groups in behavioural studies. Anim Behav 44:377–379

Goswami VR, Sridhara S, Medhi K, Williams AC, Chellam R, Nichols JD, Oli MK (2014) Community-managed forests and wildlife-friendly agriculture play a subsidiary but not substitutive role to protected areas for the endangered Asian elephant. Biol Conserv 177:74–81

Grüter CC, Zinner D (2004) Nested societies: convergent adaptations of baboons and snub-nosed monkeys? Prim Rep 70:1–98

Gupta M, Ravindranath S, Prasad D, Vidya TNC (2016) Short-term variation in sex ratio estimates of Asian elephants due to space use differences between the sexes. Gajah 44:5–15

Hess DL, Schmidt AM, Schmidt MJ (1983) Reproductive cycle of the Asian elephant (Elephas maximus) in captivity. Biol Reprod 28(4):767–773

Hodge SJ, Manica A, Flower TP, Clutton-Brock TH (2008) Determinants of reproductive success in dominant female meerkats. J Anim Ecol 77(1):92–102

Hrdy SB (1976) Care and exploitation of nonhuman primate infants by conspecifics other than the mother. Adv Stud Behav 6:101–158

Isbell LA (1991) Contest and scramble competition: patterns of female aggression and ranging behaviour among primates. Behav Ecol Sociobiol 2:143–155

Isbell LA, van Vuren D (1996) Differential costs of locational and social dispersal and their consequences for female group-living primates. Behaviour 133:1–36

Jainudeen MR, McKay GM, Eisenberg JF (1972a) Observations on musth in the domesticated Asiatic elephant (*Elephas maximus*). Mammalia 36(2):247–261

Jainudeen MR, Katongole CB, Short RV (1972b) Plasma testosterone levels in relation to musth and sexual activity in the male Asiatic elephant, *Elephas maximus*. J Reprod Fert 29:99-103

Janson C (1985) Aggressive competition and individual food consumption in wild brown capuchin monkeys (*Cebus apella*). Behav Ecol Sociobiol 18(2):125–138

Janson CH, Goldsmith ML (1995) Predicting group size in primates: foraging costs and predation risks. Behav Ecol 6:326–336

Janson CH, Van Schaik CP (1988) Recognizing the many faces of primate food competition: methods. Behaviour 105(1–2):165–186

Johnsingh AJT, Prasad SN, Goyal SP (1990) Conservation of the Chilla-Motichur corridor for elephant movement in Rajaji—Corbett National Parks area, India. Biol Conserv 51:125–138

Keerthipriya P (2018) Associations, dominance interactions, and musth in male Asian elephants in Nagarahole and Bandipur National Parks, southern India. Ph.D. thesis, Jawaharlal Nehru Centre for Advanced Scientific Research, Bengaluru

Keerthipriya P, Vidya TNC (2019) A comparison of the occurrence of musth in the Kabini elephant population with other populations. Gajah 50:4–16

Keerthipriya P, Vidya TNC (2021) Kabini Elephant Project: a long-term programme for understanding Asian elephant behaviour and ecology. Trumpet 2(2):14–21

Keerthipriya P, Nandini S, Gautam H, Revathe T, Vidya TN (2020) Musth and its effects on male–male and male–female associations in Asian elephants. J Mammal 101:259–270

Keerthipriya P, Nandini S, Vidya TNC (2021) Effects of male age and female presence on male associations in a large, polygynous mammal in southern India: the Asian elephant. Front Ecol Evol 9:616666

Koenig A (2002) Competition for resources and its behavioural consequences among female primates. Int J Primatol 23(4):759–783

Konrad CM, Frasier TR, Whitehead H, Gero S (2019) Kin selection and allocare in sperm whales. Behav Ecol 30(1):194–201

Kummer H (1971) Primate societies: group techniques of ecological adaptation. Aldine-Atherton, Inc., Chicago

Lancaster JB (1971) Play mothering: the relations between juvenile females and young infants among free-ranging vervet monkeys (*Cercopithecus aethiops*). Folia Primatol 15:161–182

Lee PC (1986) Early social development among African elephant calves. Natl Geogr Res 2:388–401

Lee PC, Moss CJ (2011) Calf development and maternal rearing strategies. In: Moss CJ, Croze H, Lee PC (eds) The Amboseli elephants: a long-term perspective on a long-lived mammal. University of Chicago Press, Chicago, pp 224–237

Lee PC, Poole JH, Njiraini N, Sayialel CN, Moss CJ (2011) Male social dynamics: independence and beyond. In: Moss CJ, CrozeH, Lee PC (eds) The Amboseli Elephants: a long-term perspective on a long-lived Mammal. University of Chicago Press, Chicago and London, pp 260–271

Leimgruber P, Gagnon JB, Wemmer C, Kelly DS, Songer MA, Selig ER (2003) Fragmentation of Asia's remaining wildlands: implications for Asian elephant conservation. Anim Conserv 6:347–359

Mar KU, Lahdenperä M, Lummaa V (2012) Causes and correlates of calf mortality in captive Asian elephants (*Elephas maximus*). PLoS One 7(3):e32335

McComb K, Shannon G, Durant SM, Sayialel K, Slotow R, Poole J, Moss C (2011) Leadership in elephants: the adaptive value of age. Proc R Soc B Biol Sci 278(1722):3270–3276

McKay GM (1973) Behavior and ecology of the Asiatic elephant in southeastern Ceylon. Smithsonian Contrib Zool 125:1–113

Mumby HS, Chapman SN, Crawley JAH, Mar KU, Htut W, Soe AT, Aung HH, Lummaa V (2015) Distinguishing between determinate and indeterminate growth in a long-lived mammal. BMC Evol Biol 15:214

Murphy D, Mumby HS, Henley MD (2020) Age differences in the temporal stability of a male African elephant (*Loxodonta africana*) social network. Behav Ecol 31(1):21–31

Nair PV (1989) Development of nonsocial behaviour in the Asiatic elephant. Ethology 82:46–60

Nandini S (2016) Social structure, genetic relatedness, and dominance relationships in female Asian elephants in Nagarahole and Bandipur National Parks, southern India. Ph.D. thesis, Jawaharlal Nehru Centre for Advanced Scientific Research, Bengaluru, India

Nandini S, Keerthipriya P, Vidya TNC (2017) Seasonal variation in female Asian elephant social structure in Nagarahole-Bandipur, southern India. Anim Behav 134:135–145

Nandini S, Keerthipriya P, Vidya TNC (2018) Group size differences may mask underlying similarities in social structure: a comparison of female elephant societies. Behav Ecol 29:145–159

Olivier RCD (1978) On the ecology of Asian elephant. Ph.D. thesis, University of Cambridge, Cambridge

Pascal JP (1986) Explanatory booklet on forest map of South India: Belgaum-Dharwar-Panaji, Shimoga, Mercara-Mysore. Institut Français de Pondichéry, Pondichéry, India

Pascal JP (1988) Wet evergreen forests of the Western Ghats of India: ecology, structure, floristic composition and succession. Travaux de la Section Scientifique et Technique: tome 20 bis. Institut Français de Pondichéry, Pondichéry, India

Plotnik JM, Lair R, Suphachoksahakun W, Frans de Waal BM (2011) Elephants know when they need a helping trunk in a cooperative task. Proc Natl Acad Sci U S A 108(12):5116–5121

Pokharel SS, Seshagiri PB, Sukumar R (2017) Assessment of season-dependent body condition scores in relation to faecal glucocorticoid metabolites in free-ranging Asian elephants. Conserv Physiol 5(1):cox039

Pokharel SS, Singh B, Seshagiri PB, Sukumar R (2019) Lower levels of glucocorticoids in crop-raiders: diet quality as a potential 'pacifier' against stress in free-ranging Asian elephants in a human-production habitat. Anim Conserv 22(2):177–188

Poole JH (1989) Announcing intent: the aggressive state of musth in African elephants. Anim Behav 37:140–152

Queller DC, Goodnight KF (1989) Estimating relatedness using genetic markers. Evolution 43:258–275

Revathe T (2022) Calf development, calf-conspecific interactions, and the effect of calves on female social structure in the Kabini Asian elephant population, southern India. Ph.D. thesis, Jawaharlal Nehru Centre for Advanced Scientific Research, Bengaluru, India

Revathe T, Anvitha S, Vidya TNC (2020) Development of motor control and behaviour in Asian elephants in the Kabini elephant population, southern India. Int J Dev Biol 64:377–392

Ross C, MacLarnon A (2000) The evolution of non-maternal care in anthropoid primates: a test of the hypotheses. Folia Primatol 71:93–113

Salguero-Gómez R, Evans DM, Gaillard J-M, Lancaster L, Sanders NJ, Scandrett K, Meyer J (2022) Time counts in animal ecology. J Anim Ecol 91:2154–2157

Sheldon BC, Kruuk LEB, Alberts SC (2022) The expanding value of long-term studies of individuals in the wild. Nat Ecol Evol 6:1799–1801

Smith JE, Kolowski JM, Graham KE, Dawes SE, Holekamp KE (2008) Social and ecological determinants of fission-fusion dynamics in the spotted hyaena. Anim Behav 76:619–636

Sterck EH, Watts DP, van Schaik CP (1997) The evolution of female social relationships in nonhuman primates. Behav Ecol Sociobiol 41:291–309

Stokke S, du Toit JT (2000) Sex and size related differences in the dry season feeding patterns of elephants in Chobe National Park, Botswana. Ecography 23:70–80

Sukumar R (1989) The Asian elephant: ecology and management. Cambridge University Press, Cambridge

Sukumar R (2003) The living elephants: evolutionary ecology, behaviour, and conservation. Oxford University Press, New York

Sukumar R, Joshi NV, Krishnamurthy V (1988) Growth in the Asian elephant. Proc Anim Sci 97:561–571

Sukumar R, Krishnamurthy V, Wemmer C, Rodden M (1997) Demography of captive Asian elephants (*Elephas maximus*) in southern India. Zoo Biol 16:263–272

Terborgh J, Janson CH (1986) The socioecology of primate groups. Annu Rev Ecol Syst 17:111–136

van Schaik CP (1989) The ecology of social relationships amongst female primates. In: Standen V, Foley RA (eds) Comparative socioecology: the behavioural ecology of humans and other mammals. Blackwell Publishers, Oxford, pp 195–218

van Schaik CP (1999) The socioecology of fission-fusion sociality in orangutans. Primates 40:69–86

Vancuylenberg BWB (1977) Feeding behaviour of the Asiatic elephant in south-east Sri Lanka in relation to conservation. Biol Conserv 12:33–54

Vehrencamp SL (1983) A model for the evolution of despotic versus egalitarian societies. Anim Behav 31:667–682

Vidya TNC (2016) Evolutionary history and population genetic structure of Asian elephants in India. Indian J Hist Sci 51(2):391–405

Vidya TNC, Sukumar R (2005) Social organization of Asian elephants (*Elephas maximus*) in southern India inferred from microsatellite DNA. J Ethol 23:205–210

Vidya TNC, Fernando P, Melnick DJ, Sukumar R (2005) Population genetic structure and conservation of Asian elephants (*Elephas maximus*) across India. Anim Conserv 8:377–388

Vidya TNC, Sukumar R, Melnick DJ (2009) Range-wide mtDNA phylogeography yields insights into the origins of Asian elephants. Proc R Soc B 276:893–902

Vidya TNC, Prasad D, Ghosh A (2014) Individual identification in Asian elephants. Gajah 40:3–17

Vijayakrishnan S, Kumar MA, Umapathy G, Kumar V, Sinha A (2018) Physiological stress responses in wild Asian elephants *Elephas maximus* in a human-dominated landscape in the Western Ghats, southern India. Gen Comp Endocrinol 266:150–156

Watts DP (1985) Relations between group size and composition and feeding competition in mountain gorilla groups. Anim Behav 33(1):72–85

Williams AC, Johnsingh AJT, Krausman PR (2001) Elephant-human conflict in Rajaji National Park, northwestern India. Wildl Soc Bull 29:1097–1104

Williams C, Tiwari SK, Goswami VR, de Silva S, Kumar A, Baskaran N, Yoganand K, Menon V (2020) *Elephas maximus*. The IUCN Red List of threatened species 2020: e.T7140A45818198. https://doi.org/10.2305/IUCN.UK.20203.RLTS.T7140A45818198.en

Wilson G, Gray RJ, Sofyan H (2020) Identifying the variation in utilization density estimators and home ranges of elephant clans in Aceh, Sumatra, Indonesia. Eur J Wildl Res 66(6):1–20

Wilson G, Gray RJ, Radinal R, Hasanuddin H, Azmi W, Sayuti A, Muhammad H, Abdullah A, Nazamuddin BS, Sofyan H, Riddle HS, Stremme C, Desai AA (2021) Between a rock and a hard place: rugged terrain features and human disturbance affect behaviour and habitat use of Sumatran elephants in Aceh, Sumatra, Indonesia. Biodivers Conserv 30(3):597–618

Wittemyer G, Douglas-Hamilton I, Getz WM (2005) The socioecology of elephants: analysis of the processes creating multitiered social structures. Anim Behav 69:1357–1371

Wittemyer G, Getz WM (2007) Hierarchical dominance structure and social organization in African elephants, *Loxodonta africana*. Anim Behav 73(4):671–681

Wrangham RW (1980) An ecological model of female-bonded primate groups. Behaviour 75:262–300

Zhang D, Xia DP, Wang X, Zhang QX, Sun BH, Li JH (2018) Bridging may help young female Tibetan macaques *Macaca thibetana* learn to be a mother. Sci Rep 8:1–8

25 Patterns of Crop Foraging by Asian Elephants (*Elephas maximus*) in Central Kerala, India

Suresh K. Govind and E. A. Jayson

Abstract

Crop foraging by wild Asian elephants (*Elephas maximus*) is a growing concern in Central Kerala, India. This study was carried out in Thrissur district, from April 2009 to March 2013, with the aim to assess the pattern of crop foraging by wild elephants and to find the environmental factors associated with these patterns. Quadrats were laid in the damaged crop fields for quantifying the area, and logistic regression analysis was run for identifying factors associated with the patterns of encountering the species in human habitations. Plantain (*Musa paradisiaca*) (74.11%) was the main food of elephants in the fringe areas of the forest. It was cultivated in between young rubber plants (*Hevea brasiliensis*) (below 4 years). Crop foraging was at its peak during the northeast monsoon (September to November). Marginal farmers harvest plantains in the above-mentioned period due to the Onam festival in Kerala. While consuming plantains, the unpalatable young rubber plants were also destroyed. As the market price of rubber increased, farmers were expanding the cultivation of rubber, followed by cultivating plantain in between the young rubber plants. The cultivation of plantains in the immediate fringe areas of the forest is the cause of increased human-elephant interactions in the area.

Keywords

Human-elephant interaction · Crop foraging · Wildlife conservation · Wildlife management

S. K. Govind (✉)
Department of Psychology, Christ College, Irinjalakuda, Thrissur, Kerala, India

E. A. Jayson
Department of Wildlife, Kerala Forest Research Institute, Peechi, Thrissur, Kerala, India

S. Sridhara et al. (eds.), *Animal Behavior in the Tropics*,
https://doi.org/10.1007/978-981-97-7324-4_25

25.1 Introduction

Asian elephant is listed under Schedule I of the Indian Wildlife (Protection) Act, 1972, and categorized as 'endangered' by International union for conservation of nature (IUCN) (Williams et al. 2020). The conflict between humans and wild elephants (*Elephas maximus*) is a contentious issue in India and other tropical nations (Saaban et al. 2011; Zhang and Wang 2003). Increasing human population, developmental activities in the periphery of forests, nutritive value of crops, scarcity of water within the forests, encroaching habitat of wildlife for agriculture, lack of maintenance of physical barriers and stress in elephants are the major causes of human-elephant conflict (Sukumar 1985; Hoare 2000; Nigam 2002; Leimgruber et al. 2003; Choudhury 2004; Gubbi 2012). Gubbi (2012) postulates that human-elephant conflict is high in the places where human settlements and undisturbed forest areas are confronted. In Malaysia, oil palm plantations were preferred by elephants and incurred huge losses to farmers (Blair et al. 1979; Saaban et al. 2011). Rice (*Oryza sativa*) and plantain (*Musa paradisiaca*) were mainly damaged in Simao, China (Zhang and Wang 2003). A significant economic loss was reported due to crop depredation in Assam (Nath et al. 2015) and Karnataka (Venkataramana et al. 2017). In Kerala, studies were conducted on human-elephant conflict (Veeramani and Jayson 1995; Jayson and Christopher 2008; Prasad et al. 2011). Crop foraging was the main problem faced by marginal people, and, in that, coconut (*Cocos nucifera*) and areca nut (*Areca catechu*) palms were mainly consumed (Veeramani and Jayson 1995). The cultivation of crops in the immediate fringe areas of the forest is the cause of the conflict (Sukumar 1990). Recently, Govind and Jayson (2018) estimated the economic loss incurred by the farmers due to elephant raids in Thrissur district, Central Kerala (₹17,00,000/- per annum; one US dollar = ₹60). It negatively affected the rural agricultural economy. To ameliorate the deteriorating situation, the Kerala Forest and Wildlife Department is disbursing ex gratia payments to the farmers for damaging crops. Here, an attempt has been made to understand the pattern of crop foraging by wild elephants in Thrissur district, Kerala, India, and to find out the environmental factors associated with these patterns.

25.2 Study Area

Thrissur district (10° 46′ to 10° 7′ N and 75° 57′ to 76° 55′ E) is in the central part of Kerala, India, spanning 3032 km^2 (Fig. 25.1). The district has a tropical humid climate and gets plentiful rainfall from south-west monsoon (June to August) and northeast monsoon (September to November). Different types of soil, namely laterite, sandy loam, alluvial, clayey and black soil, are found. The district comprises 11 forest ranges within three forest divisions, namely Thrissur (210.64 km^2), Chalakudy (279.71 km^2) and Vazhachal (413.94 km^2), and three wildlife sanctuaries (213.44 km^2). Vegetation types are moist deciduous (52.86%), riverine (10%) and plantations (37.14%) including teak (*Tectona grandis*), rubber (*Hevea brasiliensis*)

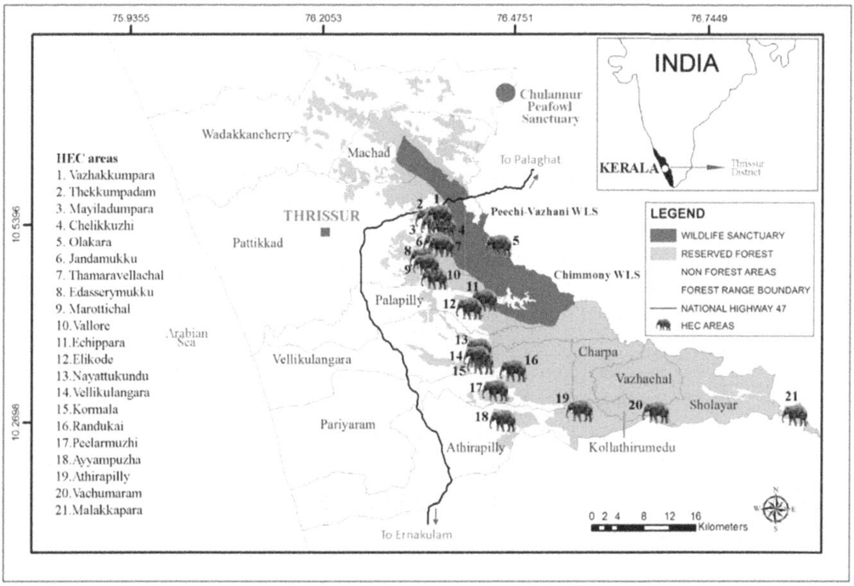

Fig. 25.1 Human-elephant conflict (HEC) areas of Thrissur district, Kerala, India

and cashew (*Anacardium occidentale*). Agriculture is the main occupation of people in the fringe areas of the forest. Coconut, areca nut, rubber, cocoa (*Theobroma cacao*) and plantain (*Musa paradisiaca*) are the major cultivated crops. Multiple crops are cultivated in the private farms adjacent to the forest. Asian elephant (*E. maximus*), wild pig (*Sus scrofa*), leopard (*Panthera pardus*), chital (*Axis axis*), sambar (*Rusa unicolor*) and Indian giant squirrel (*Ratufa indica*) are the common wild animals found in the forest.

25.3 Methods

Case studies were conducted from April 2009 to March 2013 ($n = 31$). Crop foraging incidents were collected from the newspaper media, and the areas were visited by forest officials. Discussion was done with the owners and neighbours of the raided plots ($n = 52$). Age of damaged crop plants, mode of feeding on crop plants, distance to the area of crop foraging from the boundary of the forest, time of encounter in the crop fields, herd composition of elephants and previous reports of crop foraging by elephants were recorded from each location. Quadrats were laid in the affected crop fields to assess the area of damage, and the number of each damaged crop species within the quadrats was quantified (Jayson 1999). A total of 238 running quadrats were taken during the period. The encounter of elephants in the crop fields in different months was calculated by the monthly encounter index (MEI)

Table 25.1 Environmental variables selected for the logistic regression model

Variables	Source
Altitude (m)	Measured using GPS
Slope (angle)	Recorded from digital elevation model (DEM)
Aspect value	Recorded from (DEM)
Canopy density	Calculated by preparing a normalized difference vegetation index (NDVI) map of the study area from Landsat 7 satellite images using ArcGIS 10 software
Distance to the river (m)	Measured from toposheets
Distance to the reserve forest (m)	Measured from toposheets
Coconut	Recorded from the field
Areca nut	Recorded from the field
Plantain	Recorded from the field

(mean encounters in different months were divided by the overall mean, which was multiplied by 100).

Pearson's correlation (r) was used to determine the relationship between climatic parameters (rainfall and temperature) and MEI, and with this, the influence of the climatic parameters on the incidence of crop foraging was calculated.

For identifying the variables affecting the pattern of crop foraging, logistic regression analysis was performed, using SPSS v.17. The intensive study area was divided into grid cells with a size of 2 km × 2 km, and all the independent variables were superimposed onto the grid cells, covering the range of elephants. The grid cell containing the human-elephant conflicts was taken as a conflict cell. The non-conflict cells were selected by a simple random method, and within each cell, the midpoint was identified. The values of environmental variables recorded from these midpoints were taken for the analysis. The entry and exit of environmental variables were determined by Wald test statistic with P-values of 0.05 and 0.1, respectively. Spearman's rank correlation (r_s) was used to examine the relationship between the predicted probability of occurrence of crop damage by elephants from the logistic regression model and the number of crop-raiding incidents in the grid cells (Sitati et al. 2003) (Table 25.1).

Several environmental factors are responsible for human-elephant conflict (Hoare 1999; Smith and Kasiki 2000; Sitati et al. 2003). The variables selected here are the possible contributors and have been taken for predicting the conflict in Kerala (Prasad et al. 2011).

25.4 Results

Asian elephants consumed seven species of crops, viz., plantain, areca nut palm, coconut palm, rubber, nutmeg (*Myristica fragrans*), oil palm (*Elaeis guineensis*) and cocoa (*Theobroma cacao*). The conflict was reported from Pattikkad, Peechi,

Palappilly, Pariyaram, Athirappilly, Charpa, Kollathirumedu and Malakkapara forest ranges (Fig. 25.1). The forest ranges namely Vazhachal, Kollathirumedu and Sholayar have not experienced crop foraging. But, damage to property and houses was recorded from these ranges. The National Highway-47 passing from Ernakulam to Palakkad blocks the movement of elephants to the northern side of the district like Machad and Wadakkanchery forest ranges (Fig. 25.1).

Thirty-one incidences of crop foraging were recorded during the study period. Twenty-six cases were recorded in the early midnight (20.00–00.00 h), and an exceptional case of day-time feeding (15.00 h) was reported from Vallore of Pattikkad forest range. Twenty-three cases occurred during the northeast monsoon season (September–November). The encounter of elephants in the crop fields in different months is shown in Fig. 25.2. Eleven cases occurred in the immediate fringe areas, whereas ten cases were observed within 100 m of the reserve forest. The remaining encounters were in between 100 and 200 m (n = 5), 200 and 300 m (n = 3) and 300 and 500 m (n = 2). While raiding crops, on average, 5.42 ± 1.68 elephants were recorded per herd, and 65% of herds had juveniles. In all the encounters, a male elephant was present.

Elephants fed on plantains (74.11%), followed by areca nut palm (11.38%) and coconut palm (3.37%). Plantains were cultivated between young rubber plants (below 4 years), and while consuming plantains, the unpalatable young rubber plants (10.44%) were destroyed. This cropping pattern was found in 30% of the total area raided (independent sample t-test, $t = 3.49$, n = 6, $P < 0.05$), namely in Pattikkad (43.43%), Palappilly (27.91%), Pariyaram (23.21%) and Peechi (22.22%) forest ranges (Table 25.2). No rubber plants were destroyed in the Palappilly forest range, as the age of the plants was more than 4 years, and all immature trees survived while consuming plantains. The pseudo-stem of plantain was preferred as food. Forty-nine per cent of areca nut palms were trampled during the immature phase, and the core portion of the trunk was consumed. Seventy-six per cent of

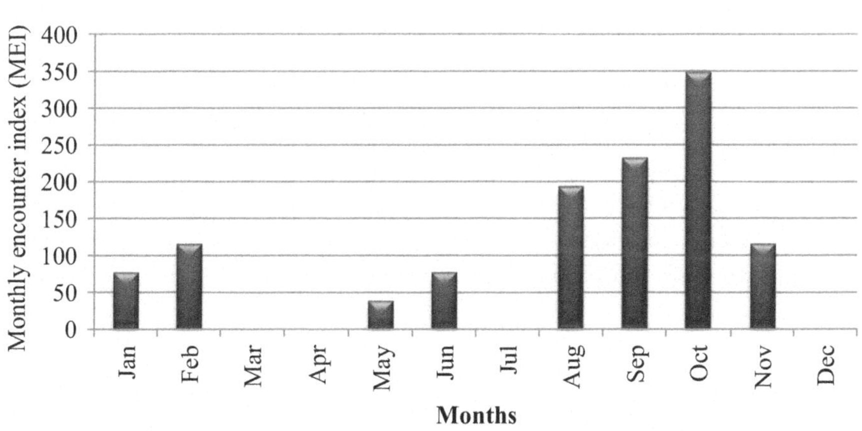

Fig. 25.2 Monthly variations in the incidence of crop foraging by Asian elephants in Thrissur district ($n = 31$)

Table 25.2 Area of crops damaged in different forest ranges of Thrissur district (n = 6)

Sl. no.	Forest ranges	Total area of crops damaged (ha)	Damaged area with inter-crops (ha)
1	Pattikkad	0.99	0.43
2	Palappilly	0.43	0.12
3	Pariyaram	0.56	0.13
4	Charpa	0.18	0
5	Athirappilly	0.04	0
6	Peechi	0.18	0.04
	Total	**2.38**	**0.72**

coconut palms were uprooted during the productive phase (76.09%), and the fresh leaves including central rachis were consumed. Nutmeg plants (0.14%), oil palms (0.18%) and cocoa plants (0.37%) were also destroyed, as they were planted as an inter-crop in the plantations.

Discussion with the people at marginal studies revealed that plantain (40.38%) was the most vulnerable crop to foraging, followed by coconut (21.15%), rubber (19.23%), tubers (17.31%) and areca nut (1.92%). Foraging crops by elephants was not previously reported in the regions of Vazhukumpara, Thekkumpadam, Mayiladumpara, Chelikuzhy, Jandamukku, Vallore, Edasserymukku and Marottichal of Pattikkad forest range; Elikode of Palappilly forest range; and Nayattukundu, Vellikulangara, Randukai and Peelarmuzhi of Pariyaram forest range. In the year 2010, the first incident of crop foraging by elephants was reported from the above-mentioned areas. The cultivation of rubber was preferred by the farmers in the fringe areas of the forest, and this trend commenced in 2009. The high price of rubber was the cause for cultivating rubber. The price of rubber in Kerala in 2008 was ₹100.54 ± 30.05/- per kg, whereas in 2011, it increased to ₹207.96 ± 17.77/- (Rubber Board, Kottayam, Kerala). No significant relationship was observed between the monthly encounter index (MEI) and climatic parameters, namely temperature (Pearson's correlation coefficient, $r = -0.15$, $n = 12$, $P > 0.05$) and rainfall ($r = 0.275$, $n = 12$, $P > 0.05$).

Nine environmental variables were included in the logistic regression analysis, and only five independent variables are significant ($P < 0.1$) (Table 25.3). Increasing the slope of landscape and plantain cultivation were the predictor variables influencing crop foraging. The model predicted that the chances for the occurrence of crop foraging were 24.09 times higher in the steep slopes (68.4%) (Chi-square, $\chi^2 = 9.29$, $P < 0.05$) and 15.66 times higher for the plantain cultivation (82.4%) ($\chi^2 = 18.416$, $P < 0.001$). Plantains were cultivated on the steep slopes of the hilly areas, and the elephants preferred this crop. A positive

Table 25.3 Environmental variables in the logistic regression model

Sl. no.	Variables	Regression coefficients (β)	Standard error (β)	Wald test statistic value	df	P values	Odds ratio Exp (β)
1	Slope	3.182	1.437	4.9	1	0.03	24.09
2	Distance to the reserve forest (m)	−0.859	1.832	0.22	1	0.64	0.42
3	Plantain cultivation	2.751	1.404	3.836	1	0.05	15.66
4	Areca nut plantation	2.178	1.718	1.607	1	0.21	8.83
5	Coconut plantation	0.171	1.52	0.013	1	0.91	1.19
	Constant	−7.673	4.597	2.786	1	0.10	0

Low slope = 1 (reference category); high slope = 2
Distance to the reserve forest (low) = 1 (reference category) (<100 m); high = 2 (>100)
Presence of plantain cultivation = 1; absence = 0
Presence of areca nut plantation = 1; absence = 0
Presence of coconut cultivation = 1; absence = 0

Fig. 25.3 The relationship between the predicted probability of occurrence of crop raiding from the model and number of crop-raiding incidents in the grid cells (rs = 0.80, $n = 39$, $P < 0.001$)

relationship was observed between predicted probabilities and number of crop-raiding incidents in the grid cells (Fig. 25.3).

25.5 Discussion

Asian elephants mainly consumed plantain, areca nut, rubber and coconut. The mode of feeding on these crops was already reported by Jayson and Christopher (2008). In the study area, the mean number of elephants in a herd was five, and in Peppara Wildlife Sanctuary, Kerala, it was ten (Jayson and Christopher 2008). In Africa, the crop foraging elephants in the herds ranged from 1 to 40 (Hoare 1999). In each crop foraging incident, one male elephant was present in a herd, and the same was found in the Peppara Wildlife Sanctuary. The severity of crop damage was influenced by the presence of a male elephant in a herd because male elephants will take risks more than female elephants while consuming crops in human habitation (Sukumar 1991). Sukumar (1985) stated that, during the dry season (January–April), browsing is important for elephants, and the rainy season (May–August) is the time for consuming freshly growing tall grass. When the tall grass becomes unpalatable, they will consume protein-rich fodder during the northeast monsoon season (September–December) and elephants move to low-elevation areas. The movement pattern of elephants showed the highest after the rainy season because the raiding behaviour was observed only during the matured phase of crops (Hoare 1995). The same is found true in this study.

The history of crop foraging incidents in the study area revealed that marginal areas of the forest ranges, namely Pattikkad, Palappilly, Pariyaram and Peechi, were not reported as a core range of elephant movement. As the market price of rubber increased, the farmers were cultivating rubber plants in remote hilly areas, with plantain as an inter-crop. This inter-crop pattern (30%) in the marginal areas increased the human-elephant conflict. The availability of matured plantains during the northeast monsoon season is boosting the foraging because the movement of elephants to low elevated areas is high during this period. The price of bananas was also the highest during August (₹33.3 ± 3.05/- per kg) and September (₹30.67 ± 3.21/- per kg), due to the Onam festival in Kerala (Fig. 25.4). Previous studies have reported that less availability of water within forests triggered elephants to feed on crops (Desai and Baskaran 1996; Eisenberg and Lockhart 1972). In this study, no significant relationship was observed with variations in temperature and rainfall. Besides, the encounter was the highest during the northeast monsoon, after receiving heavy rainfall in the southwest monsoon. As the availability of water was high

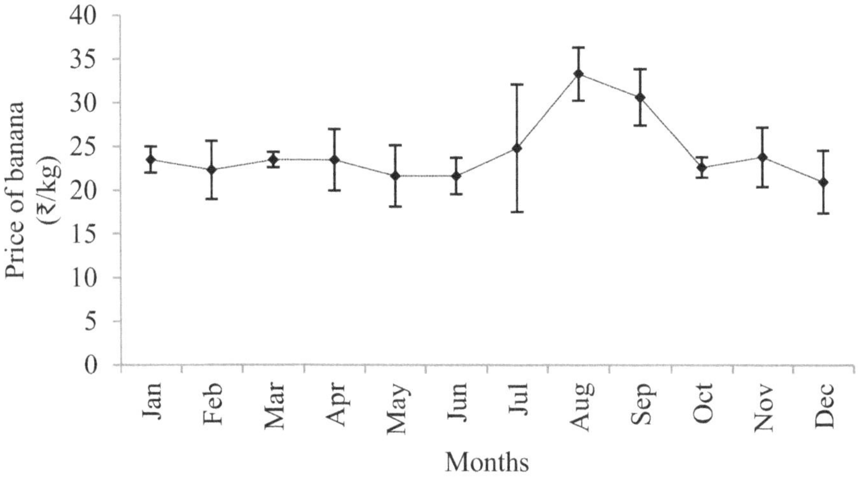

Fig. 25.4 Price of bananas during the study period (April 2009 to March 2013). (Farm Information Bureau, Kerala)

in the forest areas and the reservoirs after the southwest monsoon, the unavailability of water within the forest was not a reason for crop preference.

Grass is the main food of elephants in the forest (Baskaran et al. 2010), and the shortage of grass within the forest limits them to select other foods (White et al. 1993). Sukumar (1985) reported that higher palatability and nutritive value are the reasons for selecting crops in human habitations. According to the optimal foraging theory (MacArthur and Pianka 1966; Stephens and Krebs 1986), animals prefer more profitable food than less profitable one. When the elephants reach low elevated areas, they feed on forest plants, only if the availability of crops in human habitation is less. As the nutritive value of the crops is high, elephants always prefer crops to forest plants. The study area has tropical evergreen forest, west coast tropical semi-evergreen forest and south Indian moist deciduous forest, spanning 888.89 km^2 (Menon and Sasidharan 2005). The status of the population of elephant is found stable in the district (Elephant Census Reports, Kerala Forest and Wildlife Department), and it showed that the food within the forest limits was adequate. So, the chance of increasing the search time for food within the protected areas is less. Therefore, the 'zero-one rule' was confirmed when the elephants reached the fringe areas of the forest. In Africa, the home range of elephants was recorded as 24,000 km^2 to feed on nutritious food (Blake et al. 2003). Frequent crop foraging in certain areas is influenced by the availability and abundance of crops (Charnov 1976) and no human disturbances (Hoare 1995). Increasing the slope of landscape and plantain cultivation were the two environmental variables that influenced the human-elephant interaction. As the marginal farmers were cultivating plantain on the steep slopes of the hilly areas adjacent to the forest boundary, it attracted the elephants and increased the conflict. The logistic regression analysis confirmed this observation.

25.6 Conclusion

Asian elephants feed on seven species of crops in the fringe areas of forests in the Thrissur district of Kerala, India. Plantain (74.11%) was the main food of elephants when they entered into human habitations. Marginal farmers planned to harvest plantain during the northeast monsoon season (September–November), and the incidences of crop foraging were recorded the highest during this period.

Acknowledgement We are thankful to the Director, Dr. K.V. Sankaran (Retd.), Kerala Forest Research Institute for providing the necessary facilities for the study. Forest officials of Thrissur District and farmers in the fringe areas of the forest who gave all the cooperation needed for the study are acknowledged.

References

Baskaran N, Balasubramanian M, Swaminathan S, Desai AA (2010) Feeding ecology of Asian elephant Elephas maximus Linnaeus in the Nilgiri Biosphere Reserve, southern India. J Bombay Nat Hist Soc 107:3–13

Blair JAS, Boon GG, Noor NM (1979) Conservation or cultivation: the confrontation between the Asian elephant and land development in peninsular Malaysia. Land Dev Digest 2:27–59
Blake S, Bouche P, Rasmussen H, Orlando A, Douglas-Hamilton I (2003) The last Sahelian elephants: ranging behaviour, population status and recent history of the desert elephants of Mali. Kenya: Save the elephants. Nairobi, Kenya
Charnov EL (1976) Optimal foraging, the marginal value theorem. Theor Popul Biol 9:129–136
Choudhury A (2004) Human-elephant conflicts in north-east India. Hum Dimens Wildl 9:261–270
Desai AA, Baskaran N (1996) Impact of human activities on the ranging behaviour of elephants in the Nilgiri Biosphere, south India. J Bombay Nat Hist Soc 93:559–569
Eisenberg JF, Lockhart M (1972) An ecological reconnaissance of Wilpattu National Park, Ceylon. Smithsonian contribution of zoology, number 101. Smithsonian Institution Press, Washington
Govind SK, Jayson EA (2018) Crop damage by wild animals in Thrissur District, Kerala, India. In: Sivaperuman C, Venkataraman K (eds) Indian hotspots. Springer, Singapore. https://doi.org/10.1007/978-981-10-6983-3_18
Gubbi S (2012) Patterns and correlates of human-elephant conflict around a south Indian reserve. Biol Conserv 148:88–95
Hoare R (1995) Options for the control of elephants in conflict with people. Pachyderm 19:54–63
Hoare R (1999) Determinants of human-elephant conflict in a land-use mosaic. J Appl Ecol 36:689–700
Hoare R (2000) African elephants and humans in conflict: the outlook for co-existence. Oryx 34:34–38
Jayson EA (1999) Studies on crop damage by wild animals in Kerala and evaluation of control measures. KFRI Research Report 169, Kerala, India
Jayson EA, Christopher G (2008) Human-elephant conflict in the southern Western Ghats: a case study from Peppara wildlife sanctuary, Kerala, India. Indian For 134:1309–1325
Leimgruber P, Gognon JB, Wemmer C, Kelly DS, Songer MA, Selig ER (2003) Fragmentation of Asia's remaining wildlands: implications for Asian elephant conservation. Anim Conserv 6:347–359
MacArthur R, Pianka E (1966) On optimal use of a patchy environment. Am Nat 100:603–609. https://doi.org/10.1086/282454
Menon ARR, Sasidharan N (2005) Biodiversity characterization at landscape level using remote sensing and GIS in Kerala. KFRI Research Report No. 274, Kerala, India
Nath NK, Dutta SK, Das JP, Lahkar BP (2015) A quantification of damage and assessment of economic loss due to crop raiding by Asian elephant Elephas maximus (Mammalia: Proboscidea: Elephantidae): a case study of Manas National Park, Assam, India. J Threat Taxa 7:6853–6863
Nigam BC (2002) Elephants of Jharkhand—increasing conflicts with man. Indian For 128:189–196
Prasad G, Shiny R, Reghunath R, Prasannakumar V (2011) A GIS-based spatial prediction model for human-elephant conflict (HEC). Wildl Biol Pract 7:30–40
Saaban S, Othman NB, Yasak MNB, Nor BM, Zafii A, Compose-Arceiz A (2011) Current status of Asian elephants in Peninsular Malaysia. Gajah 35:67–75
Sitati NW, Walpole MJ, Smith RJ, Williams NL (2003) Predicting spatial aspects of human-elephant conflict. J Appl Ecol 40:667–677
Smith RJ, Kasiki SM (2000) A spatial analysis of human-elephant conflict in the Tsavo ecosystem, Kenya. A report to the African Elephant Specialist Group, Human-elephant conflict Task Force of IUCN, Gland
Stephens DW, Krebs JR (1986) Foraging theory. Princeton Univ. Press, Princeton
Sukumar R (1985) Ecology of the Asian elephant (Elephas maximus) and its interaction with man in South India. Ph.D. Thesis, vol 1 and 2. Indian Institute of Science, Bangalore, India
Sukumar R (1990) Ecology of the Asian elephant in southern India: II. Feeding habits and crop raiding pattern. J Trop Ecol 6:33–53
Sukumar R (1991) The management of large mammals in relation to male strategies and conflict with people. Biol Conserv 55:93–102
Veeramani A, Jayson EA (1995) A survey of crop damage by wild animals in Kerala. Indian For 121:949–953

Venkataramana GV, Sreenivasa, Lingaraju HG (2017) An assessment of crop damage and economic loss caused by elephants in Harohalli and Kodihalli ranges of Bannerghatta National Park, Karnataka, India. Curr Sci 113:161–167

White LJT, Tutin EGC, Fernandez M (1993) Group composition and diet of elephants Loxodonta Africana cyclotis Matschie 1900 in the Lope Reserve, Gabon. Afr J Ecol 31:181–199

Williams C, Tiwari SK, Goswami VR, de Silva S, Kumar A, Baskaran N, Yoganand K, Menon V (2020) Elephas maximus. The IUCN Red List of threatened species 2020: e.T7140A45818198. https://doi.org/10.2305/IUCN.UK.2020-3.RLTS.T7140A45818198.en

Zhang L, Wang N (2003) An initial study on habitat conservation of Asian elephant (Elephas maximus), with a focus on human–elephant conflict in Simao, China. Biol Conserv 112:453–459

Asiatic Lions: Behaviour in the Wild and Captivity

26

Shakunthala Sridhara, Akshay Kumar Chakravarthy, and A. Usha Nandini

Abstract

Lions forage in large ranges of grasslands and deciduous forests. Lion is a sociable animal. Lions live in groups called pride. Prides mostly consist of females and cubs. Males lions have solitary lives. There are over 600 lions in the Gir forest, India. Today, worldwide lions are facing several threats like habitat loss, climate change and human interference. One of the main risks lion populations are facing today is inbreeding due to restricted areas. Conservation measures have contributed to the increase in lion population in Asia and Africa.

Keywords

Lions · Behaviour · Captivity · Conservation

26.1 Introduction

Lion is a large, powerful cat belonging to the family Felidae. It is the second-largest carnivore after tiger. Popularly called as the 'king of beasts', the lion has been one of the most awe-inspiring animals, written and talked about since times immemorial. Unfortunately, it became extinct in North America, 2000 years ago in the Balkans, and during the crusades in Palestine. Still, over one million lions lived

S. Sridhara
Department of Zoology, University of Agricultural Sciences, Bengaluru, Karnataka, India

A. Kumar Chakravarthy (✉)
Division of Entomology and Nematology, Indian Institute of Horticultural Research, Bengaluru, Karnataka, India

A. Usha Nandini
Krishi Sai Ektha Pvt Ltd., Bengaluru, Karnataka, India

S. Sridhara et al. (eds.), *Animal Behavior in the Tropics*,
https://doi.org/10.1007/978-981-97-7324-4_26

during the nineteenth century. Tragically, since two centuries, 60% of lions have declined, and today only about 20,000 lions in the wild exist due to the loss and degradation of forests and expanding agriculture. Currently, there are two species of lions, *Panthera leo leo* (Linn, 1758), the Asiatic lion spread over once in India, Central and Western Africa but at present limited to Gir National Park in Gujarat, and *Panthera leo melanochaita* (C.H. Smith, 1842), the African lion found in Eastern and Southern Africa.

Three biotic factors help the lion survive in their ecosystem, namely grass, trees and prey. Three environmental factors that help them are rocks because they form caves where the pride sleeps. The second abiotic factor is mud, which protects the lion's skin from harsh sun and cools its body. The next factor is the water bodies such as wetlands, ponds, rivulets and rivers. A male lion can reach a 2 m (6.8 ft) height, weigh up to 217 kg and have strong muscular bodies (Anon 2015). Their mighty hind legs aid them to leap up to 36 ft. Their sharp claws are powerful enough to kill larger-sized animals like bison. The roar of a lion is loud, helps in communication amongst themselves and acts as a warning signal to other animals. The roar is very loud and fierce and can travel up to 5 miles. It can reach up to 114 decibels, quite close to 120 decibels emanating when jet planes take off.

Lions prey on a wide types of animals varying in size from rodents to buffalo and hippopotamuses. But they prefer to kill and consume medium- to large-sized ungulates such as wild beasts, zebra and antelopes. Prey preferences vary geographically and also between neighbouring prides. What is interesting is that they scavenge on any meat, including carrion and fresh kills. Equally interesting is their penchant to forcefully steal the hunted prey of hyenas, cheetahs and even from wild dogs. Lionesses inhabiting in savannas do most of the hunting. The lazy males consume the female's kills. However, male lions are also skilful hunters.

Asian lions are among the most charismatic of large cats. Lions forage in large grasslands and deciduous forests. The Asian lion is a population of *Panther leo leo* (Linn, 1758) that today survives in the wild in India only. The first scientific description of the Asian lion was published in 1826 by the Austrian zoologist Johann N. Meyer, who named it *Felis leo persica*, as the lion skin was from Persia (Asiatic Lions 2019), Asian lion existed in Saudi Arabia, Turkey, Iran, Mesopotamia, Pakistan and from east of the Indus River to Bengal and the Narmada River in Central India until the nineteenth century. The Gir forest in Gujarat is the major habitat for the Asian lions in India (Anonymous 1975). It has been extirpated from the Gambia, Mauritania, Mali, Sierra Leone, Ivory Coast and Togo. The lion is among five Pantherine cats native to India along with the Bengal tiger *Panthera tigris tigris* (Linn 1758), Indian leopard *Panthera pardus fusca* (Linn 1758), snow leopard *Panthera unica* (Schreber 1775) and clouded leopard *Neofelis nebulosa* (Griffith 1821).

26.2 Life History, Habits and Habitat

Male and female lions are polygamous. However, females mate with one or two males. Although lions breed every year in captivity, in nature they breed once in 2 years. The gestation time is 108 days, and the litter size ranges from one to six cubs. Newborn cubs are helpless and blind. Fifty per cent of cubs do not survive due to painful teething, which weakens them, diseases and predators. Further, when a male from outside the pride takes over the resident pride, he will kill all the cubs less than a year old. But females prevent this infanticide by hiding or directly defending their cubs. In the wild, lions live between 8 and 10 years; in captivity, they live up to 25 years. Lions sleep 16–20 h a day. Lions forage and hunt often at dawn or dusk. Lions communicate using vocalizations, hood rubbing, licking, facial expressions, chemical marking and visual marking.

Lions live commonly in clusters called pride. The cluster size may vary from 4 to 37 members, but about 15 is the average. Hunting and sharing the kill may be carried out by several members of the pride. Many times a pride consists of several generations of lionesses, one or two breeding males and cubs. The females may be related also. Male lions protect pride, while lionesses are hunters. They ensure food availability for the pride. Each pride has a well-defined territory consisting of a core area that is its own and a fringe area where other prides are tolerated. There seems to be some sort of territorial attachment as some prides are known to use the same territory for decades and then pass on the area to females (https://animalia.bio/asian.lion).

Asian lions are quite like the African lions in appearance and size. But Asian lions are less in size than African relatives; tassels on the end of their tails and a bunch of hair on their elbows are longer than those of African lions. Lion is a sociable animal. The lions gather into small units called prides. The prides usually have two females. Males are not as sociable as females, joining the pride mainly for mating and on a large kill. Adult males tend not to take part in hunting. In habitats such as Serub, with little vegetation, lions go hunting at night. In areas with thick vegetation, lions can hunt during the day. Asian lions rest for most of the day, with 20 h per day. Lions have a large repertoire of vocalizations. Most of them are variations of growling, snarling and roaring. Roaring is used for advertising their presence. Lions often roar at night and can be heard from several miles away. Asian lions are carnivores feeding on wild boars, buffaloes, antelopes and deer. They also regularly forage specific sites within the protected area to scavenge on dead livestock dumped by Maldhari livestock herders in the Gir, Gujarat. Asian lions are polygynous, with no definite mating period. Lions can mate throughout the year. The season lasts from 100 to 119 days with 18–26 months between births. A female gives birth to 1–6 cubs. Young start eating meat when they reach 3 months. But young ones keep suckling for another 3 months. The cubs learn hunting for 9 months, becoming independent at 1 year. Males reach maturity at 5 years, while females reach maturity at 3–4 years.

26.2.1 Gir National Park

There are more than 600 lions in Gir. Although historically Asiatic lions were abundant in India, their populations declined to just 31 in the nineteenth century due to periodical famines and intensive hunting by Indian kings and Britishers during colonial rule. When the Nawab of Junagadh invited Lord Curzon for hunting, they later agreed on the condition that there should be no hunting of lions. Not only that Lord Curzon refused the request of Maharajah of Gwalior to send lions to him but also prohibited hunting of lions. This is perhaps the first effort to conserve Asiatic lions. The Gir National Park covers 259 km^2, is devoid of human habitation and is exclusively available for free-ranging lions. In about −13,000 km^2 around this core area, humans and lions coexist (Kumar and Meena 2014).

Research on Asiatic lions began in late 1960 (Joslin 1973). It continued from 1986 to 2018 and is considered as the longest ecological project in India (Jhala et al. 2011). Habitat improvement and protective measures by the Gujarat Forest Department resulted in an increase in lion population to 287 by 1965, to 500 by 2015 and to nearly 730 by 2022. This includes 300 adult lions within the Gir National Park Wildlife Sanctuary and several populations of <50 lions in each in the surrounding agropastoral, human-dominated landscapes. As compared to the 19.2% increase in the growth rate of human population in Gujarat (www.census2011.co.in), the lions in Gir National Park and Wildlife Sanctuary have grown by 126% during the past 20 years (Singh 2007).

The social life of Asiatic lions has been studied only recently. Their prides are reported to comprise only females and their dependent cubs, while adult males are solitary or form separate all-male groups. Female prides are territorial with their territory remaining unchanged for several years. The male prides do move into the home ranges of female prides. Males exhibit a social hierarchy with a dominant male who has the greatest access to females and food.

A major factor contributing to the increase in lion numbers is the expanded prey base (Talbot 1960). Earlier in 1970, the main food of lions in Gir was wild herbivores such as ungulates, nilgai and porcupines in the forest, which was inadequate for forcing lions to kill livestock. But with improved habitat in later years, forest herbivores became major food of lions, accounting for 76.4% of their food requirements. Livestock contributed 25–42% of lion's food, 16% of which was by predation and the rest was scavenged. Still, annually 3000–4000 livestock are killed. The food patterns were seasonal with livestock being hunted more in summer when they visit waterholes and wild herbivores during other seasons (Anon 2021).

To conserve lions in Gir, the forest department has dedicated employees, which include 800 forest personnel and 175 trackers. It is the trackers who collect information on different prides, their activities, birth of the young, their survival and death of lions if any. Their surveillance is spread throughout the day. The excess lion population is maintained by habitat management of the Gir protected area and providing dispersal corridors to the scattered small populations living in the surrounding grasslands and agro-pastoral landscape. Judiciary too is proactive in lion conservation, which gave the ruling that lions should be allowed to live in peace and interactions between people and lions should be reduced in the interest of both. The State Government should have a policy on this matter.

Gir forest is inhabited by *Maldharis* inhabiting the forest for over two centuries. They have been co-existing with lions all their life with mutual benefit. They are vegetarian, with their livelihood coming from rearing cattle and collecting forest resources permitted by the government. With free grazing rights, collection of minor forest produce and adequate compensation for cattle predated by ions, *Maldharis* do not disturb lions. On the other hand, lions get a considerable quantum of food from the livestock reared by *Maldharis*. An additional source of food for lions comes from dead livestock dumped by *Maldharis*. These forest-dwelling cattle herders do not hunt or kill lions but revere nature and lions and totally avoid encounters with lions. It is a unique case of coexistence between humans and a large carnivore with mutual benefit. Thus the relationship between *Maldharis* and lions is maintained by a delicate balance of cultural practices, vigorous enforcement of laws, adequate monetary compensation for livestock killed and the inherent tolerance and respect of this tribe towards lions.

As far as conflict with humans is concerned, the lion is not a problematic animal; panthers and crocodiles kill more people than lions. Human deaths occur due to accidental encounters, mistaken identities and while rescuing cattle from lions. Only 40 attacks on humans by lions were reported between 2007 and 2017, with a fatality rate of 1.3 attacks on humans per year (Berwick and Jordan 1971; Chellam 1993; Dharaiya et al. 1998).

Lions contribute two important economic benefits in Gir, namely (1) their predation on wild herbivores like nilgai and porcupines reduces considerable damage to cultivated crops, and (2) lions have enhanced tourism as 'Lion Shows' are organized for a hefty fee wherein lions are lured to carcases on private land. Tourists watch lions at close quarters and take pictures. Such human proximities affect lions. The Government of Gujarat is implementing strict measures to curb 'Lion Shows'.

A large number of lions in Gir is a problem for themselves. Today, there are 694 lions in Gir, which is beyond the productivity of the sanctuary. The sanctuary can support a maximum of 300 lions, with the result that the excess members are pushed out to the surrounding grasslands. Some are known to have been living 150 km away from the core area. The inbreeding of lions in the confined area of Gir has resulted in several adverse health factors like reduced immunity, dental abnormalities and decreased heterozygosity leading to declined reproductive vigour (Jhala et al. 2011).

To ease the burden on the ecosystem and make it more sustainable, an attempt was made to translocate ten lions to Chandra Prabha Wildlife Sanctuary, but these lions did not survive in the new environment either due to habitat quality, the belligerent attitude of locals or both. With the Asiatic lion population increasing, the Supreme Court of India asked the Union government in 2013 to translocate some lions to the Kuno-Palpur Wildlife Sanctuary. The proposal was aimed at protecting the Gir population from possible extinction due to forest fires, epidemics, infighting, inbreeding, etc. However, the Gujarat government rejected the idea of translocating lions and insisted that they are 'completely safe' in Gir. Alternately, the Gujarat government is developing a suitable habitat with lion-familiar biodiversity near Jamnagar in the state.

One of the major factors for the upward trend in the Gir population of lions is the attitude of people. The people in and around Gir are highly tolerant of lions; they are not afraid of the ferocious carnivore and do not unnecessarily provoke the giant. There is mutual respect between lions and people. The empathy of villagers is so great that when lions fall into wells during summer in the night, not only forest personnel are informed immediately, but by the time the experts arrive, the villagers are ready with ropes and other help needed to extricate the lions from the well.

After the lion was stripped of as a national animal, paving the way for the tiger, the Gujarat government proclaimed the lion as a state animal, making it an iconic symbol of the state. The people of Gujarat are proud of this iconic status. No doubt this passionate identity has enhanced conservation efforts, leading to increased lion numbers. At the extreme level of this passion, the Gujarat government is hesitant to provide a stock of founder lions to establish a second sanctuary for Asiatic lions in Madhya Pradesh, inspired by the directive from the Supreme Court. The implementation of the judgement is still pending, mired in endless discussions amongst the forest departments of Guajrat, Madhya Pradesh and the Union Ministry of Environment, Forests and Climate Change.

The lion population dwindled throughout the nineteenth century due to intensive hunting by British colonial officers and Indian rulers. By the nineteenth century, only about 12 Asiatic lions remained in India. The residue populations of Asiatic lions were found only in the Gir forest. Gir forest was once personal property and has now been converted into a forest, national park and wildlife sanctuary. Asiatic lions are endangered species as per the IUCN Red List. Their population is now 674 according to the census of 2023.

Lions in Gir National Park prefer hunting chital (*Axis axis*), sambar deer (*Rusa unicolon*), Nilgai (*Boselaphus tragocamelus*), cattle and wild boar (*Sus scrofa* Linn). There have been a few livestock kills within the sanctuary because of changes in prey community distribution. Male Asiatic lions tend to roam, rest, hunt and feed with a pride of up to three, while female prides comprise up to 12 individuals. Lions share the carcasses with each other and seldom do they do so with males. Home ranges for males can be as large as 230 km^2 (89 mL^2), and those for females can be up to 85 km^2 (33 mL^2). Males form a coalition of three to four individuals who hold and defend territories that contain one or two female prides. When males choose to coexist in a coalition, one will tend to dominate among the others. During mating season (October and November), males and females tend to associate for a couple of days but do not feed or live together.

26.3 Lions in Protected Rangelands

The most recent conservation activity is the reintroduction of Asiatic lions to the Kuna Wildlife Sanctuary, Madhya Pradesh. In the early 1990s, an ecological assessment of protected areas within the historical range of lions was conducted by the

Wildlife Institute of India. This was a crucial move as Asiatic lion numbers have increased to a point where about 50% of the country's lion population currently lives outside a protected area. The steady rise in populations of Asiatic lions showed that protecting rangelands of endangered animals may result in extraordinary conservation efforts.

Large carnivores like lions are territorial and require large landscapes and a good prey base to survive. A good understanding of carnivore ecology is fundamental for conservation planning. When carnivores, particularly lions, disperse out of designated protected areas, it becomes crucial to reassess their range and habitat requirements specific to these heterogeneous land uses. In their range, lions are known to occupy a wide range of habitat types. Within their current range, the endangered Asiatic lions utilize landscapes of varied terrain, vegetation and land use.

Endangered Asiatic lions are noted for adaptations and as a flagship of successful conservation and management. Lions dispersing out of the Gir forest have established in the east of Gujarat about 25 years back. The home range and spatial distribution of lions inhabiting the coastal habitats are distinct from the forested habitats of the protected areas. Ram et al. (2022) monitored each individual for an average of 367.2 ± 99.05 days from 2019 to 2021. The mean core area was 33.8 km^2 (50% FK, SE8.7 km^2), and the overall average range was 171.8 km^2 (90% FK, AE 40.5 km^2). The lion distribution model was built on MaxEnt, and inputs included location fixes of lions and variables, including 18 land-use categories and categories and Euclidean distance to linear infrastructures and human settlements. Lions were shown to use forest habitat patches extensively. The study highlights the importance of corridors connecting to the Gir protected area and the importance of coastal forest patches for lion conservation.

Terrestrial wild quadruped mammals like lions are one of the most threatened groups and 57% of the <250 ungulate species. Habitat loss, poaching and grazing are the important factors for the decline in their population. Additionally, wild ungulates exhibit remarkable diversity in terms of body size, foraging zone, etc. Body size plays an important role in ungulates diet and movements. Ram et al. (2023a, b, c) assessed the population density, structure and population change of nine wild prey species in the semi-arid landscape of Saurashtra, Gujarat. The workers deployed the road transect/methodology under a distance sampling framework to achieve objectives. The study revealed that deer species exhibited higher density and biomass in woodlands compared to grasslands and coastal forests. Antelopes showed higher density and biomass in grasslands and coastal forests compared to woodlands. The density gradient of wild prey species was influenced by factors, such as habitat, structure, social organization, grouping tendencies and topography. Over the last four decades, the population of wild prey species in Gir showed minimal changes.

26.4 Behavioural in Captivity-Threats, Human–Lion Conflict and Conservation

The only animal that can cause the death of a lion is porcupine, whose quills pierce the heart or major arteries resulting in death or seriously wounding the lion. Such wounded lions are too weak to hunt agile or bulky prey and slowly become killers of livestock and in rare cases human beings. Still, during severe droughts, lions do hunt porcupines (Douglas Main 2019).

As for mortality, diseases killed 30% of lions, while natural causes accounted for 60% of mortality. The other causes were human associated such as coming in contact with a lethal electrical current illegally drawn by villagers and falling in open wells while trying to drink water (Wielebnowski 1999).

Forest settlements, state highways and railway lines are all inside Gir. All of these certainly affect the movement and behaviour of lions. In addition, there is a considerable disturbance to lions as many resorts have come up inside the forests, which attract tourists. Further, the surrounding grasslands where lions living outside the core area rest and hide are disappearing fast.

The Asiatic lion now exists as a single subpopulation and is vulnerable to extinction from a disease epidemic or a large forest fire. In recent years, lions have also been poached, and reports indicate that organized gangs have switched poaching from tigers to lions.

Various studies have shown tremendous habitat recovery and increases in wild ungulate populations during the last four decades. Even though their habitat is a preserve, farmers still graze cattle and goats within the boundaries, and people regularly enter the part to collect firewood. Domestic cattle have historically been part of the Asian lion's diet, but now the most common prey species is the chital deer. The establishment of at least one other wild population is recommended, maximizing genetic diversity. However, there are problems in attempting this, a previous attempt to establish a second subpopulation in the Chandraprabha Wildlife Sanctuary in eastern Uttar Pradesh. But Palpur-Kuno Wildlife Sanctuary in northern Madhya Pradesh has now been selected as the best candidate area. Communities will require resettlement to create dens for lions, but this time great care is being taken to make the process participatory and attempting to satisfy local needs, and not engender hostility towards lion conservation (Gosling 2008; Hansen and Jeppesen 2006).

African lions face various threats, such as climate change and habitat loss, but the biggest threat comes from humans. Humans kill lions out of fear or to use in illegal trade and also indulge in poaching. The demand for lion bone has increased in recent times as it is used increasingly in traditional Asian medicine. These factors have resulted in a more than 50% decrease in the African lion populations over the past 20 years. Asiatic lions also face threats such as poaching, habitat loss, and conflict with humans. The major risk for lions is inbreeding leading a single population living in a restricted area in Gujarat. However, various conservation measures have resulted in an increase in lion populations in both Africa and India (Pavlov 1906; Powell and Gartner 2011; Ravi 1993; Singh and Kamboj 1996).

26.5 Human–Lion Conflict

Asiatic lions now exist only in the Gir protected areas and surrounding landscape in Saurashtra, Gujarat. Scientifically monitoring their activity and behaviour is critical for their conservation. Ram et al. (2023a, b, c) used GPS radio collars to study the movement and activity of 19 individual lions. This pilot study's results revealed how technology-driven scientific monitoring facilitates conserving endangered species. The workers discussed the use of virtual geofence functions and alert generation to reduce the chances of endangered species mortality on linear infrastructures. They also presented a pilot case study on how GPS radio-telemetry helped in securing the safety of dispersing sub-adult coalition males extensively moving in a multi-use land matrix.

Human-wildlife interactions that were earlier reported at protected area interfaces are increasingly being reported even in distant cities, as animals forage away from protected areas (Woodroffe and Frank 2005; Athreya et al. 2013). Monitoring animal movement, particularly endangered species, outside protected areas in multi-use landscapes is a great challenge, with possibilities of retaliatory killing, poaching, hunting and accident-related mortality (Roelke-Parker et al. 1996; Williams et al. 2017). As species traverse rural and urban settings, navigating across a matrix of land uses and continuously monitoring their health and status is a difficult feat to accomplish without a good understanding of the concerned species' ecology (Meena et al. 2014). Additionally, conservation managers have to mitigate costs for people caused by wildlife and reconcile damage to life, property and livelihoods in order to gain support for and safeguard wild species' survival in the multi-use land matrix (Dickman et al. 2013; Redpath et al. 2013). Real-time monitoring of wildlife movement through the Global Positioning System (GPS) telemetry, further strengthened by improved communication networks, is proving to be a great aid for coping with these novel conservation challenges (Wall et al. 2014). Thus, it has become inevitable to complement traditional wildlife management methods with technology and artificial intelligence (Wall et al. 2014; Shivik 2004; Weise et al. 2019).

Asiatic lions presently range extensively in multi-use land matrices in varying administrative and legal land-use areas (Meena et al. 2014; Ram et al. 2022). Previously, it was shown that this movement has considerable intraspecific variation, with adult and sub-adult males having a greater tendency to make long-distance movements (Ram et al. 2022). The study established that movements of this nature are important for fulfilling life-history requirements and, therefore, crucial for the continued survival of lions (Ram et al. 2022; Meena and Jhala 2010). It is known that ranging behaviour makes lions vulnerable to accidents on linear infrastructures and increases the risk of their falling into open wells, accidental electrocution and other causes of unnatural mortality (Singh 2017). It is essential that managers need to heed two categories of alerts while remotely monitoring them—one is when a lion is not moving (immobility), and the other is when it is traversing through risk-prone areas (movement). The study demonstrated how these dual signals can be interpreted and managed using the Asiatic lion as a case study. A renewed conservation strategy was implemented by deploying GPS radio collars on lions so that

Table 26.1 Enrichment device usage with behaviour performed (N = 109,238 scans)

Enrichment device	Feed (%)	Manipulable (%)	Sensory (%)
Aggression	2.94	11.51	0
Chase	11.54	1.69	0
Explore	3.05	5.95	11.77
Fear	0	1.96	0
Forage	11.48	0	0
Play	0	35.16	1.21
Social	1.16	0.35	0
Total	30.20	56.8	12.99

Source: Sitendu Goswami et al. (2020)

individual lions are remotely monitored, almost in real time, and prompt management responses to threat situations could be triggered, thereby strengthening on-ground protection measures.

Goswami et al. (2020) tested the efficacy of feeding, sensory and manipulable enrichment interventions on the welfare of Asiatic lions at the conservation breeding centre of Sakkarbaug Zoological Garden, Gujarat. We adopted a holistic approach by measuring the physiological and behavioural responses of 35 captive Asiatic lions, divided into control (N = 16) and test (N = 19) groups. The test subjects approached feeding devices first and used manipulable devices for a longer duration. Manipulable devices were used homogenously with two significant time peaks, but sensory controls were deployed throughout the day. The control animals remained the same in welfare parameters compared to pre-treatment levels. Post-enrichment behavioural assessments revealed greater behaviour diversity, reduced enclosure zone bias (40.25% reduction) and aberrant repetitive behaviours (80.68%) in samples. The faecal corticosterone measures showed lower stress levels in test samples (58% decrease), confirming the significant improvement in welfare indices than control groups (Table 26.1).

References

Anonymous (1975) The Gir Lion Sanctuary Project, Government of Gujarat, p 131

Anon (2015) Achievements, biodiversity conservation and rural livelihood project. Asiatic lion landscape, Forest Department, Sasan, p 37

Anon (2021) Lions should be allowed to live in peace with minimal human interaction: Gujarat. HC Vibes of India 9:22

Asiatic Lions (2019). https://wildcatconservation.org/wild-cats/asia/asiatic-lion

Athreya VM, Odden JD, Linnell JK, Karanth U (2013) Big cats in our backyards: persistence of large carnivores in a human-dominated landscape in India. PLoS One 8:e57872

Berwick SH, Jordan PA (1971) First report of the Yale–Bombay Natural History Society—studies of wild ungulates at the Gir forest, Gujarat, India. J Nat Hist Soc 68(2):412–423

Chellam R (1993) Ecology of the Asiatic lion (*Panthera leo persica*). Dissertation/Ph.D. thesis. Saurashtra University, Rajkot

Dharaiya N, Singh M, Raval PP (1998) The feeding pattern of big cats in Gir protected area. Gujarat Forest Department

Dickman AJ, Marchini S, Manfredo M (2013) The importance of the human dimension in addressing conflict with large carnivores. In: Macdonald DW, Willis KJ (eds) Key topics in conservation biology. Oxford University Press, Oxford, pp 110–226
Douglas Main (2019) When lions attack porcupines, humans suffer unexpected consequences. https://nationalgeographic.com/animals/lions, July 6 2019
Gosling SD (2008) Personality in non-human animals. Soc Personal Psychol Compass 2(2):985–1001
Goswami S, Patel SK, Kadivar R, Tyagi PC, Malik PK, Mondol S (2020) Effects of a combined enrichment intervention on the behavioural and physiological welfare of captive Asiatic lions (*Panthera leo persica*). Appl Anim Behav Sci 236:105222. www.elsevier.com/locate/applanim
Hansen SW, Jeppesen LL (2006) Temperament, stereotypies and anticipatory behaviour as measures of welfare in mink. Appl Anim Behav Sci 99(1–2):172–182
Jhala YV, Ravi C, Meena V, Banerjee K, Basu P (2011) Ecology of lion in Greater Gir landscape. Social organisation and dispersal of Asiatic lions. Technical Report, Wildlife Institute of India, Dehradun
Joslin P (1973) The Asiatic lion: a study of ecology and behavior (Dissertation/Ph.D. thesis). Department of Forestry and Natural Resources University of Edinburgh UK, 249 p
Kumar S, Meena RL (2014) Conservation of Asiatic lion in Gir. Forester's diary, Asiatic lion landscape. Gujarat Forest Department, 3–18, p 113
Meena V, Jhala YV (2010) Radio-telemetry studies on lions in India. In: Telemetry in wildlife science. ENVIS bulletin: wildlife & protected areas
Meena V, Macdonald DW, Montgomery RA (2014) Managing success: Asiatic lion conservation interface problems and peoples' perceptions in the Gir protected area. Biol Conserv 174:120–126
Pavlov IP (1906) The scientific investigation of the psychical faculties or processes in the higher animals. Science 24(620):613–619
Powell DM, Gartner MC (2011) Applications of personality to the management and conservation of non-human animals. In: Inoue Murayama M, Kawamura S, Weiss A (eds) From genes to animal behavior: social structures, personalities, communication by color. Springer, Tokyo, pp 185–199
Ram M, Vasavada D, Tikadar S, Jhala LS, Zala Y, Meena V (2022) Movement and activity of endangered Asiatic lions in relation to land-use, season and group characteristics. J Zool 319:23–31
Ram M, Sahu A, Tikadar S, Jayawant H, Jhala L, Zala Y, Venkataraman M (2023a) Living to the sea-cost: ranging and habitat distribution of Asiatic lions. Sci Rep 12:19235. www.nature.com/scientific reports
Ram M, Sahu A, Srivastava N, Jhala L, Zala Y, Venkataraman M (2023b) Conservation management of the endangered Asiatic lions in Gujarat, India, using GPS satellite telemetry. Animals 13(1):125. https://doi.org/10.3390/ani13010125
Ram M, Sahu A, Srivastava N, Chaudhary R, Jhala L, Zala Y (2023c) The semi-arid ecosystem of Asiatic lion landscape in Saurashtra, Gujarat: population density, biomass and conservation of nine wild prey species. PLoS One 18(9):e0292048
Ravi C (1993) The ecology of the Asiatic lion (Panthera leo persica), PhD thesis submitted to Saurashtra University, Rajkot
Redpath SM, Young J, Evely A, Adams WM, Sutherland WJ, Whitehouse A, Amar A, Lambert R, Linnell JDC, Watt A (2013) Understanding and managing conservation conflicts. Trends Ecol Evol 28:100–109
Roelke-Parker ME, Munson L, Packer C, Kock R, Cleaveland S, Carpenter M, O'Brien SJ, Pospischil A, Hofmann-Lehmann R, Lutz H et al (1996) A canine distemper virus epidemic in Serengeti lions (*Panthera leo*). Nature 379:441–445
Shivik JA (2004) Non-lethal alternatives for predation management. Sheep Goat Res J 19:64–71
Singh HS (2007) The Gir lion—Panthera leo persica: a natural history, conservation status and future prospect. Pugmark Qmulus, Ahmedabad, pp 163–194

Singh HS (2017) Dispersion of the Asiatic Lion Panthera leo persica and its Survival in Human-Dominated Landscape Outside the Gir Forest, Gujarat, India, Current Science, 112(5):933–940. https://doi.org/10.18520/cs/v112/i05/933-940

Singh HS, Kamboj RD (1996) Biodiversity conservation plan for Gir—a management plan for Gir National Park and Sanctuary. Gujarat Forest Department, p 242

Talbot LM (1960) A look at the threatened species. A report on some animals of the Middle East and southern Asia which are threatened with extinction. Oryx 5:153–293

Wall J, Wittemyer G, Klinkenberg B, Douglas-Hamilton I (2014) Novel opportunities for wildlife conservation and research with real-time monitoring. Ecol Appl 24:593–601

Weise FJ, Hauptmeier H, Stratford KJ, Hayward MW, Aal K, Heuer M, Tomeletso M, Wulf V, Somers MJ, Stein AB (2019) Lions at the gates: trans-disciplinary design of an early warning system to improve human-lion coexistence. Front Col Evol 6:242

Wielebnowski NC (1999) Behavioral differences as predictors of breeding status in captive cheetahs. Zoo Biol 18(4):335–349

Williams ST, Williams KS, Lewis BP, Hill RA (2017) Population dynamics and threats to an apex predator outside protected areas: implications for carnivore management. R Soc Open Sci 4:161090

Woodroffe R, Frank LG (2005) Lethal control of African lions (*Panthera leo*): local and regional population impacts. Anim Conserv 8:91–98

Further Reading

Banerjee K, Jhala YV, Chauhan KS, Dave CV (2013) Living with lions: the economics of coexistence in the Gir forests, India. PLoS One 8(1):e49457. https://doi.org/10.1371/journal.pone.0049457

Kays RW (2020) Lion. Encyclopedia Britannica. 29 Jul 2020

https://www.britannica.com/animal/lion. Accessed 24 Jan 2022

https://www.frontiersin.org/articles/10.3389/fevo.2019.00312/full

The manufacturer's authorised representative in the EU is Springer Nature Customer Service Centre GmbH, Europaplatz 3, 69115 Heidelberg, Germany. If you have any concerns regarding our products, please contact ProductSafety@springernature.com

Printed and bound by CPI Group (UK) Ltd, Croydon, CR0 4YY
01/10/2026
02225956-0017